Interacting Multiagent Systems

Interacting Multiagent Systems

Kinetic equations and Monte Carlo methods

Lorenzo Pareschi
Department of Mathematics and Computer Science,
University of Ferrara, Italy

Giuseppe Toscani
Department of Mathematics,
University of Pavia, Italy

OXFORD
UNIVERSITY PRESS

OXFORD
UNIVERSITY PRESS

Great Clarendon Street, Oxford, OX2 6DP,
United Kingdom

Oxford University Press is a department of the University of Oxford. It furthers the University's objective of excellence in research, scholarship, and education by publishing worldwide. Oxford is a registered trade mark of Oxford University Press in the UK and in certain other countries

First Edition published in 2014
Impression: 1

Published in the United States of America by Oxford University Press
198 Madison Avenue, New York, NY 10016, United States of America

British Library Cataloguing in Publication Data
Data available

Library of Congress Control Number: 2013937550

ISBN 978–0–19–965546–5

Printed in Great Britain by
Clays Ltd, St Ives plc

To Roberta and Roberta

Preface

> I am conscious of being only an individual struggling weakly against the stream of time. But it still remains in my power to contribute in such a way that, when the theory of gases is again revived, not too much will have to be rediscovered.
>
> Ludwig Boltzmann, Foreword to the second volume of *Vorlesungen über Gastheorie*, 1898.

The description of emerging collective phenomena and self-organization in systems composed of large numbers of individuals has gained increasing interest from various research communities in biology, ecology, robotics and control theory, as well as sociology and economics. This challenging research activity aims to find methods that allow us to build complex systems composed of autonomous agents who, as a result of their mutual interactions, exhibit a well-defined collective behaviour.

Although the definition of multiagent systems is clear and unambiguous, the treatment of such systems by mathematicians, physicists and engineers varies greatly, and some approaches are distinctly incompatible with others. A well-developed line of research deals with the analysis of multiagent systems using tools drawn from game theory, economics and biology. It supplements these with concepts and algorithms arising from artificial intelligence research, like planning, reasoning methods, search methods and machine learning.[1]

A second approach to multiagent systems has its roots in statistical physics. The first attempts made in this direction date back nearly twenty years, when the term econophysics was coined to describe an interdisciplinary research field which aims to solve problems in economics by means of well-established physical methods. The main idea here is that the description of emerging phenomena could be obtained by assuming that the collective behaviours of a group composed of a sufficiently large number of individuals could be treated using the laws of statistical mechanics, as happens in a physical system composed of many interacting particles.

Certainly, the basic entities in these fields differ from physical particles in that they already have an intermediate complexity themselves. Therefore, these entities are commonly denoted as *agents*. This term means a subunit of the system that may already have internal degrees of freedom to allow certain activities, such as (active) movement, and interaction with other agents.

Part of this approach related to statistical physics makes use of concepts and methods borrowed from kinetic theory of rarefied gases. Similarly to the Boltzmann description of the tendency of a rarefied gas in a vessel to reach statistical equilibrium,

[1] See, for example, http://multiagent.com.

collisional kinetic equations have been fruitfully employed to describe the emergence of universal structures through their equilibria.

In fact, kinetic equations play a major role in several applications where the multiscale nature of the phenomena cannot be described by a standard macroscopic approach. They are particularly useful in the study of emergent behaviours in complex systems characterized by the spontaneous formation of spatio-temporal structures as a result of simple local interactions between agents.

The mathematical modelling based on collisional kinetic theory furnishes a useful link between the different approaches to multiagent systems. In principle, since any local description of the modification of a certain attribute of agents (wealth, opinion, direction, etc.) can be treated as an interaction with an external subject, or as an internal interaction with another agent, the mathematical description of this interaction can serve as a building block for the construction of the kinetic equation in terms of the density of agents which have the same quantity of this attribute.

Most speculative markets at national and international level share a number of stylized facts, like volatility clustering and fat tails of returns, for which a satisfactory explanation is still lacking in standard theories of financial markets. Such stylized facts are nowadays almost universally accepted among economists and physicists, and it is clear that financial market dynamics give rise to some kind of universal scaling laws. Other effects, in particular the emergence of power laws, are present in wealth and income distributions of different countries as postulated more than a century ago by Vilfredo Pareto [257]. Such overpopulated tails are a manifestation of the existence of an upper class of very rich agents, i.e. an indication of social inequality. Showing similarities with scaling laws for other systems with many interacting particles, a description of financial and economic markets as multiagent interacting systems appeared to be a natural consequence. This topic has been pursued by quite a number of contributions appearing in both the physics and economics literature in recent years.

Starting from microscopic dynamics, kinetic models can be derived with the tools of the classical kinetic theory of fluids. In contrast with microscopic dynamics, where behaviour can often be studied only empirically through computer simulations, kinetic models based on partial differential equations allow us to derive analytically general information on the model and its asymptotic behaviour. For example, knowledge of the tails behaviour for the income distribution allows us to calculate characteristic features, like the Pareto index of the wealth distribution over a long period, directly from the model parameters.

On the social side, the modelling of opinion formation, based on two-body interactions involving both compromise and diffusion properties in exchanges between individuals, has attracted the interest of an increasing number of researchers. The binary interactions among agents are quantified here in terms of compromise and diffusion, which are mainly responsible for the behaviour of the model, and allow for a rigorous asymptotic analysis.

Lastly, different biological systems can be treated from the point of view of statistical mechanics. Swarming is a well-known behaviour in many animal species from swarming locusts to schools of fish to flocks of birds. Emergent structures are a common strategy found in many animal groups: colonies of ants, mounds built by termites, swarms of bees, shoals/schools of fish, flocks of birds and herds/packs of mammals. They share several common aspects with human crowd dynamics and road traffic flow. Emergent behaviours may also apply to the cells inside our bodies, like the rules by which cancer cells work together to build tumours or migrate through tissues. In this context the leading example can be considered the original Luria–Delbrück mutation problem [214].

These topics enlighten the common methodological background: to identify similar modelling approaches, similar analytical and numerical techniques for systems consisting of a large number of *individuals* which show a *collective behaviour*, and how to obtain *average* information from them. We identify this common background in kinetic theory, where the underlying equations can be derived in a systematic way, on the basis of the identification of microscopic binary interactions.

This book is written both for researchers in applied mathematics who want to apply these methodologies to interdisciplinary problems, and for scientists from other fields (physics, biology, social sciences, economics and others) interested in mathematical methods developed for interacting particle systems and multiagent systems.

Plan of the book

The book essentially consists of two parts, the first being devoted to a guided explanation of the fundamental tools related to kinetic equations, together with an introduction to Monte Carlo methods for the efficient numerical solution of the models. Because of their introductory purpose, the first four chapters are supplemented by exercises. The second part is devoted to the modelling and analysis of economic, social and biological systems whose common denominator is to be constituted of many agents that interact through binary interactions.

In the first part, particular attention is paid both to modelling aspects and to the mathematical background necessary to rigorously justify the consistency of the models. While modelling is treated in Chapter 1, the essential mathematical tools are listed in Chapter 2. The material of the first two chapters is organized in such a way that the reader may learn as much as possible about the aims and consequences of kinetic modelling and its mathematical treatment in a self-contained way.

The next two chapters are devoted to a detailed presentation of Monte Carlo methods. In Chapter 3 a general overview of Monte Carlo methods is given together with a direct approach based on Monte Carlo methods for solving several types of partial differential equation. Next, in Chapter 4, using the kinetic models introduced in the first chapter and the Monte Carlo tools presented in Chapter 3, we derive different Monte Carlo methods for kinetic equations. Similarly to the first two chapters, the treatment of Monte Carlo methods in Chapters 3 and 4 is substantially self-contained.

The second part of the book is aimed at applications. We have necessarily had to make a selection of suitable topics, and have focused on applications of kinetic modelling and simulation to the fields of economics, social sciences and biology.

Chapter 5 is concerned with the description of wealth distribution in western society. This basic aspect of mathematical modelling in economics has been extensively treated in the last twenty years from the point of view of statistical mechanics, and perfectly fits into the framework of multiagent systems which undergo binary interactions (trades). We discuss three main aspects of wealth distribution. The first is concerned with the properties of the stationary distribution profile, including the presence of the so-called Pareto tails, which are one of the indicators of the health of an economy. The second aspect deals with the modelling of wealth distribution in the presence of a taxation policy. The third aims to understand and describe the possibility that anomalous behaviour of agents in trading can modify the wealth distribution curve in a sensible way. This last aspect is connected, for example, to the recent crisis in Argentina, which has seen a dramatic reduction of the middle class. Where possible, benchmark examples and explicit solutions are collected and discussed. Modelling is accompanied by a thorough mathematical treatment, in which many results are rigorously justified.

Chapter 6 aims to describe the formation of consensus in a society as a consequence of exchanges of information among individuals. Many aspects of modern life are influenced by this type of phenomena. While the formation of opinion around the choice of government is a very important, though rare, event (occurring periodically every few years), advertising campaigns for products are now ubiquitous. We will deal with the description of these problems by introducing kinetic models for three main typologies of problem. First, we will consider the modelling of opinion formation, in which opinion in society distributes according to a certain profile in consequence of the allowed rules of exchange of information. Second, we will consider the related problem of the formation of choice. In this second case, the effect of the information exchanges is to move opinion towards the various available options. Lastly, we will discuss the importance and the consequences of the presence of leaders in addressing opinion, for example in the choice of spokesperson in an advertising campaign.

While the modelling described in Chapters 5 and 6 is restricted to describing statistically the evolution of a one-variable distribution, say wealth or opinion, Chapter 7 aims to describe more realistic situations, in which more than one variable can enter into the game. Each section of Chapter 7 provides a description of a model of wealth distribution in which wealth exchange in trades follows very precise realistic rules. With respect to the elementary ones introduced in Chapter 5, these rules take into account new aspects, like personal utility, opinion, propensity and others. Clearly, the increasing complexity of these models, which are extremely useful in understanding phenomena such as the formation of prices, booms and crashes in the stock markets and shifts of opinion in the markets, makes rigorous mathematical analysis difficult and requires the use of numerical simulations.

Finally, Chapter 8 provides a short introduction to the potential applications of statistical mechanics to the life sciences. Two main examples are dealt with. The first one deals with the classic topic of mutation in genetics, which in its original formulation goes back to the middle of the last century. The second example deals with the description of the collective behaviour of a set of organisms, animals or devices. The pertinent literature in both cases is very wide, and these examples are intended as significant prototypes for an introduction to the subjects.

Acknowledgements

As often happens when doing research, we discovered the potential application of kinetic theory to the field of multiagent systems in a purely random fashion. In the early years of the new century, we were part of a large European project in applied mathematics, the HyKE project: Hyperbolic and Kinetic Equations [13]. One of the main tasks of this project was to investigate the time evolution of dissipative gases by means of the study of the dissipative Boltzmann equation. In particular, the problem of cooling of a spatially homogeneous dissipative gas attracted the interest of a number of researchers in view of an interesting conjecture by Ernst and Brito [141,142] concerning the fat tails (in terms of the velocity variable) of the self-similar solution in the case of Maxwell molecules. The phenomenon of fat tails was new in the kinetic theory of rarefied gases, since for the classical elastic Boltzmann equation the equilibrium configuration is represented by a Maxwellian function, which decays exponentially fast at infinity.

Looking at recent papers on the subject on *arXiv*,[2] through a research based on the word 'inelastic', we became aware of a work by Frantisek Slanina [287], whose title *Inelastically scattering particles and wealth distribution in an open economy* tied dissipation and wealth. Moved by curiosity, we began to enter into the diverse world of econophysics, remaining fascinated by the many analogies between statistical physics and economics and, at the same time, intrigued by the lack of rigorous mathematical theories behind them. In 2005, we published a first paper, *On a kinetic model for a simple market economy* [105], where we developed, by analogy with dissipative gases, the general methodology for the study of the asymptotic behaviour of such systems. In 2006, in *Self-similarity and power-like tails in non-conservative kinetic models* [254], we characterized the general properties of nonlinear collision-like models with respect to the formation of fat tails.

The founding of the HyKE project, made it possible to start seriously with this new field of application of mathematical kinetic theory. After that it was clear that many ideas and methodologies have analogies in other fields, such as sociology and biology, where non-Gaussian behaviours are originated by the interaction of many agents.

The results of these new applications of classical kinetic theory were announced and disseminated in a series of lectures and conferences, where the keen interest of the participants motivated the writing of this monograph, which deliberately addresses only a few relatively simple models and simulation methods. Indeed, mathematical research in the fields covered in this book is growing rapidly, leading to more and more models, with an increasing degree of difficulty. It is our opinion, however, that the knowledge of a few basic elementary facts may contribute significantly to the understanding of the underlying phenomena, and make them accessible to a larger audience.

The research in kinetic theory and numerical methods that yielded many of the results and some of the new proofs for old results in this book was partially supported by the Italian Ministry of Education, Universities and Research, and the Italian Institute of Higher Mathematics (INDAM).

[2]http://arxiv.org/archive/cond-mat

Of particular importance for the development of our research in this new field has been our continuous and valuable collaboration with Giacomo Albi, Federico Bassetti, José Antonio Carrillo, Stephane Cordier, Bertram Düring, Giacomo Dimarco, Massimo Fornasier, Peter Markowich, Daniel Matthes, Giovanni Naldi, Cyrille Piatecki, Giovanni Russo, Cedric Villani, Bernt Wennberg and many others.

The final content of the book benefited greatly from comments from numerous students and colleagues. We would like to extend our sincere thanks to all of them.

Ferrara, Pavia, September 2013

Lorenzo Pareschi, Giuseppe Toscani

Contents

PART I KINETIC MODELLING AND SIMULATION

1 A short introduction to kinetic equations 3
1.1 Boltzmann's legacy 3
1.2 Notation 8
1.3 Some linear kinetic models 11
1.4 Binary interaction models on the real line 26
1.5 Binary interaction models on the half-line 36
1.6 Some classical results 40

2 Mathematical tools 48
2.1 How to be certain of the predictions of a model? 48
2.2 Some mathematical tools 49
2.3 The drift equation and Dirac delta functions 57
2.4 Dissipative models and the drift equation 61
2.5 Growth processes 69

3 Monte Carlo strategies 79
3.1 Why Monte Carlo methods? 79
3.2 Generating random variables 83
3.3 Monte Carlo techniques 96
3.4 Applications to evolutionary PDEs 110

4 Monte Carlo methods for kinetic equations 123
4.1 The general framework 123
4.2 Relaxation problems 125
4.3 Binary interaction models 131
4.4 Asymptotic preserving Monte Carlo 139
4.5 Kinetic approximation of diffusion equations 143
4.6 Remarks on multi-dimensional problems 149

PART II MULTIAGENT KINETIC EQUATIONS

5 Models for wealth distribution 159
5.1 Wealth, trades and kinetic equations 159
5.2 Economic and kinetic dictionaries 162
5.3 Kinetic market models for conservative economies 165
5.4 Non-conservative kinetic market models 181
5.5 Exact solutions 189
5.6 Modelling heterogeneous traders 199
5.7 Individual preferences 205

5.8 Taxation and wealth redistribution 210

6 Opinion modelling and consensus formation 225
6.1 Opinion formation 225
6.2 Kinetic models of opinion formation 227
6.3 Other Fokker–Planck models of opinion formation 238
6.4 Choice formation and influence of external factors 244
6.5 Opinion formation in the presence of leaders 252

7 A further insight into economics and social sciences 259
7.1 Towards more realistic models 259
7.2 A kinetic model for trading goods 261
7.3 Modelling speculative financial markets 272
7.4 A model for different groups of traders 281
7.5 Inhomogeneous models for the evolution of wealth 295

8 Modelling in life sciences 308
8.1 The Luria–Delbrück distribution 308
8.2 The quasi-invariant limit of the growth of mutant cells 315
8.3 Self-organized systems and swarming models 322
8.4 Systems interacting with few individuals 335
8.5 Final remarks 340

Appendix A Basic arguments on Fourier transforms 341
A.1 Definitions 341
A.2 Properties of the Fourier transform 342

Appendix B Important probability distributions 346
B.1 Uniform distribution 346
B.2 Beta distribution 346
B.3 Normal distribution 347
B.4 Exponential distribution 347
B.5 Gamma distribution 348
B.6 Bernoulli distribution 348
B.7 Poisson distribution 349

List of figures 350

References 356

Index 374

Part I

Kinetic modelling and simulation

1 A short introduction to kinetic equations

> I think a strong claim can be made that the process of scientific discovery may be regarded as a form of art. This is best seen in the theoretical aspects of Physical Science. The mathematical theorist builds up on certain assumptions and according to well understood logical rules, step by step, a stately edifice, while his imaginative power brings out clearly the hidden relations between its parts. A well constructed theory is in some respects undoubtedly an artistic production. A fine example is the famous Kinetic Theory of Maxwell.
>
> Sir E. Rutherford at the Royal Academy of the Arts, 1932.

1.1 Boltzmann's legacy

The systematic study of economic, social and biological systems by means of methods typical of statistical mechanics, and, more particularly, of kinetic theory, is a relatively new field of applied mathematics.

The founding idea is that in systems composed of a huge number of living agents one can reasonably identify universal interaction rules which allow for the construction of a mathematical model following the well understood methodology of kinetic theory of rarefied gases. This idea has its roots in the recent history of kinetic modelling, which has been successfully applied to the study of interacting particle systems that come from both physical and engineering applications: electrons in micro-devices, ions in the plasma of fusion reactors, atoms in a Bose–Einstein condensate, and gas flowing over the wings of aircraft are leading examples.

Fig. 1.1 Daniel Bernoulli[1]

The experience gained from the study of those problems is proof enough that applied mathematics plays a substantial role in two basic steps of the realization

[1]**Daniel Bernoulli**: born on 8 February 1700 in Groningen (the Netherlands), died on 17 March 1782 in Basel (Switzerland). In 1738 he published *Hydrodynamica*, which laid the basis for the kinetic theory of gases. In this work, Bernoulli posited the argument that gases consist of great numbers of molecules moving in all directions, that their impact on a surface causes the gas pressure, and that what we experience as heat is simply the kinetic energy of their motion.

process: modelling and numerical simulation. On the one hand, it is important that these models must be solvable on computers at reasonable cost (or in reasonable time); on the other hand they must retain sufficient physical accuracy. These two goals are clearly opposed. Hence, the derivation of physical models with the right balance between complexity and efficiency becomes a key issue, vital to continuously improving the physical description of the relevant phenomena.

The kinetic theory of rarefied gases provided a successful attempt to describe the relevant macroscopic properties of a gas composed of an enormously large number of particles obeying the laws of classical mechanics. According to the molecular theory of matter, a macroscopic volume of gas, say one cm^3, is a system composed of a number, say 10^{20}, of molecules moving in a rather irregular way. It is evident that solving the equations of motion for any single particle is an impossible and useless task. The choice of kinetic theory was to consider the evolution, not of a system, but of a set of identical systems whose initial data differ from each other by amounts corresponding to the accepted errors [80]. The consequence of such a choice is that the only significant and useful results are those describing the behaviour of many systems in the form of statistics, that is information about probable distributions. In this statistical description, a given particle will not have a definite position and velocity, but instead certain probabilities of having different positions and velocities.

1.1.1 The Boltzmann equation

The Boltzmann equation is the most famous kinetic model, both due to its important current applications and for historical reasons. For this model, L. Boltzmann [54] proved his celebrated H-theorem about the increase of entropy, which was the first analytical proof ever of the second principle of thermodynamics. Boltzmann's analysis was subsequently adapted for numerous variants of the Boltzmann equation which are used in the description of collective dynamics in physics, engineering and biology, and in the description of the temporal transition from non-equilibrium to equilibrium in thermodynamic processes [13].

Fig. 1.2 Ludwig Eduard Boltzmann[2]

The Boltzmann equation is a partial integro-differential equation which describes the evolution of the particle density of a rarefied monatomic gas. This equation reads [82, 84]

$$\frac{\partial f(x,v,t)}{\partial t} = -v \cdot \nabla_x f(x,v,t) + Q(f,f)(x,v,t), \tag{1.1}$$

where $f(x,v,t)$ is assumed to be a non-negative function that represents the density of particles at position $x \in \mathbb{R}^3$ with velocity $v \in \mathbb{R}^3$ at time $t > 0$.

[2]**Ludwig Eduard Boltzmann**: born on 20 February 1844 in Vienna (Austria), died on 5 September 1906 in Duino, near Trieste (Italy). Boltzmann developed statistical mechanics, which uses probability to describe how the properties of atoms determine the properties of matter. In 1872 he introduced the Boltzmann equation and stated the H-Theorem. His ideas were not accepted by many scientists at that time. His tombstone bears the inscription $S = k\log(W)$.

The Boltzmann equation contains terms accounting for the two ways that the density can change. The $-v \cdot \nabla_x f(x, v, t)$ term represents the effects of free particle transport or *streaming*; that is, the motion $x_0 \mapsto x_0 + (t - t_0)v_0$ of molecules of velocity v_0 between collisions. The $Q(f, f)(x, v, t)$ term represents the effects of binary collisions and describes relaxation to the local Maxwellian equilibrium [82, 84] as a function of mass $\rho(x, t)$, velocity $U(x, t)$ and temperature $T(x, t)$:

$$M(x, v, t) = \rho(x, t) \frac{1}{(2\pi T(x, t))^{3/2}} \exp\left\{ -\frac{|v - U(x, t)|^2}{2T(x, t)} \right\}. \tag{1.2}$$

These macroscopic quantities are computed as moments of the particle density:

$$\rho = \int f\, dv, \quad U = \frac{1}{\rho} \int f v\, dv, \quad T = \frac{1}{3\rho} \int f (v - U)^2\, dv. \tag{1.3}$$

The collision term Q accounts for all kinematically possible (those that conserve both momentum and energy) binary collisions. Given a pair of particles with velocities v and w, the post–collisional velocities v^* and w^* are given by

$$v^* = \frac{1}{2}(v + w + |v - w| n), \qquad w^* = \frac{1}{2}(v + w - |v - w| n), \tag{1.4}$$

in which n is a unit vector.

The collision term depends upon the type of binary interactions. One of the most-used interaction models is furnished by the so-called Maxwellian molecules [44, 47, 77]:

$$Q(f, g)(v) = \int_{\mathbb{R}^3 \times S^2} B\left((v - w) \cdot n\right) \left[f(v^*)g(w^*) - f(v)g(w)\right] dw dn, \tag{1.5}$$

in which the collision frequency $B(\cdot)$ is independent of the modulus of the relative velocity, which leads to various simplifications. In (1.5), dn denotes the *normalized* surface measure on the unit sphere S^2. It can be easily checked by direct inspection that, if M is defined as in (1.2), $Q(M, M) = 0$.

The Boltzmann description relies on one main *ansatz*. While the number of gas molecules is normally enormously high, so that collisions between molecules do not retain memory of the incoming velocities, for rarefaction only binary collisions are important for the evolution.

In all cases in which these assumptions can be considered appropriate to describing the underlying system of interacting particles, the bilinear structure of the Boltzmann collision operator can be considered the starting point for building up the operator describing the interaction process.

1.1.2 Interacting multiagent systems

An important case in which this analogy can be fruitfully proposed, and that will be described in detail in the forthcoming chapters, is the natural description of emergent phenomena in a system of interacting agents.

Among others, trading markets composed of a large number of agents fit this analogy. Starting from the pioneering works of Mandelbrot [223] on distribution of income in a trading society, it is now commonly accepted that in many aspects a socio-economic system can be described using the laws of statistical mechanics.

In fact, as far as wealth distribution is concerned, there is an almost literal translation of concepts: *molecules* are identified with the *agents*, the *particles' energy* corresponds to the *agents' wealth*, and binary *collisions* translate into *trade interactions*.

Fig. 1.3 James Clerk Maxwell[3]

This modelling is clearly rather ad hoc, but if one is willing to accept the proposed analogies between interacting agents and colliding particles, then various well-established methods from kinetic theory and statistical physics are ready for application to these socio-economic fields. Most notably, the numerous tools originally devised for the study of the energy distribution in a rarefied gas can now be used to analyze the underlying steady distributions. In this way, the kinetic description of market models via a Boltzmann-type equation provides one possible explanation for the development of universal profiles in the wealth distributions of real economies.

It is clear from the previous discussion that the most important aspect in this new modelling of socio-economic phenomena is the relaxation towards equilibrium, or, more generally, the evolution of density due to binary interactions. Hence, the main legacy of Boltzmann's ideas are related both to the modelling of the interaction operator which follows from the microscopic interactions between agents, and to the consequent evolution of the density.

In the kinetic theory of rarefied gases, and for Maxwellian molecules, this problem has been intensively studied for many years now [44, 83]. Typically, one investigates the behaviour of the (spatially uniform) density $f(v,t)$ which satisfies the equation

$$\frac{\partial f(v,t)}{\partial t} = Q(f,f)(v,t), \tag{1.6}$$

or, more simply, its linear version

$$\frac{\partial f(v,t)}{\partial t} = Q(f,M)(v,t), \tag{1.7}$$

where Q is the collision operator (1.5), while M is the Maxwellian function (1.2).

The meanings of the two equations (1.6) and (1.7) are deeply different. In the former, particles of the gas interact with each other, and the relaxation process is due to the internal collision rule. In the latter, particles of the gas interact with an external background (the Maxwellian M), and the relaxation process is caused by an external collision rule.

The strong link between the Boltzmann equation for the spatially homogeneous Maxwell molecules (1.6) for dissipative interactions and its application to the study of

[3]**James Clerk Maxwell**: born on 13 June 1831 in Edinburgh (Scotland), died on 5 November 1879 in Cambridge (United Kingdom). In 1859, after reading a paper by Clausius, he formulated the Maxwell distribution of molecular velocities. In 1867 he introduced a kinetic equation which is formally equivalent to the Boltzmann equation. He showed that the Maxwellian densities are equilibria of the kinetic equation and observed that the equation simplifies for the case of what are now called Maxwell molecules.

non-conservative economies based on binary trades has been outlined, among others, in [287]. There, a clear parallelism between the evolution of wealth in a simple economy and the evolution of the particle density in a one-dimensional dissipative gas has been established. This paper motivated us to eventually adapt more and more of the ideas developed in the studies of dissipative Maxwell gases to the economic framework.

It should be emphasized, however, that there are substantial differences between the collision mechanism for classical gas molecules and the modelling of human interactions. In the new framework, interactions can lack the usual microscopic conservation laws for (the analogues of) momentum and energy; moreover, random effects play a crucial role. For example, the key step in establishing a reasonable kinetic market model is the definition of sensible rules on the *microscopic* level, i.e., the prescription of how wealth is exchanged in trades. Such rules are usually derived from plausible assumptions in a largely questionable manner (this is clearly in contrast to the original Boltzmann equation, where the microscopic collisions are governed by the laws (1.4)).

Nevertheless, strong analogies remain. Maybe the most important one is related to the fact that, as happens in the classical Boltzmann equation [82, 84], where relaxation to Maxwellian equilibrium is shown to be a universal behaviour of the solution, in the description of a socio-economic phenomenon through a density function, the corresponding outputs of the model are the *macroscopic* statistics of the underlying distribution in the society, to which the solution is shown to relax.

Most importantly, while relaxation to equilibrium in the Boltzmann equation is achieved by looking at the monotonicity of the entropy functional, relaxation to the steady wealth profile can be achieved by looking instead at the monotonicity of new convex functionals.

In general, the richness of the steady states for socio-economic models is another remarkable difference from the theory of Maxwell molecules [44]. For example, while the Maxwell distribution (1.2) is the universal steady profile for the velocity distribution of molecular gases, the stationary profiles for wealth can be manifold, and are in general not explicitly known analytically. In fact, they depend heavily on the precise form of the microscopic modelling of trade interactions. Consequently, in investigations of the large-time behaviour of the wealth distribution, one is typically limited to describing a few analytically accessible properties (e.g. moments and smoothness) of the latter.

The first goal of our analysis will be the step-by-step construction of a kinetic model starting from a precise type of binary interaction. This construction will emphasize the link between the kinetic description of the interactions in the spirit of Boltzmann's ideas and some basic arguments from probability theory. In fact, there are many analogies between the relaxation processes in linear or nonlinear interactions of Maxwell type and the various forms of the central limit theorem. Likewise, the use of Fourier transform methods is a natural extension of the pioneering works of Bobylev [43, 44], and their links with probabilistic applications.

Before entering into the new fields of the kinetic description of socio-economic properties of multiagent systems, we will introduce with some details various well-known kinetic models, starting from models with a linear interaction operator. By analogy with the classical kinetic theory of rarefied gases, in which the Boltzmann equation describes the one-particle density of particles which interact through binary collisions, in our approach we will always deal with a system of agents which are forced to interact according to well-defined laws for binary interactions. In particular, we will emphasize the probabilistic aspects of the construction of the operators, which will be useful when trying to approximate them by numerical methods.

1.2 Notation

In this book we will be mainly concerned with the study of the large-time behaviour of the solution to various partial differential and/or integro-differential equations similar in structure to the Boltzmann equations (1.6) and (1.7). These equations describe the evolution of the one-particle density in a system of interacting particles, or, more generally, in a system of interacting agents. In most of our applications the independent variable, say v, is one-dimensional, so that this particle density, which we will denote by $f(v,t)$, will satisfy, for $v \in \mathbb{R}$ and $t \in \mathbb{R}^+$, the equation

$$\frac{\partial f(v,t)}{\partial t} = I(f)(v,t). \tag{1.8}$$

The operator $I(f)$ describes the variation of the particle density due to interactions. As introduced in Section 1.1, the kinetic theory of rarefied gases, where interactions are given by (1.4), is described in the Boltzmann model by an interaction operator like (1.5).

In analogy with the Boltzmann equation, the interaction operator on the right-hand side of the kinetic equation (1.8) always satisfies two natural properties. First, for each time $t \geq 0$,

$$\int_{\mathbb{R}} I(f)(v,t)\, dv = 0. \tag{1.9}$$

Equality (1.9) is the mathematical translation of a natural requirement on our physical or socio-economic system: the mass density conservation, which ensures that particles (or agents) cannot disappear. In consequence of condition (1.9), taking the integral on both sides of (1.8) shows, in fact, that

$$\frac{d}{dt}\int_{\mathbb{R}} f(v,t)\, dv = 0,$$

or, alternatively, if $f_0(v)$ denotes the initial one-particle density,

$$\int_{\mathbb{R}} f(v,t)\, dv = \int_{\mathbb{R}} f_0(v)\, dv. \tag{1.10}$$

Second, even if this condition is not expressed by an explicit formula, the operator $I(\cdot)$ is *positivity* preserving. Together with condition (1.10), this property guarantees that the kinetic equation is well posed, in the sense that it maps a non-negative density function with a certain mass to a non-negative density function with equal mass at any subsequent time.

A typical case of a positivity-preserving operator which we will deal with in the remainder of this book is

$$I(f) = I_+(f) - \sigma f,$$

where $I_+(f) \geq 0$ if $f \geq 0$. In this situation, writing the solution to equation (1.8) via the so-called Duhamel formula,

$$f(v,t) = f_0(v)e^{-\sigma t} + \int_{\mathbb{R}} e^{-\sigma(t-s)} I_+(f(v,s))\,ds, \tag{1.11}$$

shows that, starting from a non-negative initial density, the solution at any subsequent time remains non-negative.

Mass conservation relates kinetic theory with probability theory. If we choose the initial density $f_0(v)$ of mass equal to one, the solution to equation (1.8) remains a density of mass equal to one for any $t \geq 0$. Thus, equation (1.8) describes the time evolution of a probability density, and one can fix this density to be that of a random process $X(t)$ [165, 200].

1.2.1 Observable quantities

This analogy suggests the introduction of a number of operating instruments that will form the basis of the methodology used in this book. The natural use of a probability density relies on the computation of averages. If we know the probability density $f(v)$ of a random variable X, we can compute the average value of any given function $\varphi(v)$. As a matter of fact, we can define averages as

$$\langle \varphi(X) \rangle = \langle \varphi, f \rangle = \int_{\mathbb{R}} \varphi(v) f(v)\,dv, \tag{1.12}$$

where brackets are conventional notation for averages. We will also say that $\varphi(\cdot)$ is an *observable* quantity. Among the various observables, some are of particular importance. The case $\varphi(v) = v$ and $\varphi(v) = |v|^\alpha$, with $\alpha > 0$, define the mean value and the moment of order α, respectively, of the random variable X. As we shall see, in most applications the knowledge that some moment of the solution to equation (1.8) remains uniformly bounded in time will be of paramount importance in the analysis of the evolution of its solution.

To fix notations, we will denote by $\mathcal{P}(\mathbb{R})$ the space of all probability densities in $\mathbb{R}$, and by

$$\mathcal{P}_\alpha = \left\{ f \in \mathcal{P} : \int_{\mathbb{R}} |v|^\alpha f(v)\,dv < +\infty, \alpha \geq 0 \right\} \tag{1.13}$$

the space of all probability densities of finite momentum of order α.

Observables will also play an important role in the following definition of a solution for equation (1.8), corresponding to an initial density $f_0 \in \mathcal{P}_\alpha$ (initial value problem). By a weak solution of the initial value problem for equation (1.8), corresponding to the initial probability density $f_0(v) \in \mathcal{P}_\alpha, \alpha \geq 0$, we shall mean any probability density $f(v,t)$ satisfying the weak form of the equation

$$\frac{d}{dt}\int_{\mathbb{R}} \varphi(v) f(v,t)\,dv = \int_{\mathbb{R}} \varphi(v) I(f)(v,t)\,dv, \tag{1.14}$$

for $t > 0$ and all (smooth) observable functions φ, and such that for all φ

$$\lim_{t\to 0} \int_{\mathbb{R}} \varphi(v) f(v,t)\, dv = \int_{\mathbb{R}} \varphi(v) f_0(v)\, dv. \tag{1.15}$$

The physical meaning of this definition is clear. One substitutes the complete knowledge of the solution to equation (1.8) with the knowledge of the action this solution has on observables. This notion most importantly allows us to include certainty among random phenomena. If the random variable X is known with complete certainty, in that it has the definite value $X = \bar{v}$, then it means that its probability density, say $\bar{f}(v)$, has to be equal to zero for any value $v \neq \bar{v}$. On the other hand, the equation

$$\int_{\mathbb{R}} \bar{f}(v)\, dv = 1$$

has to be satisfied. No ordinary function can satisfy these requirements, and if one wants to include certainty as a particular case of probability one needs to enlarge the previous concept of probability density. The required generalization has been achieved by resorting to the concept of *generalized* functions or *distributions* [211]. Note that this inconsistency does not hold at the level of observables. If X is known with complete certainty, and has the definite value $X = \bar{v}$,

$$\langle \varphi(X) \rangle = \langle \varphi, \bar{f} \rangle = \varphi(\bar{v}) \tag{1.16}$$

for any (smooth) observable $\varphi(\cdot)$. In the case in which $\bar{v} = 0$, we will define the *generalized density* of the variable X which assumes the value 0 with certainty as the Dirac delta function (concentrating at $v = 0$), and we will denote it by $\varepsilon_0(v)$. Hence

$$\langle \varphi(X) \rangle = \langle \varphi, \delta_0 \rangle = \int_{\mathbb{R}} \varphi(v) \varepsilon_0(v)\, dv = \varphi(0).$$

We will postpone a more exhaustive discussion on generalized functions to Chapter 2, Section 2.3. The Dirac delta function allows us to denote in a unified way the probability densities associated with both continuous and discrete random variables. If X is the random variable who can assume the values a and b with probabilities p and q respectively ($p + q = 1$), we will denote its probability density by

$$g(v) = p\, \varepsilon_0(v - a) + q\, \varepsilon_0(v - b),$$

and we will treat it as a normal density concerning the evaluation of observables, so that

$$\langle \varphi(X) \rangle = \langle \varphi, g \rangle = p\, \varphi(a) + q\, \varphi(b).$$

In addition to moments, other observables will be intensively used in this book. For any given $\xi \in \mathbb{R}$, we will denote by

$$\mathcal{F}_\xi(v) = e^{-i\xi v}$$

the observable which defines the Fourier transform [59]. If the random variable X has probability density $f(v)$,

$$\langle \mathcal{F}_\xi(X)\rangle = \langle e^{-i\xi X}\rangle = \int_{\mathbb{R}} e^{-i\xi v} f(v)\, dv. \tag{1.17}$$

The use of Fourier transforms will be ubiquitous in this book. Given $f \in \mathcal{P}(\mathbb{R})$, we will often denote (cf. Appendix A)

$$\hat{f}(\xi) = \int_{\mathbb{R}} e^{-iv\xi}\, f(v)\, dv. \tag{1.18}$$

Since the Fourier transform is defined starting from observables, it follows that it is well defined even for generalized functions. Also, any time the weak form (1.14) can be expressed in closed form in terms of the Fourier transform, the representation of equation (1.8) in terms of the Fourier transform will be preferred to other weak forms. As we shall see, most of the kinetic models we will be concerned with in this book satisfy this property.

Last, since we will often deal with kinetic models in which the independent variable can assume only non-negative values (the wealth of an agent is the leading example), we will also consider the Laplace transform of a probability density $f(v)$, with $v \in \mathbb{R}^+$. For any given complex number $s \in \mathbb{C}$, we will denote by

$$\mathcal{L}_s(v) = e^{-sv}$$

the observable which defines the Laplace transform [59]

$$\langle \mathcal{L}_s(X)\rangle = \langle e^{-sX}\rangle = \int_0^{+\infty} e^{-sv} f(v)\, dv. \tag{1.19}$$

Given $f \in \mathcal{P}(\mathbb{R})$, we will often use the notation

$$\tilde{f}(s) = \int_{\mathbb{R}+} e^{-s\xi}\, f(v)\, dv.$$

Analogously to the Fourier transform, the Laplace transform is defined starting from observables, and it is well defined even for generalized functions. Both the Fourier and Laplace transforms satisfy a number of properties, which are assumed to be known hereafter. Nevertheless, for reasons of clarity and completeness, in many cases the derivation of some of these properties will be illustrated in detail.

1.3 Some linear kinetic models

1.3.1 The Goldstein–Taylor model

In order to understand the principal ingredients which lead to the construction of a linear kinetic equation, we will start by considering one of the best-known models with a finite number of allowed velocities, which was introduced independently by Taylor in 1921 [303] and Goldstein in 1951 [160].

The model describes the time evolution of the two particle densities $f^+(x,t)$ and $f^-(x,t)$, with $x \in \mathbb{R}$ and $t \in \mathbb{R}^+$, where $f^+(x,t)$ (respectively $f^-(x,t)$) denotes the density of particles at time $t > 0$ which travel along a straight line with velocity $+c$

(respectively $-c$), and can at the same time change and assume the opposite velocity at random.

In Goldstein's original description, at time $t = 0$ a large number of non interacting particles start from an origin and move with a uniform velocity of absolute value c along a straight line for an interval of time t. To begin with, half move in each direction. Thereafter, and at the end of each successive interval of time t, each particle starts a new partial path; it still moves with speed c, and there is a probability α that it will continue to move in the same direction as its previous path, and a probability $\beta(= 1-\alpha)$ that the direction of its velocity will be reversed. Consequently, the directions in any two consecutive intervals are correlated with a correlation coefficient $\alpha - \beta$. The partial correlations for non-consecutive intervals are zero. The limit behaviour of the particle density functions as the number of particles tends to infinity was shown to obey the so-called Goldstein–Taylor system

Fig. 1.4 Sir Geoffrey Ingram Taylor[4]

$$\begin{aligned} \frac{\partial f^+(x,t)}{\partial t} + c\frac{\partial f^+(x,t)}{\partial x} &= \sigma\left(f^-(x,t) - f^+(x,t)\right), \\ \frac{\partial f^-(x,t)}{\partial t} - c\frac{\partial f^-(x,t)}{\partial x} &= \sigma\left(f^+(x,t) - f^-(x,t)\right). \end{aligned} \tag{1.20}$$

The time variation of the densities in (1.20) is due to the fact that the particles possess a velocity of absolute value c which transports them in time (the left-hand side operators), and also a random variation of their velocities (the right-hand side operators). If at each time step we consider sequentially the transport and interaction operators, during this short time interval the evolution of the density is given by the joint action of the transport

$$\begin{aligned} \frac{\partial f^+(x,t)}{\partial t} + c\frac{\partial f^+(x,t)}{\partial x} &= 0 \\ \frac{\partial f^-(x,t)}{\partial t} - c\frac{\partial f^-(x,t)}{\partial x} &= 0, \end{aligned} \tag{1.21}$$

and interaction

$$\begin{aligned} \frac{\partial f^+(x,t)}{\partial t} &= \sigma\left(f^-(x,t) - f^+(x,t)\right) \\ \frac{\partial f^-(x,t)}{\partial t} &= \sigma\left(f^+(x,t) - f^-(x,t)\right). \end{aligned} \tag{1.22}$$

The transport equations (1.21) can be solved explicitly. If $f_0^+(x)$ and $f_0^-(x)$ denote the initial densities, the densities at time $t > 0$ are given by

[4]**Sir Geoffrey Ingram Taylor**: born on 7 March 1886 in London (United Kingdom), died on 27 June 1975 in Cambridge (United Kingdom). He was one of the most notable scientists on fluid dynamics and wave theory of the 20th century. Several physical phenomena are named after him, like Taylor cone, Taylor dispersion, Taylor–Couette flow, Rayleigh–Taylor instability, Taylor–Green vortex and many others.

$$f^+(x,t) = f_0^+(x - ct); \quad f^-(x,t) = f_0^-(x + ct). \tag{1.23}$$

In fact, the number of particles moving to the right (respectively to the left) found at time $t > 0$ in position x is given by the same number of particles which at time $t = 0$ are located at $x - ct$ (respectively $x + ct$).

Let us now discuss in some detail the meaning of the interaction operator (1.22). Taking the sum of both equations shows that, for any given x, the total density $\rho(x,t) = f^+(x,t) + f^-(x,t)$ is preserved in time. Thus, unless $f^+(x,t) = f^-(x,t) = 0$, the total density $\rho(x,t) = \rho_0(x)$ is different from zero, and we can set (dropping the dependence on x)

$$p(t) = \frac{f^+(x,t)}{\rho_0(x)}, \quad q(t) = \frac{f^-(x,t)}{\rho_0(x)}.$$

Since $p(t) + q(t) = 1$ for all $t \geq 0$, these quantities can be interpreted as the proportion of particles with velocities $+c$ and $-c$ which at time t are located at position x. Thus, we can define, at each time $t \geq 0$, a discrete random process $X(t)$ which assumes the values $+c$ and $-c$ with probabilities $p(t)$ and $q(t)$ respectively, so that

$$P(X(t) = +c) = p(t); \quad P(X(t) = -c) = q(t).$$

The evolution of $p(t)$ and $q(t)$, starting from $p(t = 0) = p_0$ and $q(t = 0) = q_0$, can be easily derived from system (1.22), to give

$$\begin{aligned} \frac{dp(t)}{dt} &= \sigma\left(q(t) - p(t)\right), \\ \frac{dq(t)}{dt} &= \sigma\left(p(t) - q(t)\right). \end{aligned} \tag{1.24}$$

Derivation from discrete random processes

In order to understand how system (1.24) can be derived from kinetic considerations, let us introduce a related problem, in which the proportion of particles with velocities $\pm c$ is described by $X(t)$, and the variation of $X(t)$ is due to interaction with an external background. Let M be a discrete random variable (the background), which characterizes the variation of the velocity of particles through interaction with it. To preserve the set of allowed velocities, this random variable can assume the two values $+1$ and -1, with constant-in-time probabilities α and β, respectively,

$$P(M = +1) = \alpha; \quad P(M = -1) = \beta.$$

If $v = \pm c$ is the velocity before the interaction, we assume that the post-interaction velocities are given by

$$v^* = wv \tag{1.25}$$

where w is sampled by the random variable M.

Next, we will introduce a rule for the time variation of the process $X(t)$ due to interactions. To this end we will assume that, in a short time interval Δt, there is a probability of interacting with the background proportional to Δt. The simplest way to describe this mechanism is to introduce a random variable T_μ, independent of both $X(t)$ and M, that takes values 0, 1 with probabilities

$$P(T_\mu = 1) = \mu\Delta t, \quad P(T_\mu = 0) = 1 - \mu\Delta t. \tag{1.26}$$

In (1.26) the positive constant μ furnishes a measure of the interaction frequency. Note that, for a given μ, the time interval Δt has to be chosen so that $\mu\Delta t < 1$. If this is the case, the value of $X(t + \Delta t)$ is related to the value of $X(t)$ as follows:

$$X(t + \Delta t) = (1 - T_\mu)X(t) + T_\mu MX(t). \tag{1.27}$$

For any given observable quantity $\varphi(\cdot)$, let us evaluate the mean value of $\varphi(X)$, which we will denote by $\langle\varphi(X(t))\rangle$. Owing to (1.27), we can first calculate the mean value with respect to T_μ, to obtain

$$\begin{aligned}\langle\varphi(X(t + \Delta t))\rangle &= \langle\varphi((1 - T_\mu)X(t) + T_\mu MX(t))\rangle \\ &= (1 - \mu\Delta t)\langle\varphi(X(t))\rangle + \mu\Delta t\langle\varphi(MX(t))\rangle.\end{aligned} \tag{1.28}$$

Within this assumption on the frequency of interactions, in a small interval of time Δt we can describe the law of variation of the observable as

$$\frac{\langle\varphi(X(t + \Delta t))\rangle - \langle\varphi(X(t))\rangle}{\Delta t} = \mu\left(\langle\varphi(MX(t))\rangle - \langle\varphi(X(t))\rangle\right). \tag{1.29}$$

Equation (1.29) is a discrete-in-time approximation to the differential equation

$$\frac{d}{dt}\langle\varphi(X(t))\rangle = \mu\left(\langle\varphi(MX(t)) - \varphi(X(t))\rangle\right). \tag{1.30}$$

Hence, equation (1.30) describes the instantaneous variation of the mean value of the observable due to the interaction (1.25) with the background.

The evaluation of the mean value $\langle\varphi(MX(t))\rangle$ is immediate:

$$\begin{aligned}\langle\varphi(MX(t))\rangle = \varphi(+c)&\left[P(X(t) = +c,\ M = +1) + P(X(t) = -c,\ M = -1)\right] \\ +\varphi(-c)&\left[P(X(t) = +c,\ M = -1) + P(X(t) = -c,\ M = +1)\right].\end{aligned}$$

Here comes the main assumption which leads to system (1.24). If the random variables which describe both the particles and the background are independent from each other, then

$$\begin{aligned}\langle\varphi(MX(t))\rangle &= \varphi(+c)\left[P(X(t) = +c)P(M = +1) + P(X(t) = -c)P(M = -1)\right] \\ &\quad +\varphi(-c)\left[P(X(t) = +c)P(M = -1) + P(X(t) = -c)P(M = +1)\right] \\ &= \varphi(+c)\left[\alpha p(t) + \beta q(t)\right] + \varphi(-c)\left[\beta p(t) + \alpha q(t)\right].\end{aligned}$$

Finally

$$\langle\varphi(MX(t)) - \varphi(X)\rangle = \varphi(+c)\beta\left[q(t) - p(t)\right] + \varphi(-c)\beta\left[p(t) - q(t)\right]. \tag{1.31}$$

Substituting (1.31) into (1.30), we obtain that the probabilities $p(t)$ and $q(t)$ satisfy, for all (smooth) functions $\varphi(\cdot)$,

$$\frac{d}{dt}\left[\varphi(+c)p(t) + \varphi(-c)q(t)\right] = \varphi(+c)\mu\beta\left[q(t) - p(t)\right] + \varphi(-c)\mu\beta\left[p(t) - q(t)\right]. \tag{1.32}$$

Note that equation (1.32) is the weak form of system (1.24), with frequency $\sigma = \mu\beta$. In this case, the frequency of interactions σ equals the product of the probability of changing velocity through interaction with the background, and of the frequency with which the interactions take place.

We note that the random variable T_μ which describes the frequency at which interactions happen in a time interval Δt can be chosen in various ways. A different law can be given by choosing

$$P(T_\mu = 1) = \frac{\mu \Delta t}{1 + \mu \Delta t}, \quad P(T_\mu = 0) = \frac{1}{1 + \mu \Delta t}$$

which is such that, for any given time interval Δt, the probabilities are non-negative for any value of the frequency μ. This choice also leads to the instantaneous law of variation (1.30). This shows that equation (1.30) is the arrival point of a number of approximations at discrete times. This is particularly useful to know when one is faced with the numerical approximation of the continuous-in-time equation.

System (1.24) can also be derived from kinetic considerations in a slightly different way. In fact, since $p(t) + q(t) = 1$, $q(t) - p(t) = 1 - 2p(t)$, and system (1.24) can be equivalently written as

$$\begin{aligned} \frac{dp(t)}{dt} &= 2\sigma \left(\frac{1}{2} - p(t) \right), \\ \frac{dq(t)}{dt} &= 2\sigma \left(\frac{1}{2} - q(t) \right). \end{aligned} \tag{1.33}$$

As before, let us introduce a related problem, in which the proportion of particles with velocities $\pm c$ are described by $X(t)$, and the variation of $X(t)$ is due to interaction with an external background. This time, the background M is a discrete random variable which can still assume the two values $+c$ and $-c$ with equal probabilities $1/2$,

$$P(M = +c) = P(M = -c) = \frac{1}{2}.$$

If $v = \pm c$ is the velocity before the interaction, we assume that the post-interaction velocities are given by

$$v^* = w \tag{1.34}$$

where w is sampled by the random variable M.

With these rules of interaction, and proceeding as before with interactions that happen with a frequency μ, for a given observable quantity $\varphi(\cdot)$ we then have that its mean value satisfies

$$\frac{d}{dt}\langle \varphi(X(t)) \rangle = \mu \left(\langle \varphi(M) - \varphi(X) \rangle \right). \tag{1.35}$$

Therefore

$$\langle \varphi(M) - \varphi(X) \rangle = \varphi(+c) \left[\frac{1}{2} - p(t) \right] + \varphi(-c) \left[\frac{1}{2} - q(t) \right]. \tag{1.36}$$

Substituting (1.36) into (1.35), we obtain that the probabilities $p(t)$ and $q(t)$ satisfy, for all (smooth) functions $\varphi(\cdot)$,

$$\frac{d}{dt} \left[\varphi(+c)p(t) + \varphi(-c)q(t) \right] = \mu \left\{ \varphi(+c) \left[\frac{1}{2} - p(t) \right] + \varphi(-c) \left[\frac{1}{2} - q(t) \right] \right\}. \tag{1.37}$$

Note that equation (1.37) is the weak form of system (1.24), with $\sigma = \mu/2$.

Relaxation towards the steady state

System (1.24) can be easily solved. One obtains

$$\begin{aligned} p(t) &= \frac{1}{2} + (p_0 - q_0)\, e^{-2\sigma t} \\ q(t) &= \frac{1}{2} + (p_0 - q_0)\, e^{-2\sigma t}. \end{aligned} \tag{1.38}$$

So the random variable $X(t)$ converges exponentially (in distribution) with a rate 2σ to the centred random variable X_∞ defined by

$$P(X_\infty = +c) = P(X_\infty = -c) = \frac{1}{2}.$$

Exponential relaxation towards the universal steady state given by X_∞ can also be understood in terms of arguments typical of statistical mechanics. Let $\Phi(r)$, $r > 0$ be a (regular) convex function. Then, since $\Phi'(r)$ is not decreasing,

$$\begin{aligned} \frac{d}{dt}\left[\Phi(p(t)) + \Phi(q(t))\right] &= \Phi'(p(t))\frac{dp(t)}{dt} + \Phi'(q(t))\frac{dq(t)}{dt} \\ &= -\sigma\left[\Phi'(p(t)) - \Phi'(q(t))\right](p(t) - q(t)) \le 0. \end{aligned}$$

Therefore, for any choice of a pair of initial values p_0, q_0,

$$\Phi(p_0) + \Phi(q_0) \ge 2\Phi(\frac{1}{2}) = \Phi(p_\infty) + \Phi(q_\infty).$$

The action of the background results in driving the system towards the state in which the value of $\Phi(p) + \Phi(q)$ reaches its minimum. It is then natural, for a given convex function Φ, to associate to the discrete random variable X, defined by

$$P(X = +c) = p, \quad P(X = -c) = q,$$

the entropy functional

$$S(X) = -\Phi(X) = -\Phi(p) - \Phi(q). \tag{1.39}$$

In terms of the entropy functional, in the set of all random variables X which assume only two values, the maximum of the entropy is achieved by the steady state X_∞:

$$S(X_\infty) \ge S(X).$$

Remark 1.1 *Despite its relative simplicity, the Goldstein–Taylor model exhibits many of the characteristic properties of more complex kinetic equations, like the standard Boltzmann equation (1.1). In particular, due to the possibility of performing exact computations, one can verify that the role of the operator which describes the interactions with the background is to push the solution towards a universal steady state* *(which we denoted by X_∞), at an exponential-in-time rate proportional to the interaction frequency. The typical definition used in kinetic theory for this type of operator is* relaxation operator.

Also, it has been pointed out that for the relaxation operator relative to the Goldstein–Taylor model the convergence towards the steady state is such that any convex functional of the solution is monotonically decreasing in time. This is a general property, common to most of the relaxation operators we will introduce in the following chapters. It is remarkable that this physical property, when the analytical solution is not accessible, constitutes the main mathematical argument for obtaining rates of convergence towards the steady state.

1.3.2 Radiative transfer

Radiative transfer problems are encountered in a wide and varied range of applications such as, for instance, asymptotics of Schrödinger equations with a random potential whose characteristic scale matches that of the wave function [17, 122, 144, 290]. They are also used in climate evolution modelling [304, 331], in astrophysics, since the early works of Chandrasekhar [93], or for neutron transport phenomena [78]. In the three-dimensional physical space it reads

$$\left(\frac{\partial f}{\partial t} + v \cdot \nabla_x f\right)(x, v, t) = \sigma\left(\int_\Omega f(x, w, t)\, dw - \mathfrak{m}(\Omega) f(x, v, t)\right). \tag{1.40}$$

Fig. 1.5 Subrahmanyan Chandrasekhar[5] © Bettmann/CORBIS

The unknown is the non-negative specific radiation intensity $f(x, v, t)$ which depends on position $x \in \mathbb{R}^3$, velocity $v \in \Omega$ and time $t \in \mathbb{R}_+$. In general, Ω is a convex subset of $\mathbb{R}^3$ of finite measure $\mathfrak{m}(\Omega)$. As in the previous case, the relaxation operator is linear. Dropping the dependence on the spatial variable x, relaxation is here driven by the equation

$$\frac{\partial f(v,t)}{\partial t} = \bar{\sigma}\left(\frac{1}{\mathfrak{m}(\Omega)}\int_\Omega f(w, t)\, dw - f(v, t)\right), \tag{1.41}$$

where we denoted

$$\bar{\sigma} = \mathfrak{m}(\Omega)\sigma.$$

Taking the integral over Ω on both sides of (1.41) shows that the mass of the solution is preserved in time,

$$\int_\Omega f(w, t)\, dw = \int_\Omega f_0(w)\, dw, \tag{1.42}$$

[5]**Subrahmanyan Chandrasekhar**: born on 19 October 1910 in Lahore, Punjab (India), died on 21 August 1995 in Chicago, Illinois (USA). He worked in several areas including stellar structure and dynamics, theory of radiative transfer, quantum theory, hydrodynamic and hydromagnetic stability, general relativity, mathematical theory of black holes and theory of colliding gravitational waves. In 1983 he won the Nobel Prize in Physics for his theoretical studies of the physical processes of importance to the structure and evolution of the stars.

and the relaxation operator can be equivalently written as

$$\frac{\partial f(v,t)}{\partial t} = \bar{\sigma}\left(\frac{1}{\mathfrak{m}(\Omega)}\int_\Omega f_0(w)\,dw - f(v,t)\right). \tag{1.43}$$

Thanks to (1.43), it is then immediate to obtain the stationary solution, which appears to be the unique constant function in Ω of mass equal to the mass of the initial distribution

$$f_\infty(v) = \frac{1}{\mathfrak{m}(\Omega)}\int_\Omega f_0(w)\,dw.$$

The one-dimensional Goldstein–Taylor model studied in Section 1.3.1 appears to match the so-called *two-stream* approximation of (1.41), see [304, 331].

Derivation from a random dynamic

Driven by this remark, we will approach the derivation of the interaction operator in (1.41) by using the second kinetic description of the interaction operator in the Goldstein–Taylor model. To this end, let us consider the related problem in which the number of particles with velocity v at time t is described by the process $X(t)$, and the variation of $X(t)$ is due to interaction with the external background. Let us assume that the background M is a uniform random variable taking values on the domain Ω. Hence the probability density $f_M(w)$ of M is given by

$$\begin{aligned} f_M(w) &= \frac{1}{\mathfrak{m}(\Omega)} \quad \text{if} \quad w \in \Omega, \\ f_M(w) &= 0 \quad \text{if} \quad w \notin \Omega. \end{aligned} \tag{1.44}$$

For a given velocity $v \in \mathbb{R}^3$ let the post-interaction velocity be given by

$$v^* = w \quad \text{if} \quad w \in \Omega, \qquad v^* = v \quad \text{if} \quad w \notin \Omega, \tag{1.45}$$

where w is sampled by the random variable M. From a physical point of view this means that a particle with pre-interaction velocity $v \in \Omega$ is spread out uniformly after interaction with the background, while its velocity is left unchanged whenever $v \notin \Omega$.

In terms of the process $X(t)$ the law of change given by (1.45) can be rewritten as

$$X^*(t) = MI_\Omega(X(t)) + X(t)(1 - I_\Omega(X(t)), \tag{1.46}$$

where we denote by $I_A(x)$ the indicator function of the set A, that is $I_A(x) = 1$ if $x \in A$, while $I_A(x) = 0$ if $x \notin A$.

Let us consider now a short time interval Δt, and let us repeat the steps described in detail in Section 1.3.1 to obtain the instantaneous law of variation of the mean value of observable quantities. Using (1.46), for a given observable quantity $\varphi(\cdot)$, we have that the variation of its mean value $\varphi(X)$ satisfies

$$\frac{d}{dt}\langle\varphi(X(t))\rangle = \mu\left\{\langle\varphi(MI_\Omega(X(t)) + X(t)(1 - I_\Omega(X(t))\rangle - \langle\varphi(X(t))\rangle\right\}, \tag{1.47}$$

where the positive constant μ is the interaction frequency. Let $\pi(v, w, t)$ be the joint distribution function of the pair $(X(t), M)$, that is, for $A, B \subseteq \mathbb{R}^3$,

$$\begin{aligned} \int_{A\times\mathbb{R}^3} \pi(v, w, t)\, dv\, dw &= \int_A f(v, t)\, dv, \\ \int_{\mathbb{R}^3\times B} \pi(v, w, t)\, dv\, dw &= \int_{B\cap\Omega} f_M(w)\, dw = \frac{\mathfrak{m}(B\cap\Omega)}{\mathfrak{m}(\Omega)}. \end{aligned} \tag{1.48}$$

Then, equation (1.47) can be rewritten as

$$\frac{d}{dt}\int_{\mathbb{R}^3} \varphi(v) f(v, t)\, dv = \mu \int_{\mathbb{R}^3\times\mathbb{R}^3} \left(\varphi(v^*) - \varphi(v)\right) \pi(v, w, t)\, dv\, dw. \tag{1.49}$$

By virtue both of the definition of $f_M(w)$ and of the interaction (1.45), it is easy to conclude that

$$\int_{\mathbb{R}^3\times\mathbb{R}^3} \left(\varphi(v^*) - \varphi(v)\right) \pi(v, w, t)\, dv\, dw = \int_{\Omega\times\Omega} \left(\varphi(w) - \varphi(v)\right) \pi(v, w, t)\, dv\, dw.$$

Consequently, choosing a test function $\varphi(v)$ which is equal to zero for $v \in \Omega$ shows that $f(v, t)$ remains constant outside the set Ω.

To proceed, let us remark that the variation of the process $X(t)$ as defined in (1.46) depends on the joint action of the pair $(X(t), M)$. To obtain a closed form in the evolution of the mean, one has to additionally assume that $X(t)$ and M are independent of each other. If this is the case,

$$\begin{aligned} &\int_{\Omega\times\Omega} \left(\varphi(w) - \varphi(v)\right) \pi(v, w, t)\, dv\, dw \\ &\quad = \int_\Omega \varphi(w) \frac{dw}{\mathfrak{m}(\Omega)} \int_\Omega f(v, t)\, dv - \int_\Omega \varphi(v) f(v, t)\, dv \int_\Omega \frac{dw}{\mathfrak{m}(\Omega)} \\ &\quad = \int_\Omega \varphi(v) \left(\frac{1}{\mathfrak{m}(\Omega)} \int_\Omega f(w, t)\, dw - f(v, t)\right) dv. \end{aligned}$$

Finally, equation (1.47) coincides with the weak form of equation (1.41), with $\bar{\sigma} = \mu$. Moreover, since linearity implies that equation (1.41) holds for $cf(x, t)$, for any positive constant c, the choice of $f(v, t)$ to be a probability density does not cause loss of generality.

The relaxation process

As in the simpler case of the Goldstein–Taylor model studied in Section 1.3.1, interactions with the background produce the monotonicity of convex functionals of the solution. Given a convex (smooth) function Φ, let $H_\Phi(f)$ denote the Lyapunov functional

$$H_\Phi(f) = \int_\Omega \Phi(f(v))\, dv. \tag{1.50}$$

To evaluate the time evolution of H_Φ let us make use of the radiative transfer equation (1.41). We obtain

$$\begin{aligned}\frac{dH_\phi(f(t))}{dt} &= \int_\Omega \Phi'(f(v,t))\frac{\partial f(v,t)}{\partial t}\,dv\\ &= \sigma\int_\Omega \Phi'(f(v,t))\left[\int_\Omega f(w,t)\,dw - \mathfrak{m}(\Omega)f(v,t)\right]\,dv\\ &= \sigma\int_\Omega dw\int_\Omega dv\;\Phi'(f(v,t))\;[f(w,t)-f(v,t)] = -I_\Phi(f(t)).\end{aligned}$$

Exchanging v and w (respectively w and v) in the double integral, we can write I_Φ in the equivalent form

$$\begin{aligned}I_\Phi(f(t)) &= \sigma\int_\Omega dw\int_\Omega dv\;\Phi'(f(v,t))\;[f(v,t)-f(w,t)]\\ &= \sigma\int_\Omega dv\int_\Omega dw\;\Phi'(f(w,t))\;[f(w,t)-f(v,t)]\,.\end{aligned}$$

Using both expressions for I_Φ, we finally obtain

$$I_\Phi(f(t)) = \frac{\sigma}{2}\int_\Omega dw\int_\Omega dv\;[\Phi'(f(v,t))-\Phi'(f(w,t))]\,[f(v,t)-f(w,t)] \geq 0.$$

In fact, the integrand is non-negative as a consequence of the convexity of Φ. Moreover, $I_\Phi(f(t)) = 0$ if and only if $f(v,t)$ is constant on Ω.

Since $I_\Phi \geq 0$,

$$H_\Phi(f(t)) \geq H_\Phi(f_\infty).$$

In particular,

$$\int_\Omega \Phi(f_0(v))\,dv \geq \int_\Omega \Phi(f_\infty(v))\,dv = \int_\Omega \Phi\left(\frac{1}{\mathfrak{m}(\Omega)}\int_\Omega f_0(w)\,dw\right)\,dv.$$

Since $f_\infty(v)$ is constant on Ω, while $f_0(v)$ is arbitrary, we proved that, given a convex function Φ,

$$\Phi\left(\frac{1}{\mathfrak{m}(\Omega)}\int_\Omega f_0(w)\,dw\right) \leq \frac{1}{\mathfrak{m}(\Omega)}\int_\Omega \Phi(f_0(v))\,dv. \tag{1.51}$$

Surprisingly, the study of the time evolution of the Lyapunov functional (1.50) along the solution to the radiative transfer equation gives a new physical interpretation of Jensen's inequality [154].

Equation (1.41) can be easily solved to give, for $v \in \Omega$,

$$f(v,t) = f_0(v)e^{-\bar\sigma t} + \left(1-e^{-\bar\sigma t}\right)\frac{1}{\mathfrak{m}(\Omega)}\int_\Omega f_0(v)\,dv. \tag{1.52}$$

Since the mass of the solution is preserved in time (cf. (1.42)), we can divide both sides of (1.52) by the mass of the initial value to obtain

$$g(v,t) = g_0(v)e^{-\bar\sigma t} + \left(1-e^{-\bar\sigma t}\right)\frac{1}{\mathfrak{m}(\Omega)}, \tag{1.53}$$

where

$$g(v,t) = \frac{f(v,t)}{\int_\Omega f(v,t)\,dv}, \qquad t \ge 0$$

is a probability density. This shows that, at any time $t > 0$, the solution to the radiative transfer equation can be viewed as a convex combination of the initial and the background densities.

1.3.3 Rosenau approximation from hydrodynamics

A perhaps not so well-known linear kinetic model which presents a highly interesting relaxation operator has recently been proposed by Rosenau [278, 285] to produce a regularized version of the Chapman–Enskog expansion from hydrodynamics [83]. The model is given by the scalar equation

$$\frac{\partial f}{\partial t} + \frac{\partial \Psi(f)}{\partial v} = \mathcal{F}^{-1}\left[\frac{-\varepsilon\xi^2}{1+\varepsilon^2 m^2 \xi^2}\hat{f}(\xi)\right], \tag{1.54}$$

where $\hat{f}(\xi,t)$ denotes the Fourier transform of the particle density $f(v,t)$, which depends on the velocity $v \in \mathbb{R}$ and time $t \in \mathbb{R}_+$, while $\mathcal{F}^{-1}$ denotes the inverse Fourier transform. Finally, $\varepsilon \ll 1$ is a small positive parameter.

As in the case of the Goldstein–Taylor model, we can split equation (1.54) into two parts. The operator on the left-hand side describes a hyperbolic scalar conservation law [114, 205],

$$\frac{\partial f}{\partial t} + \frac{\partial \Psi(f)}{\partial v} = 0.$$

The relaxation operator reads

$$\frac{\partial f}{\partial t} = \mathcal{F}^{-1}\left[\frac{-\varepsilon\xi^2}{1+\varepsilon^2 m^2 \xi^2}\hat{f}(\xi)\right]. \tag{1.55}$$

Even if it is not immediately recognizable, the operator on the right-hand side looks like the usual viscosity term $\varepsilon\partial^2 f/\partial v^2$ at low wave numbers ξ, while for higher wave numbers it is intended to model a bounded approximation of a linearized interaction operator, thereby avoiding the artificial instabilities that occur when the Chapman–Enskog expansion for such an operator is truncated after a finite number of terms.

Let us briefly describe how the Rosenau relaxation equation (1.55) can be derived by suitable manipulations of the linear diffusion equation [143]

$$\frac{\partial g(v,t)}{\partial t} = \varepsilon \frac{\partial^2 g(v,t)}{\partial v^2}. \tag{1.56}$$

In Fourier variables, equation (1.56) takes the form

$$\frac{\partial \hat{g}(\xi,t)}{\partial t} = -\varepsilon\xi^2 \hat{g}(\xi,t). \tag{1.57}$$

The suggestion of Rosenau was to modify the right-hand side of (1.57), and to consider instead the equation

$$\frac{\partial \hat{f}(\xi,t)}{\partial t} = \frac{-\varepsilon\xi^2}{1+\varepsilon^2 m^2\xi^2}\hat{f}(\xi,t), \tag{1.58}$$

where m is a positive constant. From the modelling point of view, one of the features of this modification is that it is equivalent for small frequencies to the heat equation (and then regularizes the initial data), whereas high frequencies behave as an absorption term. In particular, information will travel at finite speed, and tools from hyperbolic equations can be used.

Reformulation as a linear kinetic equation

The interaction operator on the right-hand side of (1.55) can be more conveniently written in the usual physical space as a linear relaxation operator, analogous in structure to the ones studied in Sections 1.3.1, 1.3.2,

$$\begin{aligned}\mathcal{F}^{-1}\left[\frac{-\varepsilon\xi^2}{1+\varepsilon^2 m^2\xi^2}\hat{f}(\xi)\right] &= \frac{\varepsilon}{(\varepsilon m)^2}\mathcal{F}^{-1}\left[\frac{1}{1+\varepsilon^2 m^2\xi^2}\hat{f}(\xi)-\hat{f}(\xi)\right] \\ &= \frac{1}{m\bar{\varepsilon}}\left[M_{\bar{\varepsilon}} * f - f\right],\end{aligned} \tag{1.59}$$

where $\bar{\varepsilon} = m\varepsilon$, $*$ denotes convolution and

$$M_\gamma(v) = \frac{1}{2\gamma}\, e^{-|v|/\gamma} \tag{1.60}$$

is a non-negative exponentially distributed probability density function.

In other words, the Rosenau approximation consists in substituting the linear diffusion equation

$$\frac{\partial g(v,t)}{\partial t} = \sigma^2\frac{\partial^2 g(v,t)}{\partial v^2} \tag{1.61}$$

with the linear kinetic equation

$$\frac{\partial f(v,t)}{\partial t} = \frac{\sigma^2}{\varepsilon^2}\left[\mathcal{M}_{\varepsilon\sigma} * f(v,t) - f(v,t)\right] \tag{1.62}$$

in which $\mathcal{M}_{\varepsilon\sigma}$, given by (1.60), has the usual role of the background distribution.

Proceeding as in Section 1.3.2, let us consider the related problem in which the number of particles with velocity v at time t is described by the process $X(t)$, and the variation of $X(t)$ is due to interaction with the external background. The background M is described by a random variable with probability density $\mathcal{M}(w)$, which we will assume to be of finite variance γ^2 (M is exponentially distributed in Rosenau model). Let us assume that the interaction process of a particle with velocity v with a background particle with velocity w generates a post-interaction velocity v^* given by

$$v^* = v + w. \tag{1.63}$$

In terms of the process $X(t)$ the law of change given by (1.63) can be rewritten as

$$X^*(t) = X(t) + M, \tag{1.64}$$

which implies, in the case in which M and $X(t)$ are independent, that the law of $X^*(t)$ is the convolution of the laws of M and $X(t)$. Assuming that $X(t), M$ are independent of each other, for a given observable quantity $\varphi(\cdot)$, we then have that the mean value of $\varphi(X)$ satisfies

$$\begin{aligned}\frac{d}{dt}\int_{\mathbb{R}} \varphi(v) f(v,t)\, dv &= \frac{d}{dt}\langle \varphi(X(t))\rangle \\ &= \mu \int_{\mathbb{R}\times\mathbb{R}} \left(\varphi(v^*) - \varphi(v)\right) f(v,t)\mathcal{M}(w)\, dv\, dw,\end{aligned} \tag{1.65}$$

where the constant $\mu > 0$ denotes as usual the interaction frequency. Note that choosing $\varphi(v) = 1$ shows that, independently of the background distribution, $f(v,t)$ remains a probability density if it is initially:

$$\int_{\mathbb{R}} f(v,t)\, dv = \int_{\mathbb{R}} f_0(v)\, dv = 1. \tag{1.66}$$

This is, in general, the unique conservation law associated with equation (1.65).

From now on, let us assume in addition that the probability distribution of the background is centred, and its variance depends on a small parameter $\varepsilon > 0$. To emphasize this dependence, we will denote this distribution by

$$\mathcal{M}_\varepsilon(w) = \varepsilon^{-1}\mathcal{M}(\varepsilon^{-1}w). \tag{1.67}$$

Then $M_\varepsilon(w)$ satisfies

$$\int_{\mathbb{R}} \mathcal{M}_\varepsilon(w)\, dw = 1, \qquad \int_{\mathbb{R}} w\mathcal{M}_\varepsilon(w)\, dw = 0, \qquad \int_{\mathbb{R}} w^2\mathcal{M}_\varepsilon(w)\, dw = \varepsilon^2\gamma^2. \tag{1.68}$$

Equation (1.65) yields immediately the time evolution of the moments of f. Taking $\varphi(v) = v$ one obtains

$$\frac{d}{dt}\int_{\mathbb{R}} v f(v,t)\, dv = \mu \int_{\mathbb{R}} w\mathcal{M}_\varepsilon(w)\, dw \int_{\mathbb{R}} f(v,t)\, dv = 0. \tag{1.69}$$

Moreover, if $\varphi(v) = v^2$

$$\frac{d}{dt}\int_{\mathbb{R}} v^2 f(v,t)\, dv = \mu \int_{\mathbb{R}^2} \left[(v^*)^2 - v^2\right] f(v,t)\mathcal{M}_\varepsilon(w)\, dv\, dw = \mu\varepsilon^2\gamma^2. \tag{1.70}$$

Thus, the second moment of f grows linearly with respect to time and depends on ε. One way to avoid this dependency is to scale the time properly. Setting $\bar{t} = \varepsilon^2 t$ and introducing a new particle distribution function g_ε such that $g_\varepsilon(v,\bar{t}) = f(v,t)$ gives, according to (1.70),

$$\frac{d}{d\bar{t}}\int_{\mathbb{R}} v^2 g_\varepsilon(v,\bar{t})\, dv = \frac{\mu}{\varepsilon^2}\frac{d}{dt}\int_{\mathbb{R}} v^2 f(v,t)\, dv = \mu\gamma^2,$$

and the second moment of g_ε does not depend on ε.

Taking $\bar{t} = t$, the distribution g_ε is a weak solution of

$$\frac{d}{dt}\int_{\mathbb{R}} g_\varepsilon(v,t)\varphi(v)\, dv = \frac{\mu}{\varepsilon^2}\int_{\mathbb{R}^2} \left[\varphi(v+w) - \varphi(v)\right] g_\varepsilon(v,t)\mathcal{M}_\varepsilon(w)\, dv\, dw, \tag{1.71}$$

with $g_\varepsilon(v, t = 0) = g_0(v) = f_0(v)$.

Note that, since

$$\int_{\mathbb{R}^2} \varphi(v+w) g_\varepsilon(v,t) \mathcal{M}_\varepsilon(w)\, dv\, dw$$
$$= \int_{\mathbb{R}^2} \varphi(z) g_\varepsilon(v,t) \mathcal{M}_\varepsilon(z-v)\, dv\, dz = \int_{\mathbb{R}} \varphi(z) \mathcal{M}_\varepsilon * g_\varepsilon(z)\, dz,$$

equation (1.71) can be rewritten as

$$\frac{d}{dt} \int_{\mathbb{R}} g_\varepsilon(v,t)\varphi(v)\, dv = \frac{\mu}{\varepsilon^2} \int_{\mathbb{R}} \varphi(v) \left(\mathcal{M}_\varepsilon * g_\varepsilon(v) - g_\varepsilon(v)\right) dv. \tag{1.72}$$

Hence, if the background distribution is given by (1.60), equation (1.71) coincides with the Rosenau approximation (1.62). Of course, other choices of the background are possible, and, provided conditions (1.68) are satisfied, the evolution of the moments of the solution to (1.71) up to the second order do not depend on the background distribution.

It is then easy to see that, in the case in which $\mathcal{M}_\varepsilon$ coincides with (1.60), g_ε is an approximate (at the leading order in ε) weak solution of the heat equation (1.61). Indeed, given that ε is small enough, one can Taylor expand φ in $v^* = v + w$ to obtain

$$\varphi(v^*) = \varphi(v) + w\varphi'(v) + \frac{w^2}{2}\varphi''(v) + \frac{1}{3!}\varphi^{(3)}(\tilde{v})w^3.$$

Using this relation in (1.71), one obtains

$$\frac{d}{dt} \int_{\mathbb{R}} g_\varepsilon(v,t)\varphi(v)\, dv = \frac{\mu\gamma^2}{2} \int_{\mathbb{R}} g_\varepsilon(v,t)\varphi''(v)\, dv + R(\varepsilon), \tag{1.73}$$

where the remainder $R(\varepsilon)$ satisfies

$$|R(\varepsilon)| \le \frac{\mu}{3!}\|\varphi^{(3)}\|_{L^\infty} \frac{1}{\varepsilon^2} \int_{\mathbb{R}} w^3 \mathcal{M}_\varepsilon(w)\, dw = \varepsilon \frac{\mu}{3!}\|\varphi^{(3)}\|_{L^\infty} \int_{\mathbb{R}} w^3 \mathcal{M}_1(w)\, dw. \tag{1.74}$$

One can easily notice that equation (1.71) is an approximation to the heat equation (1.61) (with diffusion coefficient $\sigma^2 = \mu\gamma^2/2$) any time that the remainder converges to zero as $\varepsilon \to 0$. In order to have this convergence, it is enough that the distribution of the background is such that some moment of order greater than two remains bounded. This can be fruitfully used to obtain various approximations of the heat equation, just by changing the distribution of the background.

1.3.4 Analysis of the relaxation process

The Rosenau-type kinetic equation (1.71) is such that the mass, momentum and energy of its solution have the same evolution as the corresponding moments of the solution to the heat equation (1.61). In agreement with the discussion in Sections 1.3.1 and 1.3.2, a further interesting analogy with the heat equation is found by studying the evolution of convex functionals along the solution. Let $\Phi(r)$, $r \ge 0$ be a (smooth) convex function of r. Then, using equation (1.72) we obtain

$$\frac{d}{dt}\int_{\mathbb{R}} \Phi(g_\varepsilon(v,t))\,dv = \int_{\mathbb{R}} \Phi'(g_\varepsilon(v,t))\frac{\partial g_\varepsilon(v,t)}{\partial t}\,dv$$
$$= \frac{\mu}{\varepsilon^2}\int_{\mathbb{R}} \Phi'(g_\varepsilon(v,t))\,(\mathcal{M}_\varepsilon * g_\varepsilon(v) - g_\varepsilon(v))\,dv.$$

Thanks to the convexity of $\Phi(\cdot)$, for $r, s \ge 0$

$$\Phi'(s)(r-s) \le \Phi(r) - \Phi(s),$$

and one obtains

$$\frac{d}{dt}\int_{\mathbb{R}} \Phi(g_\varepsilon(v,t))\,dv \le \frac{\mu}{\varepsilon^2}\int_{\mathbb{R}} (\Phi(\mathcal{M}_\varepsilon * g_\varepsilon(v)) - \Phi(g_\varepsilon(v)))\,dv.$$

Now, use the fact that $\mathcal{M}_\varepsilon$ is a probability distribution, so that by Jensen's inequality

$$\int_{\mathbb{R}} \Phi(\mathcal{M}_\varepsilon * g_\varepsilon(v))\,dv = \int_{\mathbb{R}} \Phi\left(\int_{\mathbb{R}} g_\varepsilon(v-w))\mathcal{M}_\varepsilon(w)\,dw\right)\,dv$$
$$\le \int_{\mathbb{R}^2} \Phi(g_\varepsilon(v-w))\mathcal{M}_\varepsilon(w)\,dv\,dw = \int_{\mathbb{R}} \Phi(g_\varepsilon(v))\,dv,$$

to conclude that

$$\frac{d}{dt}\int_{\mathbb{R}} \Phi(g_\varepsilon(v,t))\,dv \le 0.$$

Thus, any convex functional is non-increasing along the solution to the Rosenau-type kinetic equation (1.62), in agreement with the analogous property of the heat equation.

Example 1.1 (Approximation to the heat equation). A leading example of Rosenau type-approximation to the heat equation is given by assuming that the background distribution M_ε is a discrete random variable that can assume the two values $\varepsilon\gamma$ and $-\varepsilon\gamma$, with equal probabilities 1/2,

$$P(M = +\varepsilon\gamma) = P(M = -\varepsilon\gamma) = \frac{1}{2},$$

which satisfies conditions (1.68). In this case

$$\mathcal{M}_\varepsilon(v) = \frac{1}{2}[\delta_0(v+\varepsilon\gamma) + \delta_0(v-\varepsilon\gamma)], \tag{1.75}$$

where as usual $\delta_0(v)$ denotes a Dirac mass concentrated at $v = 0$.

Then, according to (1.72), the distribution g solves

$$\frac{\partial g_\varepsilon}{\partial t}(v,t) = \frac{\mu}{2\varepsilon^2}[g_\varepsilon(v-\varepsilon\gamma,t) - 2g_\varepsilon(v,t) + g_\varepsilon(v+\varepsilon\gamma,t)]. \tag{1.76}$$

Now, let us fix $\gamma = \sigma$ and $\mu = 2$. Then, given a parameter $\Delta v > 0$ and a uniform grid $v_i = i\Delta v$, $i \in \mathbb{Z}$ on $\mathbb{R}$, by setting $g_i(t) = g_\varepsilon(v_i,t)$ and $\varepsilon\sigma = \Delta v$, equation (1.76) reads

$$g_i'(t) = \sigma^2 \frac{g_{i+1}(t) - 2g_i(t) + g_{i-1}(t)}{\Delta v^2}.$$

This is exactly the classical second order central differences scheme for the heat equation (1.61). It follows that this scheme furnishes an approximation to the heat equation which satisfies all the previously listed properties, including the monotonicity of Lyapunov functionals. ■

Remark 1.2

- *It is important to remark that, contrary to what happens with the linear models studied in Sections 1.3.1 and 1.3.2, the solution to the Rosenau model does not converge in time to a steady state [272].*
- *The kinetic model presented in this section enlightens various interesting features which are not present in the two cases discussed in Sections 1.3.1 and 1.3.2. The main aspect to be outlined is that the kinetic model constructed starting from the Rosenau argument gives a* physical *way to approximate the linear diffusion equation (1.61) through a kinetic model in which the density is modified by interaction with a background, and the essential properties of this background are limited to a few assumptions on the moments of the corresponding density function.*
- *Among others, this construction allows us to select numerical methods which are consistent with Rosenau approximation, namely numerical methods which preserve all the physical properties typical of this approximation, including the monotonicity of entropy functionals. As illustrated before, the central difference scheme enters as a Rosenau-type approximation to the linear diffusion equation.*

1.4 Binary interaction models on the real line

The linear relaxation operators studied in the previous sections describe various types of interactions of particles with an external background. This allows us to describe a number of physical systems in which this type of interaction is the dominant one. In many cases, however, one cannot neglect that the variation of particle density is essentially due to particles that interact by themselves.

1.4.1 General binary interaction random processes

In the one-dimensional situation, the most general binary interaction between particles which is linear in the entering velocities (v, w) can be described by assigning the exchange rules

$$v^* = p_1 v + q_1 w, \quad w^* = p_2 v + q_2 w; \quad p_i, q_i > 0, \;\; i = 1, 2. \tag{1.77}$$

In (1.77) (p_i, q_i), for $i = 1, 2$ have the role of mixing parameters, and can be either fixed constants or random variables, depending on the problem under study. Interactions of type (1.77) are very general, and include, when the mixing parameters are random, most of the known one-dimensional Boltzmann-type models of Maxwell type, including the famous Kac model [186].

Let us describe the number of particles with velocity v at time t with the random process $X(t)$, and let us define its density by $f(v,t)$. With respect to the process studied in Section 1.3.3, where the variations were due to interactions with an external background, the variation of the law of $X(t)$ is now generated by interactions with other particles with the same law. Let us denote by $Y(t)$ the second random process with law $f(v,t)$. In terms of the processes $X(t)$ and $Y(t)$ the laws of change given by (1.77) can be rewritten as

$$X^* = p_1 X + q_1 Y, \quad Y^* = p_2 X + q_2 Y; \quad p_i, q_i > 0, \;\; i = 1,2, \tag{1.78}$$

which implies, in the case in which $X(t)$ and $Y(t)$ are independent, that the laws of $X^*(t)$ and $Y^*(t)$ are suitable (generalized) convolutions of the laws of $X(t)$ and $Y(t)$.

Assuming that $X(t), Y(t)$ are independent processes, so that the joint law of $X(t)$ and $Y(t)$ is equal to the product of the laws of $X(t)$ and $Y(t)$, one can easily evaluate the variation of the mean value of $\varphi(X(t))+\varphi(Y(t))$ due to interactions of type (1.77). To simplify computations, let us assume from now on an interaction frequency $\mu = 1$. Since $X(t)$ and $Y(t)$ are identically distributed, the strategy adopted in Section 1.3.1 shows that

$$\begin{aligned}\frac{d}{dt}\langle \varphi(X(t)) + \varphi(Y(t))\rangle &= 2\frac{d}{dt}\int_{\mathbb{R}} \varphi(v) f(v,t)\, dv \\ &= \left\langle \int_{\mathbb{R}\times\mathbb{R}} \left(\varphi(v^*) + \varphi(w^*) - \varphi(v) - \varphi(w)\right) f(v,t) f(w,t)\, dv\, dw \right\rangle,\end{aligned}$$

which is equivalent to

$$\begin{aligned}&\frac{d}{dt}\int_{\mathbb{R}} \varphi(v) f(v,t)\, dv \\ &\quad = \frac{1}{2}\left\langle \int_{\mathbb{R}\times\mathbb{R}} \left(\varphi(v^*) + \varphi(w^*) - \varphi(v) - \varphi(w)\right) f(v,t) f(w,t)\, dv\, dw \right\rangle.\end{aligned} \tag{1.79}$$

In (1.79) the expectation is relative to the eventually random parameters (p_i, q_i), for $i = 1,2$. It is important to remark that, while in the linear situation the independence hypothesis between the process $X(t)$ and the background variable M appears very reasonable, here the requested independence between the processes $X(t)$ and $Y(t)$ appears to be forced by the desire to obtain an equation in closed form in terms of $f(v,t)$, but not fully motivated from a physical point of view. However, as always happens with the modelling of a physical process, one must mediate between the complexity of the real phenomenon and the possibility of obtaining predictions from the necessarily approximate mathematical model. The assumption of independence appears to be a good compromise, which allows for a fairly complete analysis of the time behaviour of the underlying probability density. We will return later on to this point.

Remark 1.3 *For a given $v \in \mathbb{R}$, taking $\varphi(\cdot) = \delta_0(v - \cdot)$ in (1.79), where $\delta_0(\cdot)$ denotes the Dirac delta, and switching the notation for the integration variables to v_1 and v_2, we recover the following expression for the kinetic equation:*

$$\frac{\partial f(v,t)}{\partial t} = \frac{1}{2}\Big\langle \int_{\mathbb{R}\times\mathbb{R}} f(v_1,t)f(v_2,t)\Big(\delta_0(v-v_1^*)+\delta_0(v-v_2^*)\Big)\,dv_1\,dv_2\Big\rangle - f(v,t), \quad (1.80)$$

where now

$$v_1^* = p_1v_1 + q_1v_2, \quad v_2^* = p_2v_1 + q_2v_2; \quad p_i, q_i > 0, \ \ i = 1, 2.$$

We remark that the above form is often used in the physics literature [30].

From this representation, one recognizes that the time variation of the one-particle density depends on the balance between gain and loss operators. This can be made evident by writing the kinetic equation (1.80) in the form

$$\frac{\partial f(v,t)}{\partial t} = Q_+(f,f)(v,t) - f(v,t), \quad (1.81)$$

where the bilinear operator $Q_+(f,f)$ *is the gain operator given by*

$$Q_+(f,f)(v,t) = \frac{1}{2}\Big\langle \int_{\mathbb{R}\times\mathbb{R}} f(v_1,t)f(v_2,t)\Big(\delta_0(v-v_1^*)+\delta_0(v-v_2^*)\Big)\,dv_1\,dv_2\Big\rangle. \quad (1.82)$$

The case of symmetric interactions

The richness of this model has been made evident by a recent study [30] which refers to the apparently toy situation with constant mixing parameters, given by $p_1 = q_2 = p$, $p_2 = q_1 = q$. In this case the binary interactions (1.77) simplify to

$$v^* = pv + qw, \quad w^* = qv + pw. \quad (1.83)$$

Without loss of generality, one can assume $p > q > 0$. The interaction (1.83) is such that the exchange of the pre-interaction velocities v and w implies the exchange of v^* and w^*. Thus, since the mixing parameters are fixed constants, we can rewrite (1.79) as

$$\frac{d}{dt}\int_{\mathbb{R}} \varphi(v)f(v,t)\,dv = \int_{\mathbb{R}\times\mathbb{R}} \left(\varphi(v^*) - \varphi(v)\right)\, f(v,t)f(w,t)\,dv\,dw. \quad (1.84)$$

Despite its apparent simplicity, the kinetic model (1.84) has a very rich variety of possible behaviours, which depend on the possible choices of the parameters p and q. This can be shown by evaluating the moments of the solution. Without loss of generality, we can fix the initial density to satisfy

$$\int_{\mathbb{R}} f_0(v)\,dv = 1; \quad \int_{\mathbb{R}} v f_0(v)\,dv = 0; \quad \int_{\mathbb{R}} v^2 f_0(v)\,dv = 1. \quad (1.85)$$

Choosing $\varphi(v) = v$ shows that

$$m(t) = \int_{\mathbb{R}} v f(v,t)\,dv = m(0)\exp\left\{(p+q-1)t\right\}. \quad (1.86)$$

Hence, since the initial density f_0 satisfies (1.85), $m(0) = 0$ and $m(t) = 0$ for all $t > 0$. Consequently, if $\varphi(v) = v^2$,

$$E(t) = \int_{\mathbb{R}} v^2 f(v,t)\, dv = \exp\left\{(p^2 + q^2 - 1)t\right\}. \tag{1.87}$$

Higher-order moments can be evaluated recursively, remarking that the integrals $\int v^n f$ obey a closed hierarchy of equations [32].

Note that the second moment of the solution is not conserved, unless the interaction parameters satisfy

$$p^2 + q^2 = 1.$$

If this is not the case, the energy can grow to infinity or decrease to zero, depending on the sign of $p^2 + q^2 - 1$. In neither case, however, do stationary solutions of finite energy exist.

Unlike the cases studied in Sections 1.3.1 and 1.3.2, for the kinetic model (1.84) one cannot immediately find the asymptotic behaviour of the solution. On the other hand, when constructing a kinetic model which describes the time evolution of a particle density, one of the main objectives, if not *the* main objective, is exactly the knowledge of this behaviour. Indeed, as briefly discussed in the introduction when discussing the analogies between the kinetic theory of rarefied gases and wealth distribution in a multiagent society, the knowledge of the large-time distribution of wealth is one of the validation points of the model itself.

For these reasons, the study of some of the accessible properties of the solution to the nonlinear kinetic model (1.84), which is, among all possible interactions of type (1.77), the easiest case to be studied, deserve to be analyzed in some detail. We remark that the case in which the second moment is increasing appears to be quite similar to the Rosenau model (1.65) discussed in Section 1.3.3. Owing to this similarity, and remarking that the Rosenau model is an approximation of the linear diffusion equation, one is tempted to assume that the asymptotic behaviour of the solution is described well by a self-similar solution [21]. The physical idea is that the solution to the kinetic equation will approach, after a certain relaxation time, a universal profile (the so-called self-similar one), for a large class of initial data. This is a very well-known tool of nonlinear diffusion equations [312], which can be fruitfully adapted to the present situation. The standard way to look for self similarity is to scale the solution according to the rule

$$g(v,t) = \sqrt{E(t)}\, f\left(v\sqrt{E(t)}, t\right). \tag{1.88}$$

This scaling implies that $g(v,t)$ remains a probability density, while $\int v^2 g(v,t) = 1$ for all $t \geq 0$. Hence, one can make use of classical arguments of probability theory [200] to prove that the scaled solution converges in time towards a universal profile $g_\infty(v)$.

A second interesting and widely used method for studying the behaviour of the solution to kinetic equations like (1.84) is related to the possibility of investigating the behaviour of the solution in the case in which the mixing parameter p is close to 1, while at the same time the parameter q is close to zero. This situation describes a well-known regime in which the post-interaction velocities are approximately equal to the pre-interaction ones. In the kinetic theory of rarefied gases these collisions are called *grazing* collisions. The grazing collision regime for the Boltzmann collision operator has been deeply investigated by C. Villani [315,316], who showed in this regime the passage from the Boltzmann collision operator to the Boltzmann–Landau operator [201].

1.4.2 The grazing interaction limit

The *grazing interaction* regime allows us to obtain a more regular operator, in general of Fokker–Planck type [274], from which it is possible to reckon explicitly the stationary solution. This is a very powerful technique, which is an alternative to other modelling approaches, in particular the mean field limit approach, where physical systems with many interacting bodies (which are generally very difficult to solve exactly, except for extremely simple cases) are replaced by a 1-body problem with a well-chosen external field [86].

Fig. 1.6 Cedric Villani[6]
© Pierre Maraval

Let us illustrate how this procedure works for the kinetic model (1.84). The starting point is to recover the kinetic equation satisfied by the scaled function g given by (1.88). Elementary computations show that $g = g(v,t)$ satisfies the equation

$$\begin{aligned}\frac{d}{dt}\int_{\mathbb{R}} \varphi(v)g(v,t)\,dv - \frac{1}{2}\left(p^2+q^2-1\right)\int_{\mathbb{R}}\varphi(v)\frac{\partial}{\partial v}(vg)\,dv \\ = \int_{\mathbb{R}^2} g(v)g(w)(\varphi(v^*)-\varphi(v))dvdw.\end{aligned} \tag{1.89}$$

Assuming that φ vanishes at infinity, we can integrate by parts the second integral on the left-hand side of (1.89) to obtain

$$\begin{aligned}\frac{d}{dt}\int_{\mathbb{R}} \varphi(v)g(v,t)\,dv + \frac{1}{2}\left(p^2+q^2-1\right)\int_{\mathbb{R}}\varphi'(v)vg(v)\,dv \\ = \int_{\mathbb{R}^2} g(v)g(w)(\varphi(v^*)-\varphi(v))dvdw.\end{aligned} \tag{1.90}$$

With respect to equation (1.84), the equation (1.90) satisfied by g, as given by (1.88), contains the additional operator

$$D(g) = \frac{1}{2}\left(p^2+q^2-1\right)\frac{\partial(vg(v))}{\partial v}. \tag{1.91}$$

When the coefficient $\frac{1}{2}\left(p^2+q^2-1\right) > 0$, the operator $D(\cdot)$ is known as the *drift* operator (*anti-drift* whenever the coefficient is less than zero). The physical meaning

[6]**Cedric Villani**: born on 5 October 1973 in Brive-la-Gaillarde (France). At the centre of much of his work is his profound mathematical interpretation of the physical concept of entropy, which he has applied to solve major problems inspired by physics. In 2010 he was awarded the Fields Medal for his work on Landau damping and the Boltzmann equation.

of this operator appears clearly if we examine the definition of the scaled function g. Indeed, the sign of the coefficient $\frac{1}{2}\left(p^2+q^2-1\right)$ is linked to the time evolution of the second moment of f. Thanks to (1.87), the second moment increases when the coefficient is positive, while it decreases when the coefficient is negative. By means of the scaling (1.88), we force the second moment of g to be constant. Hence, operator (1.91) acts as a balance for the second moment of g. When the bilinear equation (1.84) is such that the second moment of the solution increases, the drift operator acts against the growth to maintain the second moment of g constant. This implies that the solution to the initial value problem

$$\frac{\partial f}{\partial t} = \alpha \frac{\partial (vf(v))}{\partial v}, \tag{1.92}$$

when $\alpha = \frac{1}{2}\left(p^2+q^2-1\right) > 0$ and conditions (1.85) hold for the initial value, is such that the second moment of the solution decays exponentially in time. In the opposite situation, namely when the second moment of the solution decreases, the balance is due to the anti-drift operator, which consequently is such that, starting with an initial density satisfying conditions (1.85), the second moment of the solution to (1.92) increases exponentially with time. We will present more details on the drift equation (1.92) in the next chapter.

Going back to equation (1.90), since v^* is given by the first of (1.83),

$$v^* - v = (p-1)v + qw.$$

Hence, if p is close to 1, while q is close to 0, the difference $v^* - v$ is close to zero. Let us use a second-order Taylor expansion of $\varphi(v^*)$ around v:

$$\varphi(v^*) - \varphi(v) = ((p-1)v + qw)\,\varphi'(v) + \frac{1}{2}\left((p-1)v + qw\right)^2 \varphi''(\tilde{v}),$$

where, for some $0 \le \theta \le 1$,

$$\tilde{v} = \theta v^* + (1-\theta)v.$$

Inserting this expansion in the interaction operator, we obtain the equality

$$\begin{aligned}\int_{\mathbb{R}^2} g(v)g(w)(\varphi(v^*) - \varphi(v))dvdw = &\int_{\mathbb{R}^2} g(v)g(w)\left((p-1)v + qw\right)\varphi'(v)dvdw \\ &+ \frac{1}{2}\int_{\mathbb{R}^2} g(v)g(w)\left((p-1)v + qw\right)^2 \varphi''(v)dvdw \\ &+ R(p,q),\end{aligned} \tag{1.93}$$

where

$$R(p,q) = \frac{1}{2}\int_{\mathbb{R}^2} \left((p-1)v + qw\right)^2 \left(\varphi''(\tilde{v}) - \varphi''(v)\right) g(v)g(w)dv\,dw. \tag{1.94}$$

Recalling that $g(v,t)$ satisfies (1.85), we can simplify (1.93) to obtain

$$\int_{\mathbb{R}^2} g(v)g(w)(\varphi(v^*) - \varphi(v))dvdw = (p-1)\int_{\mathbb{R}} vg(v)\varphi'(v)dv \\ + \frac{1}{2}\int_{\mathbb{R}} g(v)\left((p-1)^2v^2 + q^2\right)\varphi''(v)dv + R(p,q). \tag{1.95}$$

Substituting (1.95) into (1.90), and grouping similar terms, we conclude that $g(v,t)$ satisfies

$$\frac{d}{dt}\int_{\mathbb{R}} \varphi(v)g(v,t)\,dv + \frac{1}{2}\left((p-1)^2 + q^2\right)\int_{\mathbb{R}} \varphi'(v)vg(v)\,dv \\ = \frac{1}{2}\int_{\mathbb{R}} g(v)\left((p-1)^2v^2 + q^2\right)\varphi''(v)dv + R(p,q). \tag{1.96}$$

Hence, if we set

$$\tau = q^2 t, \quad h(v,\tau) = g(v,t), \tag{1.97}$$

which implies $g_0(v) = h_0(v)$, $h(v,\tau)$ satisfies

$$\frac{d}{d\tau}\int_{\mathbb{R}} \varphi(v)h(v,\tau)\,dv + \frac{1}{2}\left(\left(\frac{p-1}{q}\right)^2 + 1\right)\int_{\mathbb{R}} \varphi'(v)vh(v)\,dv \\ = \frac{1}{2}\int_{\mathbb{R}} h(v)\left(\left(\frac{p-1}{q}\right)^2 v^2 + 1\right)\varphi''(v)dv + \frac{1}{q^2}R(p,q). \tag{1.98}$$

Suppose now that the remainder in (1.98) is small for small values of q and $p-1$. Then equation (1.98) gives the behaviour of $g(v,t)$ for large values of time. Moreover, taking $p = p(q)$ such that, for a given constant $\lambda > 0$

$$\lim_{q\to 0}\frac{p(q)-1}{q} = \lambda, \tag{1.99}$$

equation (1.98) is approximated well by the equation (in weak form)

$$\frac{d}{d\tau}\int_{\mathbb{R}} \varphi(v)h(v,\tau)\,dv + \frac{1}{2}\left(\lambda^2 + 1\right)\int_{\mathbb{R}} \varphi'(v)vh(v)\,dv \\ = \frac{1}{2}\int_{\mathbb{R}} h(v)\left(\lambda^2 v^2 + 1\right)\varphi''(v)dv. \tag{1.100}$$

Equation (1.100) corresponds to the weak form of the Fokker–Planck equation

$$\frac{\partial h}{\partial \tau} = \frac{1}{2}\left(\frac{\partial^2}{\partial v^2}\left((1+\lambda^2 v^2)h\right) + \left(1+\lambda^2\right)\frac{\partial}{\partial v}(vh)\right), \tag{1.101}$$

which has a unique stationary state of unit mass, given by

$$M_\lambda(v) = c_\lambda\left(\frac{1}{1+\lambda^2 v^2}\right)^{\frac{3}{2}+\frac{1}{2\lambda^2}}. \tag{1.102}$$

The previous computations illustrate in a simple case the mathematical procedure which allows us to connect, in a suitable regime of the mixing parameters, a classical kinetic model written in bilinear interaction form with another model for which it is possible to obtain the explicit form of the steady solution. This procedure, when applicable, permits the extraction of information on the large-time behaviour of the solution of a kinetic model with binary interactions. It is remarkable that, while the derivation of the Fokker–Planck equation (1.101) presented here is largely formal, the mathematical details can be rigorously proven [254]. The main objective here was to illustrate in a relatively simple and instructive way how to obtain information on various properties of the solution, and that the solution itself is closely dependent on the choice of the mixing parameters.

Special cases

Among the possible choices of the parameters p and q, two are clearly distinguishable. First, if the pair satisfies the constraint $p+q=1$, by (1.86) the mean value of the solution to the kinetic equation is preserved, while by (1.87) its variance is non-increasing (in fact, $p+q=1$ implies $p^2+q^2 \le 1$). This constraint identifies a *dissipative interaction*, which leads the system towards a solution concentrated in the mean value. Second, if the pair satisfies the constraint $p^2+q^2=1$, (1.49) implies that the second moment is preserved. We refer to this case as the *conservative* case.

Let us examine in some detail these distinguished cases, starting with the second one. If one wants to maintain the constraint $p^2+q^2=1$ in the limit procedure illustrated previously, one is forced to choose $p=\sqrt{1-q^2}$, which gives $\lambda=0$ as the only possible value in (1.99). In the limit, one then obtains the linear Fokker–Planck equation [274]

$$\frac{\partial h}{\partial \tau} = \frac{1}{2}\left(\frac{\partial^2 h}{\partial v^2} + \frac{\partial}{\partial v}(vh)\right). \tag{1.103}$$

In this case the stationary solution of (1.103) is the Gaussian density (Maxwell-type distribution)

$$M(v) = \frac{1}{\sqrt{2\pi}} e^{-v^2/2}. \tag{1.104}$$

Note that $M(v)$ also solves equation (1.65) for any other pair of values satisfying the conservative constraint. In fact,

$$\begin{aligned}\int_{\mathbb{R}\times\mathbb{R}} \varphi(v^*)\, M(v)M(w)\, dv\, dw &= \int_{\mathbb{R}\times\mathbb{R}} \varphi(pv+qw)\, M(v)M(w)\, dv\, dw \\ &= \int_{\mathbb{R}\times\mathbb{R}} \varphi(v+w)\, M_p(v)M_p(w)\, dv\, dw,\end{aligned}$$

where M_p is the scaled distribution defined as in (1.67). Proceeding as in the proof of (1.72), we then obtain

$$\int_{\mathbb{R}\times\mathbb{R}} \varphi(v^*)\, M(v)M(w)\, dv\, dw = \int_{\mathbb{R}} \varphi(v)\, M_p * M_q(v)\, dv = \int_{\mathbb{R}} \varphi(v)\, M(v)\, dv. \tag{1.105}$$

The last equality in (1.105) follows from the stability property of Gaussian densities with respect to convolution [200].

An alternative proof of this identity can be obtained by resorting to Fourier transforms [59], exactly as happens in classical probability theory [200]. Indeed, equations of type (1.84) are fruitfully studied by resorting to Fourier transforms. It is enough to choose $\varphi(v) = e^{-i\xi v}$ to obtain

$$\int_{\mathbb{R}\times\mathbb{R}} e^{-i\xi v^*} f(v,t)f(w,t)\,dv\,dw$$
$$= \int_{\mathbb{R}\times\mathbb{R}} e^{-i\xi pv} e^{-i\xi qw} f(v,t)f(w,t)\,dv\,dw = \hat{f}(p\xi,t)\hat{f}(q\xi,t),$$

where $\hat{f}(\xi,t)$ is the Fourier transform of $f(v,t)$ with respect to the variable v. This leads us to write equation (1.84) in the equivalent form

$$\frac{\partial \hat{f}(\xi,t)}{\partial t} = \hat{f}(p\xi,t)\hat{f}(q\xi,t) - \hat{f}(\xi,t). \tag{1.106}$$

It is now clear that the passage to Fourier transform leads to a marked simplification of the equation, and represents an alternative way to look at this type of interaction model. A one-line argument does indeed show that a Gaussian density $M(v)$ is a stationary solution of (1.106). In fact, since $\hat{M}(\xi) = e^{-\xi^2}$, the constraint $p^2 + q^2 = 1$ immediately implies

$$\hat{M}(p\xi)\hat{M}(q\xi) = \hat{M}(\xi).$$

Let us now analyze in some detail the case in which $p + q = 1$, which describes a binary interaction model which dissipates energy. This type of kinetic equation has been intensively studied in recent years, to give an answer to the conjecture posed by Ernst and Brito [141, 142], relative to the fact that the self-similar solution to the dissipative Boltzmann equation for Maxwell molecules has overpopulated tails. The kinetic equation (1.84) is, in fact, close to the model for granular dissipative collisions introduced and studied in [18, 32, 233] as a one-dimensional caricature of the Maxwell–Boltzmann equation [45, 47].

The constraint $p + q = 1$ implies $\lambda = -1$ in (1.99), so that the stationary solution to the Fokker–Planck equation (1.103) is

$$M_{-1}(v) = \frac{2}{\pi}\left(\frac{1}{1+v^2}\right)^2. \tag{1.107}$$

Let

$$\hat{g}_\infty(\xi) = (1+|\xi|)e^{-|\xi|}. \tag{1.108}$$

Then

$$\hat{g}_\infty''(\xi) = (-1+|\xi|)e^{-|\xi|} = -2e^{-|\xi|} + \hat{g}_\infty(\xi). \tag{1.109}$$

Now, recalling that $e^{-|\xi|}$ is the Fourier transform of the Cauchy density $\left(\pi(1+v^2)\right)^{-1}$, (1.109) implies

$$(1+v^2)g_\infty(v) = \frac{2}{\pi(1+v^2)},$$

so that $\hat{g}_\infty$ is the Fourier transform of M_{-1}. Now, consider that $p + q = 1$ implies

$$\frac{1}{2}(p^2 + q^2 - 1) = pq,$$

and equation (1.89) leads to the Fourier-transformed equation

$$\frac{\partial \hat{g}(\xi, t)}{\partial t} = \hat{g}(p\xi, t)\hat{g}(q\xi, t) - \hat{g}(\xi, t) + pq\xi \frac{\partial \hat{g}(\xi, t)}{\partial t}. \tag{1.110}$$

It is immediate to verify that the function $\hat{g}_\infty(\xi) = (1+|\xi|)e^{-|\xi|}$ solves equation (1.110). This solution solves the kinetic equation (1.89), for any value of the parameters subject to the constraint $p+q = 1$. Thus, the kinetic equation (1.84) has an explicit self-similar solution.

Remark 1.4 *The analysis of the two cases $p + q = 1$ and $p^2 + q^2 = 1$, which can be treated exactly to obtain analytical results, shows that even structurally simple binary interactions can lead to bilinear equations with completely different behaviours. Likewise, they give a clear indication of the importance of the structure of binary interactions in connection with the properties of steady and/or self-similar profiles. This is a relatively new field of research, which takes its* raison d'être *from the study of new types of interaction which are typical of systems constituted by living agents. Consequently, the arguments presented in this section will be of paramount importance in the forthcoming chapters.*

1.4.3 The Kac model and variants

As explicitly mentioned in Section 1.4, the linear binary interactions (1.77) are very general, and within them one can recognize various well-studied kinetic Boltzmann models. Among them, the Kac model [186] is undoubtedly the most famous, both for its importance in the mathematical development of innovative analytical tools, and for physical reasons, linked to the possibility of exploring the fundamental aspects of chaos propagation. The Kac model corresponds to a binary interaction of type (1.77) with random mixing parameters given by

$$p_1 = q_2 = \cos\theta, \quad p_2 = -q_1 = \sin\theta, \tag{1.111}$$

where θ is a random variable uniformly distributed on the interval $[0, 2\pi]$.

Consequently, the binary interaction modifies the incoming velocity pair (v, w) into the post-interaction pair by

Fig. 1.7 Mark Kac[7]
© MFO

$$v^* = \cos\theta\, v - \sin\theta\, w, \quad w^* = \sin\theta\, v + \cos\theta\, w. \tag{1.112}$$

[7]**Mark Kac**: born on 3 August 1914 in Krzemiemiec (Russian Empire, now Ukraine), died on 26 October 1984 in California (USA). He pioneered the modern development of mathematical probability, in particular its applications to statistical physics.

Hence, the result of the interaction is a rotation of the plane vector of components (v, w) by an angle θ. Clearly, this rotation does not modify the modulus of the vector, which implies the conservation (of energy)

$$(v^*)^2 + (w^*)^2 = v^2 + w^2.$$

The Jacobian of the transformation (1.112) is in this case equal to 1, and this permits us to write the gain operator Q_+ in (1.82) in the form

$$Q_+(f,f)(v,t) = \left\langle \int_{\mathbb{R}} f(v^*,t) f(w^*,t)\, dw \right\rangle,$$

or, to maintain the usual notations of classical kinetic theory [186],

$$Q_+(f,f)(v,t) = \int_{\mathbb{R}} \int_0^{2\pi} \frac{1}{2\pi} f(v\cos\theta - w\sin\theta, t) f(v\sin\theta + w\cos\theta, t)\, dw\, d\theta. \quad (1.113)$$

The Kac operator (1.113) is a one-dimensional operator which exhibits most of the physical features of the three-dimensional Boltzmann operator for Maxwell molecules (1.5). Among others, it can be easily verified that the solution to the Kac equation is such that Boltzmann's H-functional

$$H(f)(t) = \int_{\mathbb{R}} f(v,t) \log f(v,t)\, dv \quad (1.114)$$

is monotonically decreasing in time.

By virtue of these physical characteristics, which make the Kac model very complete, binary interactions of type (1.112) have often been used to verify or discard conjectures difficult to verify for the full Boltzmann equation. Two generalizations deserve to be mentioned from the modelling point of view. The first one refers to the choice of the random parameter θ. Indeed, one can choose θ with a distribution different from the uniform one, without affecting the conservation laws of the interaction. Exploiting this idea, Desvillettes [124] studied the regularizing properties of the Kac operator in a grazing regime obtained by choosing for θ a distribution concentrated at the value 0.

Second, the choice

$$p_1 = q_2 = \cos\theta |\cos\theta|^p, \quad p_2 = -q_1 = \sin\theta |sin\theta|^p, \quad p > 0,$$

which is such that energy is dissipated,

$$(v^*)^2 + (w^*)^2 < v^2 + w^2,$$

has been introduced in [268] to investigate the Ernst–Brito conjecture [141, 142], relative to the formation of fat tails in the self-similar solution of the dissipative Boltzmann equation for Maxwell molecules introduced in [47].

1.5 Binary interaction models on the half-line

Section 1.4 describes systems in which particles move with velocities that take values on the whole real line. In many applications, however, the allowed variable is forced to assume only non-negative values. This is the case, for example, when modelling the wealth distribution in a multiagent system, or the number of mutated cells in a biological system. This apparently small difference leads to completely different behaviour of the solution for large times. To demonstrate this, just observe that if the independent variable is constrained to $\mathbb{R}_+$ then the symmetry of the self-similar or of the stationary solution is lost.

To outline the main differences between this new situation and the previous one, let us consider the case of constant interaction parameters described in (1.83), and let us describe the *grazing* case in which both $p-1$ and q are close to zero.

The kinetic equation (1.84) is replaced by

$$\frac{d}{dt}\int_{\mathbb{R}_+} \varphi(v) f(v,t)\, dv = \int_{\mathbb{R}_+^2} \left(\varphi(v^*) - \varphi(v)\right) f(v,t) f(w,t) dv dw. \tag{1.115}$$

We assume that the initial density satisfies

$$\int_{\mathbb{R}_+} f_0(v)\, dv = 1\,; \quad \int_{\mathbb{R}_+} v f_0(v)\, dv = 1. \tag{1.116}$$

We observe that, while the mass is preserved, choosing $\varphi(v) = v$ shows that, analogously to (1.86),

$$m(t) = \int_{\mathbb{R}_+} v f(v,t)\, dv = m(0) \exp\left\{(p+q-1)t\right\}. \tag{1.117}$$

Thus, unless $p+q = 1$, the mean value of the solution is not preserved (in fact, $m(0) = 1$). In particular, it is not necessary, as in Section 1.4, to proceed to the study of the evolution of the second moment. The main novelty is that the role of the energy is now supplied by the mean $m(t) = \int v f(v,t)\, dv$.

To look for self similarity we scale our solution according to

$$g(v,t) = m(t) f\left(m(t)v, t\right), \tag{1.118}$$

which implies that $g(v,t)$ remains a probability density and $\int v g(v,t) = 1$ for all $t \geq 0$. Moreover, $g = g(v,t)$ satisfies the equation

$$\begin{aligned} &\frac{d}{dt}\int_{\mathbb{R}_+} \varphi(v) g(v,t)\, dv - (p+q-1)\int_{\mathbb{R}_+} \varphi(v) \frac{\partial}{\partial v}(vg)\, dv \\ &\quad = \int_{\mathbb{R}_+^2} g(v) g(w) (\varphi(v^*) - \varphi(v)) dv dw. \end{aligned} \tag{1.119}$$

Let us assume that p is close to 1, while q is close to 0. Performing the same computations as Section 1.4, *mutatis mutandis*, we conclude that $g(v,t)$ satisfies

$$\begin{aligned} &\frac{d}{dt}\int_{\mathbb{R}_+} \varphi(v) g(v,t)\, dv + q\int_{\mathbb{R}} \varphi'(v)(v-1) g(v)\, dv \\ &\quad = \frac{1}{2}\int_{\mathbb{R}} g(v)\left((p-1)^2 v^2 + q^2 w^2 + 2(p-1)qvw\right)\varphi''(v) dv + R(p,q). \end{aligned} \tag{1.120}$$

The form of the remainder $R(p,q)$ is analogous to that of (1.94). It is clear that the correct scaling for small values of the parameter q is now

$$\tau = qt, \quad h(v,\tau) = g(v,t), \tag{1.121}$$

which implies that $h(v,\tau)$ satisfies the equation

$$\begin{aligned} \frac{d}{d\tau}\int_{\mathbb{R}_+} \varphi(v)h(v,\tau)\,dv + \int_{\mathbb{R}} \varphi'(v)(v-1)h(v)\,dv \\ = \frac{1}{2}\int_{\mathbb{R}} h(v)\frac{(p-1)^2}{q} v^2\varphi''(v)dv + R_1(p,q), \end{aligned} \tag{1.122}$$

where the remainder R_1 is given by

$$R_1(p,q) = \frac{1}{2}\int_{\mathbb{R}_+} \left(qw^2 + 2(p-1)vw\right)\varphi''(v)dv + \frac{1}{q}R(p,q).$$

Let us consider a parameter $p = p(q)$ such that, for a given constant $\lambda > 0$,

$$\lim_{q\to 0} \frac{(p(q)-1)^2}{q} = \lambda. \tag{1.123}$$

Then equation (1.122) is well approximated by the equation (in weak form)

$$\frac{d}{d\tau}\int_{\mathbb{R}_+} \varphi(v)h(v,\tau)\,dv + \int_{\mathbb{R}_+} \varphi'(v)(v-1)h(v)\,dv = \frac{\lambda}{2}\int_{\mathbb{R}_+} h(v)v^2\varphi''(v)dv. \tag{1.124}$$

Integration by parts then shows that equation (1.124) coincides with the weak form of the Fokker–Planck equation

$$\frac{\partial h}{\partial \tau} = \frac{\lambda}{2}\frac{\partial^2}{\partial v^2}\left(v^2 h\right) + \frac{\partial}{\partial v}\left((v-1)h\right), \tag{1.125}$$

provided the boundary terms produced by integration vanish. In particular, on the boundary $v = 0$, one obtains the conditions

$$v^2 h(v)\,|_{v=0} = 0, \tag{1.126}$$

and

$$(v-1)h(v) + \frac{\lambda}{2}\frac{\partial}{\partial v}\left(v^2 h(v)\right)|_{v=0} = 0. \tag{1.127}$$

While condition (1.126) is automatically satisfied for a sufficiently regular density $h(v)$, condition (1.127) requires an exact balance between the so-called advective and diffusive fluxes on the boundary $v = 0$. This condition is usually referred as the *no-flux* boundary condition.

Remark 1.5 *As this example clearly shows, boundary conditions are present any time the space variable in the one-dimensional differential problem is ranging on a bounded interval, or, more generally, on a half-line. We will encounter this type of problem* *many times in this book; each time, we will resort to an asymptotic procedure to pass from a kinetic description in terms of a Boltzmann-type collision operator to a kinetic description in terms of a Fokker–Planck-type operator, which involves derivatives with respect to the spatial variable. For the sake of simplicity, in most of the forthcoming cases we will assume that the solution to the various Fokker–Planck equations satisfies no-flux boundary conditions of type (1.127).*

It is immediately recognizable that equation (1.125) has a unique stationary solution of unit mass, given by the Γ-like distribution [55, 105]

$$M_\lambda(v) = \frac{(\mu-1)^\mu}{\Gamma(\mu)} \frac{\exp\left(-\frac{\mu-1}{v}\right)}{v^{1+\mu}}, \tag{1.128}$$

where

$$\mu = 1 + \frac{2}{\lambda} > 1.$$

This stationary distribution exhibits a power-law tail for large v's.

It is remarkable that the Fokker–Planck equation (1.125) describes the limiting behaviour of different systems [55, 221, 289].

Remark 1.6 *If we assume that the variable v has the meaning of* wealth, *and that the interaction (1.83) mimics* binary trades *between agents, then the analysis of this section describes a toy model for the evolution of the wealth distribution $f(v,t)$ of a system of traders. The study of the kinetic model (1.115) in the grazing regime then shows that the Fokker–Planck equation (1.124) follows from the kinetic model independently of the sign of the quantity $p+q-1$, which can produce exponential growth of wealth (when positive), or exponential dissipation of wealth (when negative). Hence, fat tails are produced in both situations, as soon as the compatibility condition (1.123) holds. Note that condition (1.123) is always admissible if $p+q-1<0$, while one has to require $p>1$ if $p+q-1>0$. This is quite remarkable since it shows that an uneven distribution of wealth may not only be produced as the effect of a growing mean wealth but also under critical economical circumstances.*

Remark 1.7 *Among the possible choices of the mixing parameters, one can consider the pair $(q, p = 1-q-\varepsilon)$, where ε is a small parameter. This case has been studied by Slanina in [287], who outlined the analogies between the evolution of the densities in a dissipative gas based on binary collisions and the evolution of the wealth density in a system of agents based on binary trades. Note that the Fokker–Planck equation (1.124) follows as soon as $\varepsilon = \varepsilon(q)$ satisfies*

$$\lim_{q\to 0} \frac{\varepsilon^2(q)}{q} = \lambda.$$

The particular choice

$$\varepsilon = -2\sqrt{q} + 2q,$$

which implies $\mu = 3/2$ and thus $\lambda = 4$, leads to the stationary state [287]

$$M_4(v) = \frac{1}{\sqrt{2\pi}} \frac{\exp\left(-\frac{1}{2v}\right)}{v^{5/2}}, \tag{1.129}$$

which solves the kinetic equation (1.119) for all values of the scaling parameter $q < 1/4$. We will return to this solution in the following chapters.

1.6 Some classical results

1.6.1 Kinetic equations as limit behaviour of particle systems

In the pertinent literature, kinetic equations for the one-particle distribution function with a bilinear interaction term are often derived in a slightly different way, which takes its origin in the kinetic theory of rarefied gases [82, 83, 140], and that can be easily adapted to a general system composed of a large number of interacting agents. Let us describe this derivation in the case in which the variable takes values in $\mathbb{R}_+$. Given a fixed number N of agents in a system, the interaction rules (1.77) give rise to a random process $(v_1(\tau), \dots, v_N(\tau))$ which evolves as follows: at each time τ a pair of agents (i, j) is chosen randomly and interact according to (1.77). The resulting process is clearly a discrete-time Markov process with state space $(\mathbb{R}_+)^N$ [165].

Processes of this type have been thoroughly studied, e.g. in the context of kinetic theory of ideal gases. The full information about the process in time τ is contained in the N-particle joint probability distribution $P_N(v_1, v_2, \dots, v_N, \tau)$.

However, one can write a kinetic equation for one-marginal distribution function

$$P_1(v, \tau) = \int P_N(v, v_2, \dots, v_N, \tau) dv_2 \cdots dv_N,$$

involving only one- and two-particle distribution functions [82, 84],

$$\begin{aligned} P_1(v, \tau+1) - P_1(v, \tau) = \Big\langle \frac{1}{N} \Big[\int P_2(v_i, v_j, \tau) \Big(\delta_0(v - (p_1 v_i + q_1 v_j)) \\ + \delta_0(v - (p_2 v_i + q_2 v_j)) \Big) dv_i \, dv_j - 2P_1(v, \tau) \Big] \Big\rangle, \end{aligned} \tag{1.130}$$

which may be continued to eventually give an infinite hierarchy of equations of BBGKY type [82, 140]. We recall that the BBGKY hierarchy, so called from the initials of the names of Bogoliubov, Born, Green, Kirkwood and Yvon, is a set of equations describing the dynamics of a system of a large number of interacting particles. In (1.130), $\delta(v)$ denotes a Dirac mass concentrated at $v = 0$. The standard approximation, which neglects the correlations between the wealth of the agents induced by the trade, gives the factorization

$$P_2(v_i, v_j, \tau) = P_1(v_i, \tau) P_1(v_j, \tau),$$

which implies a closure of the hierarchy at the lowest level.

Scaling the time as $t = 2\tau/N$ in the thermodynamic limit $N \to \infty$, one obtains for the one-particle distribution function $f(t, w)$ the kinetic equation

$$\begin{aligned} \frac{\partial f(v,t)}{\partial t} = \frac{1}{2} \Big\langle \int f(v_1, t) f(v_2, t) \Big(\delta_0(v - (p_1 v_1 + q_1 v_2)) \\ + \delta_0(v - (p_2 v_1 + q_2 v_2)) \Big) dv_1 \, dv_2 \Big\rangle - f(v, t), \end{aligned} \tag{1.131}$$

which describes the process (1.77) in the limit $N \to \infty$. A rigorous derivation of equation (1.131), based on the so-called propagation of chaos, can be derived following the classical argument of Kac [185,186]. In this case one needs, for example, to assume that in the discrete model the interactions take place at random times which correspond to the jumps of a Poisson process of rate N [163]. However, the rigorous derivation of bilinear equations in the limit $N \to \infty$ remains very difficult to follow, and we do not exploit it further. We address the interested reader to the paper of Carlen, Carvalho, Le Roux, Loss and Villani [70] for recent references to the subject, and for detailed information on the mathematical questions involved.

1.6.2 The central limit theorem revisited

There are deep analogies between the convergence towards the Gaussian law in the central limit theorem [177,200] and the large-time asymptotic of the nonlinear kinetic equation described in Section 1.4, in the case in which the second moment is preserved. We recall that the central limit theorem for a centred Gaussian law Y with density given by (1.104) consists of finding the set of distributions F such that, when dealing with the normalized sum

$$S_n = \frac{X_1 + \cdots + X_n}{n^{1/2}}$$

of the independent, centred and identically distributed random variables X_i with common distribution function F, the distribution F_n of the sum S_n converges to the Gaussian distribution. Let us denote with $f(x), x \in \mathbb{R}$, the probability density function of the random variables X_i, where f is normalized to satisfy (1.85), and with $f_n(x)$ the density of S_{2^n}, $n \geq 1$. Then, since

$$S_{2^{n+1}} = \frac{1}{\sqrt{2}} S_{2^n} + \frac{1}{\sqrt{2}} S^*_{2^n},$$

where S_{2^n} and $S^*_{2^n}$ are independent and identically distributed,

$$f_{n+1}(x) = \int_{\mathbb{R}} \sqrt{2} f_n\left(\sqrt{2}(x-y)\right) \sqrt{2} f_n\left(\sqrt{2}y\right)\, dy. \tag{1.132}$$

Let us change variable in the integral in (1.132), setting $x - y = \frac{1}{2}(x+z)$ (which implies $y = \frac{1}{2}(x-z)$). We obtain

$$f_{n+1}(x) = \int_{\mathbb{R}} f_n\left(\frac{x+z}{\sqrt{2}}\right) f_n\left(\frac{x-z}{\sqrt{2}}\right) dz. \tag{1.133}$$

Since $f_n(x)$ has unit mass, we can rewrite (1.132) as

$$f_{n+1}(x) = f_n(x) + \int_{\mathbb{R}} dz \left\{ f_n\left(\frac{x+z}{\sqrt{2}}\right) f_n\left(\frac{x-z}{\sqrt{2}}\right) - f_n(x) f_n(z) \right\}. \tag{1.134}$$

The recursive relation (1.134) can be viewed as the explicit Euler scheme (at discrete times $\Delta t = 1$) of the kinetic equation (in weak form)

$$\frac{d}{dt} \int_{\mathbb{R}} f(x)\varphi(x)\, dx = \frac{1}{2} \int_{\mathbb{R}^2} f(x) f(z) (\varphi(x^*) + \varphi(z^*) - \varphi(x) - \varphi(z)) dx\, dz, \tag{1.135}$$

where the post-interaction values (x^*, z^*) are related to (x, z) by the relations

$$x^* = \frac{1}{\sqrt{2}}(x + z) \tag{1.136}$$

$$z^* = \frac{1}{\sqrt{2}}(x - z). \tag{1.137}$$

Note that

$$(x^*)^2 + (z^*)^2 = x^2 + z^2,$$

and the operator is mass and energy preserving.

Hence, the evolution of densities in the central limit theorem, at least for the subsequence with indices which are powers of 2, obey a nonlinear evolution equation similar to (1.84). This analogy can be fruitfully taken into account to pass results from probability theory into kinetic theory. It is interesting to remark that this analogy is not restricted to the conservative case in which the second moment is preserved in time. As an example, it can also be shown that the dissipative case (in which the second moment decreases) has its counterpart in the central limit theorem for a stable law [200].

The central limit theorem for a centred p–stable law with distribution θ is related to finding the set of distributions F such that, when dealing with the normalized sum

$$S_n = \frac{X_1 + \cdots + X_n}{n^{1/s}}$$

of the independent, centred and identically distributed random variables X_i with common distribution function F, the distribution F_n of the sum S_n converges to θ. Note that, provided $s < 2$, the second moment of S_n is decreasing with respect to n. As before, let us denote with $f(x), x \in \mathbb{R}$, the probability density function of the random variables X_i, where f is normalized to satisfy (1.85), and with $f_n(x)$ the density of S_{2^n}, $n \geq 1$. In this case

$$S_{2^{n+1}} = \frac{1}{2^{1/s}} S_{2^n} + \frac{1}{2^{1/s}} S^*_{2^n},$$

where S_{2^n} and $S^*_{2^n}$ are independent and identically distributed, and

$$f_{n+1}(x) = \int_{\mathbb{R}} dy\, 2^{1/s} f_n\left(2^{1/s}(x - y)\right) 2^{1/s} f_n\left(2^{1/s} y\right). \tag{1.138}$$

With a change of variables $x - y = \frac{1}{2}(x + z)$ (which implies $y = \frac{1}{2}(x - z)$) in the integral (1.138) we obtain

$$f_{n+1}(x) = \int_{\mathbb{R}} dz\, 2^{\frac{2-s}{s}} f_n\left(\frac{x+z}{2^{\frac{s-1}{s}}}\right) f_n\left(\frac{x-z}{2^{\frac{s-1}{s}}}\right), \tag{1.139}$$

which, owing to mass conservation, can be written as

$$f_{n+1}(x) = f_n(x) + \int_{\mathbb{R}} dz \left\{ 2^{\frac{2-s}{s}} f_n\left(\frac{x+z}{2^{\frac{s-1}{s}}}\right) f_n\left(\frac{x-z}{2^{\frac{s-1}{s}}}\right) - f_n(x) f_n(z) \right\}. \tag{1.140}$$

This leads to the nonlinear evolution equation (1.135), where now the post-interaction values (x^*_s, z^*_s) are given by the relations

$$x_s^* = \frac{1}{2^s}(x+z) \tag{1.141}$$
$$z_s^* = \frac{1}{2^s}(x-z). \tag{1.142}$$

Note that, for $s < 2$,

$$(x_s^*)^2 + (z_s^*)^2 = 2^{\frac{s-2}{s}}\left(x^2+z^2\right) < x^2+z^2,$$

and the equation is mass preserving while it dissipates the second moment (energy).

1.6.3 Essential literature on Maxwell molecules

The kinetic modelling of this chapter is largely inspired by the representation of the collision integral for the Boltzmann equation for Maxwell molecules. In the kinetic theory of rarefied gases, the spatially homogeneous Boltzmann equation for elastic Maxwell molecules [84] is one of the most intensively studied models, because of the simplification consequent to the property that the collision rate is independent of the relative velocity of the colliding pair.

The investigation both of the spatially homogeneous Boltzmann equation for Maxwell molecules and of its simplified models made it possible to achieve essential progress and to verify or discard conjectures. To help the interested reader who wants to learn more about physical and mathematical advances in this topic, we give below a brief list of the main references, which will show how the study of relaxation to equilibrium has been a source of discovery for various methods which are now in widespread use in the mathematical and physical communities.

Fig. 1.8 Carlo Cercignani[8]

The first of these theoretical studies is due to McKean [231], who was able to find explicit rates of convergence towards the Maxwellian equilibrium for the Kac caricature of a Maxwell gas, the one-dimensional model introduced in the fifties by Mark Kac [186], briefly presented in Section 1.4.3. The pioneering paper by McKean contains a lot of enlightening remarks, and introduces into the matter the role of entropy production and of Fisher information [199], fruitfully used later on in different contexts [75]. In the early seventies, motivated by the problem of convergence to equilibrium for the Kac equation, Tanaka [301] introduced into kinetic theory the concept of a metric nowadays known by the German version of the name of the Russian mathematician L.N. Vasershtein, who introduced it independently in a different field [325]. Tanaka's work, however, was the first to contain the notable idea of obtaining results for the large-time behaviour of a nonlinear equation as a consequence of the contractivity of the metric. The properties of the Wasserstein metric were subsequently dealt with in a later paper by Tanaka [302], who added to the previous ideas the interesting

[8]**Carlo Cercignani**: born on 17 June 1939 in Teulada (Italy), died on 7 January 2010 in Milan (Italy). He was one of the founders of the modern mathematical kinetic theory. Cercignani devoted particular attention to the H-theorem, and to the problem of how macroscopic, irreversible evolution equations can follow from microscopic, reversible motion equations. His contributions to the study of Boltzmann's equation include the proof of the H-theorem for polyatomic gases. He received the Humboldt Prize in 1994.

connection between the problem of convergence to equilibrium for the Boltzmann equation for Maxwell molecules and the central limit problem of probability theory [269]. While innovative in the methodology introduced into kinetic theory, these papers do not contain results about the rate of convergence to equilibrium. This represents a point of weakness, because McKean [231] proved that, at least for a certain class of initial data, the solution to the Kac equation converges to equilibrium exponentially in time.

The same result was conjectured to hold by Cercignani [81] for the spatially homogeneous Boltzmann equation in any dimension and with any kernel, provided a suitable lower bound on the entropy production holds true (for an exhaustive discussion on this conjecture see [318]). It is now clear that, while powerful in getting convergence to equilibrium for the Boltzmann equation, the Wasserstein metric is not suitable for obtaining precise rates of convergence to equilibrium [52].

In 1995, exponential convergence towards equilibrium for both the Kac equation and the Boltzmann equation for Maxwell molecules was proven to hold in [156]. This result takes advantage of the possibility, discovered by Bobylev [43], of writing the Boltzmann equation for Maxwell molecules by changing to Fourier transforms, and makes use of a new metric for probability measures which is flexible enough to obtain precise rates of (exponential) convergence towards the Maxwellian equilibrium. In the same paper, various relationships of this metric with other, better-known metrics (including the Wasserstein one) allowed the determination of convergence in physical space. The same metric was subsequently used in [310] to prove the uniqueness of the solution to the Boltzmann equation for Maxwell molecules without cut-off, as well as in [69], again in connection with the spatially homogeneous Boltzmann equation for Maxwell molecules and its representation in Wild sums.

Hence, the study of the Boltzmann equation for elastic Maxwell molecules was responsible for the introduction of new mathematical tools, and among them, two metrics for probability measures, which are at the basis of most of the results concerned with the large-time behaviour of kinetic equations of Maxwell type.

In recent years, interest in the kinetic theory of dissipative systems, such as granular gases and fluids, has caused a revival in the study of the Boltzmann equation. Not surprisingly, the work of Bobylev, Carrillo and Gamba [47], who introduced a dissipative Maxwell model with its energy-independent collision rate which simplifies the nonlinear collision term to a convolution product, has had a great impact in that revival. A second main fact which was responsible for a noticeable increase of interest was the discovery of an exact scaling solution for a freely cooling one-dimensional Maxwell model [18].

The application of the techniques based on the use of the probability metrics, Wasserstein metric and Fourier-based ones, allowed the generalization of results of convergence to the equilibrium in classical elastic kinetic theory to the new problem of convergence to the self-similar solution (homogeneous cooling state) both in simpler one-dimensional models [254, 268] and in Maxwell models [41, 45, 48–50, 52].

These studies are very close to the modelling of socio-economic and biological systems we are going to describe in the following chapters. Indeed, contrary to what happens for the classical Boltzmann equation for Maxwell molecules which undergo elastic interactions, where the Maxwellian equilibrium is known, the self-similar solution of the inelastic Boltzmann equation is not known explicitly, and one is faced with the additional problem of characterizing it by means of some accessible properties, like regularity, presence of fat tails and others [152, 153].

Problems

Exercise 1.1 Let us denote with

$$\rho(x,t) = f^+(x,t) + f^-(x,t)$$

the mass density at time $t > 0$ of the solution to the Goldstein–Taylor system (1.20). Determine the partial differential equation satisfied by $\rho(x,t)$.

Let us substitute in system (1.20) $c \to c/\varepsilon$ and $\sigma \to \sigma/\varepsilon^2$ (diffusive scaling). Find the formal limit as $\varepsilon \to 0$ of the equation for the density.

Exercise 1.2 In Section 1.3.1, it was shown that the entropy functional (1.39)

$$S(t) = -\left[\Phi(p(t)) + \Phi(q(t))\right],$$

where $p(t)$ and $q(t)$ solve system (1.24), is increasing in correspondence to any convex function $\Phi(\cdot)$. Denote by $I(t)$ the (positive) derivative of $S(t)$ with respect to time. $I(t)$ is usually named *entropy production*. By evaluating the time derivative of $I(t)$, show that $S(X_\infty) - S(t)$ decays exponentially towards zero, and find the rate of exponential convergence.

Exercise 1.3 In 1957, Carleman [68] proposed a nonlinear kinetic model analogous to that of Goldstein and Taylor described in Section 1.3.1. The two-velocity Carleman model is

$$\begin{aligned}
\frac{\partial f_+(x,t)}{\partial t} + c\frac{\partial f_+(x,t)}{\partial x} &= \sigma\left(f_-^2(x,t) - f_+^2(x,t)\right), \\
\frac{\partial f_-(x,t)}{\partial t} - c\frac{\partial f_-(x,t)}{\partial x} &= \sigma\left(f_+^2(x,t) - f_-^2(x,t)\right).
\end{aligned} \tag{1.143}$$

The quadratic relaxation operator in (1.143) is intended to mimic the quadratic collision operator (1.5) of the Boltzmann equation. Following the procedure of Section 1.3.1, give a consistent derivation of the relaxation process

$$\begin{aligned}
\frac{dp(t)}{dt} &= \sigma\left(q^2(t) - p^2(t)\right), \\
\frac{dq(t)}{dt} &= \sigma\left(p^2(t) - q^2(t)\right).
\end{aligned}$$

Exercise 1.4 Let $H_\Phi(t)$ denote the Lyapunov functional (1.50) relative to the solution to the relaxation operator (1.43) for the radiative transfer equation, and let $I_\Phi(t)$ be the corresponding entropy production, i.e.

$$I_\Phi(t) = -\frac{d}{dt}H_\Phi(t).$$

Evaluate the time derivative of $I_\Phi(t)$, and discuss its consequences on the large-time behaviour of $H_\Phi(t)$ (see [154]).

Exercise 1.5 Let $M_\varepsilon(v)$ be the normal density

$$M_\varepsilon(v) = \frac{1}{\sqrt{2\pi\varepsilon}} e^{-v^2/(2\varepsilon)}. \tag{1.144}$$

Find the solution to equation (1.62) corresponding to a background given by (1.144). Compare this solution with the solution to the heat equation (1.61).

Exercise 1.6 Let us consider equation (1.106) when $p = q = 1$. Find the solution to equation (1.106) corresponding to a general initial value of mass equal to 1. Analyze the large-time behaviour of this exact solution.

Exercise 1.7 A relevant example of a relaxation operator is given by the Bhatnagar, Gross and Krook (BGK) [16] operator of rarefied gas dynamics (cf. Section 4.2.2 of Chapter 4). The BGK relaxation is given by

$$\frac{\partial f(v,t)}{\partial t} = \sigma(M(v) - f(v,t)),$$

where M is the Maxwellian equilibrium (1.2), namely a multivariate normal density, and σ is the collision frequency.

Given two probability densities $f(v)$ and $g(v)$, we define their relative entropy by the formula

$$H(f|g) = \int_{\mathbb{R}} f(v) \log \frac{f(v)}{g(v)}\, dv.$$

Using Jensen's inequality (1.51), prove that $H(f|g) \geq 0$. Then, verify that the relative entropy $H(f(t)|M)$, where $f(v,t)$ is the solution to the BGK equation, is monotonically decreasing in time, and find an estimate of its time decay.

2
Mathematical tools

> I have never felt the "gap" between the mode of thinking in pure mathematics and the thinking in physics, on which many mathematicians place so much stress. Anything amenable to mental analysis was congenial for me. I do not mean the distinction between rigorous thinking and more vague "imaginings"; even in mathematics itself, all is not a question of rigor, but rather, at the start, of reasoned intuition and imagination, and, also, repeated guessing. . . . It is all a multicolored thing, not very easy to describe in a way that a reader can appreciate.
>
> S.M. Ulam, *Adventures of a Mathematician*, 1976.

2.1 How to be certain of the predictions of a model?

In Chapter 1 we introduced in some detail various kinetic models, which allowed us to describe the time evolution of the (one-dimensional) density of a system of particles which interact with each other, or with an external background.

Fig. 2.1 Josiah Willard Gibbs[1]

The main physical idea behind this construction is that, because of the interactions, the system will relax towards a *universal* stationary solution. In many situations we showed that this relaxation process can be fruitfully interpreted in terms of the time decay of suitable convex functionals (the entropy). As discussed in Section 1.1, the statistical definition of entropy was originally developed by Boltzmann in the 1870s by analyzing the statistical behaviour of the microscopic components of the system. Another, related, expression of entropy is due to Gibbs; we refer to [182] for a discussion on the differences between the two notions.

In general, this universal stationary solution is not explicitly known, and the goal to be pursued is to be able, by resorting to mathematical tools, to extract some accessible properties, like its regularity, the presence or not of fat tails, and others.

[1]**Josiah Willard Gibbs**: born on 11 February 1839 in New Haven, Connecticut (USA), died on 28 April 1903 in New Haven, Connecticut, (USA). Gibbs is considered one of the founders of statistical mechanics. The importance of Gibbs findings was not immediately recognized. When his publications were read, they were considered too mathematically complex for most chemists and too scientific for many mathematicians. In 1901 Gibbs was awarded the Copley Medal of the British Royal Society, the most prestigious international science award at that time.

In order to do this, one needs to be sure that the kinetic model under study exhibits good mathematical properties, in addition to the physical ones. It is important to know if, starting from an initial datum, the solution to the model exists, and has some good features, including uniqueness. In other words, we have to be certain that the model is well posed from a mathematical point of view.

Possessing powerful mathematical instruments also allows us to justify the connections between models which present a decreasing order of difficulty, while maintaining most of the required physical properties. In Section 1.4 we discussed the importance of the simplification which comes from considering only interactions which do not vary the post-collision velocity (grazing collisions) too much. In this regime, in fact, one is led to pass from an interaction operator of Boltzmann type to a more regular operator of Fokker–Planck type, for which it is generally easier to explicitly characterize the stationary solution.

Clearly, this could be useless without knowing that what we are doing is consistent from a mathematical point of view. It would, in fact, be desirable to have an estimate of how the solutions of the two models differ from each other in terms of quantifiable distances.

In order to make usable predictions, we need mathematical instruments which are at the same time flexible and easy to handle. The recent mathematical research in kinetic theory, together with its goals in treating the aforementioned problems, gives us a variety of flexible tools for handling most of the models we will deal with in this book.

Our main proposal is to accompany the reader step by step to acquire the skills to best use these tools. For this purpose, after introducing the basic notions, we will fully handle two simple models, the first related to dissipation, and the second concerned with a growth process. These two models, which in view of their linearity are explicitly solvable, act as prototypes for more involved models that will be presented in the following chapters.

2.2 Some mathematical tools

2.2.1 Wild-type convolutions

In Chapter 1, Section 1.4, we introduced the most general one-dimensional kinetic model generated by binary collisions in which the post-interaction velocities are linear transformations of the pre-collision ones. In these interactions (cf. (1.77)) the mixing parameters $p_i, q_i, i = 1, 2$, are constant, or, in many cases, random variables.

In the language of probability theory, as described by (1.78), this interaction takes the form of a generalized convolution. Indeed, for any given pair of independent random variables X and Y, the pair (X^*, Y^*) is given by (1.78), that is a weighted sum, with random weights, of independent variables:

$$X^* = p_1 X + q_1 Y, \quad X^* = p_2 X + q_2 Y.$$

Suppose for the moment that the mixing parameters $p_i, q_i, i = 1, 2$, are positive constants. Then, proceeding as in Section 1.3.3, for any given pair of density functions f, g and for any observable $\varphi(\cdot)$,

$$\int_{\mathbb{R}^2} \varphi(v^*) f(v) g(w)\, dv\, dw = \int_{\mathbb{R}^2} \varphi(p_i v + q_i w) f(v) g(w)\, dv\, dw$$
$$= \int_{\mathbb{R}^2} \varphi(v + w) \frac{1}{p_i} f\left(\frac{v}{p_i}\right) \frac{1}{q_i} g\left(\frac{w}{q_i}\right) dv\, dw \qquad (2.1)$$
$$= \int_{\mathbb{R}} \varphi(v) f_{p_i} * g_{q_i}(v)\, dv.$$

In (2.1) the function $f_a(v)$, with $a \neq 0$, denotes as usual the function f scaled as in (1.67).

In agreement with the original definition of Wild [328], we will define Wild convolution of the two densities f and g, written $f \circ g(v)$, as the density defined through its action with any observable by the equality

$$\int_{\mathbb{R}} \varphi(v) f \circ g(v)\, dv = \frac{1}{2} \left\langle \int_{\mathbb{R}\times\mathbb{R}} (\varphi(v^*) + \varphi(w^*))\, f(v,t) g(w,t)\, dv\, dw \right\rangle. \qquad (2.2)$$

In the simpler case in which the binary collision is given by (1.83), i.e. when the mixing parameters are constant, Wild convolution takes the explicit form

$$f \circ g(v) = f_p * g_q(v), \qquad (2.3)$$

namely a standard convolution between scaled density functions. A similar explicit representation holds for the Kac model. In this case,

$$f \circ g(v) = \int_{\mathbb{R}} \int_0^{2\pi} \frac{1}{2\pi} f(v \cos\theta - w \sin\theta) g(v \sin\theta + w \cos\theta)\, dw\, d\theta. \qquad (2.4)$$

By making use of the Wild representation, we can write the bilinear kinetic model (1.79) as

$$\frac{\partial f(v,t)}{\partial t} = f \circ f(v,t) - f(v,t). \qquad (2.5)$$

Let us investigate the problem of existence and uniqueness of solutions to the initial value problem for equation (2.5) by assigning an initial density function $f(\cdot, 0) = F(\cdot)$. Since in many physical problems knowledge of some moment of the solution will be relevant for studying the large-time behaviour of the system, we will always assume that the initial density is a probability density function.

In general, we will assume that, if $v \in \mathbb{R}$, the initial datum $F(v)$ belongs to $\mathcal{P}_2$, and satisfies the normalization conditions (1.85), while, if $v \in \mathbb{R}^+$, $F(v)$ belongs to $\mathcal{P}_1$, and satisfies conditions (1.116).

The existence of a solution to the initial value problem for (2.5) can be immediately traced to a fixed point problem [328]. Define the map $f \mapsto \Phi(f)$ by

$$\Phi(f)(t) = e^{-t} F + \int_0^t e^{-(t-s)} f \circ f(s)\, ds.$$

Then, differentiating on both sides shows that $f(t)$ solves the kinetic equation (2.5) exactly when $\Phi(f) = f$. To find fixed points one considers iterations. First, put $f^{(0)} = 0$, and define, for all $j \geq 1$,

$$f^{(j+1)} = \Phi\left(f^{(j)}\right). \tag{2.6}$$

This yields

$$\begin{aligned}
f^{(1)} &= e^{-t}F \\
f^{(2)} &= e^{-t}F + e^{-t}\left(1-e^{-t}\right)F\circ F \\
f^{(3)} &= e^{-t}F + e^{-t}\left(1-e^{-t}\right)F\circ F \\
&\quad + e^{-t}\left(1-e^{-t}\right)^2\left(\frac{1}{2}F\circ(F\circ F) + \frac{1}{2}(F\circ F)\circ F\right),
\end{aligned}$$

and so on. Clearly,

$$f^{(j+1)} - f^{(j)} \geq 0$$

for all $j \geq 1$. Consequently, the limit of the monotone sequence of the $f_j(t)$ exists. Take

$$f(v,t) = \lim_{j\to\infty} f^{(j)}(t).$$

Then $f(v,t)$ is a solution to the kinetic equation (2.5). Historically, the existence argument of Wild [328] was completed by D. Morgenstern [238], who proved the uniqueness of solutions to the Boltzmann equation for Maxwell molecules three years later. We will return to uniqueness later on. Before then, let us discuss briefly the importance of Wild's argument. His idea immediately leads to the construction of a monotone sequence which approximates the solution, in which the approximations are made by iterated convolutions. Hence, the Wild approximation enters deeply into the structure of the solution to the bilinear kinetic equation. This idea has been developed in a number of papers in which the approximation has been clarified for the Kac model [232] from a probabilistic point of view.

Of course, the fixed point idea of Wild still holds for linear convolutions, such as the ones present in the linear kinetic model (1.72) of Section 1.3.3. Equations of this type can be rewritten in the form

$$\frac{\partial f(v,t)}{\partial t} = f(t)\circ M(v) - f(v,t). \tag{2.7}$$

In this case, one considers the map $f \mapsto \Phi(f)$ given by

$$\Phi(f)(t) = e^{-t}F + \int_0^t e^{-(t-s)} f\circ M\, ds.$$

As before, differentiating on both sides shows that $f(t)$ solves the kinetic equation (2.7) exactly when $\Phi(f) = f$. The linear case produces simpler iterations. Starting from $f^{(0)} = 0$, (2.6) yields

$$\begin{aligned}
f^{(1)} &= e^{-t}F \\
f^{(2)} &= e^{-t}F + te^{-t}F\circ M \\
f^{(3)} &= e^{-t}F + te^{-t}F\circ M + \frac{t^2}{2}e^{-t}(F\circ M)\circ M,
\end{aligned}$$

and so on. Clearly, also in this case

$$f^{(j+1)} - f^{(j)} \geq 0$$

for all $j \geq 1$. The function

$$f(v,t) = \lim_{j\to\infty} f^{(j)}(t),$$

the limit of the monotone sequence of the $f_j(t)$, exists, and it is a solution to the kinetic equation (2.7). Note that

$$f(v,t) = e^{-t} \sum_{k=0}^{\infty} \frac{t^k}{k!} f^{(k+1)}, \tag{2.8}$$

where the positive coefficients $f^{(k)}$, $k \geq 1$, are recursively defined by

$$f^{(k+1)} = f^{(k)} \circ M,$$

starting from $f^{(1)} = F$. It is important to remark that, at any time $t \geq 0$, $f(v,t)$, as given by (2.8), is a convex combination of the (time-independent) coefficients $f^{(k)}$.

The (unique) solution to the initial value problem for the kinetic equation (2.5) can be fruitfully described in a slightly different way by resorting to the so-called *convolution iterates*. Consider again the initial value problem for (2.5). Let us replace the time t and the density f by setting

$$\tau = 1 - e^{-t}, \qquad g(v,\tau) = f(v,t)e^t.$$

By direct inspection one can show that $g(v,\tau)$ satisfies the equation

$$\frac{\partial g(v,\tau)}{\partial \tau} = g \circ g(v,\tau), \tag{2.9}$$

and $g(\cdot,0) = F(\cdot)$. Note that, by its definition, the new time variable $\tau \leq 1$. Hence, one can expand the solution in power series of τ:

$$g(v,\tau) = \sum_{k=0}^{\infty} \tau^k f_{(k)}(v). \tag{2.10}$$

Substituting the expression (2.10) into equation (2.9) gives

$$\sum_{k=1}^{\infty} (k+1)\tau^k f_{(k+1)}(v) = \sum_{i,j=0}^{\infty} f_{(i)}(v) f_{(j)}(v) \tau^{i+j}.$$

Equating the coefficients of τ^k, the coefficients of the power series are recovered immediately. Starting from $f_{(0)} = F(v)$, the functions $f_{(k)}$, for $k \geq 0$, are given by

$$f_{(k+1)}(v) = \frac{1}{k+1} \sum_{i=0}^{k} f_{(i)} \circ f_{(k-i)}.$$

Going back to the original variables, the solution $f(t)$ can be represented by the sum

$$f(v,t) = e^{-t} \sum_{k=0}^{\infty} (1 - e^{-t})^k f_{(k)}(v). \tag{2.11}$$

The solution (2.11) to the kinetic equation (2.5) is referred to as a Wild sum. Notice that, since

$$e^{-t} \sum_{k=0}^{\infty} (1 - e^{-t})^k = 1,$$

this solution is a convex combination of the Wild coefficients $f_{(k)}$, $k \geq 0$. This property will be particularly useful when looking for the convex functionals which are monotonically decreasing along the solution.

The previous procedure can easily be generalized to obtain solutions to the kinetic-type equations which follow from multiple collisions [193]. Consider, in fact, the initial value problem for the kinetic equation

$$\frac{\partial f}{\partial t} = P_n(f)(v) - f(v),$$

where $f(\cdot, 0) = F(\cdot)$, and P is a n-linear operator from the Banach space $B^n \to B$ such that

$$|P(g_1, \ldots, g_n)| \leq |g_1| \cdots |g_n|, \qquad g_i \in B.$$

Then the unique solution $f(t)$ can be represented (in its interval of existence) by a Wild sum

$$f(v,t) = e^{-t} \sum_{k=0}^{\infty} (1 - e^{-(n-1)t})^k h_{(k)}(v),$$

where, for $k \geq 1$,

$$h_{(k+1)}(v) = \frac{1}{k+1} \sum_{i_1 + \cdots + i_n = k} \frac{b_{i_1} \cdots b_{i_n}}{(n-1) b_k} P_n(f_{(i_1)}, \ldots, f_{(i_n)}).$$

The numbers b_k are the coefficients of the Taylor expansion

$$(1 - x)^{1/(1-n)} = \sum_{k=0}^{\infty} b_k x^k.$$

Details of the proof can be found in [193].

2.2.2 Fourier-based metrics

In Section 2.2.1 it is proven that the initial value problem for the bilinear kinetic equation introduced in Section 1.4 admits a solution. Wild's idea then proves that this solution can be viewed as the limit of a monotone sequence, thus ensuring positivity. A different representation of the solution can be obtained with a suitable change of variables. It remains to prove that this solution is unique.

In Remark 1.4.2 we showed that changing to Fourier transforms leads to a marked simplification of the kinetic equations of type (1.84). In fact, for any fixed pair of constants (p, q), the Fourier-transformed equation (1.84) becomes

$$\frac{\partial \widehat{f}(\xi, t)}{\partial t} = \widehat{f}(p\xi)\widehat{f}(q\xi) - \widehat{f}(\xi). \tag{2.12}$$

Using the standard property of Fourier transforms connecting moments of f to the values of the derivatives of $\widehat{f}$ at the origin, the initial conditions (1.85) turn into

$$\widehat{f}(0) = 1, \widehat{f}'(0) = 0, \widehat{f}''(0) = -1, \tag{2.13}$$

so that $\widehat{f} \in C^2(\mathbb{R})$.

Let f_1 and f_2 be two solutions of the kinetic equation (1.84), corresponding to initial values $f_{1,0}$ and $f_{2,0}$ satisfying conditions (1.85), and let $\widehat{f_1}$, $\widehat{f_2}$ denote their Fourier transforms. Let us fix $|\xi| > 0$. Given any positive constant s, let us subtract the kinetic equation for $\hat{f}_2$ from the kinetic equation for $\hat{f}_1$. Dividing by $|\xi|^s$ we obtain

$$\frac{\partial}{\partial t} \frac{\left(\widehat{f_1} - \widehat{f_2}\right)}{|\xi|^s} + \frac{\widehat{f_1}(\xi) - \widehat{f_2}(\xi)}{|\xi|^s} = \frac{\widehat{f_1}(p\xi)\widehat{f_1}(q\xi) - \widehat{f_2}(p\xi)\widehat{f_2}(q\xi)}{|\xi|^s}. \tag{2.14}$$

Now, consider that the kinetic equation (1.84) is such that the mass is preserved. Hence, the first condition in (1.85) implies $|\widehat{f_1}(\xi)| \le 1$ and $|\widehat{f_2}(\xi)| \le 1$. Using these bounds in (2.14) we obtain

$$\begin{aligned}\left|\frac{\widehat{f_1}(p\xi)\widehat{f_1}(q\xi) - \widehat{f_2}(p\xi)\widehat{f_2}(q\xi)}{|\xi|^s}\right| &\le |\widehat{f_1}(p\xi)| \left|\frac{\widehat{f_1}(q\xi) - \widehat{f_2}(q\xi)}{|q\xi|^s}\right| q^s \\ &\quad + |\widehat{f_2}(q\xi)| \left|\frac{\widehat{f_1}(p\xi) - \widehat{f_2}(p\xi)}{|p\xi|^s}\right| p^s \le \sup \left|\frac{\widehat{f_1} - \widehat{f_2}}{|\xi|^s}\right| (p^s + q^s).\end{aligned}$$

We set

$$h(\xi) = \frac{\widehat{f_1}(\xi) - \widehat{f_2}(\xi)}{|\xi|^s}.$$

The previous computation then shows that

$$\frac{\partial h(\xi, t)}{\partial t} + h(\xi, t) \le (p^s + q^s)\|h\|_\infty(t). \tag{2.15}$$

This is equivalent to

$$\frac{\partial}{\partial t}\left(h(\xi, t)e^t\right) \le (p^s + q^s)\|h(\cdot, t)e^t\|_\infty.$$

Integrating from 0 to t, we get

$$h(\xi, t)e^t \le h(\xi, 0) + \int_0^t (p^s + q^s)\|h(\cdot, \tau)e^\tau\|_\infty \, d\tau.$$

Hence, if $H(t) = \|h(\cdot, t)e^t\|_\infty$,

$$H(t) \le H(0) + \int_0^t (p^s + q^s)H(\tau)\, d\tau.$$

Gronwall's lemma proves at once that

$$\|h(t)\|_\infty \le \exp\{(p^s + q^s - 1)t\}\, \|h_0\|_\infty. \tag{2.16}$$

Suppose at this point that, for some $s > 0$, $\|h_0\|_\infty$ is equal to zero. Then $\|h(t)\|_\infty$ is equal to zero for all subsequent times.

This remark was used in [156] to study the properties of the quantity

$$d_s(f, g) = \sup_{\xi \in \mathbb{R}} \frac{|\widehat{f}(\xi) - \widehat{g}(\xi)|}{|\xi|^s}, \tag{2.17}$$

which was shown to be a metric equivalent to the weak* convergence of measures for all $s > 0$. In our case, since the two initial data satisfy condition (1.85) (equal moments up to order two), it is clear that $\|h_0\|_\infty$ is bounded at least for all positive values of s such that $s \le 3$. This also implies that the solution is unique.

The Fourier-based metric (2.17) was introduced in [156] in connection with the study of the large-time asymptotics of the Boltzmann equation for Maxwell molecules. There, the case $s = 2+\alpha$, $\alpha > 0$, was considered. Further applications of d_s, with $s = 4$, were studied in [71], while the cases $s = 2$ and $s = 2 + \alpha$, $\alpha > 0$, have been considered in [69] in connection with the so-called McKean graphs (see Chapter 4, Section 4.3.2). The case $s = 2$ was subsequently used in [310], in connection with the uniqueness of the non-cut-off Boltzmann equation for Maxwell molecules. A further application of the general case $s > 0$ to the finding of Berry–Esseen-type bounds in the central limit theorem for a stable law has been given in [162]. Only recently, various applications to the large-time behaviour of the dissipative Boltzmann equation [40, 41, 268] showed the importance of the distance even in this case.

Let us now review the main properties of the d_s metrics, by collecting them in the following

Proposition 2.1 *The distances d_s with $s > 0$ verify the following properties:*

i) **Interpolation of metrics:** *Given any two probability measures $f, g \in \mathcal{P}_s(\mathbb{R}^n)$ with $q > 0$ with equal moments up to $[s]$ if $s \notin \mathbb{N}$, or equal moments up to $s - 1$ if $s \in \mathbb{N}$, then*

$$d_p(f, g) \le 2\left(\frac{s-p}{2p}\right)^{p/s} \frac{s}{s-p} [d_s(f, g)]^{p/s} = C_{p,s}\, [d_s(f, g)]^{p/s}$$

for any $0 < p < s$.

ii) **Scaling:** *Given any two probability measures $f, g \in \mathcal{P}_s(\mathbb{R}^n)$ with $s > 0$ with equal moments up to $[s]$ if $s \notin \mathbb{N}$, or equal moments up to $s - 1$ if $s \in \mathbb{N}$, then*

$$d_s(f_\theta, g_\theta) = \theta^{-s}\, d_s(f, g),$$

where, for $v \in \mathbb{R}^n, n \ge 1$, and a given positive constant θ,

$$f_\theta(v) = \frac{1}{\theta^n} f\left(\frac{v}{\theta}\right).$$

iii) **Convexity:** *Given f_1, f_2, g_1 and g_2 in $\mathcal{P}_s(\mathbb{R}^n)$ with $s > 0$ with equal moments up to $[s]$ if $s \notin \mathbb{N}$, or equal moments up to $s-1$ if $s \in \mathbb{N}$, and α in $[0,1]$, then*

$$d_s(\alpha f_1 + (1-\alpha) f_2, \alpha g_1 + (1-\alpha) g_2) \le \alpha d_s(f_1, g_1) + (1-\alpha) d_s(f_2, g_2).$$

iv) **Super-additivity with respect to convolution:** *Given f_1, f_2, g_1 and g_2 in $\mathcal{P}_s(\mathbb{R}^n)$ with $s > 0$ with equal moments up to $[s]$ if $s \notin \mathbb{N}$, or equal moments up to $s-1$ if $s \in \mathbb{N}$, then*

$$d_s(f_1 * f_2, g_1 * g_2) \le d_s(f_1, g_1) + d_s(f_2, g_2).$$

Proof. The first statement *i)* comes from

$$\begin{aligned} d_p(f,g) &= \sup_{\xi \in \mathbb{R}^n} \frac{|\hat{f}(\xi) - \hat{g}(\xi)|}{|\xi|^p} \le \sup_{0<|\xi|\le R} \frac{|\hat{f}(\xi) - \hat{g}(\xi)|}{|\xi|^p} + \sup_{|\xi|>R} \frac{|\hat{f}(\xi) - \hat{g}(\xi)|}{|\xi|^p} \\ &\le \sup_{0<|\xi|\le R} \frac{|\hat{f}(\xi) - \hat{g}(\xi)|}{|\xi|^s} R^{s-p} + \frac{2}{R^p} \le d_s(f,g)\, R^{s-p} + \frac{2}{R^p}, \end{aligned} \tag{2.18}$$

for any $R > 0$. Optimizing the function in the right-hand side of (2.18) over R, we obtain the desired result.

The second statement *ii)* is an easy consequence of the scaling property of the Fourier transform $\widehat{f_\theta}(\xi) = \hat{f}(\theta\xi)$ and the definition of d_s.

The third statement *iii)* follows trivially from triangular inequality and the definition of d_s.

Finally, the convolution property *iv)* is straightforward due to $\widehat{f * g} = \hat{f}\hat{g}$, the triangular inequality and the definition of d_s. ∎

Remark 2.1 *As explained in this section, the Fourier metric d_s is natural to use in connection with kinetic equations of type (2.5), any time it is possible to write these equations in terms of Fourier transforms. While it is easy to handle these equations by means of this type of metric, new problems are introduced to connect these metrics with other types of metric in more common use, and that are part of the common background of applied mathematicians.*

This connection has been partially investigated in [156], where the relationship of the d_s metric with other metrics in wide use in probability theory has been pointed out. In particular, connections of the Fourier-based metric with Prokhorov and Wasserstein metrics [330] have been investigated. The equivalence between various types of metric has subsequently been investigated in [310]. A detailed analysis of this argument is contained in [77]. We address the interested reader to these papers to learn more of this argument.

Remark 2.2 *One of the weak points of the Fourier-based distance (2.17) is that, for a given s such that $1 < s < 2$, it is not known if the space of probability measures*

$\mathcal{P}_s(\mathbb{R})$ with metric d_s is complete or not. This unpleasant fact is discussed in [77], together with a possible remedy. A further metric, however, can be introduced, which does not have the same problem, while it possesses most of the properties of the metric d_s. This metric has been introduced in [22] to characterize fixed points of convex sums of random variables with a small number of moments. For $s \in (1,2)$,

$$D_s(f_1, f_2) = \int |\xi|^{-(s+1)} |\hat{f}_1(\xi) - \hat{f}_2(\xi)|\, d\xi, \quad s > 0. \tag{2.19}$$

As proven in [22], $(\mathcal{P}_s(\mathbb{R}), D_s)$ is complete. In most cases, however, the use of the standard d_s metric leads to simpler proofs, and we will not resort to the D_s metric. The interested reader can in any case give alternative proofs of our forthcoming results in terms of D_s.

2.3 The drift equation and Dirac delta functions

Section 2.2.2 illustrates a basic fact about bilinear equations of type (1.79). These equations are easily treated by converting to Fourier transforms. This is nowadays a well-known fact, which goes back to the result by Bobylev [43] for the Boltzmann equation for Maxwell molecules. His idea introduced a new field of application for Fourier transform methods to nonlinear problems. As a matter of fact, Fourier transforms are widely used for linear problems, like, for example, linear diffusion problems [143]. In what follows, we will illustrate how this analysis works for the linear drift partial differential equation (1.92), which for simplicity we will write with the coefficient $\alpha = 1$. Hence, we will study the initial value problem for equation

$$\frac{\partial f}{\partial t} = \frac{\partial (v f(v))}{\partial v}, \tag{2.20}$$

where the initial value $f_0(v)$ is a probability density satisfying conditions (1.85). Moments of the solution to equation (2.20) can be evaluated explicitly in terms of the moments of the initial density. Integration by parts yields

$$\frac{d}{dt}\int_{\mathbb{R}} v f(v,t)\, dv = \int_{\mathbb{R}} v \frac{\partial (v f(v))}{\partial v} = -\int_{\mathbb{R}} v f(v,t)\, dv,$$

so that

$$\int_{\mathbb{R}} v f(v,t)\, dv = e^{-t} \int_{\mathbb{R}} v f_0(v)\, dv = 0, \tag{2.21}$$

while

$$\frac{d}{dt}\int_{\mathbb{R}} v^2 f(v,t)\, dv = \int_{\mathbb{R}} v^2 \frac{\partial (v f(v))}{\partial v} = -2\int_{\mathbb{R}} v^2 f(v,t)\, dv,$$

which gives

$$\int_{\mathbb{R}} v^2 f(v,t)\, dv = e^{-2t} \int_{\mathbb{R}} v^2 f_0(v)\, dv = e^{-2t}.$$

Hence, the variance of the solution is exponentially decreasing, and the drift equation reproduces a dissipative phenomenon.

If one multiplies the drift equation (2.20) by $e^{-i\xi v}$ and integrates on $\mathbb{R}$, one obtains an identity for the Fourier transform

$$\int_{\mathbb{R}} \frac{\partial f}{\partial t} e^{-i\xi v}\, dv = \frac{\partial \hat{f}(\xi,t)}{\partial t} = \int_{\mathbb{R}} \frac{\partial(vf(v))}{\partial v} e^{-i\xi v}\, dv.$$

Integrating by parts in the last integral, and considering that

$$-\int_{\mathbb{R}} vf(v) - i\xi e^{-i\xi v}\, dv = -\xi \int_{\mathbb{R}} f(v) \frac{\partial e^{-i\xi v}}{\partial \xi}\, dv = -\xi \frac{\partial}{\partial \xi} \int_{\mathbb{R}} f(v) e^{-i\xi v}\, dv,$$

yields the partial differential equation

$$\frac{\partial \hat{f}(\xi,t)}{\partial t} + \xi \frac{\partial \hat{f}(\xi,t)}{\partial \xi} = 0. \tag{2.22}$$

Since the initial value $f_0(v)$ satisfies conditions (1.85), $\hat{f}_0(\xi)$ satisfies conditions (2.13).

Equation (2.22) can be solved by the following argument. Let $\eta = \eta(\xi,t)$ be a smooth function, and let us consider the total derivative of the function $\hat{f} = \hat{f}(\eta(\xi,t),t)$. Note that by consistency one is forced to assume $\eta(\xi,t=0) = \xi$, which implies $\hat{f}(\eta(\xi,t=0),t=0) = \hat{f}_0(\xi)$. It holds that

$$\left.\frac{d\hat{f}}{dt}\right|_{\eta=\eta(t)} = \left.\left(\frac{\partial \hat{f}}{\partial t} + \frac{\partial \hat{f}}{\partial \eta}\eta'\right)\right|_{\eta=\eta(t)}.$$

Consequently, by choosing $\eta(t) = \xi e^t$, which is such that $\eta'(t) = \eta(t)$ and at the same time $\eta(\xi,t=0) = \xi$, if f satisfies (2.22) one obtains

$$\left.\frac{d\hat{f}}{dt}\right|_{\eta=\xi e^t} = \left.\left(\frac{\partial \hat{f}}{\partial t} + \frac{\partial \hat{f}}{\partial \eta}\eta\right)\right|_{\eta=\xi e^t} = 0.$$

This shows that $\hat{f}(\xi e^t,t) = \hat{f}_0(\xi)$, so that the solution to the initial value problem for equation (2.22) is

$$\hat{f}(\xi,t) = \hat{f}_0(\xi e^{-t}). \tag{2.23}$$

In the physical space, the solution to the drift equation (2.20) reads

$$f(v,t) = e^t f_0(ve^t). \tag{2.24}$$

The previous example shows that, by converting to Fourier transforms in linear problems, one gains advantages which lead to a simple way of solving the underlying equation. This technique is useful in solving linear partial differential equations. A further example will be fully presented in Section 2.5.

In the case of the drift equation, we can easily extract from the analytic expression of its solution both the main properties and the asymptotic behaviour. It is clear that expression (2.24) guarantees that the solution to (2.20) is non-negative whenever

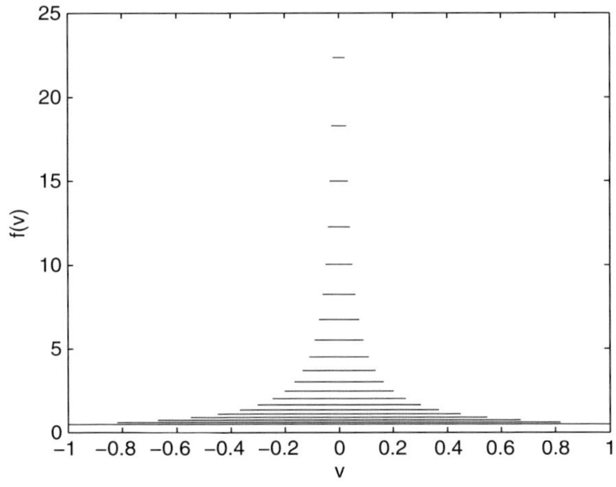

Fig. 2.2 Graph of the function (2.27) at subsequent times.

the initial datum is. Likewise, when the initial datum satisfies conditions (1.85), the solution satisfies, for all times $t \geq 0$,

$$\int_{\mathbb{R}} f(v,t)\,dv = 1\,; \quad \int_{\mathbb{R}} v f(v,t)\,dv = 0, \tag{2.25}$$

while

$$\int_{\mathbb{R}} v^2 f(v,t)\,dv = e^{-2t}.$$

Hence, if the initial value is a probability density function, the solution to the drift equation is a probability density function for all subsequent times. Likewise, if the mean value of the initial probability density is equal to zero, the mean value of the solution remains equal to zero. However, the variance of the solution decays exponentially with time. The meaning of this behaviour can be made clear by resorting to a particular initial datum, namely to a probability density which is uniform on the interval $[-1, 1]$. Hence

$$f_0(v) = \frac{1}{2} \text{ if } v \in [-1,1]; \quad f_0(v) = 0 \text{ outside.} \tag{2.26}$$

Consequently, for all $t \geq 0$,

$$f(v,t) = \frac{e^t}{2} \text{ if } v \in [-e^{-t}, e^{-t}]; \quad f_0(v) = 0 \text{ outside.} \tag{2.27}$$

Let $X(t)$ be the uniform random variable with density given by (2.27). Then at time t the variable can take values only in the interval $[-e^{-t}, e^t]$ (see Fig. 2.2). Thus, when time goes to infinity, the limit random variable will take the value $v = 0$ with complete certainty, or, equivalently, its probability density will be zero for any value $v \neq 0$. On the other hand, conditions (2.25) have to be satisfied. No ordinary function can satisfy these two requirements. If one wants to include certainty as a particular case of probability, it is necessary to enlarge the concept of probability density function.

The required generalization is achieved by resorting to *generalized functions* or *distributions* [211].

A fruitful representation of the action of the generalized limit function of the drift equation with a uniform initial datum is obtained by resorting to observables. For any given (smooth) observable $\varphi(\cdot)$ let us compute its average. Owing to the mean value theorem,

$$\langle \varphi(X(t)) \rangle = \int_{\mathbb{R}} \varphi(v) f(v,t)\, dv = \frac{e^t}{2} \int_{-e^{-t}}^{e^{-t}} \varphi(v)\, dv = \varphi(v^*(t)),$$

where $v^*(t) \in [-e^{-t}, e^{-t}]$. Finally,

$$\lim_{t\to\infty} \langle \varphi(X(t)) \rangle = \varphi(0).$$

Let us denote by $f_\infty(v)$ this *generalized* limit function. Then the integral of $\varphi(v) f_\infty(v)$ is well defined and given by

$$\int_{\mathbb{R}} \varphi(v) f_\infty(v)\, dv = \varphi(0). \tag{2.28}$$

The case described above is the simplest example of a generalized function. The limit function $f_\infty(v)$ acting on $\varphi(\cdot)$ to satisfy (2.28) defines the so-called *Dirac delta function* concentrating on the value $v = 0$. In Section 1.2 we denoted a Dirac delta function concentrated on zero by $\delta_0(\cdot)$.

As the analysis of the drift equation shows clearly, the Dirac delta function appears to be a universal limit distribution (the steady solution of equation (2.20)), independent of the choice of the initial value. Indeed, different initial values for (2.20) lead to the same limit distribution.

By choosing $\varphi(v) = e^{-i\xi v}$ one obtains that the Fourier transform of a Dirac delta function concentrated on a point $x \in \mathbb{R}$, which we denote by $\delta_x(\cdot)$, is

$$\widehat{\delta_x}(\xi) = \int_{\mathbb{R}} e^{-i\xi v} \delta_0(v - x)\, dv = e^{-i\xi x}. \tag{2.29}$$

Remark 2.3 *If x and y are real numbers, $x \neq y$, two Dirac distributions have $d_1(\delta_x, \delta_y) = |x - y|$. Indeed,*

$$d_1(\delta_x, \delta_y) = \sup_\xi \frac{|e^{-i\xi x} - e^{-i\xi y}|}{|\xi|} = \sup_\xi \frac{|e^{-i\xi(x-y)} - 1|}{|(x-y)\xi|} |x - y| = |x - y|.$$

More generally, a density $f_1(v)$ and its translation $f_2(v) = f_1(v - z)$ have Fourier distance $d_1(f_1, f_2) = |z|$. For comparison, if f_1 is supported in a small interval $[-\epsilon, +\epsilon]$, then

$$\|f_1 - f_2\|_{L^1} = \int_{\mathbb{R}} |f_1(v) - f_2(v)|\, dv = 2$$

for all $|z| > \epsilon$. Thus, the Fourier-based distances provide a more sensible notion of closeness *of densities than, e.g., the classical L^1 distance. Notice, however, that $d_s(\delta_x, \delta_y) = +\infty$ for $s > 1$ unless $x = y$.*

2.4 Dissipative models and the drift equation

The drift equation (2.20) appears to be a model which is simple to handle by resorting to Fourier transforms, and such that most of the properties of more realistic kinetic models are satisfied. Among others, we showed that the solution is mass preserving, and converges to a steady state. As we discussed in Chapter 1, Sections 1.3.1 and 1.3.2, a further property typical of kinetic models is the existence of entropy functionals (cf. Remark 1.1). Let us show that this property also holds for the drift equation, under a particular choice of the convex functional which defines the entropy.

To this end, consider that, for given positive times t and τ, the solution (2.23) satisfies

$$\hat{f}(\xi, t+\tau) = \hat{f}_0(\xi e^{-(t+\tau)}) = \hat{f}_0\left((\xi e^{-\tau})e^{-t}\right) = \hat{f}(\xi e^{-\tau}, t). \tag{2.30}$$

Since $\hat{f}_\infty(\xi) = 1$, for all positive constants s and $\xi \neq 0$,

$$\frac{\widehat{f}(\xi, t+\tau) - 1}{|\xi|^s} = \frac{\widehat{f}(\xi e^{-\tau}, t) - 1}{|\xi|^s}.$$

Thanks to the scaling property of d_s distance (property ii of Proposition 2.1), we then obtain

$$d_s(f(t+\tau), f_\infty) = \sup_\xi \frac{\left|\widehat{f}(\xi, t+\tau) - 1\right|}{|\xi|^s} = \sup_\xi \frac{\left|\widehat{f}(\xi e^{-\tau}, t) - 1\right|}{|\xi|^s} = e^{-s\tau} d_s(f(t), f_\infty).$$

Consequently, for any $s > 0$, the functionals

$$\Phi_s(f(t)) = d_s(f(t), f_\infty) \tag{2.31}$$

are monotonically decreasing along the solution to the drift equation (2.20). On the other hand, property iii of Proposition 2.1 ensures that these functionals are convex. Of course, the convex functionals defined in (2.31) are bounded only for a precise set of values of the constant s, which depends on the initial value. For example, conditions (1.85) ensure that $\Phi_s(f_0)$ is bounded for all s such that $0 < s < 2$.

The drift equation represents an idealized linear model for dissipation. From a kinetic point of view, dissipation can be easily obtained as a consequence of interactions of a system of particles with a background whose action is to slow the particles in proportion to their speed.

Proceeding as in Section 1.3.3, we can describe the evolution of the system of particles by considering the related problem in which the number of particles with velocity v at time t is produced by the process $X(t)$, and the variation of $X(t)$ is due to interaction with the external background M, described by a random variable with probability density $\mathcal{M}(w)$, which we will assume of positive mean value $\varepsilon < 1$, and variance γ^2. Let us assume that the collision process of a particle with velocity v with a background particle with velocity w generates a post-collision velocity v^* given by

$$v^* = v(1-w). \tag{2.32}$$

In terms of the process $X(t)$, the law of change given by (2.32) can be rewritten as

$$X^*(t) = X(t)(1 - M). \tag{2.33}$$

Assuming that $X(t), M$ are independent of each other, for a given observable quantity $\varphi(\cdot)$ we then have that the mean value of $\varphi(X)$ satisfies

$$\begin{aligned} \frac{d}{dt}\int_{\mathbb{R}} \varphi(v) f(v,t)\,dv &= \frac{d}{dt}\langle \varphi(X(t))\rangle \\ &= \mu \int_{\mathbb{R}\times\mathbb{R}} (\varphi(v^*) - \varphi(v))\, f(v,t)\mathcal{M}(w)\,dv\,dw, \end{aligned} \tag{2.34}$$

where the constant $\mu > 0$ denotes as usual the collision frequency. Note that choosing $\varphi(v) = 1$ shows that, independently of the background distribution, $f(v,t)$ remains a probability density if it is so initially:

$$\int_{\mathbb{R}} f(v,t)\,dv = \int_{\mathbb{R}} f_0(v)\,dv = 1.$$

If now $\varphi(v) = v$, we obtain

$$\frac{d}{dt}\int_{\mathbb{R}} v f(v,t)\,dv = -\mu \int_{\mathbb{R}\times\mathbb{R}} vw\, f(v,t)\mathcal{M}(w)\,dv\,dw = -\mu\varepsilon \int_{\mathbb{R}} v f(v,t)\,dv. \tag{2.35}$$

In an analogous way, if $\varphi(v) = v^2$ one shows that

$$\frac{d}{dt}\int_{\mathbb{R}} v^2 f(v,t)\,dv = -\mu\left[2\varepsilon - (\varepsilon^2 + \gamma^2)\right] \int_{\mathbb{R}} v^2 f(v,t)\,dv. \tag{2.36}$$

Hence, depending on the sign of the coefficient $2\varepsilon - (\varepsilon^2 + \gamma^2)$, the behaviour of the second moment of the solution can be twofold. If $2\varepsilon - (\varepsilon^2 + \gamma^2) > 0$, the second moment of the solution decays exponentially to zero (as in the drift equation). In the opposite case, the second moment of the solution increases exponentially, and the solution does not concentrate at all, so behaving differently from the solution to the drift equation.

The simplest case in which the solution to the kinetic equation (2.34) does concentrate corresponds to the choice of a background which is described by a random variable which is equal to ε with probability one. In this case, as discussed in Section 1.2, the density function of the background M is a Dirac delta function concentrating on $v = \varepsilon$, say $\delta_0(v - \varepsilon)$. Within this choice, the kinetic equation reads

$$\frac{d}{dt}\int_{\mathbb{R}} \varphi(v) f(v,t)\,dv = \mu \int_{\mathbb{R}} (\varphi(v(1-\varepsilon)) - \varphi(v))\, f(v,t)\,dv. \tag{2.37}$$

Note that in this case, while equation (2.35) for the evolution of the mean value still holds unchanged, equation (2.36) becomes

$$\frac{d}{dt}\int_{\mathbb{R}} v^2 f(v,t)\,dv = -\mu\left(2\varepsilon - \varepsilon^2\right) \int_{\mathbb{R}} v^2 f(v,t)\,dv, \tag{2.38}$$

and, since $\varepsilon < 1$, the second moment is exponentially decreasing in time.

To establish the eventual connection of equation (2.37) with the weak form of the drift equation (2.20), we first set the frequency $\mu = 1$. After this simplification, we observe that, while the mean value of the solution to the drift equation decays exponentially fast at a rate equal to 1 (cf. (2.21)), the mean value of the solution to the kinetic equation, as given by (2.35), decays exponentially at a rate equal to ε. A way to avoid the dependence on the parameter ε is to scale time to obtain for the solution to (2.37) an evolution of the mean value which does not depend on ε. It is enough to set $\bar{t} = \varepsilon t$ and $g(v, \bar{t}) = f(v, t)$. Then $g(v, \bar{t})$ satisfies the equation

$$\frac{d}{d\bar{t}} \int_{\mathbb{R}} \varphi(v) g(v, \bar{t})\, dv = \frac{1}{\varepsilon} \int_{\mathbb{R}} \left(\varphi(v(1-\varepsilon)) - \varphi(v) \right) g(v, \bar{t})\, dv, \tag{2.39}$$

which is such that, as for the drift equation,

$$\frac{d}{d\bar{t}} \int_{\mathbb{R}} v g(v, \bar{t})\, dv = - \int_{\mathbb{R}} v g(v, \bar{t})\, dv.$$

The physical idea behind this scaling is to render the macroscopic effect of the microscopic collision (at the level of mean value) independent of the size of the action of the background. It is clear that by choosing $\varepsilon \ll 1$, any interaction produces a small modification of the velocity of particles, and a bigger time is required to observe an appreciable change of the mean value. Note that, as extensively discussed in Section 1.4, the interactions corresponding to a choice of $\varepsilon \ll 1$ are close to being *grazing* collisions, namely collisions in which the post-collision velocities are approximately equal to the pre-collision ones. Thus, by letting $\varepsilon \to 0$, and at the same time by scaling time to preserve the behaviour of the mean value, we are investigating the *grazing interaction limit* of equation (2.37).

In the rest of this section, we will introduce the main tools which allow us to study the limiting behaviour (as $\varepsilon \to 0$) of the solution to equation (2.39). For reasons of simplicity, and to highlight the role of ε, we fix $t = \bar{t}$, and $g(v, \bar{t}) = g_\varepsilon(v, t)$. In the Fourier transform, equation (2.39) then takes the form

$$\frac{\partial \hat{g}_\varepsilon(\xi, t)}{\partial t} = \frac{1}{\varepsilon} \left(\hat{g}_\varepsilon((1-\varepsilon)\xi, t) - \hat{g}_\varepsilon(\xi, t) \right). \tag{2.40}$$

If the initial value of the drift equation (2.20) satisfies conditions (1.85), so that the second moment of the solution is bounded, the relationship of the drift equation with (2.40) can be shown immediately. In fact, by Taylor expansion,

$$\hat{g}((1-\varepsilon)\xi) = \hat{g}(\xi) - \varepsilon\xi \frac{\partial \hat{g}(\xi)}{\partial \xi} + \frac{1}{2} (\varepsilon\xi)^2 \frac{\partial^2 \hat{g}(\bar{\xi})}{\partial \xi^2},$$

where $\bar{\xi} \in ((1-\varepsilon)\xi, \xi)$. Therefore

$$-\xi \frac{\partial \hat{g}(\xi)}{\partial \xi} = \frac{1}{\varepsilon} \left(\hat{g}((1-\varepsilon)\xi, t) - \hat{g}(\xi, t) \right) - \frac{1}{2} \varepsilon \xi^2 \frac{\partial^2 \hat{g}(\bar{\xi})}{\partial \xi^2},$$

and the drift equation for g can be written as

$$\frac{\partial \hat{g}(\xi,t)}{\partial t} = \frac{1}{\varepsilon}\left(\hat{g}((1-\varepsilon)\xi,t) - \hat{g}(\xi,t)\right) - \frac{1}{2}\varepsilon\xi^2 \frac{\partial^2 \hat{g}(\bar{\xi})}{\partial \xi^2}. \tag{2.41}$$

Equations (2.40) and (2.41) differ only by the presence of a term of size proportional to ε (the last term in (2.41)). We remark that, by construction, the solutions to both equations are such that the mass is preserved (equal to one), while the first moment has the same law of decay. Hence, by choosing initial data in both equations with equal moments of order one (the same mean value), the two solutions have the same mean value at any subsequent time. This implies that the Fourier-based metric d_2 of the solutions is bounded. Bearing this in mind, for any given $\xi \neq 0$, let us subtract equation (2.41) from equation (2.40), and divide both sides by $|\xi|^2$ to obtain

$$\begin{aligned}\frac{\partial}{\partial t}\frac{g_\varepsilon(\xi,t) - g(\xi,t)}{|\xi|^2} &= \frac{1}{\varepsilon}\frac{g_\varepsilon((1-\varepsilon)\xi,t) - g((1-\varepsilon)\xi,t)}{|\xi|^2} \\ &\quad - \frac{g_\varepsilon(\xi,t) - g(\xi,t)}{|\xi|^2} + \frac{1}{2}\varepsilon\frac{\partial^2 \hat{g}(\bar{\xi})}{\partial \xi^2}.\end{aligned} \tag{2.42}$$

Now, consider that the solution to the drift equation (2.20), with $\mu = 1$, where the initial density satisfies (1.85), is such that

$$\int_{\mathbb{R}} v^2 g(v,t)\, dv = e^{-2t}\int_{\mathbb{R}} v^2 g_0(v)\, dv = e^{-2t}.$$

Therefore, by a classical property of Fourier transforms,

$$\left|\frac{\partial^2 \hat{g}(\bar{\xi})}{\partial \xi^2}\right| \le \left|\frac{\partial^2 \hat{g}(0)}{\partial \xi^2}\right| = e^{-2t}.$$

Using this upper bound in (2.42) we obtain that, by setting

$$h_\varepsilon(\xi,t) = \frac{g_\varepsilon(\xi,t) - g(\xi,t)}{|\xi|^2},$$

$h_\varepsilon(t)$ satisfies the inequality

$$\begin{aligned}\frac{\partial h_\varepsilon(\xi,t)}{\partial t} &\le \frac{1}{\varepsilon}\sup_\xi \frac{g_\varepsilon((1-\varepsilon)\xi,t) - g((1-\varepsilon)\xi,t)}{|\xi|^2} - \frac{1}{\varepsilon}h_\varepsilon(\xi,t) + \frac{\varepsilon}{2}e^{-2t} \\ &= \frac{1-\varepsilon}{\varepsilon}\|h_\varepsilon\|_\infty(t) - \frac{1}{\varepsilon}h_\varepsilon(\xi,t) + \frac{\varepsilon}{2}e^{-2t}.\end{aligned} \tag{2.43}$$

Note that in (2.43) we made use of property ii of Proposition 2.1. Let us repeat step by step the computations done in Section 2.2.2. Inequality (2.43) is equivalent to

$$\frac{\partial}{\partial t}\left(h(\xi,t)e^{t/\varepsilon}\right) \le \frac{1-\varepsilon}{\varepsilon}\|h(\cdot,t)e^{t/\varepsilon}\|_\infty + \frac{\varepsilon}{2}e^{[(1-2\varepsilon)t]/\varepsilon}.$$

Integrating from 0 to t, we get

$$h(\xi,t)e^{t/\varepsilon} \le h(\xi,0) + \int_0^t \frac{\varepsilon}{2}e^{[(1-2\varepsilon)\tau]/\varepsilon}\, d\tau + \int_0^t \frac{1-\varepsilon}{\varepsilon}\|h(\cdot,\tau)e^{\tau/\varepsilon}\|_\infty\, d\tau.$$

Hence, if $H(t) = \|h(\cdot,t)e^{t/\varepsilon}\|_\infty$, and

$$\psi(t) = H(0) + \int_0^t \frac{\varepsilon}{2} e^{[(1-2\varepsilon)\tau]/\varepsilon}\, d\tau, \tag{2.44}$$

$H(t)$ satisfies

$$H(t) \le \psi(t) + \int_0^t \frac{1-\varepsilon}{\varepsilon} H(\tau)\, d\tau.$$

Now, by the generalized Gronwall inequality,

$$u(t) \le \psi(t) + \int_0^t \lambda(\tau) u(\tau)\, d\tau$$

implies

$$u(t) \le \psi(0) \exp\left\{\int_0^t \lambda(\tau)\, d\tau\right\} + \int_0^t \exp\left\{\int_\tau^t \lambda(\tau)\, d\tau\right\} \frac{d\psi}{d\tau}\, d\tau.$$

Applying this inequality with $\lambda(\tau) = (1-\varepsilon)/\varepsilon$ and $\psi(t)$ given by (2.44) we obtain

$$H(t) \le \left[H(0) + \frac{\varepsilon}{2}(1 - e^{-t})\right] e^{[(1-2\varepsilon)t]/\varepsilon}.$$

Going back to $h_\varepsilon(\cdot, t)$, we finally obtain

$$\|h(\cdot,t)\|_\infty \le \left[\|h(\cdot,0)\|_\infty + \frac{\varepsilon}{2}(1-e^{-t})\right] e^{-t}.$$

Recalling now that $\|h(\cdot,t)\|_\infty = d_2(g_\varepsilon, g)(t)$, we conclude with the bound

$$d_2(g_\varepsilon, g)(t) \le \left[d_2(g_\varepsilon, g)(0) + \frac{\varepsilon}{2}(1-e^{-t})\right] e^{-t}, \tag{2.45}$$

which holds uniformly with respect to ε and t, provided the distance between the initial data is finite. In particular, by taking the same initial density for both the drift equation (2.40) and the kinetic equation (2.41) one concludes with the bound

$$d_2(g_\varepsilon, g)(t) \le \frac{\varepsilon}{2}(1-e^{-t})e^{-t}, \tag{2.46}$$

which proves that as soon as $\varepsilon \to 0$ the solution to the kinetic equation converges towards the solution to the drift equation for all times $t > 0$.

Remark 2.4 *Apart from the amount of computation, conversion to Fourier transforms and the use of the Fourier-based metric gives a powerful tool for comparing kinetic equations with their limiting equations. This tool is of concrete use any time one can resort to the natural properties of the d_s metric. This can be done both for linear kinetic equations as well as for bilinear ones. Naturally, one can ask whether this strategy can be reached by using different metrics, or the same result can be obtained without passing to Fourier. Doubtless, other approaches are possible. Concerning metrics for probability measures, many of them possess the same properties that have* *been used in the previous computations. Among others, one can resort to the paper of Zolotarev [330] for an almost complete view of probability metrics grouped in terms of their characteristics. We remark that the main properties required to get precise estimates are precisely the ones listed in Proposition 2.1.*

Formula (2.45) has an important counterpart in the numerical approximation of equations of the type of the drift. It shows, in fact, that, in terms of a weak metric, the approximation via the kinetic system is uniform both with respect to the choice of the initial density and to the time, giving an explicit estimate of the relative error with respect to both ε *and time.*

Similar results can be obtained by choosing a different background, provided its density is such that the coefficient $2\varepsilon - (\varepsilon^2 + \gamma^2)$ *which drives the evolution of the second moment is positive. In this case, one is led to treat a more complex error term in equation (2.42). We leave to the reader the proof of this result, following the steps illustrated here.*

2.4.1 A proof for general densities

The proof of (2.46), which ensures that the solutions to the kinetic model of dissipation and the drift equation are close to each other in terms of the d_2 metric, requires that the initial density for the kinetic model satisfies conditions (1.116), or, in other words, that this initial density possesses a certain number of moments that are bounded. From the physical point of view, however, there are no reasons why this condition would be necessary. In fact, to obtain the solution to the drift equation (2.23) we only used the hypothesis that the initial density is a probability density, so that one can resort to its Fourier transform.

Owing to the properties of the d_s metric, we can easily prove that the result still holds, provided that the initial probability density for the kinetic model, say $g_0(v)$, belongs to $\mathcal{P}_\alpha(\mathbb{R})$, for some $\alpha > 0$, that is,

$$\int_{|v|\le R} |v|^\alpha f_R(v)\, dv = m_\alpha < \infty.$$

Let us show in some detail how this can be proven. First of all, for any given probability density function $f(v)$, we consider, for any given constant $R > 0$, its approximation $f_R(v)$ given by

$$f_R(v) = \frac{1}{\mathfrak{m}_R} f(v) \quad \text{if } |v| \le R; \quad f_R(v) = 0 \quad \text{if } |v| > R. \tag{2.47}$$

In (2.47) we denoted

$$\mathfrak{m}_R = \int_{|v|\le R} f(v)\, dv.$$

By construction, $f_R(v)$ is a probability density function. Moreover, since its support is bounded, all moments are bounded.

By definition,

$$|\hat{f}(\xi) - \hat{f}_R(\xi)| = \left| \int_{|v|\le R} f(v)e^{-i\xi v}\,dv \left(1 - \frac{1}{\mathfrak{m}_R}\right) + \int_{|v|>R} f(v)e^{-i\xi v}\,dv \right|$$

$$= \left| \int_{|v|\le R} f(v)(e^{-i\xi v} - 1)\,dv \left(1 - \frac{1}{\mathfrak{m}_R}\right) + \int_{|v|>R} f(v)(e^{-i\xi v} - 1)\,dv \right|$$

$$\le \left(\frac{1}{\mathfrak{m}_R} - 1\right) \int_{|v|\le R} f(v) \left|e^{-i\xi v} - 1\right| dv + \int_{|v|>R} f(v) \left|e^{-i\xi v} - 1\right| dv.$$

Consequently, for $s > 0$,

$$\frac{|\hat{f}(\xi) - \hat{f}_R(\xi)|}{|\xi|^s} \le \left(\frac{1}{\mathfrak{m}_R} - 1\right) \int_{|v|\le R} |v|^s f(v) \frac{\left|e^{-i\xi v} - 1\right|}{|\xi v|^s} dv + \int_{|v|>R} |v|^s f(v) \frac{\left|e^{-i\xi v} - 1\right|}{|\xi v|^s} dv.$$

If $s \le 1$,

$$\frac{\left|e^{-ix} - 1\right|}{|x|^s} \le \left|e^{-ix} - 1\right|^{1-s} \left(\frac{\left|e^{-ix} - 1\right|}{|x|}\right)^s \le 2^{1-s} \le 2.$$

Grouping together these inequalities, if $s = \alpha/2$ we obtain

$$\frac{|\hat{f}(\xi) - \hat{f}_R(\xi)|}{|\xi|^{\alpha/2}} \le 2\left(\left(\frac{1}{\mathfrak{m}_R} - 1\right) \int_{|v|\le R} |v|^{\alpha/2} f(v)\,dv + \int_{|v|>R} |v|^{\alpha/2} f(v)\,dv\right)$$

$$\le 2\left(\left(\frac{1}{\mathfrak{m}_R} - 1\right) \int_{|v|\le R} |v|^{\alpha/2} f(v)\,dv + \frac{1}{R^{\alpha/2}} \int_{|v|>R} |v|^{\alpha} f(v)\,dv\right)$$

$$\le 2\left(\left(\frac{1}{\mathfrak{m}_R} - 1\right) m_\alpha^{1/2} + \frac{1}{R^{\alpha/2}} m_\alpha\right).$$

Hence, if $f \in \mathcal{P}_\alpha(\mathbb{R})$ for some $\alpha > 0$, for any fixed $\varepsilon > 0$, we can find $R(\varepsilon) > 0$ such that, for $R \ge R(\varepsilon)$,

$$sup_{|\xi|} \frac{|\hat{f}(\xi) - \hat{f}_R(\xi)|}{|\xi|^{\alpha/2}} = d_{\alpha/2}(f, f_R) \le \varepsilon,$$

and f_R is of bounded support.

Having in mind this result, let us denote by $\tilde{g}_0(v)$ a probability density with finite second moment, which is such that, for $s \le \alpha/2$,

$$d_s(\tilde{g}_0, g_0) \le \varepsilon,$$

and let $\tilde{g}_\varepsilon(v,t)$ be the solution to the kinetic equation (2.40) corresponding to $\tilde{g}_0(v)$ as initial datum.

If $g_\varepsilon(v,t)$ denotes as usual the solution to the kinetic equation (2.40) corresponding to $g_0(v)$, for any given time $t > 0$

$$d_s(g_\varepsilon(t), g(t)) \le d_s(g_\varepsilon(t), \tilde{g}_\varepsilon(t)) + d_s(\tilde{g}_\varepsilon(t), \tilde{g}(t)) + d_s(\tilde{g}(t), g(t)), \tag{2.48}$$

where $g(t)$ (respectively $\tilde{g}(t)$) are the solutions to the drift equations with initial data $g_0(v)$ (respectively $\tilde{g}_0(v)$).

Let us consider the first term in (2.48). By adapting the computations given to obtain (2.42) we obtain

$$\frac{\partial}{\partial t}\frac{\tilde{g}_\varepsilon(\xi,t)-g(\xi,t)}{|\xi|^s}=\frac{1}{\varepsilon}\frac{\tilde{g}_\varepsilon((1-\varepsilon)\xi,t)-g((1-\varepsilon)\xi,t)}{|\xi|^s}-\frac{\tilde{g}_\varepsilon(\xi,t)-g(\xi,t)}{|\xi|^s}. \tag{2.49}$$

Now, setting

$$h_s(\xi,t)=\frac{\tilde{g}_\varepsilon((1-\varepsilon)\xi,t)-g((1-\varepsilon)\xi,t)}{|\xi|^s},$$

we obtain that h_s satisfies the inequality

$$\frac{\partial h_s(\xi,t)}{\partial t}\le\frac{(1-\varepsilon)^s}{\varepsilon}\|h_\varepsilon\|_\infty(t)-\frac{1}{\varepsilon}h_\varepsilon(\xi,t).$$

Proceeding as in the proof of (2.45), we conclude by Gronwall's inequality that

$$\|h_\varepsilon\|_\infty(t)\le\|h_\varepsilon\|_\infty(0)\exp\left\{-\frac{1-(1-\varepsilon)^s}{\varepsilon}t\right\}\le\varepsilon\exp\left\{-\frac{1-(1-\varepsilon)^s}{\varepsilon}t\right\}. \tag{2.50}$$

The second term in (2.49) can be easily evaluated resorting to both (2.45) and to property i of Proposition 2.1. Indeed,

$$d_s(\tilde{g}_\varepsilon(t),\tilde{g}(t))\le C_{s,2}d_2(\tilde{g}_\varepsilon(t),\tilde{g}(t))^{s/2}\le C_{s,2}\left[d_2(\tilde{g}_\varepsilon,\tilde{g})(0)+\frac{\varepsilon}{2}(1-e^{-t})\right]^{s/2}e^{-st/2}. \tag{2.51}$$

Finally, owing to the explicit solution of the drift equation (2.23) and to the scaling property ii of Proposition 2.1, the last term in (2.49) can be bounded to obtain

$$d_s(\tilde{g}(t),g(t))\le d_s(\tilde{g}(0),g(0))e^{-st}\le\varepsilon e^{-st}. \tag{2.52}$$

Grouping together (2.50), (2.51) and (2.52), we obtain that, as soon as some moment of the initial density is bounded, the solution to the kinetic equation (2.40) converges towards the solution to the drift equation in d_s norm, and the following estimate holds:

$$\begin{aligned}d_s(g_\varepsilon(t),g(t))\le{}&\varepsilon\exp\left\{-\frac{1-(1-\varepsilon)^s}{\varepsilon}t\right\}\\&+C_{s,2}\left[d_2(\tilde{g}_\varepsilon,\tilde{g})(0)+\frac{\varepsilon}{2}(1-e^{-t})\right]^{s/2}e^{-st/2}+\varepsilon e^{-st}.\end{aligned} \tag{2.53}$$

Remark 2.5 *In view of Remark 2.4, it is important to outline that the linear drift equation is the prototype of more complex nonlinear models, which have been used in recent years to describe the cooling of a dissipative gas [233]. These equations, which are known in the pertinent literature as nonlinear friction equations, take the form [36, 37, 66]*

$$\frac{\partial f(v,t)}{\partial t}=\frac{\partial}{\partial v}\left[f(v,t)\int_{\mathbb{R}}|v-w|^\gamma(v-w)f(w,t)\,dw\right], \tag{2.54}$$

where the parameter $-1<\gamma\le 2$ describes the dissipation properties of the medium. For this class of equations, it has been shown in [209] that the right metric to be used *is the Wasserstein metric. The Wasserstein metric represents, in fact, a powerful instrument for handling nonlinear problems, and it is an alternative to the Fourier-based metric even in most of the linear kinetic equations treated in this chapter. We address the interested reader to the notes in [77] for a detailed description of the relationships between these two metrics.*

2.5 Growth processes

Section 2.4 considers the case of a system of travelling particles which continuously interact with an external background, constructed so as to slow down their speed (in the mean). In particular, it was shown that in the *grazing interaction* regime, this dissipative process can be described by a drift equation, which is explicitly solvable (cf. (2.24)). Because of the simplicity of the underlying interaction, it was possible to check in a clear manner the way in which the Fourier-based metric allows us to prove rigorously the connection between the kinetic model based on collisions and the continuous one. This example can be assumed as a prototype for studying other types of interactions and the corresponding grazing interaction limit. Among others, we will discuss a model which can be referred to a biological context, and describes, for example, the growth of a population of bacteria. In this type of process, one aims to describe the behaviour of a population of particles which is subject to an external control unit which acts on them on the basis of some well-defined rules.

In our example the control will be twofold. At first, it will act on the total number, by preventing the population from exceeding a fixed value. Second, if the number is below the allowed maximal bound, the control will decide either to add or to subtract a unit, or to leave the population unchanged. This choice will be taken at random. In our example we will assume a constant birth rate and a mortality rate proportional to the number of bacteria.

To fix notations, let $X(t)$ denote the number of bacteria present in the system at a given time t, and let the variation of $X(t)$ due to the action of the external control unit be described by a discrete random variable B that, according to the previous requirements, can assume only the values -1, 0 and $+1$ with certain probabilities to be defined later. Independently of the assigned values of these probabilities, the number $v \in \mathbb{N}$ of particles in the system is modified according to

$$v^* = v + w(v), \tag{2.55}$$

where w is sampled by the control variable $B(X(t))$. In terms of the process $X(t)$, this law of change is given by

$$X^*(t) = X(t) + B(X(t)). \tag{2.56}$$

Since B can attain only the values -1, 0 and $+1$, after a control operation the number of bacteria can remain unchanged, or at most increase or decrease by one unit. Therefore, if $v \in \mathbb{N}$, so also $v^* \in \mathbb{N}$.

Since the variable $v \in \mathbb{N}$, the probability density $f(v,t)$ of the solution is a sum of point masses; in general,

$$f(v,t) = \sum_{k=0}^{+\infty} p_k(t)\delta_0(v-k), \tag{2.57}$$

where $\sum_{k=0}^{+\infty} p_k(t) = 1$ for all $t \geq 0$. For a given observable quantity $\varphi(\cdot)$, fixing without loss of generality a collision frequency equal to one, we then have that the mean value of $\varphi(X)$ satisfies

$$\frac{d}{dt}\int_{\mathbb{R}+} \varphi(v) f(v,t)\, dv = \left\langle \int_{\mathbb{R}+} \left(\varphi(v + B(v)) - \varphi(v)\right) f(v,t)\, dv \right\rangle. \tag{2.58}$$

In equation (2.58), the mean $\langle\cdot\rangle$ has to be taken with respect to the control variable B. The evolution of observables will depend on the complete knowledge of the discrete variable B, namely from the knowledge of the probabilities of transition of the population from the number v to $v+1$, or from v to $v-1$.

Let $\nu > 0$ be a fixed constant. For a given positive constant $\varepsilon \ll 1$, we define $p = p(v,\nu,\varepsilon) = \varepsilon(\nu + v)$. Moreover, let $v_{\max}(\varepsilon)$ be the biggest natural number for which $p(v,\nu,\varepsilon) \leq 1$. Then, let us assume the following:

$$P(B(v) = 0) = 1, \quad \text{if} \quad v > v_{\max}(\varepsilon), \tag{2.59}$$

while, if $v = v_{\max}(\varepsilon)$,

$$\begin{aligned} &P(B(v_{\max}) = -1) = \varepsilon v_{\max}; \\ &P(B(v_{\max}) = +1) = 0; \\ &P(B(v_{\max}) = 0) = 1 - \varepsilon v_{\max}. \end{aligned} \tag{2.60}$$

Finally, if $v < v_{\max}(\varepsilon)$,

$$\begin{aligned} &P(B(v) = -1) = \varepsilon v; \\ &P(B(v) = +1) = \varepsilon \nu; \\ &P(B(v) = 0) = 1 - (v + \nu)\varepsilon. \end{aligned} \tag{2.61}$$

Conditions (2.59) and (2.60) imply that the population cannot exceed a certain value. In particular, condition (2.60) asserts that when the population arrives at its maximal allowed value, it can only decrease or remain stationary. When the number of the population is below its maximal value, then condition (2.61) assigns to birth a constant probability of happening, and to death a probability proportional to the population itself. It is important to observe that condition (2.61) also implies that population cannot reach negative values. Indeed, the possibility of decreasing the number by one is forbidden as soon as $v = 0$. Note that the total number in the population is inversely proportional to the value ε, so that the upper bound on the total number disappears as ε tends to infinity. Likewise, as $\varepsilon \to 0$ the rates of birth and death tend to zero, so producing a *grazing interaction effect.*

According to (2.58), choosing $\varphi(v) = 1$ shows that, independently of the background distribution, $f(v,t)$ remains a probability density if it is so initially. A further consequence follows from (2.59) and (2.60). Let us consider an initial population whose number is distributed according to an initial density $f_0(v)$ of support contained in the interval $[0, \bar{v}]$, where $\bar{v} < v_{\max}(\varepsilon)$. This implies that the size of the population at time $t = 0$ is smaller than the maximal allowed value. Hence, by conditions (2.59) and

(2.60), the maximal number $v(t)$ of the population at any subsequent time $t > 0$ will not exceed the value $v_{\max}(\varepsilon)$. In other words, by starting from an initial density whose support is contained in the interval $[0, v_{\max}(\varepsilon)]$, the support of the solution to equation (2.58) will remain in the interval $[0, v_{\max}(\varepsilon)]$ at any subsequent time $t > 0$. Note that, in the light of the preceding explanation, (2.58) can be explicitly written as

$$\begin{aligned}\frac{d}{dt}\int_{\mathbb{R}^+} \varphi(v) f(v,t)\, dv &= \varepsilon v_{\max} \left(\varphi(v_{\max} - 1) - \varphi(v_{\max})\right) p_{v_{\max}}(t) \\ &\quad + \varepsilon \int_{v \le v_{\max}-1} (\varphi(v-1)v + \varphi(v+1)\nu - \varphi(v)(v+\nu)) f(v,t) dv.\end{aligned} \tag{2.62}$$

In the previous formula, as in other formulas that will follow, we will use either the integral or the sum representation of the observables of the density f. For example, if $\varphi(v) = v$ we obtain

$$\begin{aligned}\frac{d}{dt}\int_{\mathbb{R}^+} v f(v,t)\, dv &= -\varepsilon v_{\max} p_{v_{\max}}(t) + \varepsilon \int_{v \le v_{\max}-1} (-1\cdot v + 1 \cdot \nu)\, f(v,t)\, dv \\ &= \varepsilon \left(\nu - \int_{\mathbb{R}^+} v f(v,t)\, dv \right) - \varepsilon \nu p_{v_{\max}}(t).\end{aligned} \tag{2.63}$$

From equation (2.63) one can easily obtain an upper bound for the mean value. In fact, equation (2.63) implies

$$\frac{d}{dt}\int_{\mathbb{R}^+} v f(v,t)\, dv \le \varepsilon \left(\nu - \int_{\mathbb{R}^+} v f(v,t)\, dv \right),$$

which can integrated to give

$$\int_{\mathbb{R}^+} v f(v,t)\, dv \le e^{-\varepsilon t} \int_{\mathbb{R}^+} v f_0(v)\, dv + \nu(1 - e^{-\varepsilon t}). \tag{2.64}$$

Once again, it is remarkable that $\varepsilon \ll 1$ implies a small variation of the mean value of the population. Along the line of the grazing collision procedure, if we want to observe the action of the control as $\varepsilon \to 0$, we need to act on time, by scaling it according to $\bar{t} = \varepsilon t$. Within this usual scaling, and using the same notation for the new time, we are led to study the equation

$$\begin{aligned}\frac{d}{dt}\int_{\mathbb{R}^+} \varphi(v) g_\varepsilon(v,t)\, dv &= v_{\max} \left(\varphi(v_{\max} - 1) - \varphi(v_{\max})\right) p_{v_{\max}}(t) \\ &\quad + \int_{v \le v_{\max}-1} (\varphi(v-1)v + \varphi(v+1)\nu - \varphi(v)(v+\nu))\, g_\varepsilon(v,t)\, dv,\end{aligned} \tag{2.65}$$

which depends on ε through the maximal value $v_{\max}$. Let us observe that the mean value of g_ε satisfies a bound independent of ε, and

$$\int_{\mathbb{R}^+} v g_\varepsilon(v,t)\, dv \le e^{-t} \int_{\mathbb{R}^+} v g_0(v)\, dv + \nu(1 - e^{-t}) \le \max\left\{\nu, \int_{\mathbb{R}^+} v g_0(v)\, dv\right\}. \tag{2.66}$$

Likewise, if $\varphi(v) = v^2$, we obtain the inequality

$$\frac{d}{dt}\int_{\mathbb{R}+} v^2 g_\varepsilon(v,t)\,dv \le -2\int_{\mathbb{R}+} v^2 g_\varepsilon(v,t)\,dv + \nu + (1+2\nu)\int_{\mathbb{R}+} v g_\varepsilon(v,t)\,dv,$$

which, using the bound (2.66), implies a uniform bound for the second moment:

$$\int_{\mathbb{R}+} v^2 g_\varepsilon(v,t)\,dv \le \max\left\{\int_{\mathbb{R}+} v^2 g_0(v), \frac{1}{2}(\nu + (1+2\nu))\int_{\mathbb{R}+} v g_0(v)\,dv\right\} = m_2. \quad (2.67)$$

By Chebyshev's inequality [246], (2.66) implies that

$$p_{v_{\max}}(t) \le P(X(t) \ge v_{\max} \le \frac{1}{v_{\max}^2}\langle X^2(t)\rangle.$$

Hence, thanks to (2.67),

$$v_{\max} p_{v_{\max}}(t) \le \frac{1}{v_{\max}} m_2.$$

Since, by virtue of its definition,

$$v_{\max} \ge \frac{1}{\varepsilon} - (1+\nu),$$

if the second moment of the population density is initially bounded,

$$v_{\max} p_{v_{\max}}(t) \le \frac{\varepsilon}{1-\varepsilon(1+\nu)} m_2 \to 0, \quad (2.68)$$

as $\varepsilon \to 0$ uniformly with respect to time.

Choosing $\varphi(v) = e^{-i\xi v}$ in (2.65) shows that, in Fourier transforms, g_ε satisfies the equation

$$\begin{aligned}\frac{\partial \hat{g}_\varepsilon(\xi,t)}{\partial t} &= e^{-i\xi v_{\max}} v_{\max}\left(e^{i\xi} - 1\right) p_{v_{\max}}(t) \\ &\quad + \int_{v \le v_{\max}-1} e^{-i\xi v}\left(e^{i\xi} v + e^{-i\xi}\nu - (v+\nu)\right) g_\varepsilon(v,t)dv \\ &= \nu e^{-i\xi v_{\max}}\left(1 - e^{-i\xi}\right) p_{v_{\max}}(t) + \int_{\mathbb{R}+} e^{-i\xi v}\left(v(e^{i\xi}-1) + \nu(e^{-i\xi}-1)\right) g_\varepsilon(v,t)dv \\ &= \nu e^{-i\xi v_{\max}}\left(1 - e^{-i\xi}\right) p_{v_{\max}}(t) + \nu(e^{-i\xi}-1)\hat{g}_\varepsilon(\xi,t) + i(e^{i\xi}-1)\frac{\partial \hat{g}_\varepsilon(\xi,t)}{\partial \xi}.\end{aligned} \quad (2.69)$$

By virtue of (2.68), the remainder term

$$R_\varepsilon(t) = \nu e^{-i\xi v_{\max}}\left(1 - e^{-i\xi}\right) p_{v_{\max}}(t) \to 0,$$

as $\varepsilon \to 0$ uniformly with respect to t. By making use of the same techniques as Section 2.4, that is by introducing the Fourier-based distance (2.17), with $s = 1$, one proves that the solution $g_\varepsilon(v,t)$ is weakly convergent to $g(v,t)$, where $g(v,t)$ solves

$$\frac{\partial \hat{g}(\xi,t)}{\partial t} = \nu(e^{-i\xi}-1)\hat{g}(\xi,t) + i(e^{i\xi}-1)\frac{\partial \hat{g}(\xi,t)}{\partial \xi}. \tag{2.70}$$

Because of (2.63), after scaling the mean value of the limit process satisfies the equation

$$\frac{d}{dt}\int_{\mathbb{R}+} vf(v,t)\,dv = \left(\nu - \int_{\mathbb{R}+} vf(v,t)\,dv\right),$$

which is solvable to give

$$\int_{\mathbb{R}+} vf(v,t)\,dv = e^{-t}\int_{\mathbb{R}+} vf_0(v)\,dv + \nu(1-e^{-t}). \tag{2.71}$$

Note that, as time tends to infinity, the mean value tends towards the value ν. This is a clear indication that the solution to equation (2.70) should converge towards a stationary solution of bounded mean (equal to ν). The eventual stationary solution solves the equation

$$\nu(e^{-i\xi}-1)\hat{g}(\xi) + i(e^{i\xi}-1)\frac{\partial \hat{g}(\xi)}{\partial \xi} = 0. \tag{2.72}$$

Using the identity

$$(e^{-i\xi}-1) = -e^{-i\xi}\left(e^{i\xi}-1\right),$$

we can write equation (2.72) as

$$\frac{1}{\hat{g}(\xi)}\frac{\partial \hat{g}(\xi)}{\partial \xi} = -i\nu e^{-i\xi},$$

which can be integrated to obtain

$$\hat{g}(\xi) = \hat{g}(\xi=0)\exp\left\{\nu\left(e^{-i\xi}-1\right)\right\}.$$

Among all possible choices, $\hat{g}(\xi=0)=1$ is the unique one which is compatible with mass conservation, and corresponds to the mass of the stationary solution being equal to one. Therefore, the unique stationary solution to equation (2.70) of mass one is

$$\hat{g}_\infty(\xi) = \exp\left\{\nu\left(e^{-i\xi}-1\right)\right\}, \tag{2.73}$$

which is the Fourier transform of a Poisson distribution of mean value equal to ν [165, 246]. It follows that equation (2.70) drives the system of particles from an initial distribution of bounded mean value towards a steady state which is a random variable X_∞ distributed according to a Poisson distribution of mean ν, so that

$$P(X_\infty = k) = \frac{\nu^k}{k!}e^{-\nu}.$$

Note that the constant ν corresponds to the probability of growth we assigned to the control. An explicit solution of equation (2.70) can be obtained by resorting to the argument first presented in Section 2.3. To this extent, let us rewrite equation (2.70) in terms of

$$\hat{k}(\xi,t) = \log \hat{g}(\xi,t) = \nu\left(1-e^{-t}\right)\left(e^{-i\xi}-1\right),$$

to obtain

$$\frac{\partial \hat{k}(\xi,t)}{\partial t} = i(e^{i\xi}-1)\frac{\partial \hat{k}(\xi,t)}{\partial \xi} + \nu(e^{-i\xi}-1). \tag{2.74}$$

First, let us look for a solution to the homogeneous equation

$$\frac{\partial \hat{k}(\xi,t)}{\partial t} - i(e^{i\xi}-1)\frac{\partial \hat{k}(\xi,t)}{\partial \xi} = 0. \tag{2.75}$$

We remark that equation (2.75) has the same features as the drift equation (2.22). Consequently, we apply the same strategy.

Let $\eta = \eta(\xi,t)$ be a smooth function, and let us evaluate the total derivative of the function $\hat{k} = \hat{k}(\eta(\xi,t),t)$, with the constraint $\eta(\xi,t=0) = \xi$, which implies $\hat{k}(\eta(\xi,t=0),t=0) = \hat{k}_0(\xi)$. Since

$$\left.\frac{d\hat{k}}{dt}\right|_{\eta=\eta(t)} = \left.\left(\frac{\partial \hat{k}}{\partial t} + \frac{\partial \hat{k}}{\partial \eta}\eta'\right)\right|_{\eta=\eta(t)},$$

it follows that $\eta(t)$ has to solve the differential equation

$$\eta'(t) = -i(e^{i\eta(t)}-1), \tag{2.76}$$

with $\eta(\xi,t=0)=\xi$. Separating variables, equation (2.76) takes the form

$$\frac{d\eta}{-i(e^{i\eta}-1)} = dt.$$

By multiplying both numerator and denominator of the left-hand side by $ie^{-i\eta}$ we get

$$\frac{d\eta}{-i(e^{i\eta}-1)} = \frac{ie^{-i\eta}d\eta}{(1-e^{-i\eta})} = d\log(1-e^{-i\eta}).$$

This shows that equation (2.76) can be integrated, and its solution satisfies, for all times $t \geq 0$,

$$1 - e^{-i\eta(t)} = (1-e^{-i\xi})e^{t}. \tag{2.77}$$

Making use of identity (2.77), we can write equation (2.74) in the form

$$\left.\frac{d\hat{k}}{dt}\right|_{\eta=\eta(t)} = \left.\left(\frac{\partial \hat{k}}{\partial t} + \frac{\partial \hat{k}}{\partial \eta}\eta'\right)\right|_{\eta=\eta(t)} = \nu(e^{-i\eta}-1)\big|_{\eta=\eta(t)} = \nu(e^{-i\xi}-1)e^{t}.$$

Integrating with respect to time we obtain

$$\hat{k}(\eta(\xi,t),t) = \hat{k}_0(\xi) + \nu(e^{-i\xi}-1)(e^{t}-1).$$

At this point, using (2.77) again to express ξ in terms of $\eta(t)$, that is

$$e^{-i\xi} - 1 = (e^{-i\eta(t)} - 1)e^{-t},$$

we conclude with

$$\hat{k}(\eta(\xi,t),t) = \hat{k}_0(\xi) + \nu(e^{-i\eta(t)} - 1)(1 - e^{-t}). \tag{2.78}$$

The simplest case in which (2.78) can be explicitly solved to find the density function $g(v,t)$ is when $\hat{k}_0(\xi) = 0$. Suppose that at time $t = 0$ the system has no particles, so that

$$P(X_0 = 0) = 1.$$

Then the initial distribution is a Dirac delta function concentrated at $v = 0$, and $\hat{g}_0(\xi) = 1$. Therefore $\hat{k}_0(\xi) = \log \hat{g}_0(\xi) = 0$. In this case,

$$\hat{g}(\xi,t) = \exp\left\{\nu(1 - e^{-t})(e^{-i\xi} - 1)\right\}, \tag{2.79}$$

which is the Fourier transform of a Poisson distribution of mean value equal to $\nu(1 - e^{-t})$. Hence, starting from a population of zero bacteria, the population at time $t \geq 0$ is distributed according to a Poisson distribution of mean $\nu(1-e^{-t})$. In all the other cases, in order to make the solution explicit, at least with regards to its Fourier transform, one needs to express the initial data $\hat{g}_0(\xi)$ in terms of $\eta(t)$. This can be done by means of (2.77), which implies

$$e^{-i\xi v} = \left(1 + e^{-t}(e^{-i\eta} - 1)\right)^v.$$

Hence

$$\hat{g}_0(\xi) = \int_{\mathbb{R}+} e^{-i\xi v} g_0(v)\, dv = \int_{\mathbb{R}+} \left(1 + e^{-t}(e^{-i\eta} - 1)\right)^v g_0(v)\, dv = \widehat{h}(\eta,t).$$

We recall that, since $g_0(v)$ is a sum of point masses,

$$g_0(v) = \sum_{k=0}^{\infty} p_k(0)\delta_0(v-k),$$

$$\widehat{h}(\eta,t) = \sum_{k=0}^{\infty} p_k(0)\left(1 + e^{-t}(e^{-i\eta} - 1)\right)^k.$$

It is then easy to see that $\widehat{h}(\eta, t = 0) = \hat{g}_0(\eta)$, while for $t \to \infty$

$$\widehat{h}(\eta,t) \to \int_{\mathbb{R}+} g_0(v)\, dv = 1.$$

Note that, returning to the original variable in (2.78) we obtain the exact form of the solution to equation (2.70). This solution takes the form

$$\hat{g}(\xi) = \widehat{h}(\xi,t) \cdot \exp\left\{\nu(1 - e^{-t})(e^{-i\xi} - 1)\right\}, \tag{2.80}$$

which corresponds, in the physical space, to a convolution between a density function which depends on the initial distribution, and a Poisson density.

Remark 2.6 *Despite its apparent simplicity, the linear problem discussed in this section exhibits most of the features common to various kinetic models we will treat in this book. On one hand, the solution to equation (2.70) has a stationary state which is easily expressed in the physical space, and corresponds to a Poisson distribution of known mean value. Likewise, in the presence of a particular initial datum, which is a Dirac delta function located at zero, one can easily recover the consequent solution, which turns out to be a Poisson distribution with mean value depending on time (cf. (2.79)). Note that the solution consequent to an initial datum which is a point mass located at zero gives rise to a source-type solution, a well-known concept in the study of linear and nonlinear diffusion equations [21, 143, 312]. On the other hand, if one starts from a general initial configuration, the knowledge of the explicit form of the solution in Fourier variables is practically useless. Therefore, if we want to describe the time evolution of the solution density towards the stationary (known) solution, we have to resort to a suitable numerical approximation.*

Problems

Exercise 2.1 As in Exercise 1.5, let us consider equation (1.62),

$$\frac{\partial f(v,t)}{\partial t} = \sigma\left[M * f(v,t) - f(v,t)\right],$$

with a background M given by the normal density

$$M(v) = \frac{1}{\sqrt{2\pi}} e^{-v^2/(2)},$$

and express its solution as a Wild sum.

Exercise 2.2 Equation (5.178) of Chapter 5 describes a model for wealth redistribution with taxation. The case $\varepsilon = \delta$ reads

$$\frac{\partial \widehat{f}(\xi,t)}{\partial t} = \hat{f}((1-\varepsilon)\xi,t)\hat{M}(\delta\xi) - \hat{f}(\xi,t).$$

Express its solution as a Wild sum. It is possible to express the solution as a Wild sum in the general case?

Exercise 2.3 Proposition 2.1 enumerates a number of properties satisfied by the Fourier-based metric d_s. Prove that these properties still hold for the metric D_s, $s \in (1,2)$, defined by (2.19). In particular, find the analogue of (2.18).

Exercise 2.4 Verify that the result of Remark 2.3 still holds for the Fourier distance D_s, $s \in (1,2)$, defined in (2.19).

Exercise 2.5 Let us consider the the initial value problem for the drift equation (2.20), written in terms of the distribution function

$$F(v,t) = \int_{-\infty}^{v} f(w,t)\,dw.$$

Then, write the drift equation in terms of the pseudo-inverse function

$$F^{-1}(v) = \inf\{w : F(w) > v\},$$

and make use of this equation to solve the original drift equation.

Exercise 2.6 Among other distances, the *Wasserstein distance* (of order one) of two density functions $f_1(w)$, $f_2(w)$ is an extremely useful instrument. This distance is given by

$$\mathbf{W}[f_1, f_2] = \int_{\mathbb{R}_+} \big|F_1(v) - F_2(v)\big|\, dv, \tag{2.81}$$

where the F_i denote the distribution functions

$$F_i(v) = \int_{-\infty}^{v} f_i(w)\, dw \quad (i = 1, 2). \tag{2.82}$$

Wasserstein and *Fourier metrics* are intimately related [134].

Assume that two probability densities f and g have first moment equal to one, and some moment of order $s \in (1, 2]$ bounded. Prove that there exists a constant $C > 0$, depending only on s and the values of the s-th moments of f and g, such that

$$\mathbf{W}[f, g] \le C(d_s[f, g])^{\frac{s-1}{s(2s-1)}}.$$

Conversely, prove that

$$d_1[f, g] \le \mathbf{W}[f, g],$$

even if no moments of f and g above the first are bounded.

3
Monte Carlo strategies

> The year was 1945. Two earthshaking events took place: the successful test at Alamogordo and the building of the first electronic computer. Their combined impact was to modify qualitatively the nature of global interactions between Russia and the West. No less perturbative were the changes wrought in all of academic research and in applied science. On a less grand scale these events brought about a renascence of a mathematical technique known to the old guard as statistical sampling; in its new surroundings and owing to its nature, there was no denying its new name of the Monte Carlo method.
>
> N. Metropolis, The beginning of the Monte Carlo method, *Los Alamos Science*, 1987.

3.1 Why Monte Carlo methods?

Kinetic models like the ones discussed in the previous chapters are very useful in order to obtain asymptotic solutions and to carry out qualitative studies on their behaviour. In general, however, it is not possible to solve these models exactly or use them directly with the aim of finding answers to practical problems. Several numerical techniques have been derived in order to solve the integro-differential equation characterizing the time evolution of the probability density function. Except for some simple models, the multi-dimensional integral in the interaction term makes numerical approximations a very difficult task. One first difficulty is related to the computational cost. Suppose that our density function $f(v, t)$ is characterized by N parameters, for example grid points in a deterministic approach, and that the time evolution is characterized by an interaction integral $Q(f)$. A quadrature formula for the evaluation of such an integral would naturally be constructed on the same set of N parameters. The overall computational effort is then $O(N^\alpha)$ with $\alpha \geq 2$. This cost would be repeated for each time step and, if the probability density depends on other variables, as in some of the models presented in the second part of this book, for each additional point characterizing such variables. Often the numerical effort for such a procedure turns out to be unpractical.

This only partially explains the popularity of Monte Carlo quadrature methods in the context of kinetic equations. Other aspects need to be outlined. The microscopic dynamics which are at the basis of the physical properties of the system under consideration are often extremely difficult to approximate on regular grids. In addition, both the large-time behaviour of a system which is not compactly supported, and the presence of fat tails for the distribution, militate in favour of a statistical description of its properties. As we will see, Monte Carlo methods, even if affected by statistical noise, are naturally suited for such purposes, and permit us to construct numerical schemes which preserve most of the physical properties of the original system.

A systematic use of Monte Carlo methods for real scientific problems has its origin in the early days of electronic computing at Los Alamos during the Second World War [235]. The main point was to use statistical sampling by adopting the newly developed electronic computing techniques. The approach seemed particularly suited to estimating neutron multiplication rates and was, therefore, used to predict the explosive behaviour of the various fission weapons being designed. We should mention that, according to [235], Enrico Fermi had independently developed the method nearly fifteen years earlier.

Fig. 3.1 John von Neumann[1]

Nowadays, Monte Carlo methods find application in many areas, including many subfields of physics, like statistical physics or high-energy physics, and ranging to areas like biology, chemistry, finance, computer graphics and video games. Classical mathematical applications of Monte Carlo involve the computation of multi-dimensional integrals, the solution of partial differential equations, Markov chains and optimization problems.

Monte Carlo methods are often used when other methods fail, since they are much less sensitive to dimensionality, which plagues deterministic methods in problems with a large number of variables. Despite the widespread use of the methods, and numerous descriptions of them in articles and monographs (see [26, 63, 148, 171, 172, 190, 213, 219, 236, 279], for example), it is virtually impossible to find a unique notion of the Monte Carlo method in the literature, and the term Monte Carlo is often used to denote any numerical technique based on some kind of stochastic simulation.

In the following we will cover a range of key topics to the depth necessary to be able to design, implement and analyze a Monte Carlo method applied to a scientific problem. We will pay particular attention to the approximation of integrals and partial differential equations, trying to cast the various Monte Carlo techniques in a standard setting of numerical analysis. This includes the derivation of the methods directly from the differential problem rather than from the fundamental properties of the system under study.

In fact, in spite of their relevance in applications, Monte Carlo methods have received little attention from numerical analysts and applied mathematicians, and most of the schemes were originally designed by practitioners, including physicists, system engineers and statisticians. The natural consequence is that Monte Carlo methods are often described in non-mathematical terms. One relevant exception to this approach to Monte Carlo methods is the introductory manuscript by Russell Caflisch [63].

[1]**John von Neumann**: born on 28 December 1903 in Budapest (Austro-Hungarian Empire, now Hungary), died on 8 February 1957 in Washington D.C. (USA). He is regarded as one of the greatest mathematicians in modern history. Von Neumann's hydrogen bomb work in the 1940s (Manhattan Project) at the Los Alamos National Laboratory was played out in the realm of computing, where he, Stanislaw Ulam and Nicholas Metropolis developed simulations for the hydrodynamic computations. During this time he contributed to the development of the Monte Carlo method.

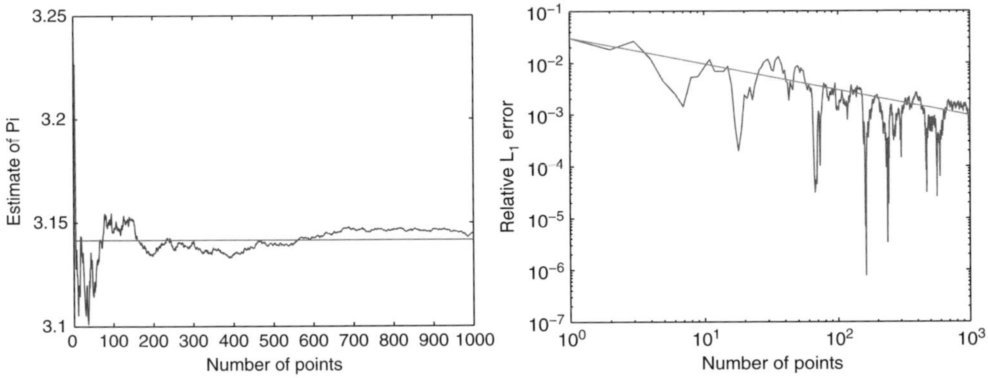

Fig. 3.2 Convergence of the Monte Carlo approximation of π.

Detailed aspects related to the construction of Monte Carlo methods for general kinetic equations will be the subject of the next chapter.

Monte Carlo methods are naturally linked to randomness, and consequently require a certain knowledge of the basic arguments of probability theory. According to this relationship, the notations of this chapter will be typical of a standard text on probability. For example, the mathematical expectation of a random variable X, which we denoted $\langle X \rangle$ in Chapters 1 and 2, will be denoted here by $E(X)$. Also, the independent variable, denoted with v or w in the first two chapters, will be denoted with x or y.

3.1.1 A simple example of a Monte Carlo method

Suppose we want to estimate π with a Monte Carlo method. We can consider a circle of diameter 2 inside the square $[-1, 1] \times [-1, 1]$. The square has area 4 and the circle π. Suppose we pick a point P uniformly at random inside the square. The probability that P lies inside the circle is $\pi/4$.

If we select N points $P_1, \dots, P_N$ independently and uniformly in the square we obtain a rough estimate of π from the formula

$$\frac{\pi}{4} \approx \frac{Z}{N},$$

where Z is the number of points inside the circle. Note that Z has a binomial distribution with parameters N and $\pi/4$, and its expectation is $E(Z) = N\pi/4$. In other words, $4Z/N$ is an unbiased estimator of π.

For example, suppose that, by taking $N = 10^5$ points, we found for Z the value 78582. Then our estimate of π would be $4 \times 78582/10^5 = 3.143280$. On the other hand, by the law of large numbers [145],

$$\lim_{N\to\infty} P(|\pi - E(4Z/N)| \geq \varepsilon) = 0,$$

and we are sure that as N gets large there is a very small probability that our estimates deviate much from π.

By extending this example it appears evident that one could use Monte Carlo to estimate areas and volumes and more in general subsets of $\mathbb{R}^d$ (and therefore also integrals).

Going back to the evaluation of π, by computing the error as $|\pi - 4Z/N|/\pi$, the observed convergence rate is $O(N^{-1/2})$ (see Figure 3.2), which is disappointing since an increased factor of 4 in the computational effort implies only an improvement of a factor of 2 in accuracy. On the other hand, the method we have constructed is extremely straightforward and general. This combination of simplicity, wide applicability and slow convergence characterizes most Monte Carlo approaches and explains why an enormous amount of computer time is nowadays spent on Monte Carlo computations [63].

Let us look further into the accuracy of the method using elementary results of probability. For each $i = 1, \ldots, N$ let us denote by X_i the random variable

$$X_i = \begin{cases} 4 & P_i \text{ lies in the circle} \\ 0 & \text{otherwise.} \end{cases}$$

Thus, X_i has a binomial distribution with $\mu = E(X_i) = \pi$ and $\sigma^2 = Var(X_i) = \pi(4 - \pi)$. Our numerical estimator $4Z/N$ coincides with the *sample mean*

$$\bar{X}_N = \frac{1}{N} \sum_{i=1}^{N} X_i. \tag{3.1}$$

When N is sufficiently large, the *central limit theorem* states that

$$\frac{\bar{X}_N - E(\bar{X}_N)}{\sqrt{Var(\bar{X}_N)}},$$

where $Var(\bar{X}_N) = \sigma^2/N$, has a standard normal distribution with zero mean and unitary variance.

Suppose we are interested in a 95% confidence interval for π. If N is sufficiently large, we have

$$P\left(\bar{X}_N - 1.96\frac{\sigma}{\sqrt{N}} \leq \pi \leq \bar{X}_N + 1.96\frac{\sigma}{\sqrt{N}}\right) \approx 0.95,$$

where the value $z \approx 1.96$ is such that $P(|X| \leq z) = 0.95$ when X has a standard normal distribution. Thus, an approximate 95% confidence interval for π is given by

$$\left[\bar{X}_N - 1.96\frac{\sigma}{\sqrt{N}}, \bar{X}_N + 1.96\frac{\sigma}{\sqrt{N}}\right].$$

To estimate σ, when unknown, one can use the *sample variance*

$$S_N^2 = \frac{1}{N-1} \sum_{i=1}^{N} (X_i - \bar{X}_N)^2, \tag{3.2}$$

which, since $E(S_N^2) = \sigma^2$, is an unbiased estimator of the variance. In our example, $\sigma = \sqrt{\pi(4 - \pi)}$ is known from the beginning. Note, however, that since we are designing a method to estimate π, we cannot assume to know its value. In addition to direct computation (3.2), there are essentially two ways to use the exact expression of σ and overcome this difficulty. First, one can use the simple bound $\sigma \leq 2$ (since $x(4-x) \leq 4$ for $x \in [0, 4]$) to have a conservative estimate of the confidence interval. Note, in particular, that in this way we can estimate the accuracy of our method for a given N before performing the simulation. In fact, the precision is measured by the width of the confidence interval, which is proportional to $\sigma/\sqrt{N}$. This confirms the estimated convergence rate.

With $N = 10^5$, the observed data give $\bar{X}_N = 3.143280$. Then, since (using 6 decimal digits) $1.96 \times 2/\sqrt{100000} = 0.012396$, we obtain the conservative confidence interval $[3.143280 - 0.012396, 3.143280 + 0.012396]$. Thus, we have a 95% probability that the interval $[3.130884, 3.155676]$ contains the true value of π. Alternatively, one can use the estimated value of π directly in the expression of σ to obtain the estimate $\sigma \approx 1.641009$. By our conservative estimate we expect this second approach to be more accurate. In this way we obtain, in fact, the approximate confidence interval $[3.133109, 3.153451]$. Note that to gain a decimal digit of precision, namely a factor of 10 in accuracy, we should increase N by a factor of 100.

From the example we can also argue the importance of the following basic issues common to most Monte Carlo methods:

- *Generation.* How do we generate random numbers with a specific distribution? We have to keep in mind that we need many random numbers and that they must be computed quickly.
- *Accuracy.* How accurate can we expect our answers to be? In the example we found that the accuracy of our estimate was proportional to $O(N^{-1/2})$. Is this a general principle of Monte Carlo methods?
- *Efficiency.* How efficient is our method? Is there another method that will give us comparably accurate answers in less time?

3.2 Generating random variables

Before entering into the description of the methods, we give a brief review of random variable generators, which are at the basis of several Monte Carlo methods. Further references can be found in [196]. As we will see, the problem is strictly related to the statistical notion of sampling, which is concerned with the selection of a subset of individuals within a statistical population.

3.2.1 Pseudo-random numbers

To generate random variables that follow a given probability distribution function, we need first to be able to generate random variables uniformly distributed in the interval $[0, 1]$. These random variables are usually called random numbers. True random numbers cannot be generated on a computer because

- floating point numbers are used as approximation of real numbers;
- a really random sequence cannot be generated even at a discrete level (it would require infinite memory).

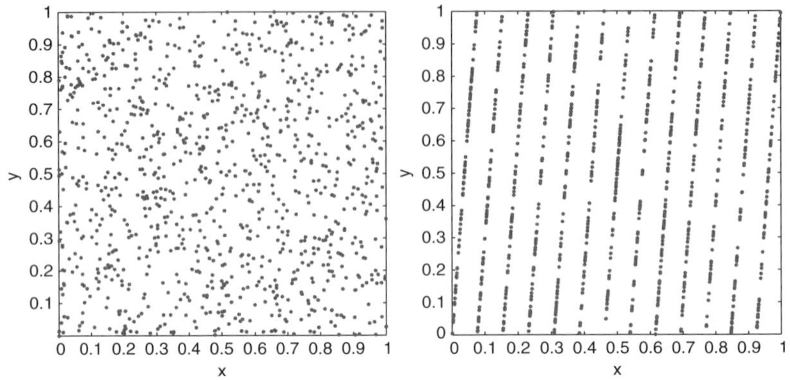

Fig. 3.3 Plot of the points (X_n, X_{n+1}). Uncorrelated pseudo-random sequence (left), and the presence of correlations (right).

An accepted alternative in the scientific community is to generate so-called *pseudo-random* numbers. Pseudo-random number generators produce a sequence of numbers which satisfy some properties of a uniform random sequence. In particular, one wishes to generate a sequence X_n which:

(a) is uniformly distributed (approximate Lebesgue measure in $[0, 1]$);

(b) has uncorrelated members (for example, absence of pairwise correlation means that (X_n, X_{n+1}) should approximate Lebesgue measure in $[0, 1]^2$); - see Figure 3.3;

(c) have to be computed quickly.

As an example, we describe the classical Linear Congruential Generators which represent one of the oldest and best-known pseudo-random number generator methods.

Linear congruential generators

The linear congruential generator (LCG) is defined by the simple recurrence relation

$$X_{n+1} = (aX_n + c) \bmod m, \quad n \geq 0,$$

with $a, c, m \in \mathbb{N}$. The start value X_0 of the sequence is usually called the *seed*. Dividing X_n by m gives an approximation of the uniform distribution in $[0, 1]$. If $c = 0$ the generator is called a *multiplicative* congruential method.

Since the sequence X_n is periodic with period at most m, m has to be large enough. If c is non-zero, it can be shown (see [196], Section 3.2.1) that:

Theorem 3.1 *The period of a LCG sequence is m for all seed values if and only if*

(i) *The values c and m are* coprime *(namely the only positive integer that evenly divides both of them is 1);*

(ii) *$a - 1$ is divisible by all prime factors of m;*

(iii) *$a - 1$ is a multiple of 4 if m is a multiple of 4.*

The quality of the result depends strongly on the choice of a, c and m. One of the first pseudo-random number generators, RANDU, in use since the 60s, was based on the values $a = 2^{16} + 3$, $c = 0$ and $m = 2^{31}$, with X_0 an odd number. The sequence generates pseudo-random integers in the interval $[0, 2^{31} - 1]$ and has a very large period of 2^{29} (see [196]). Unfortunately, the sequence contains strong correlations. This can be easily verified by the following argument. By definition,

$$X_{n+2} = (2^{16} + 3)X_{n+1} = (2^{16} + 3)^2 X_n.$$

Since $2^{32} \bmod 2^{31} = 0$, we get

$$X_{n+2} = (2^{32} + 6 \cdot 2^{16} + 9)X_n = [6 \cdot (2^{16} + 3) - 9]X_n.$$

Therefore, three points in the sequence are correlated by

$$X_{n+2} = 6X_{n+1} - 9X_n.$$

This shows that (X_n, X_{n+1}, X_{n+2}) is not uniformly distributed on $[0, 1]^3$. For this reason RANDU is an example of an ill-conceived random number generator, and due to its wide use in the early 70s many results from that time are currently being questioned. Although such correlations are characteristic of LCG, careful choices can minimize the problem, and are capable of producing pseudo-random numbers with low correlations. Among other possibilities, a *good* choice of parameters is $m = 2^{31} - 1$, $a = 7^5$, $c = 0$, which was used by the function *rand* in Matlab 4.0.

Remark 3.1

- *An exhaustive presentation of pseudo-random number generators goes beyond the scope of this chapter. The interested reader can deepen their knowledge of the topic by looking at [226, 266], where a series of reliable algorithms to generate pseudo-random numbers is presented.*
- *An alternative approach to the construction of Monte Carlo methods is based on the use of* quasi-random numbers. *Unlike pseudo-random sequences, quasi-random sequences do not attempt to mimic the behaviour of random sequences. Instead, the elements of the quasi-random sequence are correlated to make them more uniform than a pseudo-random sequence. Under suitable conditions they yield error estimates which are certain instead of probabilistic. In particular, for numerical integration they have a better convergence rate, typically $O(N^{-1}\log(N)^k)$ for some k, instead of $O(N^{-1/2})$. However, the performance of quasi-Monte Carlo methods for general high-dimensional problems often reduces to that of standard Monte Carlo methods. We refer to [63] for a recent review.*

3.2.2 Pseudo-random sequences with non-uniform distributions

We have seen how to generate uniform pseudo-random sequences. In this section we consider the task of generating a pseudo-random sequence having a given distribution.

Inverse transform methods

Let $X \in \mathbb{R}$ be a random variable and let U be a uniformly distributed random variable in $[0,1]$. Then the relation between X and U can be found using *inverse transform methods.*

Let F be the cumulative distribution function of the random variable X defined by

$$F(x) = P(X < x), \quad x \in \mathbb{R}. \tag{3.3}$$

Denote by

$$F^{-1}(x) = \inf\{y : F(y) > x\} \tag{3.4}$$

the pseudo-inverse function of the distribution function $F(x)$.

Proposition 3.1 *If X is a real-valued random variable with distribution function F, and U is a random variable uniformly distributed on $[0,1]$, $F^{-1}(U)$ has distribution function F.*

In fact, using the right-continuity of F we have, for $x \in \mathbb{R}$,

$$P\left(F^{-1}(U) < x\right) = P\left(U < F(x)\right) = F(x).$$

Note that since $1-U$ is also uniform we also have that $F^{-1}(1-U)$ is distributed like X.

Thus, to generate a pseudo-random sequence X_n distributed like X we can generate a uniformly distributed random sequence U_n and then set

$$X_n = F^{-1}(U_n). \tag{3.5}$$

The argument works for general probability distribution functions which are right-continuous [126]. If there exists a function $f : \mathbb{R} \to [0, \infty)$ such that

$$F(x) = \int_{-\infty}^{x} f(y)\,dy, \quad \forall x \in \mathbb{R}, \tag{3.6}$$

then X is a continuous random variable with probability density $f(x)$. Clearly, $f(x) \geq 0$ and $\int_{\mathbb{R}} f(x)\,dx = 1$.

Example 3.1 (Exponential distribution). Let $f(x) = \lambda \exp(-\lambda x)$, $x \geq 0$, and $f(x) = 0$, $x < 0$. Then

$$F(x) = \lambda \int_0^x \exp(-\lambda y)\,dy = 1 - \exp(-\lambda x),$$

and therefore, given a uniform pseudo-random sequence U_n, we can generate an exponentially distributed pseudo-random sequence from the relation

$$X_n = -\frac{\log(1-U_n)}{\lambda},$$

or simply $X_n = -(\log U_n)/\lambda$, since $1-U_n$ is also uniform in $[0,1]$ (see Figure 3.4). ■

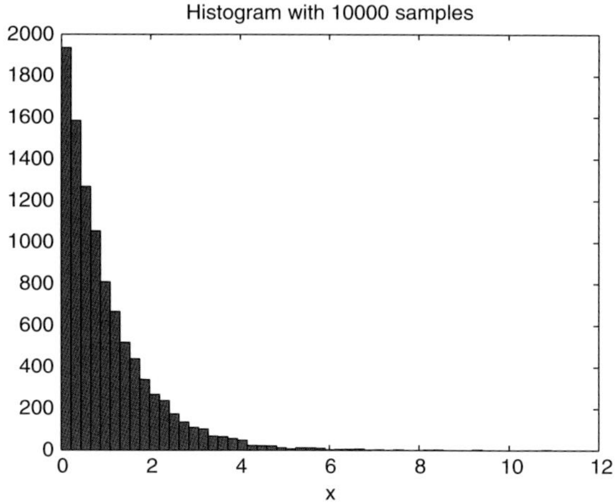

Fig. 3.4 Transformation of N uniformly distributed random numbers U_n using the relation $X_n = -\log(U_n)$.

Example 3.2 (Geometric distribution). The geometric distribution is the discrete analogue of the exponential distribution. If X is an exponentially distributed random variable with parameter λ, then

$$Y = [\![X]\!], \tag{3.7}$$

where $[\![\cdot]\!]$ denotes the integer part, is a geometrically distributed random variable with parameter $p = 1 - e^{-\lambda}$, taking values in the set $\{0, 1, 2, \ldots\}$. This can be used to generate geometrically distributed pseudo-random numbers by first generating exponentially distributed pseudo-random numbers. Since $\lambda = -\log(1-p)$, we have that

$$Y_n = \left[\!\!\left[\frac{\log(U_n)}{\log(1-p)} \right]\!\!\right]$$

is geometrically distributed with parameter p, if U_n is uniformly distributed in $[0, 1]$. ■

Example 3.3 (Pareto distribution). For positive parameters a and b the Pareto probability density is given by $f(x) = ab^a x^{-a-1}$ for $x \geq b$ (and $f(x) = 0$ for $x < b$). Then

$$F(x) = ab^a \int_b^x y^{-a-1}\, dy = -b^a x^{-a} + 1,$$

thus $1 - U_n = b^a X_n^{-a}$ which, again using the fact that $1 - U_n$ and U_n are uniformly distributed, permits the conclusion that the random sequence

$$X_n = bU_n^{-1/a}$$

has the desired Pareto distribution. ■

If $F(x)$ is known explicitly, but difficult to invert analytically, one can solve the equation $F(X_n) = U_n$ for a given U_n numerically (for example, using Newton's method). Otherwise, one can compute $F^{-1}(U_n)$ by a piecewise polynomial approximation. For example, by starting from a table of values $(F(x_i), x_i)$, $i = 0, 1, \ldots, M$, where $x_i < x_j$ for $i < j$, and choosing the points so that $F(x_{i+1}) - F(x_i)$ is small.

Example 3.4 (Sampling from a piecewise constant approximation). In general, the easiest way to sample a probability density function $f(x)$ is given by inverting its piecewise constant approximation. With this aim, given a mesh size Δx, we introduce the grid points

$$x_j = j\Delta x, \quad x_{j+1/2} = x_j + \frac{1}{2}\Delta x, \quad j = \ldots, -2, -1, 0, 1, 2, \ldots.$$

If $I_j = [x_j, x_{j+1}]$, and $f_j = f(x_j)$ are the point values of the function, the standard notation for the cell averages of the function $f(x)$, $x \in \mathbb{R}$, is

$$f_{j+1/2} = \frac{1}{\Delta x}\int_{I_j} f(x)\,dx. \tag{3.8}$$

Starting from cell averages, the corresponding piecewise constant approximation $f_{PC}(x)$ of $f(x)$ is second-order accurate. Note that, by construction,

$$\int_{\mathbb{R}} f_{PC}(x)\,dx = \Delta x \sum_j f_{j+1/2} = \int_{\mathbb{R}} f(x)\,dx = 1.$$

An algorithm to take an average of N samples X_i from $f_{PC}(x)$ is the following:

1. *Set* $k = 0$.
2. *For each interval* I_j *compute* $N_j = [\![N\Delta x f_{j+1/2}]\!]$
   (a) *In each interval* I_j *extract* $U_1, \ldots, U_{N_j}$ *uniform random numbers.*
   (b) *Set* $X_{k+i} = x_{j+1/2} + \frac{1}{2}(1 - 2U_i)\Delta x$, $i = 1, \ldots, N_j$.
3. *Set* $k = k + N_j$.

Note that the total number of samples $\sum_j N_j$ is such that $E(\sum_j N_j) = N$, and that the variance of the result depends on the choices of N and Δx. More precisely, the algorithm just described is an example of the stratified sampling method. We refer to Section 3.3.2 on variance reduction methods for further details. Similarly, one can consider a piecewise linear approximation of $f(x)$, which is also second-order accurate, starting from the pointwise values f_j [247]. ■

Acceptance–rejection methods

To compute the inverse function, in general, one has to solve a nonlinear equation. This can be achieved numerically but it can be computationally expensive, especially for multi-dimensional target distributions. A different technique is the so-called *acceptance–rejection method* due to John von Neumann. The core idea is to sample

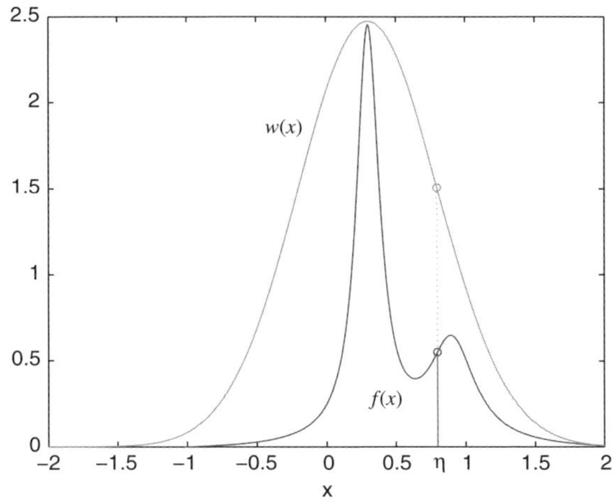

Fig. 3.5 Acceptance–rejection method: η has been sampled from $w(x)$. It will be accepted with probability equal to the ratio $f(\eta)/w(\eta)$.

from a density function which has an easily invertible (either analytically or numerically) distribution function, and to use a Monte Carlo game to accept or reject the sample.

Let X be a random variable with probability density $f(x)$, $x \in \mathbb{R}$. We search for an integrable function

$$w(x) \geq f(x) \quad \forall x \in \mathbb{R}, \tag{3.9}$$

whose primitive $W(x)$ is easily invertible. Let

$$A = \int_{\mathbb{R}} w(x)\, dx. \tag{3.10}$$

The *acceptance–rejection method* to generate one sample X_1 from the desired probability density is then based on the following three steps:

1. *Generate the pseudo-random numbers* U_1 *and* U_2 *uniform in* $[0, 1]$.
2. *Generate one sample* η *from the probability density* $g(x) = w(x)/A$, *usually referred to as the* proposal density, *by computing* η *such that* $W(\eta) = AU_1$.
3. *Generate one sample* ξ *uniform in* $[0, w(\eta)]$, *setting* $\xi = w(\eta)U_2$. *If* $\xi < f(\eta)$ *then accept the sample and set* $X_1 = \eta$, *otherwise reject the sample and repeat Step 1.*

The efficiency of the scheme depends on how easy it is to invert the function $W(x)$ and how frequently we accept the sample. The fraction of accepted samples equals the ratio of the areas below the two curves $f(x)$ and $w(x)$ (see Figure 3.5).

Let B, S be the subsets of $\mathbb{R} \times \mathbb{R}$ defined by

$$B = \{(x, y) : 0 \leq y \leq f(x)\}, \quad S = \{(x, y) : 0 \leq y \leq w(x)\}.$$

Then the probability of accepting a sample is equal to

$$\frac{\text{area(B)}}{\text{area(S)}} = \frac{1}{A}. \tag{3.11}$$

We have the following result [219]:

Theorem 3.2 *If the random variable X_1 is generated via Steps 1–3 of the acceptance–rejection method, then X_1 has probability density $f(x)$.*

Proof. First, let us remark that Steps 1–3 generate a random point (η, ξ) that is uniformly distributed on S. In fact, let us define $h(x, y)$ as the joint density of (η, ξ), and $h(y|x)$ the conditional density of ξ given $\eta = x$. Then we have

$$h(x, y) = \begin{cases} g(x)h(y|x) & \text{if } (x, y) \in S \\ 0 & \text{otherwise.} \end{cases}$$

By construction, we have $h(y|x) = w(x)^{-1}$ for $y \in [0, w(x)]$, and $h(y|x) = 0$ elsewhere. Hence, $h(x, y) = 1/A$ for every $(x, y) \in S$. Now let (η^*, ξ^*) be the accepted point. By the same arguments, (η^*, ξ^*) is uniformly distributed in B and its probability density is identically 1 in B. Therefore, the marginal probability density of $X_1 = \eta^*$ is

$$\int_0^{f(x)} 1 \, dy = f(x).$$

■

Example 3.5 (Normal distribution). Let X be a normally distributed random variable with zero mean and unit variance. Its probability density is

$$f(x) = \frac{1}{\sqrt{2\pi}} \exp\left(-\frac{x^2}{2}\right). \tag{3.12}$$

In order to sample from f one could invert the corresponding distribution function

$$\Phi(x) = (1 + \text{erf}(x/\sqrt{2}))/2, \tag{3.13}$$

where

$$\text{erf}(x) = \frac{2}{\sqrt{\pi}} \int_0^x \exp(-t^2) \, dt \tag{3.14}$$

denotes the *error function*. However, inversion of the error function may be expensive.

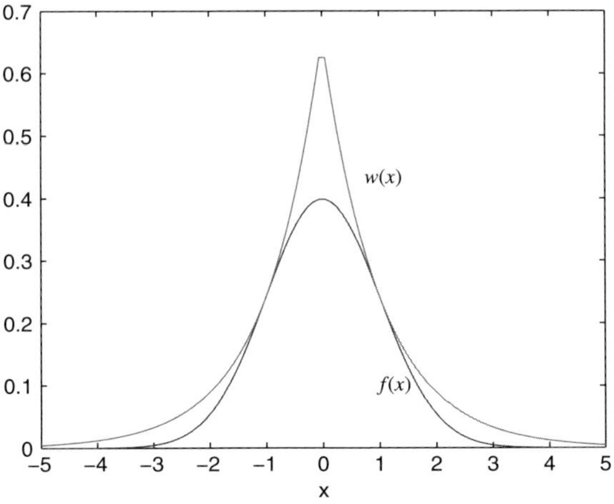

Fig. 3.6 Standard normal via rejection using the proposal density $g(x) = \frac{1}{2}e^{-|x|}$.

There are different ways to overcome this aspect. For example, one can resort to the inverse interpolation method. Here, we will see how to generate a normal sample using acceptance–rejection from the proposal density

$$g(x) = \frac{1}{2}e^{-|x|}, x \in \mathbb{R}. \tag{3.15}$$

Note that $g(x)$ can be easily sampled by generating an exponentially distributed random number multiplied randomly by ± 1. More precisely, to obtain a random sample X from $g(x)$, we can proceed as follows:

1. *Generate* U_1 *and* U_2 *uniformly in* $[0, 1]$.
2. *If* $U_1 < 1/2$ *set* $X = -\log(U_2)$ *else set* $X = \log(U_2)$.

Alternatively, one can force a random sequence of dimension n (even) to preserve exactly the zero mean of the distribution $g(x)$ simply by sampling $n/2$ random numbers X_i, $i = 1, \ldots, n/2$ from e^{-x} and then setting $X_{i+n/2} = -X_i$ for the remaining $n/2$ samples. This last approach represents an example of an *antithetic estimator* (see Example 3.11 for more details).

To apply the acceptance–rejection we must compute

$$A = \sup_{x \in \mathbb{R}} \frac{f(x)}{g(x)} = \sup_{x \geq 0} \frac{2\exp(-x^2/2)}{\sqrt{2\pi}\exp(-x)} = \sup_{x \geq 0} \sqrt{\frac{2}{\pi}} \exp(x - x^2/2).$$

Since the function $\exp(x - x^2/2)$ reaches its maximum value at $x = 1$, $A = \sqrt{2e/\pi}$ and then $w(x) = \sqrt{\frac{e}{2\pi}} e^{-|x|} \geq f(x)$, $\forall\, x \in \mathbb{R}$ (see Figure 3.6). Note that the acceptance rate in this case is approximately the 76% of the produced samples. ■

Finally, let us remark that if the function $f(x)$ can be factorized as $f(x) = Ch(x)w(x)$ where $C \geq 1$ is a constant, the primitive $W(x)$ of $w(x)$ is easily invertible and $0 \leq h(x) \leq 1$ is the weight function, we can apply a *weighted acceptance–rejection* strategy as follows:

1. *Generate the pseudo-random numbers* U_1 *and* U_2 *uniform in* $[0,1]$.
2. *Generate one sample* η *from the probability density* $g(x) = w(x)/A$ *by computing* η *such that* $W(\eta) = AU_1$.
3. *Generate one sample* ξ *uniform in* $[0,1]$, *setting* $\xi = U_2$. *If* $\xi < h(\eta)$ *then accept the sample and set* $X_1 = \eta$, *otherwise reject the sample and repeat Step 1.*

By taking $h(x) = f(x)/w(x)$ (here $C = 1$) we fall into the usual acceptance–rejection method.

Discrete rejection and convex combinations

In many cases a density function is represented as a convex combination of simpler density functions:

$$f(x) = \sum_{i=1}^{M} w_i p_i(x). \tag{3.16}$$

In (3.16) the w_i are a set of probability values, i.e.

$$w_i \geq 0, \quad \sum_{i=1}^{M} w_i = 1, \tag{3.17}$$

and the functions $p_i(x)$ are probability densities. The above decomposition is often useful in improving the efficiency of the acceptance–rejection method by constructing a set of functions $w_i(x)$ such that $w_i(x) \geq p_i(x)$, which defines the proposal densities $g_i(x) = w_i(x)/A_i$, with $A_i = \int w_i(x)\,dx$. This is the case, for example, when we decompose the original probability density function $f(x)$ on the intervals $I_i = [x_i, x_{i+1}]$ from a given set of $M+1$ points $x_i < x_{i+1}$, $i = 1, \ldots, M$.

If we know how to generate samples from each p_i we can generate a random variable X with probability density $f(x)$ as follows:

1. *Compute* $W_i = \sum_{j=1}^{i} w_j$, $i = 1, \ldots, M$. *Set* $W_0 = 0$.
2. *Generate a uniform sample* U_1 *and find the integer* k *such that* $W_{k-1} < U_1 < W_k$.
3. *Generate* X *from a random variable with density* $p_i(x)$.

Steps 1–2 of the above procedure correspond to selecting an integer $k \in \{1, \ldots, M\}$ with probability w_k. For an arbitrary set of probabilities $\{w_i\}$, once the W_i has been computed, Step 2 can be performed with a binary search in $O(\log M)$ operations.

This approach is efficient if the numbers W_i can be computed once, and then used several times, and if M is not too large. For a very large M, if we need to generate many independent random variables from the same distribution, one can use as an alternative the *alias method*, which requires a constant time (i.e. independent of M). We refer to [196, 219] for more details.

In several circumstances, however, the number M may be very large, and the numbers W_i may change, or, in other words, one is interested in sampling k with probability w_k without explicitly computing all the probabilities w_i. This can be done by using the acceptance–rejection technique, if at least an estimate $\overline{w}$ such that $\overline{w} \geq w_i,\ i = 1, \dots, M$, is known. The procedure is used as follows:

1. *Select a random integer k uniformly in $\{1, \dots, M\}$. This can be achieved by generating a uniform random number U_1 and computing*

$$k = [\![MU_1]\!] + 1,$$

where $[\![x]\!]$ denotes the integer part of $x \in \mathbb{R}$.

2. *Generate a uniform random number U_2. If $w_k U_2 < \overline{w}$ then accept the index k else reject it and repeat Step 1.*

Clearly, the procedure can be generalized to the case in which the estimate depends on k, i.e. $\overline{w}_k \geq w_k,\ k = 1, \dots, M$.

Multivariate distributions

Suppose we want to sample an n-dimensional random variable $\mathbf{X} = (X_1, \dots, X_n)$, where $\mathbf{x} = (x_1, \dots, x_n) \in \mathbb{R}^n$, whose probability density is given by $f(\mathbf{x})$. It is straightforward to extend the acceptance–rejection method if we are able to find another probability density $g(\mathbf{x})$ on $\mathbb{R}^n$ satisfying the two conditions:

1. We know how to generate a random vector $\mathbf{Y} = (Y_1, \dots, Y_n)$ from $g(\mathbf{x})$.
2. There is a constant $A \geq 1$ such that $f(\mathbf{x}) \leq Ag(\mathbf{x})$, $\forall \mathbf{x} \in \mathbb{R}^n$.

If the density can be written as a product of densities of scalar random variables (marginal probability densities), i.e. if

$$f(x_1, \dots, x_n) = p_1(x_1)p_2(x_2) \cdots p_n(x_n), \tag{3.18}$$

then the n scalar random variables $X_1, \dots, X_n$ are independent, and the problem is equivalent to sampling n monovariate random variables.

If this is not the case, then one may first look for a transformation $T : \mathbf{x} \to \mathbf{y} = T(\mathbf{x})$ such that in the new variables the probability density is factorized, i.e.

$$f(x_1, \dots x_n)dx_1 dx_2 \dots dx_n = p_{y_1}(y_1)p_{y_2}(y_2) \cdots p_{y_n}(y_n)dy_1 dy_2 \dots dy_n.$$

After sampling the random variables $Y_1, \dots, Y_n$, one can compute the random vector $\mathbf{X}$ by inverting the map T, i.e. $\mathbf{x} = T^{-1}(\mathbf{y})$.

Example 3.6 (Box–Muller method). An alternative procedure to generate a normally distributed random variable X is obtained by the so-called *Box–Muller method* described below.

Let us consider a two-dimensional normally distributed random variable. Then

$$f(x_1, x_2) = \frac{1}{2\pi} \exp\left(-\frac{x_1^2 + x_2^2}{2}\right) = p(x_1)p(x_2). \tag{3.19}$$

If we use polar coordinates,

$$x_1 = \rho\cos\theta,\ x_2 = \rho\sin\theta,$$

we have the change of variables

$$\frac{1}{2\pi} \exp\left(-\frac{x_1^2 + x_2^2}{2}\right) dx_1\, dx_2 = \frac{1}{2\pi} \exp\left(-\frac{\rho^2}{2}\right) \rho\, d\rho\, d\theta.$$

Therefore, in polar coordinates the density function is factorized as $p_\rho\, d\rho\, p_\theta\, d\theta$, with

$$p_\rho = \exp\left(-\frac{\rho^2}{2}\right) \rho, \quad \rho \geq 0$$

$$p_\theta = \frac{1}{2\pi}, \quad 0 \leq \theta < 2\pi.$$

The random variables with the corresponding densities are readily generated by inverting p_ρ and p_θ. Therefore, given U_1 and U_2 uniformly distributed we can generate

$$R = \sqrt{-2\log U_1}, \qquad \Theta = 2\pi U_2. \tag{3.20}$$

From these, switching back to Cartesian coordinates, we obtain the independent random pair (X_1, X_2). In summary, the method works as follows:

1. Generate two uniform random numbers U_1 and U_2.
2. Set $X_1 = \sqrt{-2\log U_1}\cos(2\pi U_2)$ and $X_2 = \sqrt{-2\log U_1}\sin(2\pi U_2)$.

At the end of the procedure we have two points sampled from a standard normal distribution (i.e. with zero mean and unit variance). Of course, if the random variable has mean μ and standard deviation σ, then X_1 and X_2 will be scaled accordingly as

$$X_1 = \mu + \sigma R\cos\Theta,\ X_2 = \mu + \sigma R\sin\Theta.$$

■

Example 3.7 (Surface of a sphere). Sampling multivariate distributions is useful in many circumstances. A classical situation is when the random variable is uniformly distributed on a circle or on the surface of a sphere.

Here we show how to sample a point uniformly from the surface of a unit sphere. A point (x, y, z) on a unit sphere is identified by the two polar angles (φ, θ), with $0 \leq \varphi \leq 2\pi$, $0 \leq \theta \leq \pi$.

$$\begin{aligned} x &= \sin\theta\cos\varphi, \\ y &= \sin\theta\sin\varphi, \\ z &= \cos\theta. \end{aligned} \tag{3.21}$$

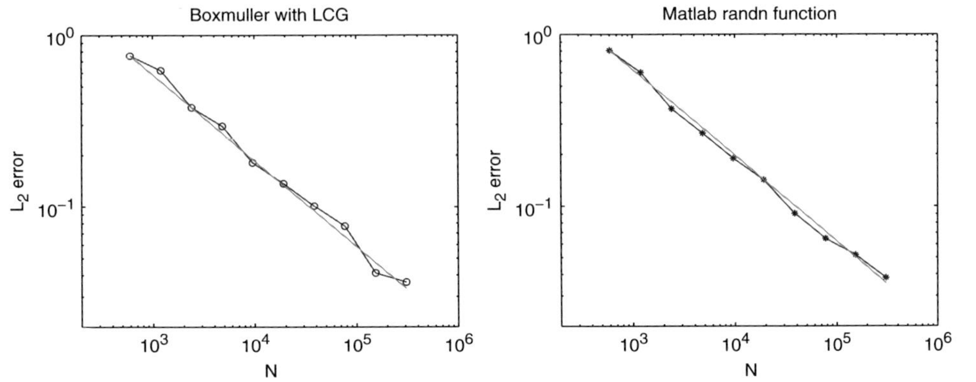

Fig. 3.7 Convergence of the Box–Muller algorithm using the LCG generator (left) against Matlab 7 `randn` function (right).

Because the distribution is uniform, the probability of finding a point in a region is proportional to the solid angle ω:

$$\frac{d\omega}{4\pi} = \frac{\sin\theta\, d\theta}{2} \cdot \frac{d\varphi}{2\pi}.$$

Hence, the density function of the solid angle is factorized as the product of the densities of two random variables, say Φ and Θ. We can easily construct the distribution functions of these random variables:

$$F_\Phi(\varphi) = \int_0^\varphi \frac{dy}{2\pi} = \frac{\varphi}{2\pi},$$

$$F_\Theta(\theta) = \int_0^\theta \frac{\sin y}{2}\, dy = \frac{1}{2}(1 - \cos(\theta)).$$

Given two uniformly distributed random numbers U_1 and U_2, and inverting the above expressions, we have

$$\Phi = 2\pi U_1, \qquad \Theta = \arccos(1 - 2U_2). \tag{3.22}$$

Reverting to the original variables gives the desired sample (see Figure 3.8). ∎

Example 3.8 (Linear combination of two independent variables). Dealing with kinetic equations, a particularly interesting situation occurs when we must generate a sample from a probability density $f(x)$ which can be expressed as an integral of the form

$$f(x) = \int_{\mathbb{R}} g(y, x - y)\, dy, \tag{3.23}$$

where $g(y, z)$ is the two-dimensional probability density of the random variables Y and Z. If the random variables are independent then $g(y, z) = g_1(y)g_2(z)$ and we have

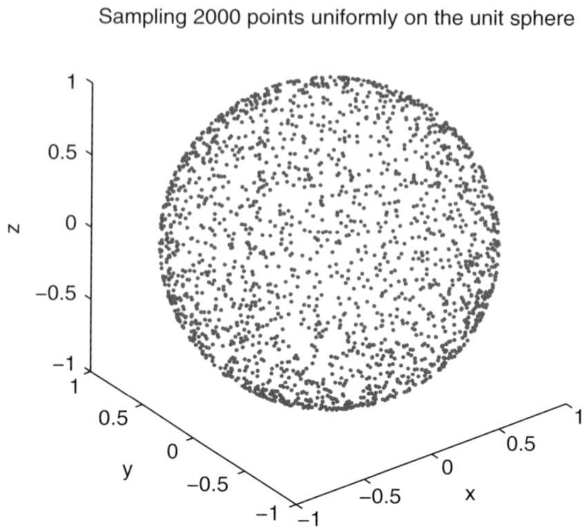

Fig. 3.8 Sampling from the surface of a sphere using $2N$ uniformly random sequences $(U_1)_i$, $(U_2)_i$ and the transformations $\Phi_i = 2\pi(U_1)_i$, $\Theta_i = \arccos(1 - 2(U_2)_i)$.

$$f(x) = \int_{\mathbb{R}} g_1(y)g_2(x-y)\,dy = \int_{\mathbb{R}} g_1(x-z)g_2(z)\,dz = (g_1 * g_2)(x). \tag{3.24}$$

We know that the probability distribution of the sum of two independent random variables is the convolution of their individual distributions [145] (see Appendix B for details). Therefore, if Y is a random variable distributed as $g_1(y)$ and Z is a random variable distributed as $g_2(z)$ then the random variable $X = Y + Z$ is distributed as $f(x)$. Thus, we can generate a sample X from $f(x)$ simply by generating two samples Y and Z distributed respectively as $g_1(y)$ and $g_2(z)$ and then setting $X = Y + Z$.

More generally, it can immediately be shown that if we consider a linear combination of the two random variables Y and Z in the form $X = aY + bZ$, $a, b > 0$, then its probability density has the form

$$f(x) = \frac{1}{b}\int_{\mathbb{R}} g_1(y)g_2\left(\frac{x-ay}{b}\right)\,dy = \frac{1}{a}\int_{\mathbb{R}} g_1\left(\frac{x-bz}{a}\right)g_2(z)\,dz. \tag{3.25}$$

■

3.3 Monte Carlo techniques

3.3.1 Monte Carlo integration

In this section we will briefly describe how to apply Monte Carlo methods to estimate the value of a definite integral (or a sum). A fundamental step in converting an approximation problem into a Monte Carlo problem is to represent the integral as the mathematical expectation of some random quantity.

Let us consider the simple integral

$$I[g] = \int_{[0,1]^d} g(x)dx, \quad d \geq 1, \tag{3.26}$$

where g is a given real-valued function in $\mathbb{R}^d$. If U is a random vector uniformly distributed in $[0,1]^d$ we have the identity $I[g] = E(g(U))$.

If U_n is a sequence of N pseudo-random vectors uniformly distributed in $[0,1]^d$ we can write

$$\bar{I}_N[g] = \frac{1}{N}\sum_{i=1}^{N} g(U_i), \quad E(\bar{I}_N[g]) = I[g]. \tag{3.27}$$

Thanks to the law of large numbers [63, 145],[2]

$$\lim_{N\to\infty} P\left(|\bar{I}_N[g] - I[g]| > \varepsilon\right) = 0, \quad \forall \varepsilon > 0.$$

Moreover, by the central limit theorem $(\bar{I}_N[g] - I[g])/(\sigma_g N^{-1/2})$, where σ_g^2 is the variance of g,

$$\sigma_g^2 = \int_{[0,1]^d} (g(x) - I[g])^2 \, dx,$$

has (approximately) a standard normal distribution.

The following theorem asserts that the error in Monte Carlo integration is $O(N^{-1/2})$, with a constant that is just the variance of the integrand function, regardless of the dimension.

Theorem 3.3 *Let us define the Monte Carlo integration error as*

$$\epsilon_N[g] = I[g] - \bar{I}_N[g]. \tag{3.28}$$

Then we have

$$E((\epsilon_N[g])^2)^{1/2} = \sigma_g N^{-1/2}. \tag{3.29}$$

Proof. Given a sequence U_n of N independent and uniformly distributed random variables, we define the sequence of random variables $X_n = (I[g] - g(U_n))/\sigma_g$. Then, since the X_i's are independent, for $i = 1, \ldots, N$ we get

$$E(X_i) = 0, \quad E(X_i^2) = \int_{[0,1]^d} \frac{(I[g] - g(x))^2}{\sigma_g^2} dx = 1, \quad E(X_i X_j) = 0 \text{ if } i \neq j.$$

Consider the sum

$$S_N[g] = \frac{1}{N}\sum_{i=1}^{N} X_i = \frac{\epsilon_N[g]}{\sigma_g}.$$

[2]The version here uses the so-called weak law of large numbers which describes convergence in probability. The same results holds true in a stronger form, the so-called strong law of large numbers, which states that $\bar{I}_N[g] \to I[g]$ almost surely. In statistical terms this says that $\bar{I}_N[g]$ is a consistent estimator of $I[g]$.

Owing to the independence of the X_i's, we obtain

$$E(S_N^2[g]) = E\left(\frac{1}{N^2}\left(\sum_{i=1}^{N} X_i\right)^2\right) = \frac{1}{N^2}\left[E\left(\sum_{i=1}^{N} X_i^2\right) + E\left(\sum_{i=1}^{N}\sum_{j\neq i} X_i X_j\right)\right]$$

$$= \frac{1}{N^2}(N+0) = \frac{1}{N}.$$

Therefore,

$$E(\epsilon_N[g]^2)^{1/2} = E((I[g] - \bar{I}_N[g])^2)^{1/2} = \sigma_g N^{-1/2}.$$

■

Clearly, the above Monte Carlo approach can be extended to the case of integrals of the form

$$I[g] = \int_\Omega g(x) f(x)\, dx, \quad \Omega \subset \mathbb{R}^d,\ d \geq 1, \tag{3.30}$$

where $f(x)$ is a probability density in $\mathbb{R}^d$. In this case we make use of the identity $I[g] = E(g(X))$, where X is a random vector with probability density $f(x)$. In particular, $g(x)$ and $f(x)$ could be identically 1. In this case we reduce to the problem of estimating the volume of Ω.

If X_n is a sequence of N samples with probability $f(x)$, a Monte Carlo estimate of $I[g]$ is obtained as

$$\bar{I}_N[g] = \frac{1}{N}\sum_{i=1}^{N} g(X_i), \qquad E(\bar{I}_N[g]) = I[g], \tag{3.31}$$

which again converges as $O(\sigma_g N^{-1/2})$. The proportionality constant is now characterized by

$$\sigma_g^2 = \int_\Omega (g(x) - I[g])^2 f(x)\, dx.$$

A typical situation of this type occurs when we evaluate some moment of a given probability density $f(x)$,

$$M^k[f] = I[x^k] = \int_{\mathbb{R}^d} f(x) x^k dx, \quad d, k \geq 1, \tag{3.32}$$

as

$$\bar{M}_N^k[f] = \frac{1}{N}\sum_{n=1}^{N} (X_n)^k. \tag{3.33}$$

Remark 3.2 *Note that the convergence rate for a deterministic grid-based quadrature method of order r is $O(N^{-r/d})$. Thus, in principle, Monte Carlo should be 'better' when $r/d \leq 1/2$. This is a surprising result and usually leaves people puzzled. How can a random method be better than a method based on a grid? First let us point out that* *for an analytic function on a periodic domain the value of r is infinite and thus the above argument fails. A second aspect concerns the fact that the Monte Carlo estimate is a probabilistic estimate which involves the expectation of the error, and thus on a single numerical run the results are affected by the presence of fluctuations, which makes Monte Carlo methods attractive only when the dimension d is significantly large. However, a crucial aspect to underline is the difference between the local nature of grid based methods and the global nature of Monte Carlo methods. In fact, in contrast to grid-based methods where each point is related to a specific evaluation of the integrand function, each point of a Monte Carlo integration formula is an estimate of the integral over the entire domain.*

3.3.2 Variance reduction strategies

In Theorem 3.3 we saw that the error ε_N and the number N of samples are related by $\varepsilon_N = O(\sigma_g N^{-1/2})$, which implies $N = O(\sigma_g^2/\varepsilon_N^2)$. Since the computational time is proportional to the sample size N, it grows like $\sigma_g^2/\varepsilon_N^2$ and thus it increases rapidly when the accuracy requirements are more stringent. Therefore, methods to accelerate the convergence of Monte Carlo techniques represent one of the central aspects in the development of effective numerical methods. Note that, for a given number of samples, estimate (3.29) shows that the only possibility of reducing the error is to lower the variance σ^2 of the statistical samples used. A related class of methods, called *quasi-Monte Carlo* methods, follows a different strategy, in the sense that it replaces the random sequence by another, more adapted, sequence, with the aim of improving the exponent 1/2. Here, we restrict ourselves to Monte Carlo methods and recall some basic techniques for variance reduction. We refer to [63] for quasi-Monte Carlo approaches.

Stratified sampling

One of the drawbacks of basic Monte Carlo methods is the large variance of the samples we obtain. This is mainly due to the fact that we sample from the whole interval of interest of the distribution function. The basic principle of stratified sampling is to divide the sampling interval into subintervals (cells), in order to improve the overall accuracy by reducing the variance of the samples. This approach presents some analogies with piecewise polynomial interpolation or composed quadrature rules in classical numerical analysis.

Let us consider again the problem of computing the integral (3.30). To obtain a stratified estimator we partition the domain Ω into a collection of disjoint sets $\Omega_1, \ldots, \Omega_M$ so that $\Omega = \cup_{i=1}^M \Omega_i$. The evaluation of the integral over Ω is consequently equivalent to the evaluation of a sum of integrals over smaller domains,

$$I[g] = \sum_{i=1}^{M} \int_{\Omega_i} g(x) f(x)\, dx = \sum_{i=1}^{M} I_i[g].$$

For $i = 1, \ldots, M$ we define with p_i the probabilities

$$p_i = \int_{\Omega_i} f(x)\, dx = P(X \in \Omega_i).$$

Observe that $p_1 + \ldots + p_M = 1$. Further, let us introduce the probability densities

$$h_i(x) = \begin{cases} \dfrac{f(x)}{p_i} & \text{if } x \in \Omega_i, \\ 0 & \text{otherwise,} \end{cases}$$

and denote by $X^{(i)}$ the random variables with probability density $h_i(x)$, $i = 1, \ldots, M$. Now fix integers $N_1, \ldots, N_M$ such that $N_1 + \ldots + N_M = N$, and generate N_i random samples $X_1^{(i)}, \ldots, X_{N_i}^{(i)}$ for each $i = 1, \ldots, M$.

We obtain that

$$T_i[g] = \frac{1}{N_i} \sum_{j=1}^{N_i} g(X_j^{(i)}), \qquad E(T_i[g]) = \frac{I_i[g]}{p_i}. \tag{3.34}$$

The stratified estimator of (3.30) is given by

$$\bar{I}_N^M[g] = \sum_{i=1}^{M} p_i T_i[g], \qquad E(\bar{I}_N^M[g]) = I[g]. \tag{3.35}$$

Note that in the above expression the choice of the N_i's is completely arbitrary except for the requirement that they sum up to N.

Let us compare the variance of the stratified estimator with that of the classical simple estimator. For the classical one we have

$$Var(\bar{I}_N[g]) = \frac{1}{N} Var(g(X)) = \frac{1}{N}\left(E(g(X)^2) - (I[g])^2\right).$$

Let us compute the variance for the stratified method. Since

$$Var(T_i[g]) = \frac{1}{N_i} Var(g(X^{(i)})) = \frac{1}{N_i}\left[E(g(X^{(i)})^2) - \left(\frac{I_i[g]}{p_i}\right)^2\right],$$

we obtain

$$Var(\bar{I}_N^M[g]) = \sum_{i=1}^{M} \frac{p_i^2}{N_i}\left[E(g(X^{(i)})^2) - \left(\frac{I_i[g]}{p_i}\right)^2\right]. \tag{3.36}$$

The main question is to know under which conditions $Var(\bar{I}_N^M[g])$ is smaller than $Var(\bar{I}_N[g])$.

The following result provides a partial answer to the question.

Theorem 3.4 *If $N_i = p_i N$, $i = 1, \ldots, M$, then*

$$Var(\bar{I}_N[g]) - Var(\bar{I}_N^M[g]) = \frac{1}{N} \sum_{i=1}^{M} p_i \left(\frac{I_i[g]}{p_i} - I[g]\right)^2. \tag{3.37}$$

Proof. Let $N_i = p_i N$. It holds that

$$\frac{1}{N}E(g(X)^2) = \frac{1}{N}\int_\Omega g(x)^2 f(x)\,dx = \frac{1}{N}\sum_{i=1}^M p_i \int_{\Omega_i} g(x)^2 \frac{f(x)}{p_i}\,dx$$
$$= \sum_{i=1}^M \frac{p_i^2}{N_i}\int_{\Omega_i} g(x)^2 \frac{f(x)}{p_i}\,dx = \sum_{i=1}^M \frac{p_i^2}{N_i} E(g(X^{(i)})^2).$$

Since

$$\sum_{i=1}^M \frac{p_i^2}{N_i}\left(\frac{I_i[g]}{p_i}\right)^2 = \frac{1}{N}\sum_{i=1}^M p_i \left(\frac{I_i[g]}{p_i}\right)^2 = \frac{(I[g])^2}{N} + \frac{1}{N}\sum_{i=1}^M p_i\left(\frac{I_i[g]}{p_i} - I[g]\right)^2,$$

(3.37) follows. ■

We can proceed in an analogous way for the integration error. We have

$$\epsilon_N[g] = I[g] - \bar{I}_N^M[g] = \sum_{i=1}^M \epsilon_N^i[g],$$

where $\epsilon_N^i[g] = I_i[g] - p_i T_i[g]$. Using (3.29) we get

$$E((\epsilon_N^i[g])^2) = p_i \frac{\sigma_i^2}{N_i},$$

with $\sigma_i^2 = Var(g(X^{(i)}))$.

Therefore, if $N_i = p_i N$, the error of the stratified method satisfies

$$\sum_{i=1}^M p_i \frac{\sigma_i^2}{N_i} = \frac{1}{N}\sum_{i=1}^M \sigma_i^2 = \frac{\sigma_T^2}{N}, \qquad \sigma_T^2 = \sum_{i=1}^M \sigma_i^2.$$

Since the variance over the whole set is always larger than the variance over a subset, $\sigma_g^2 \geq \sigma_T^2$. Thus, stratification always reduces the integration error if the distribution of points is balanced, in the sense that the number of points in the set Ω_i is proportional to its weighted size p_i. For this reason the choice $N_i = p_i N$ is called *proportional allocation.* This choice, however, generates non-integer values N_i. In practice, one can choose N_i to be an integer random variable such that $E(N_i) = p_i N$. This can be achieved using *stochastic rounding*, that is by setting

$$[\![X]\!]_S = \begin{cases} [\![x]\!], & \text{with probability } [\![x]\!] + 1 - x, \\ [\![x]\!] + 1, & \text{with probability } x - [\![x]\!]. \end{cases} \tag{3.38}$$

Clearly, $E([\![X]\!]_S) = x$. To satisfy the constraint $\sum_{i=1}^M N_i = N$ we can generate a sequence $N_1, \ldots, N_M$ where only the first $M-1$ terms are obtained by stochastic rounding and $N_M = N - \sum_{i=1}^{M-1} N_i$.

Actually, the choice $N_i = p_i N$ is not optimal, in the sense that it does not minimize the variance of the stratified estimator. A better choice is to put more points in the regions where $g(x)$ has the largest variation. It can be shown [219] that $Var(\bar{I}_N^M[g])$ is minimized by taking

$$N_i = \frac{p_i \sigma_i}{\sum_{j=1}^M p_j \sigma_j} N. \qquad (3.39)$$

Finally, let us remark that in principle all the above estimates can be used to develop adaptive Monte Carlo quadratures.

Importance sampling

Importance sampling is another way to improve the variance of Monte Carlo methods. In this approach, instead of acting on the integration domain, one acts over the distribution of the sample points in order to minimize the error.

To illustrate the idea let us consider the evaluation of the simple integral (3.26). First, let us rewrite the integral in the form

$$I[g] = \int_{[0,1]^d} \frac{g(x)}{f(x)} f(x)\, dx,$$

where $f(x)$ is an arbitrary probability density in $[0,1]^d$.

Now, given $\{X_n\}$ pseudo-random numbers distributed as $f(x)$ we can estimate

$$\bar{I}_N^f[g] = \frac{1}{N} \sum_{n=1}^{N} \frac{g(X_i)}{f(X_i)}, \qquad E(\bar{I}_N^f[g]) = I[g]). \qquad (3.40)$$

Note that the importance sampling estimator $\bar{I}_N^f[g]$ depends on the choice of f.

The error $\epsilon_N^f[g] = I[g] - \bar{I}_N^f[g]$ can be estimated from the formula

$$E((\epsilon_N^f[g])^2) = \frac{\sigma_{g/f}^2}{N},$$

where

$$\sigma_{g/f}^2 = \int_{[0,1]^d} \left(\frac{g(x)}{f(x)} - I[g] \right)^2 f(x)\, dx.$$

The problem then reduces to select a suitable probability density $f(x)$ which is easy to sample and such that $\sigma_{g/f}^2 < \sigma_g^2$. Let us observe that $\sigma_{g/f}^2$ is small if $g(x)/f(x) \approx I[g]$, that is, when $f(x)$ is approximately proportional to $g(x)$. This leads to a simple practical rule for the selection of f, namely $f(x)$ must be large where $g(x)$ is large. Hence, we must ensure that $f(x)$ gives more weight to the important values of x. One must be careful in selecting $f(x)$, since an inappropriate choice can severely reduce the efficiency of the method. The method is particularly suitable when one is dealing with rare but important events, namely small regions of the space where f is significantly large, but it can be easily modified to deal with other types of region, for example where f is very small (like the tails of the probability distribution).

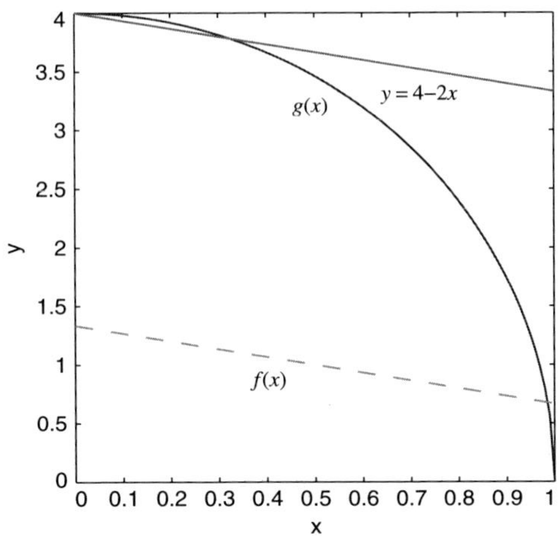

Fig. 3.9 Importance sampling to estimate the integral of $g(x) = 4\sqrt{1-x^2}$ using $f(x) = 2(2-x)/3$.

Example 3.9 Let $g(x) = 4\sqrt{1-x^2}$ and $d = 1$. In this case, $I[g] = 4\int_0^1 \sqrt{1-x^2}dx = \pi$. Given a sequence U_n of N uniformly distributed random variables in $[0,1]$, it is straightforward to evaluate the variance of the standard estimator

$$\bar{I}_N[g] = \frac{1}{N}\sum_{i=1}^{N} 4\sqrt{1-U_i^2}.$$

In fact, we have

$$\sigma_g^2 = \frac{1}{N}\left(\int_0^1 16(1-x^2)dx - \pi^2\right) = \frac{1}{N}\left(\frac{32}{3} - \pi^2\right) \approx \frac{0.797}{N}.$$

We know that $g(x)$ is a decreasing function of x in $[0,1]$ with a maximum at $g(0) = 4$. Accordingly, we can take a linear function $f(x)$ which imitates the relative height of the graph of $g(x)$. For example, we can choose $f(x) = 2(2-x)/3$ (see Figure 3.9). The resulting variance is

$$\sigma_{g/f}^2 = \frac{1}{N}\left(\int_0^1 \frac{24(1-x^2)}{2-x}dx - \pi^2\right) = \frac{1}{N}\left(60 - 72\log 2 - \pi^2\right) \approx \frac{0.224}{N},$$

which is smaller than 1/3 of the variance of the original method. ■

Other techniques

Several other techniques for the reduction of the variance in Monte Carlo methods have been developed in the literature. Many of these techniques are strictly related to a specific problem and it is beyond the scope of the present chapter to illustrate all of them. Here we limit ourselves to giving examples of some of the most common methods which are particularly useful in several problems involving kinetic equations.

Example 3.10 (Matching moments). A partial correction to the statistical sampling error is to equate the moments of the statistical samples with the true moments of the probability density. This, however, requires exact knowledge of the true moments (or an accurate estimate, for example using a deterministic numerical method) and in general should be used with some caution. In fact, the approach is based on a transformation of the statistical samples which makes them no longer independent. As a result, Monte Carlo estimates are not straightforward. For example, the central limit theorem cannot be applied directly and the method may have the wrong expectation. In some sense this method is an example of the second class of methods where we modify the properties of the random sequence with the goal of accelerating the convergence of Monte Carlo schemes.

Let us consider a set of samples $X_1, \ldots, X_N$ distributed as $f(x)$. The first two central moments of the sample are given by

$$m_1 = \frac{1}{N}\sum_{i=1}^{N} X_i, \qquad m_2 = \frac{1}{N}\sum_{i=1}^{N}(X_i)^2.$$

Knowing the exact values (or better estimates)

$$\mu_1 = \int f(x)x\,dx, \qquad \mu_2 = \int f(x)x^2\,dx,$$

one can apply the linear transformation

$$X_i^* = \mu_1 + \frac{(X_i - m_1)S_N}{\sigma}, \quad i = 1, \ldots, N, \tag{3.41}$$

where $\sigma^2 = \mu_2 - \mu_1^2$ and $S_N^2 = m_1 - m_2^2$ to get

$$\frac{1}{N}\sum_{i=1}^{N} X_i^* = \mu_1, \qquad \frac{1}{N}\sum_{i=1}^{N}(X_i^*)^2 = \mu_2.$$

In the particular case of normally distributed samples, a method to generate independent variables with a prescribed mean and variance has been developed in [267].

■

Example 3.11 (Antithetic variates). The use of antithetic variates is one of the simplest and widely used variance reduction methods. Suppose, for example, that we want to generate N samples from a symmetric density function $f(x)$ in such a way that the samples have the exact mean μ. Instead of using a moment matching method we can proceed as follows:

1. *Let N be even. Generate $N/2$ samples X_i distributed as $f(x)$.*
2. *If $X_i > \mu$ set $X_{i+N/2} = X_i - \mu$ otherwise set $X_{i+N/2} = 2\mu - X_i$.*

Note that, alternatively, we can, in the first step, generate $N/2$ random variables from the probability density $g(x) = 2f(x)$ defined only for $x \geq \mu$, and then set $X_{i+N/2} = X_i - \mu$.

Antithetic variates can also be used to evaluate integrals of the type (3.26). Let N be even. Setting

$$\bar{I}^a_N[g] = \frac{1}{N}\sum_{i=1}^{N/2}(g(U_i) + g(1-U_i)), \qquad E(\bar{I}^a_N[g]) = I[g], \tag{3.42}$$

where $U_1, \ldots, U_{N/2}$ are uniformly distributed in $[0,1]^d$ (which implies also that $1-U_1, \ldots, 1-U_{N/2}$ have the same distribution). Note that, even if the antithetic estimator $\bar{I}^a_N[g]$ uses only $N/2$ points, the number of function evaluations is N, the same as the standard estimator $\bar{I}_N[g]$. It can be shown [219] that under suitable monotonicity assumptions on $g(x)$,

$$Var(\bar{I}^a_N[g]) \leq Var(\bar{I}_N[g]). \tag{3.43}$$

Let us recall that, given two random variables X and Y,

$$Var(X+Y) = Var(X)+Var(Y)+2Cov(X,Y), \qquad Cov(X,Y) = E(XY)-E(X)E(Y).$$

Therefore, we can compute

$$\begin{aligned} Var(\bar{I}^a_N[g]) &= \frac{1}{N^2}\left[\frac{N}{2}\left(Var(g(U)) + Var(g(1-U)) + 2Cov(g(U), g(1-U))\right)\right] \\ &= \frac{1}{N}\left(Var(g(U)) + Cov(g(U), g(1-U))\right) \\ &= Var(\bar{I}_N[g]) + \frac{1}{N}Cov(g(U), g(1-U)). \end{aligned}$$

It follows that (3.43) corresponds to $Cov(g(U), g(1-U)) \leq 0$.

To illustrate this let us return to Example 3.9, where $g(x) = 4\sqrt{1-x^2}$ in dimension one. We have

$$\bar{I}^a_N[g] = \frac{1}{N}\sum_{i=1}^{N/2} 4\left(\sqrt{1-U_i^2} + \sqrt{1-(1-U_i)^2}\right).$$

Since $Var(\bar{I}_N[g]) \approx 0.797/N$, we obtain

$$Cov(g(U), g(1-U)) = 16\int_0^1 \sqrt{(1-x^2)x(2-x)}\, dx - \pi^2 \approx -0.578.$$

Hence, $Var(\bar{I}^a_N[g]) \approx 0.219/N$, which is about 1/4 the variance of the standard method. ■

Example 3.12 (Control variates). The idea of control variates is to decompose the integral we want to compute as the sum between another integral that we know exactly, and their difference. For example,

$$\int_{[0,1]^d} g(x)\, dx = \int_{[0,1]^d} f(x)\, dx + \int_{[0,1]^d} (g(x) - f(x))\, dx,$$

where $I[f] = \int_{[0,1]^d} f(x)\, dx$ is known. We have the Monte Carlo formula

$$\bar{I}_N^c[g] = I[f] + \frac{1}{N}\sum_{i=1}^{N}(g(U_i) - f(U_i)). \tag{3.44}$$

The resulting integration error is therefore $O(\sigma_{g-f} N^{-1/2})$, where

$$\sigma_{g-f}^2 = \int_{[0,1]^d} (g(x) - I[g] - f(x) + I[f])^2 \, dx.$$

Clearly, we require that $\sigma_{g-f} \leq \sigma_g$ for the control variates approach to be effective. Typically, the method considers a multiplier λ so that

$$\bar{I}_N^c[g] = \lambda I[f] + \frac{1}{N}\sum_{i=1}^{N}(g(U_i) - \lambda f(U_i)), \tag{3.45}$$

and the integration error depends on

$$\sigma_{g-\lambda f}^2 = \int_{[0,1]^d} (g(x) - I[g] - \lambda f(x) + \lambda I[f])^2 \, dx.$$

The optimal value of λ is the one that minimizes $\sigma_{g-\lambda f}^2$ and is given by

$$\bar{\lambda} = \frac{\int_{[0,1]^d}(g(x) - I[g])(f(x) - I[f])\, dx}{\int_{[0,1]^d}(f(x) - I[f])^2 \, dx}. \tag{3.46}$$

Going back to Example 3.9, we want to estimate $\int_0^1 4\sqrt{1-x^2}\, dx$. We can choose a function $f(x)$ which is easy to integrate and close to $g(x) = 4\sqrt{1-x^2}$. Let us take $f(x) = 4(1-x)$ so that $I[f] = 2$. We get, for the control variates approach:

$$\sigma_{g-f}^2 = \int_0^1 \left(4\sqrt{1-x^2} - \pi - 4(1-x) + 2\right)^2 dx = \frac{68}{3} - \pi^2 - 4\pi \approx 0.231,$$

which is less than 1/3 the value of σ_g^2. For the multiplicative method we get the optimal value $\bar{\lambda} = 3\pi/2 - 4$, which gives $\sigma_{g-\bar{\lambda} f}^2 = 16\pi - 4\pi^2 - 32/3 \approx 0.12$. This value represents a factor of 6 gain with respect to σ_g^2.

■

3.3.3 Analysis of simulation output

When dealing with Monte Carlo methods it is particularly important to have an assessment of how good the Monte Carlo estimate is. The basic problem can be formulated in this way. We observe random variables $X_1, X_2, \ldots$ produced by a Monte Carlo method and we want to estimate $\mu = E(X_i)$ and make an assessment of the error in the form of a certain percentage, say 95% for example, confidence interval. More precisely, we seek $\bar{\mu}$ and $r > 0$ such that

$$P(\bar{\mu} - r \leq \mu \leq \bar{\mu} + r) = 0.95\%. \tag{3.47}$$

We have already seen an example of this type in Section 3.1.1. In the following we will consider only two special cases, which are of particular interest since they often appear when dealing with simulations of kinetic equations. We close this section with some notions concerning post-processing of the simulation output.

The classical case

The simplest case is when the random variables $X_1, X_2, \ldots$ are independent and identically distributed with mean μ and variance σ^2. This applies to most of the situations we have described in the previous paragraphs, like Monte Carlo integration, as well as where each X_i comes from a repeated independent simulation. Since this case has been illustrated in Section 3.1.1, here we just summarize it briefly. We consider the sample mean

$$\bar{X}_N = \frac{1}{N}\sum_{i=1}^{N} X_i.$$

We know that $\bar{X}_N \to \mu$ with probability 1 as $N \to \infty$. Moreover, we have $E(\bar{X}_N) = \mu$ and $Var(\bar{X}_N) = \sigma^2/N$. We have also seen that by the central limit theorem the random variable $(\bar{X}_N - \mu)/\sqrt{\sigma^2/N}$ follows a standard normal distribution. For the sample variance

$$S_N^2 = \frac{1}{N-1}\sum_{i=1}^{N}(X_i - \bar{X}_N)^2,$$

we have $E(S_N^2) = \sigma^2$ and $S_N^2 \to \sigma^2$ with probability 1 as $N \to \infty$. By the central limit theorem the random variable $(\bar{X}_N - \mu)/\sqrt{S_N^2/N}$ is also normally distributed for large N. Thus, we find the desired confidence interval taking $\bar{\mu} = \bar{X}_N$ and $r = z\sqrt{S_N^2/N}$, where z is the number that satisfies $1 - \Phi(z) = \alpha$, with Φ the distribution function of the standard normal distribution given by (3.13) and α is such that $P(|X| \leq \alpha) = 0.95$ when X has a standard normal distribution. From tables, one finds that $\alpha \approx 0.025$ and $z \approx 1.96$.

Asymptotic stationary processes

In some cases, however, the sequence of random variables X_n is asymptotically stationary. In these situations we are interested in the mean with respect to the asymptotic equilibrium distribution:

$$\lim_{n\to\infty} X_n = \mu.$$

Examples are Markov chains starting from an arbitrary initial state. Here, we will assume that X_i has a unique asymptotic stationary distribution with mean μ and variance σ^2. In this case we must choose T large enough so that we can discard $X_1, \ldots, X_T$ and consider the remaining data $X_{T+1}, X_{T+2}, \ldots$ as outputs of a stationary process. Note that the X_{i+T}'s variables are then identically distributed but dependent.

Thus we can define

$$\bar{X}_N^T = \frac{1}{N}\sum_{i=i}^{N} X_{i+T},$$

and we have $E(\bar{X}_N^T) = \mu$ for every N and T large enough. Under suitable assumptions, the sample mean just defined satisfies the ergodic theorem

$$\lim_{N\to\infty} \bar{X}_N^T = \mu, \tag{3.48}$$

and the central limit theorem so that the random variable $(\bar{X}_N^T - \mu)/Var(\bar{X}_N^T)^{1/2}$ follows a standard normal distribution. From this we can compute the desired confidence interval as in the classical case: taking $\bar{\mu} = \bar{X}_N^T$ and $r = z_\alpha Var(\bar{X}_N^T)^{1/2}$. There is, however, a big difference since $Var(\bar{X}_N^T)$ cannot be estimated by

$$S_N^{2,T} = \frac{1}{N-1}\sum_{i=1}^{N}(X_{i+T} - \bar{X}_N^T)^2,$$

since

$$Var(\bar{X}_N^T) = \frac{1}{N^2}\sum_{i,j=1}^{N} Cov(X_{i+T}, X_{j+T}).$$

It is still true that $\lim_{N\to\infty} S_N^{2,T} = \sigma^2$ with probability 1, but in general $Var(\bar{X}_N^T)$ may be significantly larger then σ^2/N.

Let us finally remark that choosing the right T is of paramount importance since failure to discard enough initial data would result in a numerical bias. Usually one uses some empirical test to measure the convergence of X_i towards equilibrium. For example, computing the mean $\mu_j^{(k)}$ of each group of k outputs and continuing until the mean of the last group is close enough to the mean of the penultimate, namely $|\mu_j^{(k)} - \mu_{j-1}^{(k)}| \leq \delta$ for a given tolerance δ. Of course, more sophisticated methods can be constructed similarly.

Reconstruction

In many applications we need to reconstruct the probability density starting from the samples generated by the Monte Carlo method. This post-processing is usually necessary in order to produce simple graphs of the probability density, but also to construct numerical methods based on hybrid approaches where knowledge of the approximate probability density is needed by the deterministic part of our numerical method. For example, in order to evaluate an integral involving a given function which is known only through a set of samples we can first reconstruct the function from its samples and then apply a conventional quadrature formula. Of course, variance is strongly reduced using this approach, but we lose in general the efficiency of the Monte Carlo quadrature. Moreover, reconstruction itself may be quite expensive and thus should be used with caution during the simulation process. In general, the problem is strictly connected to smoothing and filtering of statistical data. Here we will sketch some basic methods which will be useful when dealing with Monte Carlo methods applied to kinetic equations.

The general problem we consider is the following. Given a set of N samples $X_1, X_2, \ldots, X_N$, the corresponding probability density is defined by

$$f_N(x) = \frac{1}{N}\sum_{k=1}^{N}\delta(x - X_k), \qquad x \in \mathbb{R}^d, \tag{3.49}$$

where $\delta(\cdot)$ is the Dirac delta function.

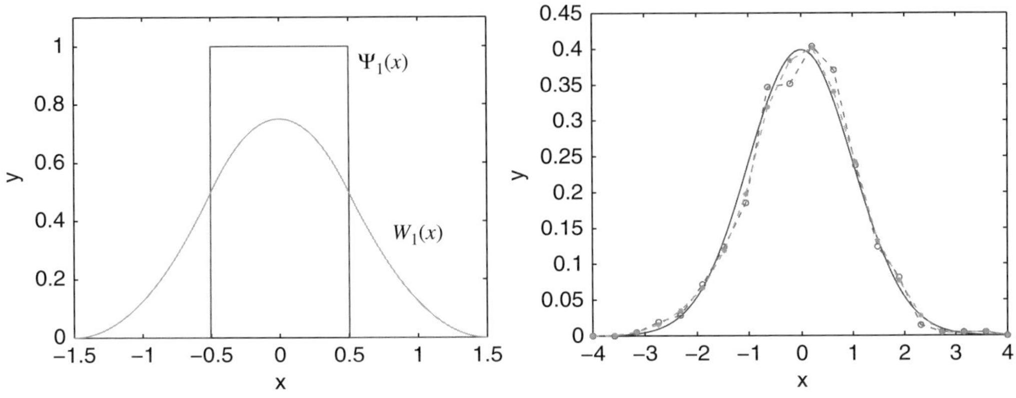

Fig. 3.10 Left: the functions $\Psi_1(x)$ and $W_1(x)$. Right: reconstruction of a standard normal density (–) from $N = 500$ samples using piecewise constant (o) and B-spline functions (∗) with $\Delta x = 0.4$.

In the one-dimensional case, the simplest method, which produces a piecewise constant reconstruction, is based on evaluating the histogram of the samples at the cell centres of a suitable grid $x_j = j\Delta x$,

$$f_N(x_{j+1/2}) = \frac{1}{N}\sum_{k=1}^{N} \Psi_{\Delta x}(x_{j+1/2} - X_k), \quad j = \ldots, -2, -1, 0, 1, 2, \ldots \tag{3.50}$$

where $\Psi_{\Delta x}(x) = 1/\Delta x$ if $|x| \leq \Delta x/2$ and $\Psi_{\Delta x}(x) = 0$ elsewhere. Note that this corresponds to approximating the Dirac delta through the function Ψ.

Smoother versions can be obtained simply by changing the function approximating the Dirac delta. This corresponds to a convolution of the samples with a suitable mollifier. More precisely, we consider

$$f_N(x_{j+1/2}) = \frac{1}{N}\sum_{k=1}^{N} W_{\Delta x}(x_{j+1/2} - X_k), \quad j = \ldots, -2, -1, 0, 1, 2, \ldots \tag{3.51}$$

where $W_{\Delta x} \geq 0$ is a smoothed approximation of the Dirac delta such that

$$\int_{\mathbb{R}} W_{\Delta x}(x)\, dx = 1.$$

For example, we can take $W_{\Delta x}$ as a B-spline. We give here the explicit expression in the case of the quadratic B-spline on $[-1.5, 1.5]$:

$$W_{\Delta x}(x) = \frac{1}{\Delta x} B_2\left(\frac{x}{\Delta x}\right), \quad B_2(x) = \begin{cases} 3/4 - |x|^2 & \text{if } |x| \leq 0.5, \\ (|x| - 3/2)^2/2 & \text{if } 0.5 < |x| \leq 1.5, \\ 0 & \text{otherwise.} \end{cases}$$

In general, the choice of Δx has a strong influence on the results and should be selected as a good compromise between fluctuations and resolution (see Figure 3.10).

Example 3.13 (Negative-valued functions). In some applications we need to construct a Monte Carlo method using functions which contain negative values on the domain of interest. How can we construct samples from these functions?

Given a function $f(x) \in \mathbb{R}$ the general idea is to construct a set of statistical values $X_1, \ldots, X_N$ such that

$$f_N(x) = \frac{1}{N} \sum_{k=1}^{N} m_k \delta(x - X_k), \tag{3.52}$$

where m_k are allowed to have negative values. There are different ways to do that, the simplest one being to introduce the positive functions $f^+(x) = \max\{f(x), 0\}$ and $f^-(x) = -\min\{f(x), 0\}$. If we denote

$$I^+ = \int_{\mathbb{R}} f^+(x)\, dx, \qquad I^- = \int_{\mathbb{R}} f^-(x)\, dx, \quad I = I^+ + I^-,$$

we can generate the sample pairs (X_i, m_i), $m_i \in \{-I, I\}$, where (X_i, I) is generated with probability I^+/I from $f^+(x)/I^+$ and $(X_i, -I)$ with probability I^-/I from $f^-(x)/I^-$. In terms of post-processing, to reconstruct (3.52) we can apply the same smoothing techniques described in the last paragraph. ■

3.4 Applications to evolutionary PDEs

Several partial differential equations (PDEs) can be solved numerically by probabilistic algorithms such as Monte Carlo methods, stochastic particle methods, ergodic algorithms, etc. Usually these probabilistic methods are not competitive in terms of efficiency with direct deterministic discretizations (finite differences/finite volumes/finite elements) of the same differential equations. However, they become efficient compared to deterministic approaches in problems where a large number of dimensions or complex geometries are involved. Moreover, thanks to the natural interpretation of the statistical sample as a physical particle, quite often Monte Carlo methods (or particle methods) allow a better description of the physical properties of the system under study. Here, we will describe different Monte Carlo approaches to evolutionary PDEs which are closely connected to the various kinetic models we consider elsewhere in the book. As mentioned in the introduction, we will try to derive the methods starting directly from the PDE as in classical numerical analysis rather than starting from the idea of performing a statistical simulation of the physics behind the model itself.

Example 3.14 (A simple ODE). We start this section with a simple case where the solution $u(x,t)$ is described by the following ordinary differential equation (ODE)

$$\frac{\partial u}{\partial t} = Cu, \tag{3.53}$$

where $C \in \mathbb{R}$ is a constant. Equation (3.53) represents the action of a linear source term at time t at the point $x \in \mathbb{R}^d$. The equation can be easily solved from the initial data $u(x,0) = u_0(x)$ to give

$$u(x,t) = u_0(x) \exp(Ct). \tag{3.54}$$

Let us assume that $u_0(x)$ is a probability density. When $C > 0$ we have exponential growth of the solution, whereas if $C < 0$ exponential decay follows. For $C = 0$ the solution is stationary.

When developing a probabilistic approach to a differential equation the goal is to find a method which transforms a given set of random variables (samples) distributed as the initial data $u_0(x)$ into a set of random variables (samples) distributed as the solution $u(x,t)$. The main difficulty, in the present case, is represented by the fact that $\int_{\mathbb{R}} u(x,t)\,dx$ is not constant in time.

Given a set of N samples $X_1^0, \ldots, X_N^0$ from the initial data, there are essentially two ways to deal with this problem. The first approach is based on the simple observation that the distribution of the samples does not change in x and so the solution can be reconstructed at all times using its representation

$$u_N(x,t) = \frac{1}{N}\sum_{k=1}^{N} m(t)\delta(x - X_k^0), \qquad E(u_N(x,t)) = u(x,t), \tag{3.55}$$

where the weight $m(t) = \exp(Ct)$ heuristically represents the 'mass' of our sample particles, the X_i's. However, this approach implies the use of samples with time-varying weights which may represent a drawback when the source term in (3.53) appears in combination with other integro-differential operators in a more general PDE problem. In this case we can use a *doubling/removing technique* based on the following first-order approximation in a small time step Δt to (3.54):

$$u^*(x, \Delta t) = u_0(x)(1 + C\Delta t). \tag{3.56}$$

Note that the above approximation corresponds to the explicit Euler scheme applied to (3.53), and that for positive C we have the identity

$$u^*(x, \Delta t) = (1 - C\Delta t)u_0(x) + C\Delta t 2u_0(x). \tag{3.57}$$

Starting from the samples $X_1^0, \ldots, X_N^0$ and a time step Δt such that $\Delta t|C| \le 1$, this originates the Monte Carlo scheme:

1. *Select uniformly an index* $k \in \{1, \ldots, N\}$.
2. *Generate a uniform random sample* U_1.
3. *If* $C > 0$ *do the following*
   - *if* $U_1 < \Delta t C$ *then create a new sample by doubling, namely set* $X_{N+1} = X_k^0$, $N = N + 1$.
4. *If* $C < 0$ *do the following*
   - *if* $U_1 < -\Delta t C$ *then destroy a sample by removing the sample* X_k^0 *and set* $N = N - 1$.
5. *Set* $X_i = X_i^0$ *for the remaining samples.*

Observe that in a time step Δt the algorithm can only create or destroy one sample. The method just described can also be written using expression (3.54) directly

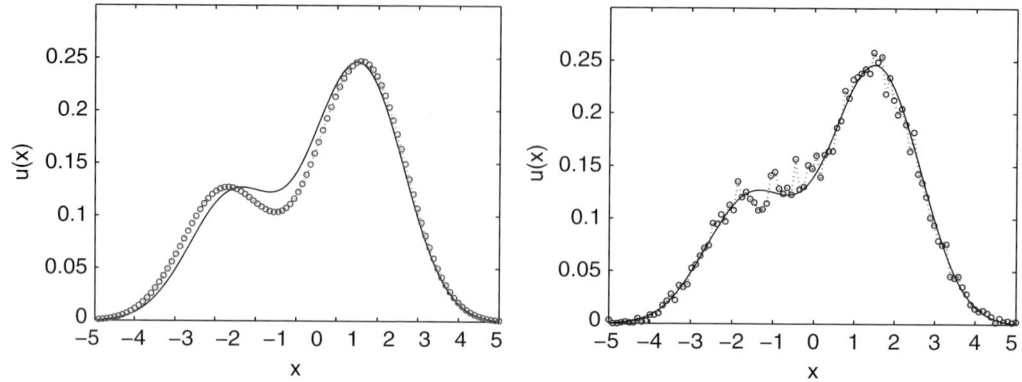

Fig. 3.11 Solution of the heat equation with $D = 1$ at $t = 0.4$ by estimating the expectation of the heat kernel with $M = 100$, $N = 120$ (left) and by the random walk method with $M = 100$, $N = 12 \times 10^3$ and simple piecewise constant reconstruction (right). The initial data is the sum of two square waves $u_0(x) = 0.5$, $x \in [-2.5, -0.5]$, $u_0(x) = 1$, $x \in [0.5, 2.5]$ and $u_0(x) = 0$ elsewhere.

instead of (3.56). However, in view of applications to nonlinear source terms, where the differential equation cannot be solved exactly, we have chosen to present it in the case of its Euler approximation (3.56). ■

3.4.1 Diffusion problems

Diffusion equations are a classical field where Monte Carlo simulations play a relevant role. Let us illustrate the basic principles of the method with the simple linear heat equation:

$$\begin{cases} \dfrac{\partial u}{\partial t} = D\triangle u, \quad x \in \mathbb{R}^d, \quad D, t > 0 \\ \\ u(x,0) = u_0(x). \end{cases} \tag{3.58}$$

The explicit exact solution is given by the integral formula

$$u(x,t) = \int_{\mathbb{R}^d} G_t(x-y)u_0(y)\,dy = (G_t * u_0)(x), \tag{3.59}$$

where $G_t(z)$ is the *heat kernel*

$$G_t(z) = \frac{1}{(4\pi Dt)^{d/2}} e^{-\frac{|z|^2}{4Dt}}. \tag{3.60}$$

If we assume that $u_0(x)$ is a probability density function, by mass conservation the solution $u(x,t)$ is also a probability density for all subsequent times $t > 0$. Let us consider a set of N samples $X_1^0, \ldots, X_N^0 \in \mathbb{R}^d$ from $u_0(x)$.

A first method is based on evaluating the integral solution as an *expected value of the heat kernel* to get

$$u_N(x,t) = \frac{1}{N}\sum_{i=1}^{N} G_t(x - X_i^0), \qquad E(u_N(x,t)) = u(x,t). \tag{3.61}$$

Note, however, that for each point $x \in \mathbb{R}^d$ we have a computational cost of $O(N)$, which implies a cost of $O(MN)$ for a grid of M points in space.

Let us now restrict the problem to the domain $\Omega \subseteq \mathbb{R}^d$, and consider the partition $\Omega = \cup_{i=1}^M \Omega_i$. Using a stratified estimator (cf. Section 3.3.2), we can write

$$u_N^M(x,t) = \sum_{i=1}^{M} \frac{p_i}{N_i} \sum_{j=1}^{N_i} G_t(x - X_j^{(i),0}), \qquad E(u_N^M(x,t)) = u(x,t), \tag{3.62}$$

where

$$p_i = \int_{\Omega_i} u_0(x)\, dx, \qquad i = 1, \ldots, M, \qquad \sum_{i=1}^{M} p_i = 1,$$

and the random variables $X_j^{(i),0}$ are distributed as $u_0(x)/p_i$ for $x \in \Omega_i$. Note that, by choosing $N_i = p_i N$, the stratified estimator improves the accuracy of the solution but not the computational cost. Indeed, at each point x of the physical space the cost is given by $O\left(\sum_{i=1}^M N_i\right) = O\left(N \sum_{i=1}^M p_i\right) = O(N)$.

This computational cost is considerable, much higher in general than the cost of a deterministic approach, which is typically $O(M)$ for a grid of M points.

A more efficient approach is based on directly sampling the solution from the integral representation. This can easily be achieved by observing that, given two independent random variables Z and Y distributed respectively as $G_t(z)$ and $u_0(y)$, the sum $X = Z + Y$ is distributed as $u(x,t) = (G_t * u_0)(x)$.

A set of samples $X_1, \ldots, X_N \in \mathbb{R}^d$ from $u(x,t)$ is obtained as follows:

1. *Generate* $X_1^0, \ldots, X_N^0$ *from* $u_0(x)$.
2. *Generate* $Z_1, \ldots, Z_N$ *from a standard normal probability.*
3. *Set*

$$X_i = X_i^0 + \sqrt{2Dt}\, Z_i, \quad i = 1, \ldots, N. \tag{3.63}$$

Note that this method coincides with the classical *random walk method* [179]. In this case the cost of the generation is $O(N)$, independent of M. By construction, the X_i's are distributed as $u(x,t)$, $\forall i$. It is straightforward to show that

$$E(X_i^2) = E((X_i^0)^2) + 2Dt.$$

Thus, the second moment of the solution exhibits the right linear growth in time.

Lastly, let us note that for a grid-based method the value of M grows with the dimension of the problem. Therefore, for high dimensions the cost of the Monte Carlo procedure may be convenient.

3.4.2 Fractional diffusion

In the case of particle transport, anomalous transport is often characterized by non-Gaussian probability distribution functions of the particles' displacements, and by the non-diffusive scaling of the corresponding statistical moments.

As example, we consider here space fractional diffusion equations of the form

$$\begin{cases} \dfrac{\partial u}{\partial t} = -D(-\triangle)^\alpha u, \quad x \in \mathbb{R}^d, \quad D, t > 0 \\ \\ u(x,0) = u_0(x). \end{cases} \tag{3.64}$$

The fractional Laplacian of order $0 < \alpha \leq 1$ is easily defined in terms of Fourier transform by the formula

$$(-\triangle)^\alpha u = \mathcal{F}^{-1}\left(|\xi|^{2\alpha}\hat{u}(\xi)\right), \tag{3.65}$$

where $\mathcal{F}^{-1}$ stands for the inverse Fourier transform. For $\alpha = 1$ the fractional diffusion equation reduces to the classical heat equation.

In Fourier variables, equation (3.64) has the unique solution

$$\hat{u}(\xi, t) = \hat{u}_0(\xi)\hat{G}_t^\alpha(\xi), \tag{3.66}$$

where

$$\hat{G}_t^\alpha(\xi) = \exp\left(-Dt|\xi|^{2\alpha}\right) \tag{3.67}$$

is the Fourier transform of the *fractional heat kernel* of the linear fractional diffusion (3.64). For each fixed $t > 0$, the function in (3.67) is also the Fourier transform of an *isotropic Lévy stable probability density* function with exponent 2α [145, 234, 280]. In physical space, these probability densities are bell-shaped functions with increasingly heavy tails as α decreases from $\alpha = 1$. If $\alpha = 1$ (linear diffusion), the corresponding fractional heat kernel is given by the Gaussian density (3.60). A relevant case in which the fractional heat kernel can be computed explicitly is obtained for $\alpha = 1/2$. With this exponent the fractional kernel is given by the *Cauchy–Lorentz density.* For other values of α, the corresponding density is not known in closed form in physical space.

For this type of problem we can immediately generalize the random walk algorithm described in the previous section. To do this, it is enough to be able to generate a sample distributed as the Lévy stable probability density. The algorithm then reads as follows:

1. *Generate* $X_1^0, \ldots, X_N^0$ *from* $u_0(x)$.
2. *Generate* $W_1, \ldots, W_N$ *from an isotropic Lévy stable probability density function with exponent* 2α.
3. *Set*

$$X_i = X_i^0 + W_i, \quad i = 1, \ldots, N. \tag{3.68}$$

We omit further details and refer the interested reader to [234, 280] for algorithms for generating Lévy stable densities. Let us finally observe that it is possible to show that

$$E(X_i^2) = E((X_i^0)^2) + (2Dt)^{\frac{1}{\alpha}}.$$

The parameter α influences the time scaling of the solution.

3.4.3 Convection–diffusion equations

First, let us consider the simple linear advection equation

$$\begin{cases} \dfrac{\partial u}{\partial t} + a \cdot \nabla_x u = 0, \quad a, x \in \mathbb{R}^d, \quad t > 0 \\ \\ u(x, 0) = u_0(x). \end{cases} \tag{3.69}$$

The exact solution reads

$$u(x, t) = u_0(x - at). \tag{3.70}$$

Again, we assume that $u_0(x)$ is a probability density function, and consider a set of N samples $X_1^0, \ldots, X_N^0 \in \mathbb{R}^d$ from $u_0(x)$. Using this set of samples corresponds to sampling directly from the exact solution. In fact, $u(x, t)$ is obtained by shifting the initial data of $|at|$ in the a direction. Thus, a set of samples $X_1, \ldots, X_N \in \mathbb{R}^d$ from $u(x, t)$ is obtained as

$$X_i = X_i^0 + at, \quad i = 1, \ldots, N. \tag{3.71}$$

Note that, apart from sampling the initial data, no additional source of randomness is present, and the method corresponds to a standard deterministic *particle transport*.

Linear convection–diffusion problems of the form

$$\begin{cases} \dfrac{\partial u}{\partial t} + a \cdot \nabla_x u = D\Delta u, \quad a, x \in \mathbb{R}^d, \quad D, t > 0 \\ \\ u(x, 0) = u_0(x), \end{cases} \tag{3.72}$$

can be treated similarly by coupling the particle transport with the random walk method. In fact, thanks to linearity, the problem can be solved exactly to give

$$u(x, t) = (G_t * u_0)(x - at). \tag{3.73}$$

Starting with a set of N samples $X_1^0, \ldots, X_N^0 \in \mathbb{R}^d$ from $u_0(x)$, a set of samples $X_1, \ldots, X_N \in \mathbb{R}^d$ from the solution $u(x, t)$ is obtained as

$$X_i = X_i^0 + at + \sqrt{2Dt}\, Z_i, \quad i = 1, \ldots, N, \tag{3.74}$$

where $Z_i \in \mathbb{R}^d$, $i = 1, \ldots, N$, is normally distributed with zero mean and unit variance.

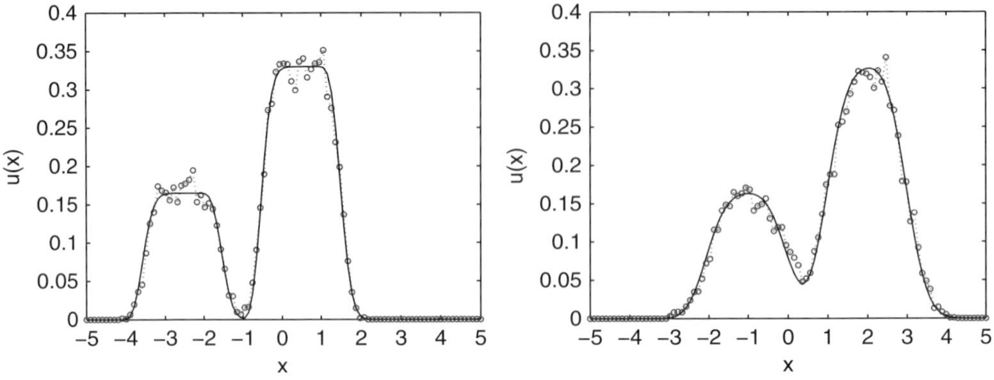

Fig. 3.12 Solution of convection–diffusion at $t = 0.5$ (left) and $t = 2$ (right) with $a = 1$ and $D = 0.05$ using $N = 12000$ samples and $M = 100$ grid points in the reconstruction. The initial data is the sum of two square waves $u_0(x) = 0.5$, $x \in [-4, -2]$, $u_0(x) = 1$, $x \in [-1, 1]$ and $u_0(x) = 0$ elsewhere.

The method just described generalizes naturally to equations of the form

$$\begin{cases} \dfrac{\partial u}{\partial t} + \nabla_x \cdot [a(x)u] = \triangle [D(x)u], \quad a(x), x \in \mathbb{R}^d, \quad D(x), t > 0 \\ \\ u(x, 0) = u_0(x). \end{cases} \tag{3.75}$$

Equation (3.75) is usually known under the name of the *Fokker–Planck equation*, and in the case of constant drift a and diffusion D reduces to the standard linear convection–diffusion case (3.72). In one dimension of the space variables we saw various examples of Fokker–Planck equations in Chapter 1, Section 1.4. For these equations, starting with a set of N samples $X_1^0, \ldots, X_N^0 \in \mathbb{R}^d$ from $u_0(x)$ we generate a set of samples from the solution $u(x, t)$ as

$$X_i = X_i^0 + a(X_i^0)t + \sqrt{2D(X_i^0)t}\, Z_i, \quad i = 1, \ldots, N, \tag{3.76}$$

where $Z_i \in \mathbb{R}^d$, $i = 1, \ldots, N$, is normally distributed with zero mean and unit variance.

Remark 3.3 *An interesting feature of Monte Carlo methods is the fact that, in the time-dependent problems just described, such as diffusion, convection and convection–diffusion, there are no stability constraints for the choice of the time step. Moreover, in contrast with most deterministic schemes, no extra dissipation due to the numerical solution of the transport part is added.*

3.4.4 Random gradient methods

A strategy for improving the variance of Monte Carlo methods for linear PDEs relies on the so-called *random gradient methods* [99, 284]. This family of algorithms is closely related to the *random vortex method* in fluid dynamics [99].

Let us illustrate the idea by treating the one-dimensional nonlinear scalar reaction–diffusion equation

$$\begin{cases} \dfrac{\partial u}{\partial t} = D\dfrac{\partial^2 u}{\partial x^2} + G(u), \quad x \in \mathbb{R}, \quad D, t > 0, \\ \\ u(x,0) = u_0(x). \end{cases} \tag{3.77}$$

Let us suppose that $u_0(x)$ is monotone and bounded from below and above. Without loss of generality, we impose $u(-\infty,t) = 0$ and $u(\infty,t) = 1$, for all $t \geq 0$. We introduce the auxiliary variable $v = \partial u/\partial x$ and we observe that it satisfies the equation

$$\begin{cases} \dfrac{\partial v}{\partial t} = D\dfrac{\partial^2 v}{\partial x^2} + G'(u)v, \quad x \in \mathbb{R}, \quad D, t > 0, \\ \\ v(x,0) = \dfrac{\partial u_0(x)}{\partial x}. \end{cases} \tag{3.78}$$

Note that, by virtue of the properties of $u_0(x)$, $v_0(x)$ is a probability density. Since $u(x,t)$ is also monotone for all subsequent times $t > 0$, by mass conservation it follows that $v(x,t)$ is a probability density for all $t > 0$. Let us first consider the case $G(u) \equiv 0$. We apply the random walk method to compute the evolution of $v(x,t)$. Note that using N samples $X_1, \ldots, X_N$ we will have

$$v_N(x,t) = \frac{1}{N}\sum_{k=1}^{N} \delta(x - X_k), \tag{3.79}$$

and since $u(x,t) = \int_{-\infty}^{x} v(y,t)\, dy$ we get

$$u_N(x,t) = \frac{1}{N}\sum_{k=1}^{N} H(x - X_k), \tag{3.80}$$

where $H(\cdot)$ is the (right-continuous) Heaviside step function

$$H(x) = 0 \text{ if } x < 0, \qquad H(x) = 1 \text{ if } x \geq 0.$$

Note that the boundary condition at the left is exactly satisfied, while the boundary condition at the right is satisfied on average. The computed solution for $u(x,t)$ contains much less fluctuation than $v(x,t)$, since all particles contribute to the solution at any point x (see Figure 3.13).

The same idea can be applied to functions $v(\cdot,t)$ of the form

$$v(x,t) = \frac{1}{N}\sum_{k=1}^{N} m_k \delta(x - X_k), \tag{3.81}$$

where the constant values m_k can also be negative. In this way, we can approximate more than monotone non-negative initial conditions. From definition (3.81) we obtain

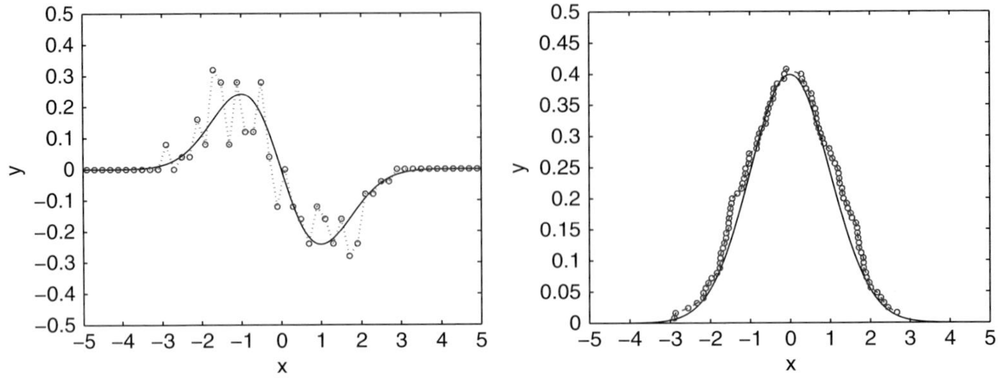

Fig. 3.13 Random gradient representation of the standard normal with zero mean and unit variance. Left: simple histogram representation using 50 cells of 100 samples from its derivative. Right: plot of the function $u_N(x)$ computed using (3.80).

$$u(x,t) = \frac{1}{N}\sum_{k=1}^{N} m_k H(x - X_k). \tag{3.82}$$

In the case where the source term $G(u)$ is present, because of the similarities between this problem and (3.53), we can apply separately the random walk method and the doubling/removal technique. The following algorithm generates the samples X_i from $v(x,t)$ starting from $X_1^0, \ldots, X_N^0$ from $v(x,0)$ and Δt such that $\Delta t C < 1$, $C = \sup_u |G'(u)|$.

1. *Compute* $X_i = X_i^0 + \sqrt{2D(X_i^0)t}\, Z_i$, *where the* Z_i*'s are standard normal samples.*
2. *Compute* $u_i = u_N(X_i)$ *using (3.80).*
3. *At those* X_i *where* $G'(u_i) > 0$ *do the following*
   - *Generate* U_1 *uniform in* $[0,1]$*. If* $U_1 < \Delta t G'(u_i)$ *then create a new sample by doubling* X_i*, set* $N = N + 1$*.*
4. *At those* X_i *where* $G'(u_i) < 0$ *do the following*
   - *Generate* U_1 *uniform in* $[0,1]$*. If* $U_1 < -\Delta t G'(u_i)$ *then destroy the sample by removing* X_i*, set* $N = N - 1$*.*

Remark 3.4 *The random gradient method has an adaptive nature, and samples are created where* $G'(u)$ *is larger; this is typically where sharp fronts appear, and where one needs to spatially resolve jump-like solutions. We note that, from the numerical point of view, the method just described is based on solving separately the two problems characterized by the linear diffusion and by the source term. As we will see in the next section, this approach in general introduces an* $O(\Delta t)$ *error in the computation of the solution.*

3.4.5 Splitting methods

Nonlinear convection–reaction–diffusion problems in the general form

$$\begin{cases} \dfrac{\partial u}{\partial t} + \nabla_x \cdot F(u) = \triangle P(u) + G(u), \quad x \in \mathbb{R}^d, \quad t > 0 \\ \\ u(x,0) = u_0(x), \end{cases} \tag{3.83}$$

where $F(u) \in \mathbb{R}^d$, $P(u), S(u) \in \mathbb{R}$, are usually much more difficult to treat using Monte Carlo methods, and most techniques developed in the literature have been designed to solve specific problems only.

One of the techniques most used is referred to as the *operator splitting method.* Let us introduce it briefly. Given a general differential problem of the type

$$\begin{cases} \dfrac{\partial u}{\partial t} = A(u) + B(u), \\ \\ u(x,0) = u_0(x), \end{cases} \tag{3.84}$$

where $A(U)$ and $B(U)$ are some integro-differential operators, the *splitting method* consists in solving separately and consecutively, in a time interval $[0, \Delta t]$, the two problems S_A and S_B defined by

$$S_A : \begin{cases} \dfrac{\partial u^*}{\partial t} = A(u^*) \\ u^*(x,0) = u(x,0) \end{cases} \qquad S_B : \begin{cases} \dfrac{\partial u^{**}}{\partial t} = B(u^{**}) \\ u^{**}(x,0) = u^*(x,\Delta t). \end{cases} \tag{3.85}$$

This idea has been used, for example, in Chapter 1 to separate the effects of particles transport and interaction processes in the Goldstein–Taylor model discussed in Section 1.3.1.

Let us denote by

$$S_A^{\Delta t}(u_0(x)) = u^*(x,\Delta t), \qquad S_B^{\Delta t}(u^*(x,\Delta t)) = u^{**}(x,\Delta t),$$

the solutions of the two subproblems characterized by A and B respectively, the first relative to the initial value $u_0(x)$, the second relative to $u^*(x,\Delta t)$ as initial value. It is well known that the method produces an approximation $u^{**}(x,\Delta t)$ of order $O(\Delta t)$ of the true solution $u(x,\Delta t)$ in the time interval Δt. This simple splitting can be written in compact notation as $S_A^{\Delta t}(S_B^{\Delta t}(u_0))$. Of course, we can exchange the operators A and B in the definition of the splitting method without affecting the accuracy of the approximation.

A second-order extension is obtained by a symmetric version of the above splitting, given by $S_A^{\Delta t/2}(S_B^{\Delta t}(S_A^{\Delta t/2}(u_0)))$, or equivalently $S_B^{\Delta t/2}(S_A^{\Delta t}(S_B^{\Delta t/2}(u_0)))$ (see [297]). The operator splitting just described can be extended, without additional difficulties, to the case where more than two operators are involved.

Thanks to the splitting, problem (3.83) can be simplified into the study of the two subproblems

$$\frac{\partial u}{\partial t} + \nabla_x \cdot F(u) = 0, \qquad \frac{\partial u}{\partial t} = \triangle P(u), \qquad \frac{\partial u}{\partial t} = G(u), \tag{3.86}$$

namely a conservation law, a nonlinear diffusion and an ordinary differential equation.

We leave to Chapter 4 the description of various Monte Carlo approaches to nonlinear conservation laws and to diffusion equations, by relating them to kinetic formulations.

Problems

Exercise 3.1 Use the simple Monte Carlo method described in Example 3.1.1 to evaluate

$$I_m = \int_0^1 x^m \, dx.$$

Since the exact solution is known to be $1/(m+1)$, evaluate the error and the convergence of the method with the number of points N sampled in $[0,1] \times [0,1]$.

Exercise 3.2 Compute the same integral I_m of the function $g(x) = x^m$ in the interval $[0,1]$, but instead of using uniform random numbers, use power-law distributed random numbers according to the distribution $p(x) = (k+1)x^k$ with $k < m$. Power-law distributed random number can be computed from uniform random numbers by transforming a uniform random sequence U_n in the following way:

$$X_n = U_n^{1/(k+1)}.$$

The importance-sampling estimate of the integral is then obtained as in Section 3.3.2 using the expression

$$I_m = \int_0^1 \frac{g(x)}{p(x)} p(x) \, dx.$$

For a given m, select $k < m$ and see how the estimate for I_m improves with increasing N, and compute the deviation from the exact result. Compare to the results of the previous problem.

Exercise 3.3 Consider the Cauchy probability density

$$f(x) = \frac{1}{\pi(1+x^2)}, \qquad x \in \mathbb{R}. \tag{3.87}$$

Use the inverse transform method described in Section 3.2.2 to generate pseudo-random numbers distributed as (3.87). Reconstruct the original distribution from the samples and compare with its analytical expression. Consider also the following algorithm:

1. *Generate Y_1 and Y_2 distributed as a standard normal.*
2. *Set $X = Y_1/Y_2$.*

It can be shown that X is distributed according to a Cauchy distribution. Can you explain why?

Exercise 3.4 Implement inverse transform and acceptance–rejection algorithms for generating random variables from the following probability densities:

$$(a)\ f(x) = \frac{1}{2}\sin(x), \quad x \in [0,\pi], \qquad (b)\ f(x) = \begin{cases} -x & -1 \le x < 0 \\ x & 0 \le x \le 1. \end{cases}$$

Exercise 3.5 Consider the Poisson distribution with parameter $\lambda > 0$,

$$f(x) = \frac{\lambda^x e^{-x}}{x!}, \qquad x = 0, 1, 2 \dots. \tag{3.88}$$

A Poisson random variable X can be interpreted as the maximal number of independent identically distributed exponential variables whose sum does not exceed 1. More precisely, given Y_n exponentially distributed with parameter λ we have that

$$X = \max \left\{ k \text{ such that } \sum_{n=1}^{k} Y_n \leq 1 \right\}, \tag{3.89}$$

is distributed according to (3.88). Now using Example 3.1 and the properties of logarithms, construct an algorithm to generate a Poisson sample starting from uniformly distributed samples.

Exercise 3.6 Consider the viscous Burgers' equation

$$\frac{\partial u}{\partial t} + u\frac{\partial u}{\partial x} = D\frac{\partial^2 u}{\partial x^2}, \tag{3.90}$$

with initial data $u(x,0) = u_0(x)$.

The Cole–Hopf transformation [178]

$$w(x,t) = \exp\left\{-\frac{1}{2D}\int_0^x u(\xi,t)\,d\xi\right\},$$

yields

$$u = -2D\frac{\partial \log(w)}{\partial x} = -2D\frac{1}{w}\frac{\partial w}{\partial x},$$

and converts the Burgers' equation into the heat equation

$$\frac{\partial w}{\partial t} = D\frac{\partial^2 w}{\partial x^2},$$

with $w(x,0) = \exp\left\{-\frac{1}{2D}\int_0^x u_0(\xi)\,d\xi\right\}$.

Use this transformation to construct a Monte Carlo method for the original Burgers' equation (3.90). Compare also with the relaxation Monte Carlo method described in next chapter, Section 4.2.1, using as initial data a standard normal.

4
Monte Carlo methods for kinetic equations

> We shall present here the motivation and a general description of a method dealing with a class of problems in mathematical physics. The method is, essentially, a statistical approach to the study of differential equations, or more generally, of integro-differential equations that occur in various branches of the natural sciences. ... Such equations are known in the kinetic theory of gases as the Boltzmann equation. In the theory of probability one has a somewhat similar situation described by the Fokker–Planck equation.
>
> N. Metropolis, S. Ulam, The Monte Carlo method, *J. Am. Stat. Ass.*, 1949.

4.1 The general framework

Fig. 4.1 Stanislaw Marcin Ulam[1]

As we have already discussed in Chapter 3, Monte Carlo methods are an effective tool for the numerical solution of kinetic equations. In fact, as we have seen in Chapter 1, the particle dynamics can be constructed in a natural way on a probabilistic rather than a deterministic basis. For its relevance to physical and engineering problems, the development of Monte Carlo tools found its natural application in rarefied gas dynamics, described by the Boltzmann equation. For nonlinear collisional kinetic equations, these studies led to the development of simulation schemes which are closely connected with the classical Direct Simulation Monte Carlo (DSMC) method developed by Bird in the early 1970s [38, 39], or with the variant proposed by Nanbu in the 1980s [243]. The fundamental assumption of the DSMC method is that the molecular movement and collision phases can be decoupled over time periods that are smaller than the mean collision time. What was new in the Nanbu method was that it does not attempt to simulate the deterministic N-body

[1]**Stanislaw Marcin Ulam**: born on 13 April 1909 Lemberg (Austro-Hungarian Empire, now Lviv, Ukraine), died on 13 May 1984 Santa Fe, New Mexico, USA. He made important contributions in many areas of mathematics: number theory, set theory, ergodic theory and algebraic topology. Together with N. Metropolis and J. Von Neumann he is considered the main inventor of Monte Carlo methods. The method was named after the Monte Carlo Casino, where Ulam's uncle often gambled away his money.

dynamics but, rather, offers a probabilistic description of the system starting from the kinetic equation. Both Bird's and Nanbu's methods are nowadays well understood, from both a physical and a mathematical viewpoint, and have been rigorously proved to converge to a solution of the Boltzmann equation provided the number of sample particles is sufficiently large [15, 323]. Several variants of the above approaches have been considered in the literature, in particular to deal with the presence of small time-scales, or to reduce the variance of the computed solution. We refer the interested reader to [85, 250, 275] for some recent reviews on these topics.

To illustrate the basic aspects of a Monte Carlo approximation to a kinetic equation we consider the simple Goldstein–Taylor model (1.20) extensively analyzed in Section 1.3.1. One can derive a Monte Carlo method by sampling directly from the exact solutions of the operator-splitting steps (1.21)–(1.22). Let us recall that in a small time interval Δt the exact solutions $\tilde{f}^{\pm}$ of the free transport (1.21) read

$$\begin{aligned} \tilde{f}^{+}(x,\Delta t) &= f_0^{+}(x - c\Delta t), \\ \tilde{f}^{-}(x,\Delta t) &= f_0^{-}(x + c\Delta t), \end{aligned} \tag{4.1}$$

and, setting $\tilde{\rho}(x,\Delta t) = \tilde{f}^{+}(x,\Delta t) + \tilde{f}^{-}(x,\Delta t)$, the solution of the relaxation step (1.22) yields the approximated values at time Δt:

$$\begin{aligned} f^{+}(x,\Delta t) &= e^{-2\sigma\Delta t}\tilde{f}^{+}(x,\Delta t) + (1 - e^{-2\sigma\Delta t})\frac{1}{2}\tilde{\rho}(x,\Delta t), \\ f^{-}(x,\Delta t) &= e^{-2\sigma\Delta t}\tilde{f}^{-}(x,\Delta t) + (1 - e^{-2\sigma\Delta t})\frac{1}{2}\tilde{\rho}(x,\Delta t). \end{aligned} \tag{4.2}$$

The probabilistic interpretation is readily obtained by introducing, as in Section 1.3.1, the new variables

$$p(x,t) = \frac{f^{+}(x,t)}{\rho(x,t)}, \qquad q(x,t) = \frac{f^{-}(x,t)}{\rho(x,t)}, \tag{4.3}$$

with $p(x,t) + q(x,t) = 1$, to have

$$\begin{aligned} p(x,\Delta t) &= e^{-2\sigma\Delta t}p_0(x - c\Delta t) + \frac{1}{2}(1 - e^{-2\sigma\Delta t}), \\ q(x,\Delta t) &= e^{-2\sigma\Delta t}q_0(x + c\Delta t) + \frac{1}{2}(1 - e^{-2\sigma\Delta t}). \end{aligned} \tag{4.4}$$

We can readily set up a Monte Carlo method in the following way. Given a set of samples $(X_1^0, V_1^0), \ldots, (X_N^0, V_N^0)$, where $V_i \in \{-c, c\}$ characterizes the samples of $f_0^{+}(x)$ from those of $f_0^{-}(x)$, a new set of samples $(X_1, V_1), \ldots, (X_N, V_N)$ is generated by the following algorithm:

1. *Compute* $X_i = X_i^0 + V_i^0 \Delta t, \quad i = 1, \ldots, N.$
2. *For each sample* (X_i, V_i^0):
   (a) *with probability* $1 - e^{-2\sigma\Delta t}$ *do the following:*
      - *with probability* $\frac{1}{2}$ *set* $V_i = c$,
      - *otherwise set* $V_i = -c$.
   (b) *otherwise set* $V_i = V_i^0$.

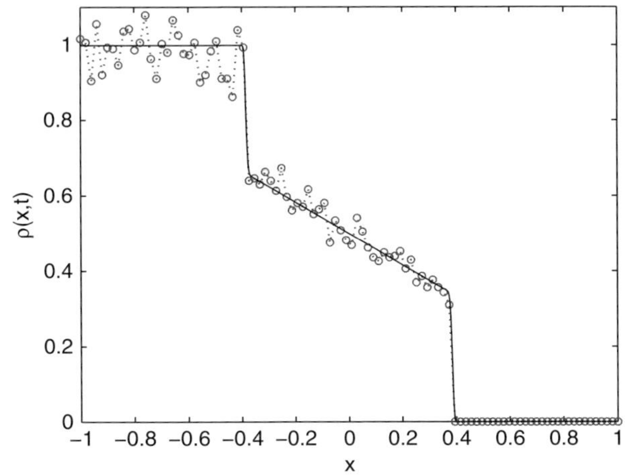

Fig. 4.2 Monte Carlo solution of the Goldstein–Taylor model for $\rho(x,t)$ at time $t = 0.38$ with $\sigma = 1$ and $c = 1$. The figure refers to $N = 15000$ samples and 100 grid points in the reconstruction. The initial data is $\rho_0(x) = 1$ for $x < 0$ and $\rho_0(x) = 0$ for $x > 0$, with $f_0^+(x) = f_0^-(x)$. The continuous line represents the reference solution obtained by finite differences.

This method can be interpreted as a free transport of sample particles (X_i, V_i) that randomly change their velocities V_i in the set $\{-c, c\}$. In particular, since the probability of velocity change is independent of the space position during the relaxation process, the resulting method is fully meshless.

Despite its simplicity this example contains several essential features that can be found in most Monte Carlo approaches to kinetic equations. First, a decoupling of free flow and collisions and, second, a probabilistic reformulation of the collision process. The latter aspect, as we will see, often characterizes the different schemes.

4.2 Relaxation problems

4.2.1 Hyperbolic relaxation systems

Let us now consider the following simple prototype of hyperbolic system with relaxation [183]:

$$\begin{cases} \dfrac{\partial \rho}{\partial t} + \dfrac{\partial j}{\partial x} &= 0, \\[2ex] \dfrac{\partial j}{\partial t} + c^2 \dfrac{\partial \rho}{\partial x} &= -\dfrac{1}{\varepsilon}(j - F(\rho)), \end{cases} \tag{4.5}$$

supplemented with the initial conditions $\rho(x,0) = \rho_0(x)$, $j(x,0) = j_0(x)$.

In the limit $\varepsilon \to 0$ one obtains the local equilibrium $j = F(\rho)$. Under the sub-characteristic condition $c^2 > F'(\rho)^2$ (see [183]), it is possible to show that solutions to (4.5) are well approximated by the scalar conservation law

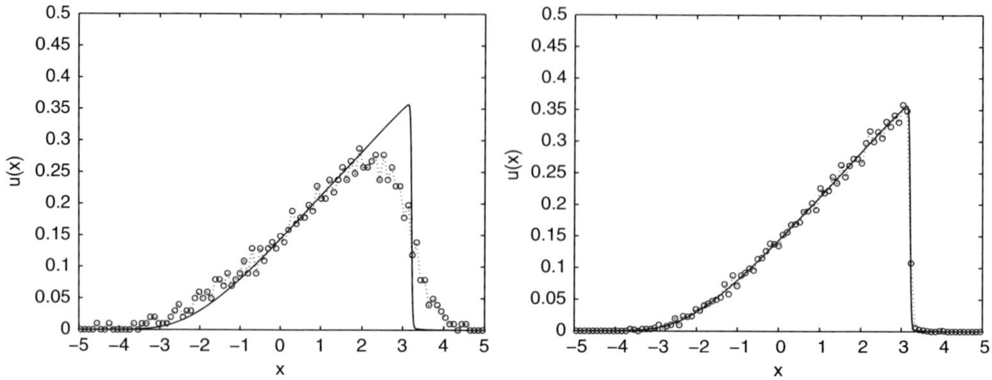

Fig. 4.3 Standard Monte Carlo. Solution at $t = 10$ for $F(\rho) = \rho^2/2$. Left: $N = 1000$ and $\Delta t = 0.1$. Right: $N = 10000$, $\Delta t = 0.01$. The limit case is $\varepsilon = 0$ with $c = 1$. The solution is reconstructed on $M = 100$ grid points. The initial data is a normal with zero mean and unit variance.

$$\frac{\partial \rho}{\partial t} + \frac{\partial F(\rho)}{\partial x} = 0. \tag{4.6}$$

For small values of ε, the study of a numerical approximation solution of system (4.5) is particularly challenging, since a standard explicit discretization will lead to time constraints of the order of ε. Given $c > 0$, let us rewrite system (4.5) introducing the kinetic variables

$$f^+ = \frac{c\rho + j}{2c}, \ f^- = \frac{c\rho - j}{2c}, \ E^+(\rho) = \frac{c\rho + F(\rho)}{2c}, \ E^-(\rho) = \frac{c\rho - F(\rho)}{2c}.$$

In these variables, we obtain the diagonal system

$$\begin{cases} \dfrac{\partial f^+}{\partial t} + c\dfrac{\partial f^+}{\partial x} = -\dfrac{1}{\varepsilon}(f^+ - E^+(\rho)), \\[2ex] \dfrac{\partial f^-}{\partial t} - c\dfrac{\partial f^-}{\partial x} = -\dfrac{1}{\varepsilon}(f^- - E^-(\rho)). \end{cases} \tag{4.7}$$

System (4.7) presents several analogies with the Goldstein–Taylor model (1.20). Note that the frequency parameter σ is equal to $1/\varepsilon$, and relaxation occurs towards a functional expression of the mass density instead of the mass density itself. The kinetic interpretation requires $E^{\pm}(\rho) \geq 0$ and this is guaranteed if $|F(\rho)| \leq c\rho$. Under this assumption one can try to apply the same Monte Carlo approach just described for the Goldstein–Taylor model.

The solution in a small time interval $[0, \Delta t]$ is approximated by means of operator splitting. The only difference lies in the relaxation process, which now reads

$$\begin{cases} \dfrac{\partial f^+}{\partial t} = -\dfrac{1}{\varepsilon}(f^+ - E^+(\rho)), \\[2ex] \dfrac{\partial f^-}{\partial t} = -\dfrac{1}{\varepsilon}(f^- - E^-(\rho)). \end{cases} \tag{4.8}$$

It is straightforward to show that the normalized densities (4.3) at the end of one splitting step provide the probabilistic solution

$$\begin{aligned} p(x,\Delta t) &= e^{-\Delta t/\varepsilon} p_0(x - c\Delta t) + (1 - e^{-\Delta t/\varepsilon}) \frac{E^+(\tilde{\rho}(x,\Delta t))}{\tilde{\rho}(x,\Delta t)}, \\ q(x,\Delta t) &= e^{-\Delta t/\varepsilon} q_0(x + c\Delta t) + (1 - e^{-\Delta t/\varepsilon}) \frac{E^-(\tilde{\rho}(x,\Delta t))}{\tilde{\rho}(x,\Delta t)}. \end{aligned} \tag{4.9}$$

We note that the probabilistic interpretation holds true independently of the choice of $\Delta t/\varepsilon$. In particular, as $\Delta t/\varepsilon \to 0$ it reduces to the simple projections

$$p(x,\Delta t) = \frac{E^+(\tilde{\rho}(x,\Delta t))}{\tilde{\rho}(x,\Delta t)}, \quad q(x,\Delta t) = \frac{E^-(\tilde{\rho}(x,\Delta t))}{\tilde{\rho}(x,\Delta t)}, \tag{4.10}$$

which represent a probabilistic approximation to the limiting equation (4.6). A Monte Carlo method can be derived by sampling directly from the exact solutions of the operator splitting steps (4.10). Note, however, that, differently from the Goldstein–Taylor model, the probability of velocity change now depends on the space position. Therefore, in order to set up a Monte Carlo method and to estimate the probability of a velocity change, we must reconstruct the mass density ρ in a neighbourhood of the particle position. The simplest way is to introduce a grid in space. In this way we obtain the following method.

Given a set of samples $(X_1^0, V_1^0), \ldots, (X_N^0, V_N^0)$, where $V_i \in \{-c, c\}$ characterize the samples of $f_0^+(x)$ from those of $f_0^-(x)$, a new set of samples $(X_1, V_1), \ldots, (X_N, V_N)$ is generated by the algorithm below [253]:

1. *Compute* $X_i = X_i^0 + V_i^0 \Delta t, \quad i = 1, \ldots, N.$
2. *At the cell centres of a space grid* $x_j = j\Delta x$, *compute*

$$\rho_j(t) = \frac{1}{N} \sum_{k=1}^{N} \Phi_{\Delta x}(x_{j+1/2} - X_k), \qquad j = \ldots, -2, -1, 0, 1, 2, \ldots$$

   where $\Phi_{\Delta x}(\cdot)$ *is defined as in (3.50).*
3. *In each space cell* j, *given a sample* (X_i, V_i^0)
   (a) *with probability* $1 - e^{-\Delta t/\varepsilon}$ *do the following:*
      - *with probability* $\dfrac{E^+(\rho_j(t))}{\rho_j(t)}$ *set* $V_i = c$,
      - *otherwise set* $V_i = -c$.
   (b) *otherwise set* $V_i = V_i^0$.

Note that there is no need for the grid to be uniform, and the choice of the grid points can easily change at any time step.

The simple method just described can be improved significantly using a random gradient strategy. In fact, by introducing the new variables $w = \partial\rho/\partial x$ and $z = \partial j/\partial x$, we can rewrite (4.5) in the form

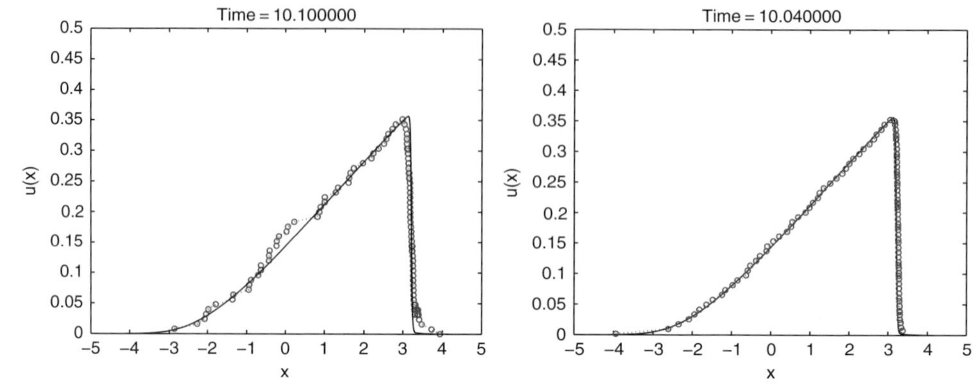

Fig. 4.4 Random gradient method. Solution at $t = 10$ for $F(\rho) = \rho^2/2$. Left: $N = 100$ and $\Delta t = 0.1$. Right: $N = 500$, $\Delta t = 0.05$. The limit case is $\varepsilon = 0$ with $c = 1$. The initial datum is a normal density with zero mean and unit variance.

$$\begin{cases} \dfrac{\partial w}{\partial t} + \dfrac{\partial z}{\partial x} = 0, \\ \dfrac{\partial z}{\partial t} + c^2 \dfrac{\partial w}{\partial x} = -\dfrac{1}{\varepsilon}(z - F'(\rho)w). \end{cases} \tag{4.11}$$

Most of the arguments just presented for system (4.5) apply again, with a substantial difference: there is no need to reconstruct the solution $\rho(x,t)$ on a grid. In fact, in terms of the diagonal variables $g^+ = (cw + z)/(2c)$ and $g^- = (cw - z)/(2c)$, we have the equilibrium states

$$D^+(\rho, w) = \frac{w(c + F'(\rho))}{2c}, \quad D^-(\rho, w) = \frac{w(c - F'(\rho))}{2c}.$$

Since $D^+(\rho, w) + D^-(\rho, w) = w$, the probabilities of a random velocity change $D^+(\rho, w)/w = (c + F'(\rho))/(2c)$ and $D^-(\rho, w)/w = (c - F'(\rho))/(2c)$ depend only on ρ. Note that, under the subcharacteristic condition $c > |F'(\rho)|$, both probabilities have the same sign of w.

Starting with a set of samples $(X_1^0, V_1^0), \ldots, (X_N^0, V_N^0)$, where $V_i \in \{-c, c\}$ separates the samples of $g_0^+(x)$ from those of $g_0^-(x)$, a new set of samples $(X_1, V_1), \ldots, (X_N, V_N)$ is generated as follows:

1. *Compute* $X_i = X_i^0 + V_i^0 \Delta t, \quad i = 1, \ldots, N$.
2. *For each sample* X_i, *evaluate* $\rho_i = \rho_N(X_i)$ *using (3.80).*
3. *For each sample pair* (X_i, V_i^0)
   (a) *with probability* $1 - e^{-\Delta t/\varepsilon}$ *do the following:*
   - *with probability* $\dfrac{c + F'(\rho_i)}{2c}$ *set* $V_i = c$,
   - *otherwise set* $V_i = -c$.

   (b) *otherwise set* $V_i = V_i^0$.

By comparing Figures 4.3 and 4.4 the remarkable improvement in variance reduction obtained by the use of the random gradient method is evident.

Remark 4.1 *As a side result of our derivation, we constructed Monte Carlo methods for a general scalar conservation law of the form (4.6). As for the construction of deterministic relaxation schemes for conservation laws [183], this is easily obtained by taking the limit case $\epsilon = 0$ in the schemes just described. Other approximations for scalar conservation laws, based on Monte Carlo, are found in the literature for rather specific situations like the Burgers' equation (see [53] for details).*

4.2.2 Kinetic relaxation systems

The direct simulation schemes presented in the previous section can easily be extended to the case of kinetic equations of relaxation type:

$$\begin{cases} \dfrac{\partial f}{\partial t} + v \cdot \nabla_x f = \dfrac{\nu}{\varepsilon}(f_\infty - f), & v, x \in \mathbb{R}^d, \quad t > 0 \\ \\ f(x, v, 0) = f_0(x, v), \end{cases} \tag{4.12}$$

where $f = f(x, v, t) \geq 0$ is the number density of particles having velocity v in position x at time t and f_∞ is the local equilibrium.

In Section 1.3.2 we discussed the case of radiative transfer, which corresponds to the local equilibrium

$$f_\infty(x, v, t) = \frac{1}{\mathfrak{m}(\Omega)} \int_\Omega f(x, w, t)\, dw, \qquad \nu = \mathfrak{m}(\Omega)\sigma, \tag{4.13}$$

with $\Omega \subset \mathbb{R}^d$. A second relevant example is given by the Bhatnagar, Gross and Krook (BGK) [16] operator of rarefied gas dynamics, for which

$$f_\infty(x, v, t) = M(x, v, t), \tag{4.14}$$

and M is the local Maxwellian equilibrium (1.2), namely a multivariate normal density with the same first three moments of f,

$$\int_{\mathbb{R}^d} f(x, v, t)\phi(v)\, dv = \int_{\mathbb{R}^d} M(x, v, t)\phi(v)\, dv, \quad \phi(v) = 1, v, |v|^2.$$

Here, $\varepsilon > 0$ is a relaxation time. In particular, for the BGK model ε is referred to as the Knudsen number. As $\varepsilon \to 0$ we have $f = M$, so that the solution depends only on the evolution of the first three moments of f. In fact, by integrating (4.12) for $f_\infty = M$ with respect to 1, v and $|v|^2$ we have

$$\frac{\partial}{\partial t} \int_{\mathbb{R}^d} f\phi(v)\, dv + \nabla_x \cdot \left(\int_{\mathbb{R}^d} v f \phi(v)\, dv \right) = 0, \quad \phi(v) = 1, v, |v|^2. \tag{4.15}$$

These equations describe the balance of mass, momentum and energy. The system is not closed since it involves higher-order moments of the distribution function f. However, for small values of ε the particle density f approaches the local Maxwellian and the higher-order moments of the distribution function can be computed as functions of ρ, U and T. Therefore, setting $f = M$ in (4.15) we obtain the closed system of *compressible Euler equations*

$$\begin{aligned} &\frac{\partial \rho}{\partial t} + \nabla_x \cdot (\rho U) = 0 \\ &\frac{\partial \rho U}{\partial t} + \nabla_x \cdot (\rho U \otimes u + p) = 0 \\ &\frac{\partial E}{\partial t} + \nabla_x \cdot (EU + pU) = 0 \end{aligned} \tag{4.16}$$

$$p = \rho T, \quad E = \frac{3}{2}\rho T + \frac{1}{2}\rho U^2, \tag{4.17}$$

where p is the gas pressure.

To approximate system (4.12) we again consider splitting the transport and relaxation steps. Hence, we solve in a small time step Δt the free transport

$$\frac{\partial f}{\partial t} + v \cdot \nabla_x f = 0,$$

and the homogeneous relaxation

$$\frac{\partial f}{\partial t} = \frac{\nu}{\varepsilon}(f_\infty - f).$$

Note that again the relaxation step can be solved exactly since f_∞ does not change during the relaxation process. Thus, we obtain the approximation

$$\begin{aligned} \tilde{f}(x, v, \Delta t) &= f_0(x - v\Delta t), \\ f(x, v, \Delta t) &= e^{-\nu \Delta t/\varepsilon} \tilde{f}(x, v, \Delta t) + (1 - e^{-\nu \Delta t/\varepsilon})\tilde{f}_\infty(x, v, \Delta t). \end{aligned}$$

The functions

$$g(x, v, t) = \frac{f(x, v, t)}{\rho(x, t)}, \quad g_\infty(x, v, t) = \frac{f_\infty(x, v, t)}{\rho(x, t)}, \tag{4.18}$$

are probability densities. In fact, we have

$$\int_{\mathbb{R}^d} g(x, v, t)\, dv = 1, \quad \int_{\mathbb{R}^d} g_\infty(x, v, t)\, dv = 1.$$

In this way we obtain the probabilistic description

$$g(x, v, \Delta t) = e^{-\nu \Delta t/\varepsilon} g_0(x - v\Delta t) + (1 - e^{-\nu \Delta t/\varepsilon}) g_\infty(x - v\Delta t). \tag{4.19}$$

Note that for the radiative transfer

$$g_\infty(v) = 1/\mathfrak{m}(\Omega), \tag{4.20}$$

where $v \in \Omega$ is the uniform distribution over the set Ω and, as for the Goldstein–Taylor model, there is no dependence on the space position. Differently, in the BGK case

$$g_\infty(v) = \frac{1}{(2\pi T)^{d/2}} \exp\left(-\frac{|v - U|^2}{2T}\right) \tag{4.21}$$

is a normal density with mean and variance depending on the mean velocity $U = U(x, t)$ and temperature $T = T(x, t)$. Of course, if $\nu = \nu(x)$, space dependence cannot be avoided. A Monte Carlo method is then obtained by direct sampling of (4.19). The resulting algorithm for a constant ν is summarized here. Given a set of samples $(X_1^0, V_1^0), \dots, (X_N^0, V_N^0)$, where $X_i, V_i \in \mathbb{R}^d$, a new set of samples $(X_1, V_1), \dots, (X_N, V_N)$ is obtained as follows:

1. *First compute* $X_i = X_i^0 + V_i^0 \Delta t, \quad i = 1, \dots, N$.
2. *For the radiative transfer:*
   (a) *with probability* $1 - e^{-\nu \Delta t/\varepsilon}$ *replace* V_i^0 *with a uniform velocity sample on* Ω,
   (b) *otherwise set* $V_i = V_i^0$.
3. *For the BGK operator:*
   (a) *In cell* j *of a* d*-dimensional grid reconstruct the desired macroscopic quantities, mean velocity* U_j *and temperature* T_j.
   (b) *In each space cell* j *given a sample* (X_i, V_i^0)
      i. *with probability* $1 - e^{-\nu \Delta t/\varepsilon}$ *replace* V_i^0 *with a normal velocity sample with mean* U_j *and variance* T_j *(for example, using the Box–Muller algorithm),*
      ii. *otherwise set* $V_i = V_i^0$.

Remark 4.2 *Note that approximation (4.19) is well defined independently of* $\nu \Delta t/\varepsilon$ *and in the limit of small relaxation times yields the projection*

$$g(x, v, t) = g_\infty(x - v\Delta t), \tag{4.22}$$

which, in the case of the BGK model, corresponds to a so-called kinetic solver for the Euler equations [106, 123]. Therefore, in this latter case, setting $\varepsilon = 0$ *in the Monte Carlo algorithm just described we obtain a Monte Carlo method for the system of Euler equations (4.17).*

4.3 Binary interaction models

For the sake of simplicity we introduce the algorithms in the case of the collision model introduced in Section 1.4. Extensions to the other models presented in Chapter 1 are also discussed later in the section.

It is convenient to start from the bilinear model expressed in Dirac delta form:

$$\frac{\partial f(v,t)}{\partial t} = \frac{1}{2} \int_{\mathbb{R}\times\mathbb{R}} f(v_1, t) f(v_2, t) \Big(\delta_0(v - v_1^*) + \delta_0(v - v_2^*) \Big) dv_1 \, dv_2 - f(v, t), \tag{4.23}$$

where $p_i, q_i > 0$, $i = 1, 2$, are given constants, and

$$v_1^* = p_1 v_1 + q_1 v_2, \quad v_2^* = p_2 v_1 + q_2 v_2. \tag{4.24}$$

Let us first observe that performing the integrations against the Dirac delta we obtain several equivalent forms for the interaction integral. By a single change of variables we get:

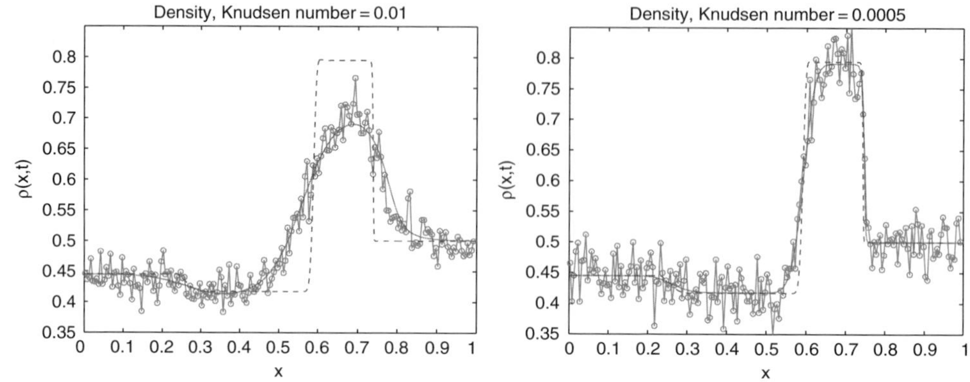

Fig. 4.5 Solution of the BGK model at $t = 0.05$ for the Lax shock tube problem ($\rho = 0.445$, $U_x = 0.698$, $U_y = U_z = 0$, $p = 3.528$ for $x < 0.5$ and $\rho = 0.5$, $U_x = U_y = U_z = 0$, $p = 0.571$ for $x > 0.5$) and Maxwellian initial data using 200 space cells and 500 particles per cell with $\Delta t = 0.1$. Left: density for $\varepsilon = 0.01$. Right: density for $\varepsilon = 0.0005$. The dashed line is the corresponding solution of the Euler equation (4.16).

$$
\begin{aligned}
\int_{\mathbb{R}\times\mathbb{R}} f(v_1)f(v_2)\delta_0(v-v_1^*)\,dv_1\,dv_2 &= \frac{1}{q_1}\int_{\mathbb{R}\times\mathbb{R}} f(v_1)f\left(\frac{v_1^*-p_1v_1}{q_1}\right)\delta_0(v-v_1^*)\,dv_1\,dv_1^* \\
&= \frac{1}{q_1}\int_{\mathbb{R}} f(v_1)f\left(\frac{v-p_1v_1}{q_1}\right)dv_1, \\
&= \frac{1}{p_1}\int_{\mathbb{R}} f\left(\frac{v-q_1v_2}{p_1}\right)f(v_2)\,dv_2.
\end{aligned}
\tag{4.25}
$$

To keep the notation as simple as possible, we have omitted the time dependence. Similarly,

$$
\begin{aligned}
\int_{\mathbb{R}\times\mathbb{R}} f(v_1)f(v_2)\delta_0(v-v_2^*)\,dv_1\,dv_2 &= \frac{1}{q_2}\int_{\mathbb{R}} f(v_1)f\left(\frac{v-p_2v_1}{q_2}\right)dv_1, \\
&= \frac{1}{p_2}\int_{\mathbb{R}} f\left(\frac{v-q_2v_2}{p_2}\right)f(v_2)\,dv_2.
\end{aligned}
\tag{4.26}
$$

Note that, under the assumption that $f(v,t)$ is a probability density and v_1 and v_2 are independent random variables, the first term (4.25) is nothing other than the probability density function of the random variable v_1^*, whereas the second term (4.26) is the probability density function of the random variable v_2^* (cf. equation (3.25) in Example 3.8). As we will see, these representations are particularly useful in deriving a Monte Carlo simulation of the kinetic equation.

Remark 4.3 *Performing the double change of variables $(v_1, v_2) \to (v_1^*, v_2^*)$ when integrating against the Dirac delta we get*

$$\begin{aligned}
&\int_{\mathbb{R}\times\mathbb{R}} f(v_1)f(v_2)\delta_0(v-v_1^*)\,dv_1\,dv_2 \\
&= \frac{1}{|J|}\int_{\mathbb{R}\times\mathbb{R}} f\left(\frac{q_2v_1^*-q_1v_2^*}{J}\right) f\left(\frac{p_1v_2^*-p_2v_1^*}{J}\right)\delta_0(v-v_1^*)\,dv_1^*\,dv_2^* \qquad (4.27)\\
&= \frac{1}{|J|}\int_{\mathbb{R}} f\left(\frac{q_2v-q_1v_2^*}{J}\right) f\left(\frac{p_1v_2^*-p_2v}{J}\right) dv_2^*,
\end{aligned}$$

where $J = p_1q_2 - p_2q_1$ is the Jacobian of the transformation. Similarly, we have, for the second term on the right-hand side of (4.23),

$$\begin{aligned}
&\int_{\mathbb{R}\times\mathbb{R}} f(v_1)f(v_2)\delta_0(v-v_2^*)\,dv_1\,dv_2 \\
&= \frac{1}{|J|}\int_{\mathbb{R}} f\left(\frac{q_2v_1^*-q_1v}{J}\right) f\left(\frac{p_1v-p_2v_1^*}{J}\right) dv_1^*.
\end{aligned} \qquad (4.28)$$

Expressions (4.27)–(4.28) yield the classical Boltzmann-like quadratic representation of the interaction integral in (4.23).

4.3.1 Monte Carlo algorithms

Starting from the initial data $f(v, 0) = f_0(v)$, with $f_0(v)$ a probability density, we can rewrite the kinetic model (4.23) as

$$\frac{\partial f(v,t)}{\partial t} = Q^+(f,f)(v,t) - f(v,t). \qquad (4.29)$$

In (4.29), $Q^+(f, f)(v, t)$ denotes the probability density function

$$Q^+(f,f)(v,t) = \frac{1}{2}(Q_1^+(f,f)(v) + Q_2^+(f,f)(v)). \qquad (4.30)$$

Owing to (4.25) and (4.26), we defined

$$Q_1^+(f,f)(v) = \frac{1}{q_1}\int_{\mathbb{R}} f(v_1) f\left(\frac{v-p_1v_1}{q_1}\right) dv_1, \qquad (4.31)$$

$$Q_2^+(f,f)(v) = \frac{1}{p_2}\int_{\mathbb{R}} f\left(\frac{v-q_2v_2}{p_2}\right) f(v_2)\,dv_2, \qquad (4.32)$$

the probability densities of v_1^* and v_2^* respectively.

Let us consider a discretization of the kinetic equation in a small time interval Δt. Using the forward Euler scheme we get

$$f(v,\Delta t) = (1-\Delta t)\,f_0(v) + \Delta t Q^+(f_0,f_0)(v) \qquad (4.33)$$

$$= (1-\Delta t)\,f_0(v) + \frac{1}{2}\Delta t Q_1^+(f_0,f_0)(v) + \frac{1}{2}\Delta t Q_2^+(f_0,f_0)(v). \qquad (4.34)$$

Under the restriction $\Delta t \le 1$, the coefficient of $f_0(v)$ is non-negative. Consequently, $f(v, \Delta t)$, a convex combination of probability densities, is a probability density. The probabilistic interpretation is straightforward and we can readily derive the following algorithm:

> 1. *Given N samples V_i^0, $i = 1, \dots, N$, from the initial distribution $f_0(v)$.*
> 2. *For each sample V_i^0*
>   (a) *with probability Δt select an index j uniformly among $1, \dots, N$;*
>   i. *with probability $1/2$ set $V_i = p_1 V_i^0 + q_1 V_j^0$.*
>   ii. *otherwise set $V_i = p_2 V_i^0 + q_2 V_j^0$.*
>   (b) *otherwise set $V_i = V_i^0$.*

The method just described is the analogue of Nanbu's method for the direct simulation of the collision step in the full Boltzmann equation [243]. Let us consider the case of symmetric interactions, that is $p_1 = q_2 = p$ and $p_2 = q_1 = q$. In this particular case $Q_1^+(f, f) = Q_2^+(f, f)$, and the method simplifies since there is no need to distinguish between sampling from $Q_1^+(f, f)$ and $Q_2^+(f, f)$. From Section 1.4 we know that the first two moments of the solution change according to

$$m(t) = \int_{\mathbb{R}} f(v, t) v \, dv = m_0 e^{(p+q-1)t}, \qquad E(t) = \int_{\mathbb{R}} f(v, t) v^2 \, dv = E_0 e^{(p^2+q^2-1)t}.$$

The two cases $p + q = 1$ and/or $p^2 + q^2 = 1$ are distinguished from the others. In these cases, indeed, the kinetic equation preserves exactly some moments of the solution in time. In such situations it may be desirable to design a Monte Carlo procedure in such a way that the conservation properties are also preserved exactly at the level of the statistical samples. This can be achieved by formulating the method in a symmetric form, by changing the velocity samples by pairs instead of individually. In fact, the conservation properties are a direct consequence of the microscopic binary interactions. A Monte Carlo procedure with the right conservation properties is described below:

> 1. *Given N samples V_i^0, $i = 1, \dots, N$, from the initial distribution $f_0(v)$.*
> 2. *Set $N_c = [\![\Delta t N/2]\!]_S$.*
> 3. *Select N_c sample pairs (V_i^0, V_j^0) uniformly without repetition among all possible pairs;*
>   (a) *Set $V_i = pV_i^0 + qV_j^0$ and $V_j = qV_i^0 + pV_j^0$.*
> 4. *Set $V_h = V_h^0$ for all the remaining $N - N_c$ samples.*

Here, $[\![\cdot]\!]_S$ denotes the stochastic rounding defined in (3.38). In classical rarefied gas dynamics this represents the analogue of the modified Nanbu's method proposed by Babovsky [14]. Note that this symmetrized approach introduces some correlation between the random samples due to the fact that we impose sampling without repetition in order to allow at most one update for each sample as in the original unsymmetric formulation of the Monte Carlo scheme.

4.3.2 Higher-order Monte Carlo methods

The kinetic model (4.29) can be written in the equivalent form

$$\partial(f(v,t)e^t)\partial t = Q^+(f,f)(v,t)e^t. \tag{4.35}$$

Taking the integral on both sides over the time interval $[0,t]$ leads to the integral representation of the solution:

$$\begin{aligned} f(v,t) &= f_0(v)e^{-t} + \int_0^t e^{-(t-s)}Q^+(f,f)(v,s)\,ds \\ &= f_0(v)e^{-t} + (1-e^{-t})\int_0^t \lambda_t(s)Q^+(f,f)(v,s)\,ds, \end{aligned} \tag{4.36}$$

where, for each $s \in (0,t)$, $\lambda_t(s) = e^{-(t-s)}/(1-e^{-t})$ is an exponential density over the interval $[0,t]$.

This means in particular that the kinetic equation describes particles which independently undergo binary interactions at time intervals that are exponentially distributed. This is the rationale behind the well-known Bird's method for direct simulation of the Boltzmann equation in rarefied gas dynamics [38, 39]. Even though the method cannot be derived directly as a sampling approach to the kinetic equation, for the sake of completeness we will briefly describe it here.

For N individuals the method is based on the introduction of a local random time counter, exponentially distributed. Given a random variable U, uniformly distributed in $[0,1]$, this random time can be computed as

$$\Delta t_c(U) = -\frac{\ln(U)}{N}. \tag{4.37}$$

Alternatively, one can consider a constant time counter Δt_c corresponding to the average time between interactions. In fact, since $N_c = N\Delta t$ is the number of average interactions in the time interval, in a time interval $[0,\Delta t]$ we have

$$\Delta t_c = \frac{\Delta t}{N_c} = \frac{1}{N}. \tag{4.38}$$

Of course, taking time averages, the two formulations (4.37) and (4.38) are equivalent.

Using the time counter (4.38) in the time interval $\Delta t = n_t \Delta t_c$, we can describe the Monte Carlo algorithm in the case of symmetric interaction as follows:

1. *Given N samples V_i^0, $i = 1,\ldots,N$, from the initial distribution $f_0(v)$.*
2. *Set $\Delta t_c = 1/N$, $V_i = V_i^0$, $i = 1,\ldots,N$.*
3. *For $k = 0,\ldots,n_t$*
   (a) *Select a random pair (V_i, V_j) uniformly from all possible pairs.*
   (b) *Set $V_i = pV_i + qV_j$ and $V_j = qV_i + pV_j$.*

Note that in the above formulation the method has much in common with the approach described in the previous section, except for the fact that multiple interactions are also allowed during the dynamics. In the limit of large numbers of samples, the method is expected to converge towards the time-continuous kinetic equation (4.29), and not to its time-discrete counterpart (4.34). This can be easily understood by observing that the local time stepping vanishes as the number of samples goes to infinity. Because of the presence of local time stepping, the method is also referred to as *time counter* direct simulation Monte Carlo, in contrast to the *no time counter* method described in the last section. We refer to [323] for detailed convergence results for the approach in the case of the full Boltzmann equation.

Recursive Monte Carlo methods

We consider here another approach, which still converges to the continuous model as the number of samples increases. In contrast to that of Bird, this method can be derived directly from the kinetic equation. Using the Wild representation introduced in Section 2.2.1, we can express the solution to (4.29) as an exponential series in the form

$$f(v,t) = f_0(v)e^{-t} + \sum_{k=1}^{\infty} e^{-t}\left(1-e^{-t}\right)^k f_{(k)}(v)\,, \tag{4.39}$$

where the functions $f_{(k)}$ are defined recursively as

$$f_{(k+1)} = \frac{1}{k+1}\sum_{i=0}^{k} Q^+(f_{(i)}, f_{(k-i)}). \tag{4.40}$$

Note that $f_{(0)} = f_0$, $f_{(1)} = Q^+(f_0, f_0)$ and, being f_0 a probability density, all terms $f_{(k)}$ are also probability densities since

$$\int_{\mathbb{R}} f_{(k)}(v)\,dv = \int_{\mathbb{R}} f_0(v)\,dv.$$

Moreover, since

$$\sum_{k=0}^{\infty} e^{-t}\left(1-e^{-t}\right)^k = 1, \qquad e^{-t}\left(1-e^{-t}\right)^k > 0, \quad \forall\, k,$$

the solution $f(v,t)$ is expressed as a convex combination of probability densities. This sum has a very clear probabilistic interpretation. Taking a sample at random from the velocity distribution of particles at time t, $f(v,t)$, it might happen that it has not interacted with another sample one single time. The probability of this event is e^{-t}, and the distribution in v, given this, is $f_0(v)$. The first term in the sum corresponds to those samples that have been involved in exactly one interaction. The probability that a sample has such a history is $e^{-t}(1-e^{-t})$, and its distribution in v is exactly $f_{(1)} = Q^+(f_0, f_0)$. In the same way, $f_{(k)}$ is the conditional distribution given that exactly $k+1$ samples have been involved in its interaction history back to the initial time. The number of samples involved, say k, is geometrically distributed. This probabilistic description of the solution is due to McKean [232].

It is instructive to describe this by *interaction trees*, which are sometimes called McKean graphs. Each of these trees describes a possible interaction history.

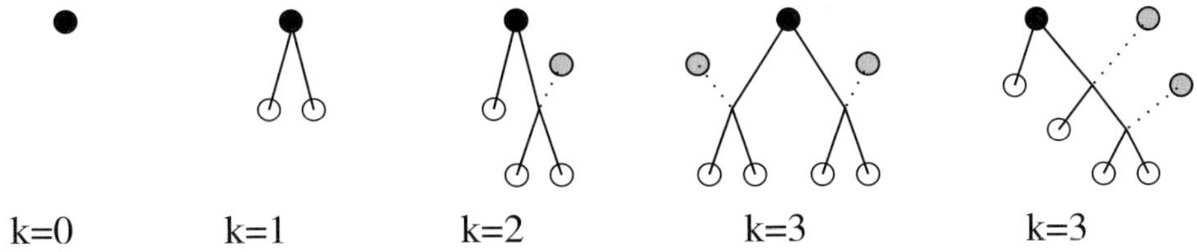

Fig. 4.6 McKean graphs for $k = 0$, $k = 1$, $k = 2$ and $k = 3$.

Figure 4.6 shows some of these graphs, where k ranges from 0 to 3. Note that, by reflections in each node, there would be two different graphs for $k = 2$ and six for $k = 3$; in general, the density $f_{(k)}$ corresponds to $k!$ trees with $k+1$ leaves. In each tree, a leaf corresponds to a particle from the initial distribution, and a node corresponds to an interaction.

In reality there are two particles entering into an interaction, and there are also two particles leaving it, but in the McKean graph only one of them is kept; the grey balls represent particles which are returned to the big bath of particles, where they are forgotten.

We first describe a simple recursive algorithm to generate a random sample V from $f(v, t)$; the method was derived in [255] and is essentially equivalent to the one given in [139].

1. *Choose k from a geometric distribution with parameter $1 - e^{-t}$ (see Example 3.2).*
2. *Take a sample V from the distribution with density $f_{(k)}$. So if $k = 0$, take V^0 from the initial density f_0 and set $V = V^0$. Otherwise do the following:*
   (a) *Choose $i \in \{0, 1, 2, 3, \ldots, k-1\}$ with equal probability.*
   (b) *Choose V^i from the density $f_{(i)}$, and V^{k-i-1} from the density f_{k-i-1}.*
   (c) *With probability 1/2, set $V^k = p_1 V^i + q_1 V^{k-i-1}$, otherwise set $V^k = p_2 V^i + q_2 V^{k-i-1}$.*
   (d) *Then V^k is a random variable distributed according to the density $f_{(k)}$. Set $V = V^k$.*

Of course, the samples from $f_{(i)}$ and $f_{(k-i-1)}$ are computed recursively using steps (a)–(d). This process always terminates after at most k steps. The algorithm is easy to program in a recursive way. Note, however, that if the kinetic model preserves some moments then the method is not able to exactly reproduce this feature. For this to be true, the method must be symmetrized, i.e. both velocities that result from an interaction must be kept.

A conservative algorithm

In [255, 256], a Monte Carlo method that attempt to solve the previous drawbacks in the case of the Boltzmann equation has been derived. The main idea is to symmetrize

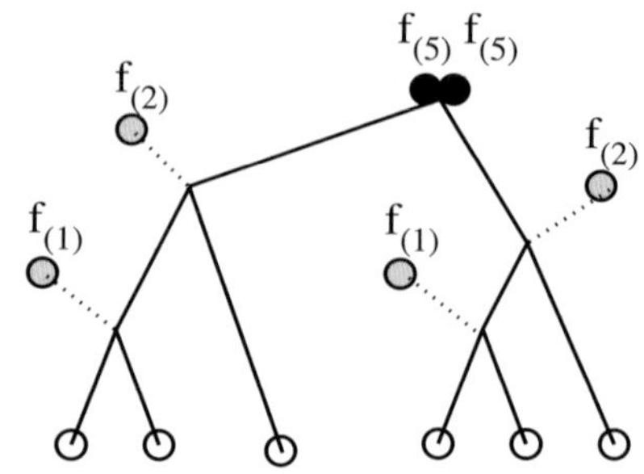

Fig. 4.7 Generation of samples in the algorithm.

the Wild sum calculation and to keep all particles used in the simulation.

Given a set of N particles distributed according to f_0, then at time t one can split the particles into sets $N_k \approx Ne^{-t}\left(1-e^{-t}\right)^k$, $k = 0, 1, 2, \ldots$, of particles sampled from $f_{(k)}$. Note that, since the number of particles N is finite and N_k must be an integer, in practice there exists a maximum index m_{max} such that $N_k \neq 0$, $k \leq m_{max}$. We can use the following procedure based on stochastic rounding:

1. *For* $k \geq 0$, *compute*

$$N_k = [\![Ne^{-t}\left(1-e^{-t}\right)^k]\!]_S \tag{4.41}$$

and stop for $k = m_{max}$, *where* m_{max} *is such that* $N_p = \sum_{k=0}^{m_{max}-1} N_k \leq N$ *and* $N_p + N_{m_{max}} > N$.

2. *Set* $N_{m_{max}} = N - N_p$.

A schematic description of the algorithm follows. In the beginning of the interaction step, time t, there are N particles distributed according to f_0. After the recursive iteration, at time t, we expect a certain number of particles distributed according to the initial distribution f_0, a certain number of particles distributed according to $f_{(1)}$, and so on. For each of these, the final number of particles is given by N_k.

The very first step in this calculation is to compute a first pair of particles (velocities) distributed according to $f_{m_{max}}$. This involves the interaction of $m_{max}+1$ particles, and hence this number of particles is drawn at random from the initial distribution. Obviously, this sets a limit on the number of terms of the Wild sum that can be estimated with a finite number of particles at the initial time.

All $m_{max}+1$ particles must be kept in order that the conserved quantities remain exact. In the particular example shown in Figure 4.7, a pair of $f_{(5)}$-particles are generated, and, on the way to doing this, two $f_{(1)}$-particles and two $f_{(2)}$-particles are also generated. We also store these particles and increment the relative c_1 and c_2 counters by two.

It is clear that if this procedure is continued until the appropriate number $N_{m_{max}}$ of $f_{(m_{max})}$-particles are generated, one risks finding that there are not enough particles at the initial step, and also that the other $f_{(k)}$'s are not properly distributed. A way out of this problem is the following. An interaction tree, such as the one shown in Figure 4.7, is evaluated from top to bottom, just as described. The difference here is that each time a particle from a given distribution is needed for an interaction, one first checks whether such a particle has already been stored. In the example above, the upper interaction involves two $f_{(2)}$-particles, and if one or two particles already exist, then those are taken as interaction partners and the corresponding counter c_2 decreased. If there are more than needed, then there is a random choice, and if there are none, or not enough, then the algorithm is called recursively, just as before.

This is repeated until all particles with distribution $f_{(1)}$ up to $f_{(m_{max})}$ are generated.

Starting with N particles distributed according to f_0, the method is conservative and can be summarized in the following algorithm:

1. *For $k = 0, \ldots, m_{max}$, compute N_k as in (4.41).*
2. *Set counters $c_0 = N$ and $c_k = 0$ for $k = 1, \ldots, m_{max}$.*
3. *For $k = m_{max}, \ldots, 0$ take N_k samples from the distribution with density $f_{(k)}$. So when $k = 0$, take the remaining c_0 samples from the initial density f_0. Otherwise, for $k \geq 1$ proceed as follows to generate two particles from $f_{(k)}$:*
   (a) *Choose $i \in \{0, 1, 2, 3, \ldots, k-1\}$ with equal probability.*
   (b) *Choose V^i from the density $f_{(i)}$, and V^{k-i-1} from the density f_{k-i-1}. This is done in the following way:*
      i. *If $c_i > 0$, use a previously stored particle with a random choice and decrease the c_i counter by one. Otherwise, take a sample from $f_{(i)}$ recursively; this will produce two particles distributed as $f_{(i)}$, so one is stored and the c_i counter increased by one.*
      ii. *If $c_{k-i-1} > 0$, use a previously stored particle with a random choice and decrease the c_{k-i-1} counter by one. Otherwise, take a sample from $f_{(k-i-1)}$ recursively; this will produce two particles distributed as $f_{(k-i-1)}$, so one is stored and the c_{k-i-1} counter increased by one.*
   (c) *Set $V_1^k = pV^i + qV^{k-i-1}$ and $V_2^k = qV^i + pV^{k-i-1}$.*
   (d) *Then V_1^k and V_2^k are random variables distributed according to the density $f_{(k)}$.*

Note that since the conservative collision rule produces particles by pairs, the number of generated particles at each interaction level k may differ from the desired number N_k by one particle.

4.4 Asymptotic preserving Monte Carlo

For some particular choices of p and q in the binary interaction terms, we have seen in Section 1.4 that we can obtain some information on the asymptotic behaviour of the system. In these cases one can look to take advantage of the analytic knowledge of

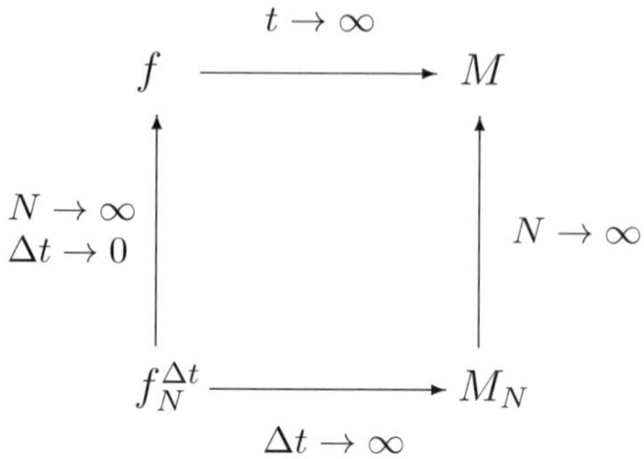

Fig. 4.8 The asymptotic preserving Monte Carlo diagram. Here, f is the solution of the kinetic equation, $f_N^{\Delta t}$ the corresponding Monte Carlo approximation characterized by Δt and by the use of N samples, M is the Maxwellian equilibrium and M_N its approximation with N samples.

the asymptotic behaviour of the model to construct more efficient numerical approximations. This idea is the basis of the so-called *asymptotic preserving* Monte Carlo methods for the Boltzmann equation constructed in [65, 127, 249, 255, 256].

Here we will consider the case of the Kac equation introduced in Section 1.4.3. In the Kac model the binary interaction is described by the random mixing parameters given by [186]:

$$p_1 = q_2 = \cos\theta, \quad p_2 = -q_1 = \sin\theta, \tag{4.42}$$

where θ is a random variable uniformly distributed on the interval $[0, 2\pi]$.

The model can be written in the form of equation (4.29), with

$$Q^+(f,f)(v,t) = \int_0^{2\pi} \frac{d\theta}{2\pi} \int_{\mathbb{R}} f(v\cos\theta + w\sin\theta, t) f(w\cos\theta - v\sin\theta, t)\, dw. \tag{4.43}$$

Expression (4.43) corresponds to the form (4.28), since here the Jacobian $J = \cos^2(\theta) + \sin^2(\theta) = 1$. Note that expressions (4.43) and (1.113) coincide, since the latter is obtained from the former by the change of variable $\theta \to -\theta$ into the integral.

The Monte Carlo algorithms described in Section 4.3 are generalized by generating a uniform angle on $[0, 2\pi]$ before computing the interaction. For example, for the Nanbu-like conservative scheme of Section 4.3.1 we replace point (a) as follows:

(a) *Generate* Θ *uniformly in* $[0, 2\pi]$. *Set* $V_i = \cos(\Theta)V_i^0 - \sin(\Theta)V_j^0$ *and* $V_j = \sin(\Theta)V_i^0 + \cos(\Theta)V_j^0$.

In an analogous way, for the conservative recursive algorithm described in Section 4.3.2, we modify point (d):

(d) *Generate* Θ *uniformly in* $[0, 2\pi]$. *Set* $V_1^k = \cos(\Theta)V^i - \sin(\Theta)V^{k-i-1}$ *and* $V_2^k = \sin(\Theta)V^i + \cos(\Theta)V^{k-i-1}$.

By direct inspection, one can verify that the one-dimensional Maxwellian density

$$M(v) = \rho \frac{1}{\sqrt{2\pi T}} e^{-\frac{|v-U|^2}{2T}} \tag{4.44}$$

is a stationary solution to the Kac equation. If we assume $f_0(v)$ is a probability density, then $\rho = 1$ and $M(v)$ is a normal distribution with mean U and variance T.

The physical idea behind the asymptotic preserving methods for the Kac model is that Boltzmann's H-functional

$$H(f)(t) = \int_{\mathbb{R}} f(v,t) \log f(v,t)\, dv \tag{4.45}$$

is monotonically decreasing in time and reaches its minimum when $f = M$. The time monotonicity of the H-functional can be fruitfully employed to prove that in the Wild sum expansion (4.39) the coefficients $f_{(k)} \to M$ as $k \to \infty$ (see [69]).

It was first suggested in [155] that this could be used to construct a class of efficient numerical schemes by replacing the higher-order terms in the Wild sum (4.39) with the corresponding Maxwellian equilibrium. These approximations over a time interval Δt are referred to as *time relaxed Monte Carlo* and can be written as

$$f(v,\Delta t) = \sum_{k=0}^{m} e^{-\Delta t}(1-e^{-\Delta t})^k f_{(k)}(v) + (1-e^{-\Delta t})^{m+1} M(v), \tag{4.46}$$

where $m \geq 1$ is an integer and the functions $f_{(k)}$ are defined by (4.40). It is easy to show that expression (4.46) is m-th order approximation of the true solution, and that it satisfies the asymptotic preserving property $f(v,\Delta t) \to M(v)$ as $\Delta t \to \infty$. Moreover, since

$$\sum_{k=0}^{m} e^{-\Delta t}(1-e^{-\Delta t})^k + (1-e^{-\Delta t})^{m+1} = 1,$$

conservation of mass and energy are guaranteed, as well as the fact that $f(v,\Delta t)$ is a convex combination of probability densities.

The above expression can be used to derive Monte Carlo methods both directly [248, 249] (see Exercise 4.5) and recursively [255]. For a given m, the recursive algorithm can be modified easily by taking into account that at most m interactions in a given time t are allowed. Of course, if $m \geq m_{max}$, where m_{max} has been computed from (4.41), the algorithm remains unchanged, since all possible admissible interactions are taken into account. On the other hand, when $m < m_{max}$ we stop the computation of (4.41) when $k = m+1$ and set $N_{m+1} = N - N_p$, where $N_p = \sum_{k=0}^{m} N_k$. Therefore, the algorithm just described is modified simply by generating N_{m+1} samples from a normal distribution at the end. Note that this choice avoids the computation of the binary interactions in the case of large times, and thus improves the overall efficiency of the algorithm. The main features of the resulting scheme are illustrated in the asymptotic preserving diagram in Figure 4.8.

We will present below some numerical results based on the use of the recursive asymptotic preserving Monte Carlo algorithm. Since exact solutions are known for the Kac equation, we can easily check the accuracy of the schemes.

An exact solution for the Kac equation has been found by Ernst [140]. This solution is in the form

$$f(v,t) = \frac{1}{2C^{3/2}}\left[\frac{3}{2}(C-1) + (3-C)\frac{v^2}{C^2}\right]\exp\left(-v^2/C\right),$$

with $C(t) = 3 - 2\exp(-\sqrt{\pi}t/16)$.

In order to compute the relative error, the density function f has been reconstructed using (3.51). All the computations have been performed with a single time step $\Delta t = t$.

Table 4.1 shows the behaviour of the relative L_2 error for t ranging from 0 to 15, where the Maxwellian equilibrium state is essentially reached, using different values for m. It is remarkable how uniform accuracy in time is essentially obtained for $m = 1000$ (for $t = 10$ the maximum value is $m_{max} = 13832$), but also the values 100 and 25 give reasonable approximations with a slight deterioration of accuracy for intermediate times. All three different truncations provide the same result for $t = 15$, since all particles are sampled from a Maxwellian.

Table 4.1 The Kac equation: relative L_2 error norm in time. The simulations are performed for $t \in [0, 15]$ by starting with $N = 5 \times 10^4$ particles.

m	1000	100	25	
$t = 0$	0.010390	0.010390	0.010390	
$t = 1$	0.007169	0.007169	0.007169	$m_{max} = 20$
$t = 2$	0.005421	0.005421	0.006588	$m_{max} = 52$
$t = 3$	0.006266	0.007841	0.010110	$m_{max} = 118$
$t = 5$	0.005971	0.006217	0.021790	$m_{max} = 550$
$t = 7$	0.008950	0.014841	0.019492	$m_{max} > 1000$
$t = 10$	0.007481	0.010583	0.010218	$m_{max} > 1000$
$t = 15$	0.006202	0.006202	0.006202	$m_{max} > 1000$

Remark 4.4 *In practical simulations the number m_{max} can be very large, depending on the time step and on the number of test particles. Clearly, small values of m make the algorithm faster, because the collision process is replaced by the projection to the local Maxwellian equilibrium, but far from the fluid regime keeping m too small can produce less accurate results. The main problem is to choose the right m, in order to have the best combination of efficiency and accuracy. In [256], an adaptive technique has been developed to select the maximum depths of the interaction trees, based on evaluating the distance of the solution from the equilibrium through a suitable indicator. This can be performed by measuring the variation of some macroscopic variables such as the fourth-order moment or the components of the shear stress tensor.*

4.5 Kinetic approximation of diffusion equations

In Chapter 3 we introduced several techniques to approximate linear diffusion problems. Nonlinear problems are usually much more difficult to treat using Monte Carlo methods, and most techniques developed in the literature have been designed for rather specific problems. Here, we first consider the Rosenau approximation for linear diffusion and then, following [251], we will illustrate a related approach based on the use of a suitable kinetic formulation of some nonlinear diffusion equations.

4.5.1 Linear diffusion

We recall that in Section 1.3.3 the linear diffusion

$$\frac{\partial g(v,t)}{\partial t} = \sigma^2 \frac{\partial^2 g(v,t)}{\partial v^2}, \quad v \in \mathbb{R},\, t > 0 \tag{4.47}$$

was approximated through the convolution-based linear kinetic equation

$$\frac{\partial f(v,t)}{\partial t} = \frac{\sigma^2}{\varepsilon^2}\left[(\mathcal{M}_{\varepsilon\sigma} * f)(v,t) - f(v,t)\right]. \tag{4.48}$$

In equation (4.48) ε is a small parameter and $\mathcal{M}_\gamma$ is given by

$$\mathcal{M}_\gamma(v) = \frac{1}{2\gamma} e^{-|v|/\gamma}. \tag{4.49}$$

We have shown in Section 1.3.3 that as $\varepsilon \to 0$ the solution $f(v,t)$ converges towards $g(v,t)$. This representation provides a further possibility for solving a linear diffusion problem with Monte Carlo methods. To this end, let us denote with $f_0(v) = g_0(v)$ the initial data, and introduce the discrete time approximation to (4.48):

$$f(v,\Delta t) = f_0(v)\left(1 - \frac{\sigma^2 \Delta t}{\varepsilon^2}\right) + \frac{\sigma^2 \Delta t}{\varepsilon^2}(\mathcal{M}_{\varepsilon\sigma} * f_0)(v). \tag{4.50}$$

If we assume that $f_0(v)$ is a probability density, for $\Delta t \le \varepsilon^2/\sigma^2$ the above expression is a convex combination of probability densities, and can be easily cast into a Monte Carlo game:

1. *Let $V_1^0, \ldots, V_N^0$ be a set of samples from $f_0(v)$.*
2. *For each sample V_i^0*
   (a) *with probability $\sigma^2\Delta t/\varepsilon^2$, replace the sample V_i^0 with a new sample V_i from the probability density $(\mathcal{M}_{\varepsilon\sigma} * f_0)(v)$. This can be done as in Example 3.8 by generating a sample W_i from the exponential density (4.49) (see Example 3.5, equation (3.15)) and setting $V_i = V_i^0 + W_i$,*
   (b) *otherwise, set $V_i = V_i^0$.*

Remark 4.5 *Since our goal is to approximate the diffusion equation (4.47) for a given Δt, we can select $\varepsilon = \sqrt{\sigma^2 \Delta t}$, namely the minimum allowed by the scheme. With this* *choice, the resulting approximation is similar to the standard random walk method introduced in Section 3.4.1, except that now the random shifts follow an exponential law instead of a normal law. Thus,*

$$V_i = V_i^0 + \sqrt{\sigma^2 \Delta t}\, Z_i, \quad i = 1, \ldots, N,$$

where the Z_i's are generated from (3.15).

In particular, as observed in Section 1.3.2, the kinetic approximation (4.48) is converging as $\varepsilon \to 0$ to (4.47) for any choice of distribution of the background, provided that some moment of order greater than two remains bounded. Taking as background a normal density, and $\varepsilon = \sqrt{\sigma^2 \Delta t}$, we recover exactly the classical random walk method with the update

$$V_i = V_i^0 + \sqrt{2\sigma^2 \Delta t}\, Z_i, \quad i = 1, \ldots, N,$$

where the Z_i's are generated from a standard normal density.

4.5.2 Nonlinear diffusion problems

This section illustrates the flexibility of Monte Carlo methods, and the possible applications to nonlinear problems. We introduce the fourth-order nonlinear degenerate diffusion equation

$$\frac{\partial g(v,t)}{\partial t} = -\frac{\partial}{\partial v}\left(g(v,t) \frac{\partial^3 g(v,t)}{\partial v^3} \right), \quad v \in \mathbb{R},\, t > 0, \tag{4.51}$$

with

$$g(v, t = 0) = g_0(v). \tag{4.52}$$

Equation (4.51) models the surface-tension-dominated motion of thin viscous films and spreading droplets [241], and it is also known as the Hele-Shaw model.

Let $M_\varepsilon(x)$ be the centred normal density of mass one and second moment equal to $\varepsilon^2 > 0$, that is

$$M_\varepsilon(v) = \frac{1}{(2\pi)^{1/2}\varepsilon} \exp\left(-\frac{v^2}{2\varepsilon^2} \right). \tag{4.53}$$

For all $t \geq 0$, let $f(v,t)$ be the solution to

$$\frac{\partial f(v,t)}{\partial t} = -\frac{\partial}{\partial v}\left(f \frac{\partial^3 f}{\partial v^3} * M_\varepsilon \right)(v,t), \quad x \in \mathbb{R},\, t > 0, \tag{4.54}$$

corresponding to the initial value $f_0(v) = g_0(v)$. As usual, $*$ denotes convolution. Since the equation preserves the total mass, without loss of generality we can fix $\int_{\mathbb{R}} f(v,t)\, dv = 1$.

Equation (4.54) was introduced in [251], where it was shown that it represents a consistent approximation to equation (4.51) as $\varepsilon \to 0$. Hence, in the following we will concentrate on the simulation of this equation.

Let us set

$$A_\varepsilon(v) = -\frac{\partial^3}{\partial v^3}\left(M_\varepsilon(v)\right).$$

Using elementary properties of the convolution operator and integrating by parts shows that equation (4.54) can be written in the equivalent form

$$\frac{\partial f}{\partial t} = Q_\varepsilon(f) = \frac{\partial}{\partial v}\left(f(v)\int_{\mathbb{R}} A_\varepsilon(w) f(v-w)\,dw\right). \tag{4.55}$$

Equation (4.55) has the same structure as some dissipative equations studied recently in connection with the cooling of granular gases [307].

Having this analogy in mind, we consider a suitable approximation of (4.55), which has the structure of a Boltzmann-like equation. Let us mention that a similar idea has been applied recently by Bobylev and Nambu in [51] to the study of the Landau–Fokker–Planck equation (cf. also [64]).

We approximate the derivative in (4.55) as

$$\frac{\partial z(v)}{\partial v} = \frac{z(v+\delta) - z(v)}{\delta},$$

where the *small* parameter δ is given by

$$\delta = \begin{cases} h\dfrac{A_\varepsilon}{|A_\varepsilon|^\alpha}, & A_\varepsilon \neq 0 \\ 0, & A_\varepsilon = 0. \end{cases} \tag{4.56}$$

In (4.56), $0 \leq \alpha \leq 1$, and the smallness of δ is guaranteed by the presence of the small parameter $h > 0$. Note that δ has the same sign as A_ε.

Substituting the derivative by its approximation in (4.55) we obtain the Boltzmann-like approximation to (4.51):

$$\frac{\partial f}{\partial t} = Q_\varepsilon^{(\alpha)}(f) = \frac{1}{h}\int_{\mathbb{R}} |A_\varepsilon(w)|^\alpha \left(f(v')f(v'-w) - f(v)f(v-w)\right) dw, \tag{4.57}$$

where the collisional velocity v' is given by

$$v' = v + h\frac{A_\varepsilon(w)}{|A_\varepsilon(w)|^\alpha}. \tag{4.58}$$

In what follows, we will consider the two limit cases $\alpha = 0$ and $\alpha = 1$. Note that for $\alpha = 0$ equation (4.57) has a constant rate function, which is typical of the Boltzmann equation for Maxwell molecules, and therefore the numerical solution can be computed with a small change of the algorithms described in the previous sections. Following standard arguments of kinetic theory it is easy to show that the collisional kinetic models (4.57) satisfy conservation of mass and momentum as well as positivity of the solution.

We recall here the following consistency result derived in [251]. For $m \geq 1$, let $C_0^m(\mathbb{R})$ be the set of m-times compactly supported continuously differentiable functions, endowed with its natural norm $\|\cdot\|_m$. Given two probability density functions $f(x)$ and $g(x)$, $x \in \mathbb{R}$, let

$$\|f - g\|_m^* = \sup\left\{ \left| \int \varphi(x)\,(f(x) - g(x))\,dx \right|, \varphi \in C_0^m, \|\varphi\|_m \leq 1 \right\}. \tag{4.59}$$

The above formula defines a distance which metrizes the weak-* topology on $P_s(\mathbb{R})$, $s > 0$, namely the class of all probability distributions F on $\mathbb{R}$, such that

$$\int_{\mathbb{R}} |x|^s \, dF(x) < \infty.$$

This distance has been extensively used in [310], where precise connections with other, better-known, metrics have been established. The following result was proven in [251]:

Theorem 4.1 *We are given a non-negative, initial condition $f_0 \in L^1(\mathbb{R}) \cap H^1(\mathbb{R})$ with unit mass and $v^2 f_0 \in L^1(\mathbb{R})$. Then there exists a strong solution $f(t, v)$ to the Cauchy problem* (4.51) *such that, for all $T > 0$, the distance $\|f f_{vvv}(\cdot, T) - f f_{vvv} * M_\varepsilon(\cdot, T)\|_3^*$ tends to zero as $\varepsilon \to 0$, and the following bound holds:*

$$\|f f_{vvv}(\cdot, T) - f f_{vvv} * M_\varepsilon(\cdot, T)\|_3^* \leq \varepsilon C(f_0, T). \tag{4.60}$$

The following theorem [251] gives similar estimates for the distance between the operator defined by (4.55) and the corresponding Boltzmann-like collision operators given in (4.57).

Theorem 4.2 *We are given a non-negative, initial condition $u_0 \in L^1(\mathbb{R}) \cap H^1(\mathbb{R})$ with unit mass. Then there exists a strong solution $f(t, v)$ to the Cauchy problem* (4.51) *such that, for all $T > 0$, the distance $\|\varepsilon^{7-4\alpha} Q_\varepsilon(f(\cdot, T)) - \varepsilon^{7-4\alpha} Q_\varepsilon^{(\alpha)}(f(\cdot, T))\|_2^*$ tends to zero as $h \to 0$, and the following bound holds:*

$$\|\varepsilon^{7-4\alpha} Q_\varepsilon(f(\cdot, T)) - \varepsilon^{7-4\alpha} Q_\varepsilon^{(\alpha)}(f(\cdot, T))\|_2^* \leq \frac{1}{2} h \int_{\mathbb{R}} |A_1(v)|^{2-\alpha} \, dv \int_{\mathbb{R}} f_0^2 \, dv. \tag{4.61}$$

4.5.3 A Monte Carlo algorithm

The kinetic approximation just introduced can be used directly for the construction of a Monte Carlo method. The kinetic form can be cast again into the general framework of (4.29) now taking

$$Q^+(f, f) = \frac{1}{h} \int_{\mathbb{R}} |A_\varepsilon(w)|^\alpha f(v') f(v' - w) \, dw. \tag{4.62}$$

Let us observe that

$$\begin{aligned}
&\int_{\mathbb{R}} |A_\varepsilon(w)|^\alpha f(v')f(v'-w)\,dw \\
&\quad = \int_{\mathbb{R}} |A_\varepsilon(w)|^\alpha f\left(v_* + h\frac{A_\varepsilon(w)}{|A_\varepsilon(w)|^\alpha}\right) f\left(v_* - w + h\frac{A_\varepsilon(w)}{|A_\varepsilon(w)|^\alpha}\right) \delta_0(v_* - v) dw dv_* \\
&\quad = \int_{\mathbb{R}} |A_\varepsilon(v'-w')|^\alpha f(v')\, f(w')\, \delta_0(v_* - v) dv'\, dw',
\end{aligned} \tag{4.63}$$

with

$$v_* = v' - h\frac{A_\varepsilon(v'-w')}{|A_\varepsilon(v'-w')|^\alpha}. \tag{4.64}$$

In (4.63), we used the fact that the Jacobian of the transformation $(v_*, w) \to (v', w')$ is one, and $w = v' - w'$. For $\alpha = 0$ the interaction process simplifies, and given v' and w' distributed as f, the above integrals define the probability density of the random variable $v_* = v' - hA_\varepsilon(v' - w')$. This can be used to adapt the Monte Carlo algorithms described in Section 4.3.1 to the present case. Let us consider Nanbu's scheme. Assuming $\Delta t \leq h$, the standard method can be easily modified by changing point (a) as follows:

(a) *with probability $\Delta t/h$ select an index j uniformly among $1, \ldots, N$; set $V_i = V_i^0 - hA_\varepsilon(V_i^0 - V_j^0)$.*

Alternatively, in its conservative variant, by setting $N_c = [\![\Delta t N/(2h)]\!]_S$ and using pairwise interactions, by modifying point (a) as

(a) *Set $V_i = V_i^0 - hA_\varepsilon(V_i^0 - V_j^0)$ and $V_j = V_j^0 + hA_\varepsilon(V_i^0 - V_j^0)$.*

Let us now consider the case $\alpha = 1$. In order to simulate the dynamic with a Monte Carlo method we must rely on the acceptance–rejection approach in this case, in fact, sampling from the interaction integral is not straightforward. However, let us observe that if we compute a constant $A_\alpha > 0$ such that $|A_\varepsilon(w)|^\alpha \leq A_\alpha$, $\forall\, w \in \mathbb{R}$, we have

$$\int_{\mathbb{R}} \frac{|A_\varepsilon(w)|^\alpha}{A_\alpha} f(v')f(v'-w)\,dw \leq \int_{\mathbb{R}} f(v')f(v'-w)\,dw,$$

and thus we can adapt the algorithm for $\alpha = 0$ to sample from the proposal density on the right-hand side. Here, however, we use a slightly different formulation of the acceptance–rejection strategy, since the standard version described in Chapter 3 would require the evaluation of the proposal density (which now corresponds to an integral) at each time step. Instead, given two initial samples (V_i^0, V_j^0) we accept the interaction sample $V_i = V_i^0 - hA_\varepsilon(V_i^0 - V_j^0)/|A_\varepsilon(V_i^0 - V_j^0)|^\alpha$ with probability given by the ratio $|A_\varepsilon(V_i^0 - V_j^0)|^\alpha / A_\alpha$. The resulting algorithm has much in common with standard direct simulation methods for rarefied gas dynamics in the case of hard spheres [38, 39, 243]. In the conservative case it can be summarized in the following way:

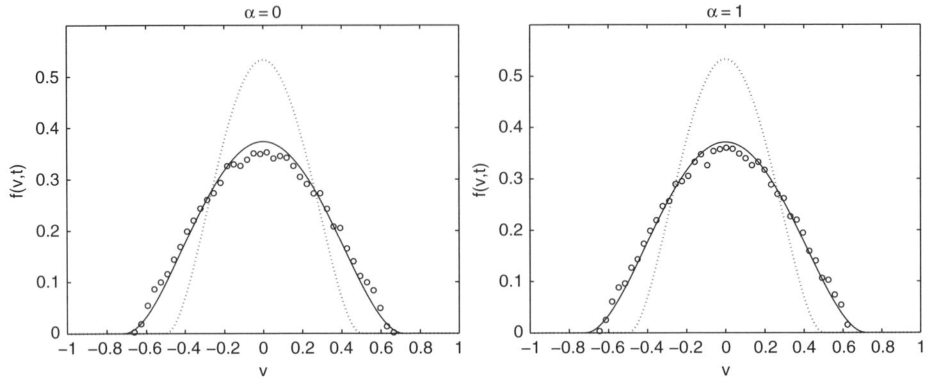

Fig. 4.9 Solution at $t = 0.005$ for $\alpha = 0$ (left) and $\alpha = 1$ (right). The dotted line represents the initial datum. The results were obtained with $N = 50000$ particles, $\delta = 0.1$ at time $t = 0.005$. The values of h used in the computations were $h = 2 \times 10^{-6}$ for $\alpha = 0$ and $h = 5 \times 10^{-3}$ for $\alpha = 1$.

1. *Given N samples V_i^0, $i = 1, \ldots, N$, from the initial distribution $f_0(v)$.*
2. *Set $N_c = [\![A_\alpha \Delta t N/(2h)]\!]_S$.*
3. *Select N_c sample pairs (V_i^0, V_j^0) uniformly without repetition among all possible pairs. For each pair:*
   (a) *generate a uniform sample U in $[0,1]$. If $UA_\alpha < |A_\varepsilon(V_i^0 - V_j^0)|$ then accept the pair and set $V_i = V_i^0 - hA_\varepsilon(V_i^0 - V_j^0)/|A_\varepsilon(V_i^0 - V_j^0)|^\alpha$ and $V_j = V_j^0 + hA_\varepsilon(V_i^0 - V_j^0)/|A_\varepsilon(V_i^0 - V_j^0)|^\alpha$;*
   (b) *otherwise, set $V_i = V_i^0$ and $V_j = V_j^0$.*

Note that in this case N_c is larger than the effective number of interacting pairs, since we take into account that a fraction A_α of the samples is discarded in the acceptance–rejection procedure.

Remark 4.6 *Since $\Delta t/h \leq 1$ in order for the probabilistic interpretation to be valid, and h has to be taken as small as possible, the choice $h = \Delta t$, the minimum value allowed by the scheme, is in some sense optimal. In this case the local update of the conservative approach becomes simply*

(a) Set $V_i = V_i^0 - \Delta t A_\varepsilon(V_i^0 - V_j^0)$ and $V_j = V_j^0 + \Delta t A_\varepsilon(V_i^0 - V_j^0)$,

which can be seen as a particle description of equation (4.55).

In Figure 4.9 we compare the numerical results obtained with the kinetic approximation for $\alpha = 0$ and $\alpha = 1$ with the similarity solution

$$f(v,t) = \frac{1}{120(t+\tau)^{(1/5)}} \left[\omega^2 - \frac{v^2}{(t+\tau)^{2/5}}\right]_+^2, \tag{4.65}$$

where $[\,\cdot\,]_+$ denotes the positive part. We have chosen $\omega = 2$, $\tau = 4^{-5}$, $v \in [0,1]$.

4.6 Remarks on multi-dimensional problems

We conclude this chapter by showing how to extend some of the Monte Carlo methods presented here to multi-dimensional kinetic equations of the type discussed in the second part of Chapter 8. With this aim we consider the general kinetic equation

$$\frac{\partial f}{\partial t} + v \cdot \nabla_x f = Q(f, f), \tag{4.66}$$

with initial data $f(x, v, 0) = f_0(x, v)$, where $v \in \mathbb{R}^d$, $x \in \mathbb{R}^d$, $d \geq 1$, and $Q(f, f)$ is some given interaction operator. As in most Monte Carlo methods, the starting point is a splitting method based on evaluating in two different steps the transport and interaction parts of the equation

$$\frac{\partial f}{\partial t} = -v \cdot \nabla_x f, \qquad \frac{\partial f}{\partial t} = Q(f, f). \tag{4.67}$$

Again, transport can be solved exactly and corresponds to free motion of particles. Therefore, if $(\mathbf{X}_i^0, \mathbf{V}_i^0)$, $i = 1, \ldots, N$, are the initial particles' position and velocity vectors in $\mathbb{R}^d$, during the transport in a time interval Δt we have the position vectors update

$$\mathbf{X}_i = \mathbf{X}_i^0 + \mathbf{V}_i^0 \Delta t.$$

The different Monte Carlo methods are then characterized by the interaction operator and by the different approaches used to approximate it. As a prototype multi-dimensional operator, we consider the Povzner–Boltzmann equation introduced in Section 8.3.1. The operator describes a non-local interaction dynamic and reads

$$Q_P(f, f)(x, v) = \int_{\mathbb{R}^d} \int_{\mathbb{R}^d} B \left(\frac{1}{J} f(x, v_*) f(y, w_*) - f(x, v) f(y, w) \right) dw\, dy, \tag{4.68}$$

where $B = B(x - y, v - w) \geq 0$ is the interaction kernel and J the Jacobian of the transformation

$$v^* = (I - A(|x - y|))v + A(|x - y|)w, \tag{4.69a}$$

$$w^* = A(|x - y|)v + (I - A(|x - y|))w, \tag{4.69b}$$

with A a $d \times d$ matrix and I the identity matrix. Equations (4.69) imply conservation of momentum since $v^* + w^* = v + w$. Here we will restrict to the simple case of $B = \sigma$, with σ a positive constant.

By decomposing the operator (4.68) in its gain and loss parts we can rewrite the interaction step as

$$\frac{\partial f}{\partial t} = Q_P^+(f, f) - \sigma \rho f, \tag{4.70}$$

where $\rho > 0$ represents the total mass and Q_P^+ the gain part of the interaction operator. Without loss of generality, in the following we assume $\sigma = 1$ and

$$\rho = \int_{\mathbb{R}^{2d}} f(x, v, t) dx dv = 1.$$

4.6.1 A Nanbu-like method

As in the one-dimensional case we start from the forward Euler scheme applied to the interaction step (4.70):

$$f(x, v, \Delta t) = (1 - \Delta t)\, f_0(x, v) + \Delta t Q_P^+(f_0, f_0)(x, v), \tag{4.71}$$

where, since f_0 is a probability density, thanks to mass conservation, also $Q_P^+(f_0, f_0)$ is a probability density. Under the restriction $\Delta t \leq 1$ then $f(x, v, \Delta t)$ is also a probability density, since it is a convex combination of probability densities.

We can readily derive the following algorithm in order to generate samples from the probability density $f(x, v, \Delta t)$:

1. *Given N samples $(\mathbf{X}_i^0, \mathbf{V}_i^0)$, $i = 1, \ldots, N$, from the initial distribution $f_0(x, v)$.*
2. *For each sample $(\mathbf{X}_i^0, \mathbf{V}_i^0)$:*
   (a) *with probability Δt select an index j uniformly among all possible individuals $(\mathbf{X}_k^0, \mathbf{V}_k^0)$ except i;*
      i. *evaluate $A(|\mathbf{X}_i - \mathbf{X}_j|)$;*
      ii. *compute the velocity change*

$$\mathbf{V}_i^* = (I - A(|\mathbf{X}_i - \mathbf{X}_j|))\mathbf{V}_i^0 + A(|\mathbf{X}_i - \mathbf{X}_j|)\mathbf{V}_j^0.$$

   (b) *otherwise set $\mathbf{V}_i^* = \mathbf{V}_i^0$.*

A symmetric version of the previous algorithm which preserves other interaction invariants at a microscopic level, like momentum, is obtained in the following way:

1. *Given N samples $(\mathbf{X}_i^0, \mathbf{V}_i^0)$, $i = 1, \ldots, N$, from the initial distribution $f_0(x, v)$.*
2. *Set $N_c = [\![\Delta t N/2]\!]_S$.*
3. *Select N_c sample pairs (i, j) uniformly without repetition among all possible pairs of individuals.*
   (a) *Evaluate $A(|\mathbf{X}_i - \mathbf{X}_j|)$.*
   (b) *Compute the velocity changes*

$$\begin{aligned} \mathbf{V}_i^* &= (I - A(|\mathbf{X}_i - \mathbf{X}_j|))\mathbf{V}_i^0 + A(|\mathbf{X}_i - \mathbf{X}_j|)\mathbf{V}_j^0 \\ \mathbf{V}_j^* &= (I - A(|\mathbf{X}_i - \mathbf{X}_j|))\mathbf{V}_j^0 + A(|\mathbf{X}_i - \mathbf{X}_j|)\mathbf{V}_i^0. \end{aligned}$$

   (c) *Set $(\mathbf{X}_i, \mathbf{V}_i) = (\mathbf{X}_i^0, \mathbf{V}_i^*)$, $(\mathbf{X}_j, \mathbf{V}_j) = (\mathbf{X}_j^0, \mathbf{V}_j^*)$.*
4. *Set $(\mathbf{X}_h, \mathbf{V}_h) = (\mathbf{X}_h^0, \mathbf{V}_h^0)$ for all the remaining $N - N_c$ individuals.*

As usual, $[\![\cdot]\!]_S$ denotes stochastic rounding.

Remark 4.7 *For non-constant interaction kernels $B = B(x - y, v - w)$, the above algorithm should be modified using and acceptance–rejection technique in order to perform the correct number of interactions. For a finite set of N particles, in fact, we can compute an upper bound Σ such that*

$$B(\mathbf{X}_i - \mathbf{X}_j, \mathbf{V}_i - \mathbf{V}_j) \leq \Sigma, \quad \forall i, j \in \{1, \ldots, N\}. \tag{4.72}$$

Point (b) of the algorithm is then performed with relative probability given by the ratio $B(\mathbf{X}_i - \mathbf{X}_j, \mathbf{V}_i - \mathbf{V}_j)/\Sigma$, as described below:

(b) *If $U \leq B(\mathbf{X}_i - \mathbf{X}_j, \mathbf{V}_i - \mathbf{V}_j)/\Sigma$, U uniform in $[0, 1]$, then compute the velocity changes*

$$\mathbf{V}_i^* = (I - A(|\mathbf{X}_i - \mathbf{X}_j|))\mathbf{V}_i^0 + A(|\mathbf{X}_i - \mathbf{X}_j|)\mathbf{V}_j^0$$
$$\mathbf{V}_j^* = (I - A(|\mathbf{X}_i - \mathbf{X}_j|))\mathbf{V}_j^0 + A(|\mathbf{X}_i - \mathbf{X}_j|)\mathbf{V}_i^0.$$

Clearly, the efficiency of the acceptance–rejection approach depends on the sharpness of the bound (4.72). Note that the computation of the optimal bound Σ as the maximum of $B(\mathbf{X}_i - \mathbf{X}_j, \mathbf{V}_i - \mathbf{V}_j)$, $\forall i, j$, at each time step may be expensive due to the $O(N^2)$ cost. For this reason, whenever possible, it is preferable to search for other kinds of bounds which can be computed at a reduced cost (typically $O(N)$) depending on the particular structure of B. For example, if B depends only on the absolute value of the relative velocity (and similarly for the relative distance) $B = |v - w|^\alpha$, $\alpha \geq 1$, we can estimate

$$|\mathbf{V}_i - \mathbf{V}_j| \leq |\mathbf{V}_i - \mathbf{U}| + |\mathbf{V}_j - \mathbf{U}| \leq 2 \max_k |\mathbf{V}_k - \mathbf{U}|,$$

where $\mathbf{U} = \frac{1}{N} \sum_k \mathbf{V}_k$ is the mean velocity.

4.6.2 A Bird-like method

Finally, we report the Bird algorithm for the interaction operator described by (4.68) with a constant B. From the above considerations, using the symmetric formulation and the local time counter (4.38), we obtain the following method in a time interval $\Delta t = n_t \Delta t_c$:

1. *Given N samples $(\mathbf{X}_i^0, \mathbf{V}_i^0)$, $i = 1, \ldots, N$, from the initial distribution $f_0(x, v)$.*
2. *Compute $\Delta t_c = 1/N$ and set $\mathbf{V}_i = \mathbf{V}_i^0$, $i = 1, \ldots, N$.*
3. *For $k = 0, \ldots, n_t$:*
   (a) *Select a random pair $(\mathbf{V}_i, \mathbf{V}_j)$ uniformly from all possible pairs.*
   (b) *Set*

$$\mathbf{V}_i = (I - A(|\mathbf{X}_i - \mathbf{X}_j|))\mathbf{V}_i + A(|\mathbf{X}_i - \mathbf{X}_j|)\mathbf{V}_j$$
$$\mathbf{V}_j = (I - A(|\mathbf{X}_i - \mathbf{X}_j|))\mathbf{V}_j + A(|\mathbf{X}_i - \mathbf{X}_j|)\mathbf{V}_i.$$

Remark 4.8

- *Note that for the binary interaction model (4.68) the resulting Monte Carlo methods are fully meshless. This is a remarkable difference when compared to the classical Boltzmann collision operator of rarefied gas dynamics (1.5). In the latter case, as in Section 4.2.1, it is necessary to introduce a space grid in order to simulate the collision dynamic. We refer to [39, 250, 275] for further details.*
- *In an analogous way it is possible to extend to the multi-dimensional case the recursive Monte Carlo method described in Section 4.3.2. On the other hand, the extension of the asymptotic preserving Monte Carlo method of Section 4.4 depends on the detailed structure of the matrix A and therefore on the equilibrium state of the interaction operator (4.68).*

Problems

Exercise 4.1 In Exercise 1.3 we introduced the Carleman model. Consider its scaled version obtained for $\sigma = 1/\varepsilon$ and compute the fluid limit as $\varepsilon \to 0$. Using the formalism of Section 4.2.1 write a Monte Carlo method for its numerical solution that works without time step restrictions for all values of the scaling parameter ε. Apply both standard Monte Carlo as well as random gradient Monte Carlo and compare the numerical solution with a standard finite difference approach to the model.

Exercise 4.2 Let us consider the Goldstein–Taylor model written using the variables $\rho = f^+ + f^-$ and $j = c(f^+ - f^-)$ as

$$\begin{cases} \dfrac{\partial \rho}{\partial t} + \dfrac{\partial j}{\partial x} &= 0, \\ \dfrac{\partial j}{\partial t} + c^2 \dfrac{\partial \rho}{\partial x} &= -\sigma j. \end{cases} \tag{4.73}$$

If we now derive the first equation with respect to time, we can write

$$\frac{\partial^2 \rho}{\partial t^2} + \frac{\partial}{\partial x}\left(\frac{\partial j}{\partial t}\right) = \frac{\partial^2 \rho}{\partial t^2} - c^2 \frac{\partial^2}{\partial x^2} - \sigma \frac{\partial J}{\partial x} = 0,$$

which, using again the first equation of (4.73) and dividing by c^2, gives the *telegrapher's equation*

$$\frac{1}{c^2}\frac{\partial^2 \rho}{\partial t^2} + \frac{\sigma}{c^2}\frac{\partial \rho}{\partial t} = \frac{\partial^2 \rho}{\partial x^2}. \tag{4.74}$$

In particular, as $\sigma \to 0$ we obtain the wave equation, whereas as $c, \sigma \to \infty$ and $\sigma/c^2 \to 1/D$ we recover the heat equation. Let us denote with $\rho(x,0) = \rho_0(x)$ the initial data, with $\partial\rho(x,0)/\partial t = 0$. We know that the solution of the wave equation for $\sigma = 0$ is given by

$$\rho(x,t) = \frac{1}{2}\left(\rho_0(x+ct) + \rho_0(x-ct)\right). \tag{4.75}$$

If we think of $\sigma/2$ as the probability per unit time of a Poisson process, then if $N(t)$ is the number of events occurring up to time t it has the distribution

$$P\{N(t) = k\} = e^{-\sigma t/2}\frac{(\sigma t/2)^k}{k!}.$$

Now, if a particle moves with velocity c for time t it travels a distance $ct = \int_0^t c\,ds$. If it undergoes random Poisson-distributed direction reversals (collisions) with probability per unit time $\sigma/2$, the distance travelled in time t is $C(t) = \int_0^t c(-1)^{N(s)}\,ds$. It can be proven [187] that if we replace the free transport distance ct in (4.75) by the effective distance $C(t)$ we obtain the solution to the initial value problem for the telegrapher's equation (4.74).

Construct a Monte Carlo method for the Goldstein–Taylor model based on the above stochastic interpretation of the telegrapher's equation and compare it with the Monte Carlo approaches described in Section 4.1.

Exercise 4.3 Consider the one-dimensional Broadwell model [60, 264]

$$\begin{cases} \dfrac{\partial f_1}{\partial t} + c\dfrac{\partial f_1}{\partial x} = \sigma(f_3^2 - f_1 f_2) \\ \dfrac{\partial f_2}{\partial t} - c\dfrac{\partial f_2}{\partial x} = \sigma(f_3^2 - f_1 f_2) \\ \dfrac{\partial f_3}{\partial t} = \sigma(f_1 f_2 - f_3^2), \end{cases} \tag{4.76}$$

where f_1, f_2 and f_3 are the densities of particles with velocity $c, -c, 0$ respectively, with $c > 0$. The collision invariants correspond to conservation of mass and momentum

$$\rho = f_1 + f_2 + 2f_3, \quad m = c(f_1 - f_2).$$

Let us focus on the rescaled form corresponding to $\sigma = 1/\varepsilon$. Introducing the additional macroscopic quantity $z = f_1 + f_2$, the model can be written as an hyperbolic relaxation system in the form

$$\begin{cases} \dfrac{\partial \rho}{\partial t} + \dfrac{\partial m}{\partial x} = 0 \\ \dfrac{\partial m}{\partial t} + c^2\dfrac{\partial z}{\partial x} = 0 \\ \dfrac{\partial z}{\partial t} + \dfrac{\partial m}{\partial x} = \dfrac{1}{2\varepsilon}\left(\rho^2 + \dfrac{m^2}{c^2} - 2\rho z\right). \end{cases} \tag{4.77}$$

In the fluid-dynamic limit $\varepsilon \to 0$ we have $z = \left(\rho^2 + m^2/c^2\right)/(2\rho)$, and the Broadwell system is well approximated by the reduced system

$$\begin{cases} \dfrac{\partial \rho}{\partial t} + \dfrac{\partial m}{\partial x} = 0 \\ \dfrac{\partial m}{\partial t} + \dfrac{c^2}{2}\dfrac{\partial}{\partial x}\left(\rho + \dfrac{m^2}{c^2\rho}\right) = 0, \end{cases} \tag{4.78}$$

which represents the corresponding Euler equations of fluid dynamics. Using the approach described in Section 4.2.1, construct a Monte Carlo method for the kinetic model and solve for $\varepsilon = 1, 0.1, 10^{-4}$ the following Riemann problem:

$$\rho = 1, \quad m = 0, \quad z = 1, \quad x < 0; \qquad \rho = 0.2, \quad m = 0, \quad z = 1, \quad x > 0.$$

Hint: Note that after splitting system (4.77) into transport and collision phases, the relaxation step can be solved exactly, since ρ and m are constants and the differential equation for z is linear. Reverting to the original kinetic variables one obtains the exact solution of the relaxation step for the Broadwell model with respect to the kinetic densities. This can be easily used to construct a probabilistic interpretation of the underlying dynamics.

Exercise 4.4 In the planar case, the Broadwell model describes a simple gas where particles can attain only four possible velocities in $\mathbb{R}^2$: $v_1 = (c, 0)^T$, $v_2 = (0, c)^T$, $v_3 = (-c, 0)^T$ and $v_4 = (0, -c)^T$, $c > 0$. Denoting the respective kinetic densities as f_i, $i = 1, 2, 3, 4$, the Broadwell equations reads [60, 264]:

$$\begin{cases} \dfrac{\partial f_1}{\partial t} + c\dfrac{\partial f_1}{\partial x} = \sigma(f_2 f_4 - f_1 f_3), \\ \dfrac{\partial f_2}{\partial t} + c\dfrac{\partial f_2}{\partial y} = \sigma(f_1 f_3 - f_2 f_4), \\ \dfrac{\partial f_3}{\partial t} - c\dfrac{\partial f_3}{\partial x} = \sigma(f_2 f_4 - f_1 f_3), \\ \dfrac{\partial f_4}{\partial t} - c\dfrac{\partial f_4}{\partial y} = \sigma(f_1 f_3 - f_2 f_4). \end{cases} \tag{4.79}$$

This model is the simplest two-dimensional discrete velocity model of the Boltzmann equation with the correct number of collision invariants (mass and momentum). Using a standard splitting between transport and collision develop a Monte Carlo method for the above system of equations. Consider the scaled case $\sigma = 1/\varepsilon$, compute the corresponding asymptotic behaviour for $\varepsilon \to 0$ and construct a Monte Carlo method that works uniformly in ε. Is it possible to apply a random gradient Monte Carlo for the Broadwell equations?

Hint: Let us define $\rho = f_1 + f_2 + f_3 + f_4$ and denote with $\mathbf{f}$ the vector with the four kinetic densities. Then the scaled homogeneous collision step can be written as

$$\frac{\partial f_i}{\partial t} = \frac{1}{\varepsilon}\left[R_i(\mathbf{f}) - \rho f_i\right], \qquad i = 1, 2, 3, 4, \tag{4.80}$$

where, using the compact notation $f_k = f_l$ if $k \equiv l \bmod 4$,

$$R_i(\mathbf{f}) = (f_i + f_{i+1})(f_i + f_{i+3}), \qquad i = 1, 2, 3, 4.$$

In particular, note that the R_i's remain constant during the collision process. Therefore, starting from the initial values f_i^0 the equations can be solved exactly to give

$$f_i = e^{-\rho t/\epsilon} f_i^0 + \left(1 - e^{-\rho t/\epsilon}\right) \frac{R_i(\mathbf{f^0})}{\rho}, \qquad i = 1, 2, 3, 4. \tag{4.81}$$

A simple calculation shows that the local Maxwellian $M_i = f_i^\infty$ is given by

$$M_i(\mathbf{f}) = \frac{1}{\rho} R_i(\mathbf{f}), \qquad i = 1, 2, 3, 4.$$

Exercise 4.5 A generalization of the asymptotic preserving (AP) schemes based on the Wild sum presented in Section 4.4 consists in taking a general convex combination of probability densities of the form

$$f(v, \Delta t) = \sum_{k=0}^{m} A_k(\Delta t) f_{(k)}(v) + A_{m+1}(\Delta t) M(v), \tag{4.82}$$

where the functions $f_{(k)}$ are defined by (4.40) and the quantities A_k are such that

$$\sum_{k=0}^{m+1} A_k(\Delta t) = 1, \quad A_k(\Delta t) \geq 0, \qquad k = 0, \ldots, m+1.$$

Moreover, if

$$\lim_{\Delta t\to\infty} A_{m+1}(\Delta t) = 1, \qquad \lim_{\Delta t\to\infty} A_k(\Delta t) = 0, \quad k = 0,\dots,m,$$

they define an AP method. We refer to [72, 127, 248] for different choices of the weights A_k.

In particular, first-order asymptotic preserving schemes take the form

$$f(v,\Delta t) = A_0(\Delta t) f_0(v) + A_1(\Delta t) f_{(1)}(v) + A_2(\Delta t) M(v). \tag{4.83}$$

Note that the non-AP choice $A_0(\Delta t) = 1 - \Delta t$, $A_1(\Delta t) = \Delta t$ and $A_2(\Delta t) = 0$ corresponds to the standard explicit Euler method characterizing Nanbu's Monte Carlo approach. Give a probabilistic interpretation of (4.83) and use it to construct a Monte Carlo method without using the recursive strategy. Is it possible to make the method exactly conservative?

Exercise 4.6 Apply the Monte Carlo approach described in Section 4.5.2 to the porous media equation

$$\frac{\partial g(v,t)}{\partial t} = \frac{\partial^2}{\partial v^2}\left(g(v,t)^2\right), \quad v \in \mathbb{R},\ t > 0. \tag{4.84}$$

Compare the computed solution with the Barenblatt [20] solution

$$g(x,t) = \begin{cases} \dfrac{1}{\lambda(t)}\left[1 - \left(\dfrac{x}{\lambda(t)}\right)^2\right] & \text{for } |x| \le \lambda(t),\ t \ge 0 \\ 0 & \text{for } |x| > \lambda(t),\ t \ge 0, \end{cases} \tag{4.85}$$

where

$$\lambda(t) = \left[12\,(t+1)\right]^{1/3} \quad \text{for } t \ge 0. \tag{4.86}$$

We note that the solution (4.85) is not a classical solution, in the sense that the transition between the region of the medium with $g > 0$ to the one with $g = 0$ is not smooth. The boundary $\lambda(t)$ in (4.85) is called the interface.

Part II

Multiagent kinetic equations

5
Models for wealth distribution

> One of Pareto's equations achieved special prominence, and controversy. He was fascinated by problems of power and wealth. How do people get it? How is it distributed around society? How do those who have it use it? The gulf between rich and poor has always been part of the human condition, but Pareto resolved to measure it. He gathered reams of data on wealth and income through different centuries, through different countries: the tax records of Basel, Switzerland, from 1454 and from Augsburg, Germany in 1471, 1498 and 1512; contemporary rental income from Paris; personal income from Britain, Prussia, Saxony, Ireland, Italy, Peru. What he found – or thought he found – was striking. When he plotted the data on graph paper, with income on one axis, and number of people with that income on the other, he saw the same picture nearly everywhere in every era. Society was not a "social pyramid" with the proportion of rich to poor sloping gently from one class to the next. Instead it was more of a "social arrow" – very fat on the bottom where the mass of men live, and very thin at the top where sit the wealthy elite. Nor was this effect by chance; the data did not remotely fit a bell curve, as one would expect if wealth were distributed randomly. "It is a social law", he wrote: something "in the nature of man".
>
> B.B. Mandelbrot, R.L. Hudson, *The (mis)Behaviour of Markets: A fractal view of risk, ruin and reward*, 2004.

5.1 Wealth, trades and kinetic equations

Typical sectors of application of the kinetic methodology are the systems of agents which interact in a market by some type of trading. A closely related field is generally known under the name of *econophysics*. This definition describes an interdisciplinary research field which aims to apply theories and methods originally developed by statistical physicists in order to solve problems in economics. It is now commonly accepted that the term econophysics was coined by H. Eugene Stanley in the early 1990s at a conference on statistical physics in Calcutta, and first appeared in its proceedings publication [291]. Among these papers, some put the roots of modelling the evolution of the distribution of wealth in a simple market economy in kinetic theory. For an introduction to econophysics we refer to the recent books [89, 98, 225, 299, 300] (and the references therein), or, for example, to the introductory articles [95, 168, 174, 321].

The analogy between binary particle interactions in kinetic theory and trades was investigated by F. Slanina [287] in 2004; there, a clear parallelism between the evolution of wealth in a simple economy and the evolution of the particle density in a one-dimensional dissipative gas was established. This paper motivated us to adapt ideas, developed in studies of dissipative Maxwell gases to the economic framework [254]. Other contributors have taken advantage of this analogy, which allows us to make exact computations in many situations [18, 32, 33, 46].

The collision mechanism for molecules and the modelling of trade interactions are clearly different. In the new framework, interactions typically lack the usual microscopic conservation laws for (the analogues of) impulse and energy; moreover, random effects play a crucial role. In fact, the key step in establishing a reasonable kinetic market model is the definition of sensible rules at the *microscopic* level, i.e. the prescription of how wealth is exchanged in trades. Such rules are usually derived from plausible assumptions in an empirical way.

Fig. 5.1 Vilfredo Pareto[1]

The corresponding outputs of the model are the *macroscopic* statistics of wealth distribution in the society. The comparison of this output with realistic data has been, so far, the only way to evaluate – *a posteriori* – the quality of a proposed model. For instance, it is commonly accepted that the wealth distribution should approach a stationary (or, in general, a self-similar) profile for large times, and that the latter should exhibit a *Pareto tail.* Such overpopulated tails are a manifestation of the existence of an upper class of very rich agents, i.e. an indication of an unequal distribution of wealth.

In general, the richness of the steady states for kinetic market models is another remarkable difference from the theory of Maxwell molecules. While the Maxwell distribution (1.2) is the universal steady profile for the velocity distribution of molecular gases which interact according to elastic collisions, the stationary profiles for wealth can be manifold, and are in general not explicitly known analytically. In fact, while subject to some conservation (for example, conservation of mean wealth), trade interactions depend heavily on the precise form of the microscopic modelling. Consequently, in investigations of the large-time behaviour of wealth distribution, one is typically limited to describing a few of its analytically accessible properties (e.g. moments and smoothness).

A variety of models have been proposed and numerically studied in view of the relation between parameters in the microscopic rules and the resulting macroscopic statistics. The features typically incorporated in kinetic trade models are saving effects and randomness. Saving means that each agent is guaranteed to retain at least a certain minimal fraction of his initial wealth at the end of the trade. This concept was

[1]**Vilfredo Pareto**: born on 15 July 1848 in Paris (France), died on 19 August 1923 in Geneva (Switzerland). He made several important contributions to economics, particularly in the study of income distribution and in the analysis of individuals' choices. He was the first to discover that income follows a power law probability distribution (the Pareto distribution). He also contributed to the fields of sociology and mathematics.

probably first introduced in [91] where a fixed saving rate for all agents was proposed. Randomness means that the amount of wealth changing hands is non-deterministic. Among others, this idea has been developed in [105], in order to include the effects of a risky market. Depending upon the specific choice of the saving mechanism and the stochastic nature of the trades, the systems studied either produce wealth curves with the desired Pareto tail or they do not.

In the first part of this chapter we introduce and discuss several of these kinetic market models, following some recent review papers [133, 134]. Roughly speaking, the models can be split into two classes, which correspond to two different types of interactions. The first type is such that the binary trade is *conservative*, either microscopically or in the statistical mean. In this situation, the mean wealth in the model Boltzmann equation is preserved, and one expects the formation of a stationary profile. In the second type of interaction, the mean wealth is *not preserved*, and therefore the long-time behaviour of the wealth distribution is not described by the formation of a stationary solution, but rather by some self-similar solution.

In the class of conservative trades, the focus is on variants of the model with *risky investments*, originally introduced by Cordier et al. in [105], and of the model designed by Chakraborti and Chakrabarti [91]. The applied analytical techniques, however, easily generalize to a broader class of conservative economic games. These techniques have been applied in the current mathematical literature [105, 132, 135, 229, 230, 254], where kinetic market models have been treated in the framework of Maxwell-type molecules. These mathematical results are briefly illustrated, before they are applied to the specific models under consideration. The interested reader who wishes to obtain a deeper understanding of the mathematical roots (and possible extensions) of the applied tools is referred, e.g., to [77, 319].

For the treatment of the class of non-conservative trades, the starting point is the simplified model introduced by Slanina [287], which has subsequently been studied in [254]. This model can easily be generalized to random mixing parameters in order to include *risky market* effects. The risky effects are designed with a certain bias to *increase* the total wealth in trade interactions; the same effect was obtained in [105], if risky gains and losses do not balance in the statistical mean.

As a further approach, we discuss how to extend these kinetic models for wealth distribution in a market which comprises a whole number of countries, or alternatively different social groups within the same country [136]. The goal is to verify analytically the existence of bimodal stationary distributions [167]. Bimodal distributions (and a polymodal distribution, in general) are, in fact, reported with real data for income distributions in various countries, such as, for example, Argentina [147]. In the proposed model, a bimodal steady state can indeed be obtained, e.g. when the saving parameter takes only two fixed values that are sufficiently widely separated.

Next, in the second part of the chapter, we consider some models for which a detailed description of exact stationary solutions is possible. This is, with respect to the classical kinetic theory of rarefied gases, a completely new topic that depends on the possibility of having a large variety of possible interactions. Exact solutions to a simplified model of binary interactions, the so-called *pure gambling* models have recently been classified in [24]. These exact steady profiles exhibit fat tails in the case in which the pure gambling interaction is not pointwise conservative, and agents can take advantage of the amount of wealth available in the market. The last part of the chapter includes the description of some possible extensions of this kinetic modelling to more realistic situations, which can include non-microscopic effects, like global taxation (and subsequent redistribution).

5.2 Economic and kinetic dictionaries

We will be interested in the distribution of wealth in a closed system of agents (more concisely, a market), which interact through binary trades. In this market, the *wealth distribution* $f(w,t)$ refers to the relative density of agents with wealth w at time $t \geq 0$.

It is commonly assumed that agents can participate in market activities only if they possess a certain amount of money. In other words, in the models considered here, debts are excluded. This corresponds to assuming that $f(w,t) = 0$ for $w < 0$. In most of the models present in the literature, in fact, trade rules are built so that the post-interaction wealth is never negative. On the other hand, agents can exit from the trade with zero wealth, which does not exclude concentration of wealth in $w = 0$.

The associated cumulative probability distribution function

$$F(w,t) = \int_w^\infty f(w,t)\, dv \leq 1 \tag{5.1}$$

provides the fraction of agents wealthier than w. Likewise, the first moment of $f(w,t)$ yields the *average wealth per agent*,

$$m(t) = m_1(t) = \int_{\mathbb{R}_+} w f(w,t)\, dw. \tag{5.2}$$

In the models we will present here, the density $f(w,t)$ stabilizes at some *stationary wealth curve* $f_\infty(w)$ in the large-time limit $t \to \infty$.

We recall that the central notion in the theory of wealth distributions is that of the celebrated *Pareto index* $\alpha \geq 1$. This number describes the size of the rich upper class in the considered ensemble of agents. Roughly, the smaller α is, the more of the total wealth is concentrated in the hands of a small group of individuals.

The stationary curve $f_\infty(w)$ satisfies the Pareto law [257] with index α, provided that f_∞ decays like an inverse power function for large w,

$$f_\infty(w) \propto w^{-(\alpha+1)} \quad \text{as } w \to +\infty. \tag{5.3}$$

More precisely, f_∞ has Pareto index $\alpha \in [1, +\infty)$ if the moments

$$m_s = \int_{\mathbb{R}_+} w^s f_\infty(w)\, dw \tag{5.4}$$

are finite for all positive $s < \alpha$, and infinite for $s > \alpha$. If all m_s are finite (e.g. for a Gamma distribution), then f_∞ is said to possess a *slim tail*.

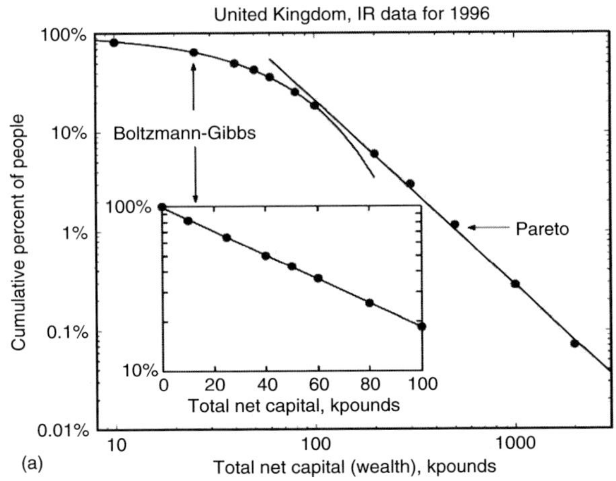

Fig. 5.2 The cumulative probability distribution of wealth[2] in the United Kingdom. Low capitals fit an exponential curve, and high capitals fit a Pareto power law. Reprinted from [130], © 2001, with permission from Elsevier.

According to empirical data from ancient Egypt until today [95,98,130], the wealth distribution among the population in a capitalist country follows the Pareto law, with an index α ranging between 1.5 and 2.5. Slim tails are typical for societies with a highly equal distribution of wealth, intuitively, one may think of socialist countries.

A seminal paper on Pareto law and the wealth curve is due to Mandelbrot [223]. Surprisingly, however, the mathematical description of the stationary wealth curve f_∞ attracted the interest of mathematicians many years before Mandelbrot's work. A description of this curve by means of a generalized Gamma distribution is due to Amoroso [7] and D'Addario [113]. If one assumes for f_∞ a unit mean, the Amoroso distribution reads

$$f_\alpha(w) = \frac{(\alpha-1)^\alpha}{\Gamma(\alpha)} \frac{\exp\left(-\frac{\alpha-1}{w}\right)}{w^{1+\alpha}}, \quad \alpha > 1. \tag{5.5}$$

Note that this stationary distribution exhibits a Pareto power-law tail of order α for large w's.

Kinetic modelling of an agent-based market is based on few universal assumptions. First, agents are indistinguishable, so that an agent's *state* at any instant of time $t \geq 0$ is completely characterized by his current wealth $w \geq 0$. Second, the time variation of the wealth distribution is entirely due to binary trades between agents. A trade represents a binary interaction in which part of the money of each agent is modified according to well-defined rules. When two agents undertake in a trade, their *pre-trade*

[2]**1996 IR data**: The distribution of wealth is not easy to measure. When a person dies, all assets must be reported for the purpose of inheritance tax. Using this data and an adjustment procedure, the British tax agency, the Inland Revenue (IR), reconstructed the wealth distribution of the whole UK population (see `http://www.hmrc.gov.uk/statistics/wealth.htm`).

wealths v, w change into the *post-trade wealths* v^*, w^* according to a linear exchange rule:

$$v^* = p_1 v + q_1 w, \quad w^* = q_2 v + p_2 w. \tag{5.6}$$

The *interaction coefficients* p_i and q_i, $i = 1, 2$, are, in general, non negative random parameters. While $q_1 v$ denotes the fraction of the second agent's wealth transferred to the first agent, the difference $(p_1 - q_2)v$ is the relative gain (or loss) of wealth of the first agent due to market risks. The coefficient p_i and q_i are assumed to have fixed laws, which are independent of v and w, and of time. This means that the amount of wealth an agent contributes to a trade is (on average) proportional to the respective agent's wealth.

Notice that a trade of type (5.6) is described by a law identical to the exchange (1.77) introduced in Chapter 1, Section 1.4. Consequently, the forthcoming kinetic equation which describes the evolution of the wealth density (assuming the independence hypothesis between agents) takes the form (1.79).

The wealth distribution $f(w,t)$ of the ensemble coincides with agent density and satisfies the associated spatially homogeneous Boltzmann equation,

$$\partial_t f + f = Q_+(f, f), \tag{5.7}$$

on the real half-line, $w \geq 0$. The collisional gain operator Q_+ acts on test functions $\varphi(w)$ as

$$\begin{aligned} Q_+(f,f)[\varphi] &= \int_{\mathbb{R}_+} \varphi(w) Q_+(f,f)(w)\, dw \\ &= \frac{1}{2} \int_{\mathbb{R}_+^2} \langle \varphi(v^*) + \varphi(w^*) \rangle f(v) f(w)\, dv\, dw. \end{aligned} \tag{5.8}$$

As the discussion of Section 1.5 clarifies, the large-time behaviour of the density in this type of equation is heavily dependent on the evolution of the average wealth. As a consequence of this remark, it appears natural to split models into two main classes.

Conservative models are such that the average wealth of the society is conserved with time, so that

$$m(t) = m_1(t) = \int_{\mathbb{R}_+} w f(w,t)\, dw = m, \tag{5.9}$$

where m is finite. In terms of the interaction coefficients, this is equivalent to

$$\langle p_1 + q_2 \rangle = \langle p_2 + q_1 \rangle = 1. \tag{5.10}$$

Non-conservative models are such that $m(t)$ is not conserved with time. Among all possible models we will restrict ourselves to describing trades for which $\langle p_1 + q_2 \rangle = \langle p_2 + q_1 \rangle \neq 1$, so that the average wealth is exponentially increasing or decreasing:

$$m(t) = m(0)\, e^{(\langle p_1 + q_2 \rangle - 1)t}. \tag{5.11}$$

According to the analysis of Sections 1.4 and 2.2.2, a useful way of writing equation (5.7) is to resort to the Fourier transform. Assuming the initial distribution of wealth in $\mathcal{P}_s$, with $s > 1$, in view of the collision rule (5.6), the transformed kernel reads

$$\widehat{Q}(\hat{f},\hat{f})(\xi) = \frac{1}{2}\left\langle \widehat{f}(p_1\xi)\widehat{f}(q_1\xi) + \widehat{f}(p_2\xi)\widehat{f}(q_2\xi)\right\rangle - \widehat{f}(\xi), \tag{5.12}$$

where, since the initial density has a bounded mean,

$$\widehat{f_0}'(0) = -im.$$

Hence, the Boltzmann equation (5.7) can be rewritten as

$$\frac{\partial \widehat{f}(\xi,t)}{\partial t} + \widehat{f}(\xi,t) = \frac{1}{2}\langle \widehat{f}(p_1\xi)\widehat{f}(q_1\xi) + \widehat{f}(p_2\xi)\widehat{f}(q_2\xi)\rangle = \langle \widehat{f}(p_i\xi)\widehat{f}(q_i\xi)\rangle_+. \tag{5.13}$$

5.3 Kinetic market models for conservative economies

As for the classical Boltzmann equation [80, 82, 83], where molecules interact through the elastic collisions (1.4), and the density is consequently shown to converge towards the equilibrium Maxwellian M given by (1.2), the hypothesis of a conservative trade will be enough to conclude with the convergence of the wealth density towards a steady profile. The aim of this section is to verify this physical idea, and to establish a precise connection between the microscopic interaction rules (5.6) and the shape of the equilibrium profile. This connection has been shown recently in [229] (cf. also the analysis of [132, 133]).

5.3.1 Trading rules and tails

The main argument in [229] was to show that the trade rules determine an evolution of the wealth density towards a steady solution with tails which depend on the function

$$\mathbf{S}(s) = \frac{1}{2}\Big(\sum_{i=1}^{2}\langle p_i^s + q_i^s\rangle\Big) - 1. \tag{5.14}$$

For almost all realistic cases, the function $\mathbf{S}(s)$ is convex in $s > 0$, with $\mathbf{S}(0) = 1$. Also, $\mathbf{S}(1) = 0$ because of the conservation property (5.10). The only exceptions are the cases in which the mixing coefficients p_i, q_i, $i = 1, 2$, are equal to the constant values 0 or 1. In these cases, in fact, $\mathbf{S}(s)$ is constant. Since condition (5.10) has to be satisfied, this choice, excluding the case in which people retain all their wealth ($p_1 = p_2 = 1$), can only describe two situations. The first corresponds to a trade in which the two agents simply exchange their wealth ($q_1 = q_2 = 1$). The second describes a trade in which the total amount of wealth which is traded is gained by one of the two agents. We call this *winner takes all* dynamics. For example, we can assign the values $p_1 = q_1 = 1$, so that

$$v^* = v + w, \quad w^* = 0. \tag{5.15}$$

In each trade, the second agent loses all of his wealth in favour of the first agent. Note that in this case $\mathbf{S}(s)$ is identically equal to zero. Due to its simplicity, this model can be studied in full detail, to illustrate, as for the model presented in Section 2.5, the power (and the limits) of the Fourier representation (5.13).

The results from [132, 229] simply associate with every behaviour the consequent properties of the solution to the kinetic equation (5.7), as well as the characteristics of its steady solution. Unless $\mathbf{S}(s) \geq 0$ for all $s > 0$, any solution $f(w,t)$ tends to a steady wealth distribution $P_\infty(w) = f_\infty(w)$, which depends on the initial wealth distribution only through the conserved mean wealth $m > 0$. Moreover, exactly one of the following is true:

(PT) if $\mathbf{S}(\alpha) = 0$ for some $\alpha > 1$, then $P_\infty(w)$ has a *Pareto tail* of index α;
(ST) if $\mathbf{S}(s) < 0$ for all $s > 1$, then $P_\infty(w)$ has a *slim tail*;
(DD) if $\mathbf{S}(\alpha) = 0$ for some $0 < \alpha < 1$, then $P_\infty(w) = \delta_0(w)$, a *Dirac Delta* at $w = 0$.

The mathematical details of these results will be postponed to Section 5.3.2. Before entering into the detailed proofs, we detail some of the driving physical ideas. Since the tails of the solution are linked to moments, as given by (5.4), one studies the evolution equation for the moments,

$$m_s(t) = \int_{\mathbb{R}_+} w^s f(w,t)\, dw, \tag{5.16}$$

which are obtained by integration of (5.7) against $\varphi(w) = w^s$,

$$\frac{d}{dt} m_s = Q_+[w^s] - m_s. \tag{5.17}$$

Then using in (5.8) the elementary inequality

$$x^s + y^s \leq (x+y)^s \leq x^s + y^s + 2^{s-1}(xy^{s-1} + x^{s-1}y), \tag{5.18}$$

valid for $x, y \geq 0$, $s \geq 1$, one calculates for the right-hand side of (5.17)

$$\mathbf{S}(s)m_s \leq Q_+[w^s] - m_s \leq \mathbf{S}(s)m_s + 2^{s-2}\sum_{i=1}^{2}\langle p_i q_i^{s-1} + p_i^{s-1}q_i\rangle m m_s^{1-1/s}. \tag{5.19}$$

Solving (5.17) with (5.19), one finds that either $m_s(t)$ remains bounded for all times when $\mathbf{S}(s) < 0$, or it diverges like $\exp[t\mathbf{S}(s)]$ when $\mathbf{S}(s) > 0$.

In case (PT), the moments $m_s(t)$ with $s > \alpha$ blow up as $t \to \infty$, giving rise to a Pareto tail of index α. We emphasize that $f(w,t)$ possesses finite moments of all orders at any finite time. The Pareto tail forms *in the limit* $t \to \infty$.

In case (ST), all moments converge to limits $m_s(t) \to m_s^*$, so the tail is slim. One can obtain additional information on the stationary wealth distribution $P_\infty(w)$ from the *recursion relation* for the principal moments,

$$-\mathbf{S}(s)m_s^* = \frac{1}{2}\sum_{k=1}^{s-1}\sum_{i=1}^{2}\binom{s}{k}\langle p_i^k q_i^{s-k}\rangle m_k^* m_{s-k}^*, \qquad s = 2, 3, \ldots \tag{5.20}$$

The latter is obtained by integration of (5.7) against $\varphi(w) = w^s$ in the steady state $\partial_t f = 0$.

In case (DD), all moments $m_s(t)$ with $s > 1$ blow up. The underlying process is a separation of wealth as time increases: while more and more agents become extremely poor, fewer and fewer agents possess essentially the entire wealth of the society. In terms of $f(w,t)$, one observes an accumulation in the pauper region $0 \leq w \ll 1$, while the density rapidly spreads into the region $w \gg 1$. The expanding support of $f(w,t)$ is balanced by a decrease in magnitude, since the average wealth is fixed. This induces a pointwise convergence $f(w,t) \to 0$ for all $w > 0$. Such a *condensation of wealth* has been observed and described in several contexts [55, 62, 90, 180] before.

The Chakraborti and Chakrabarti model

The previous classification of models based on binary trades is exhaustive, and immediately allows us to give the essential features of the steady wealth distribution. It is interesting to remark that historically the first models of exchange of wealth belong to the case (ST).

The first explicit description of a binary wealth-exchange model dates back to Angle [9], (cf. also [10]), although the intimate relation to statistical mechanics was only described about a decade later [129, 180]: in each binary interaction, winner and loser are randomly chosen, and the loser pays a random fraction of his wealth to the winner. From here, Chakraborti and Chakrabarti [91] developed the class of *strictly conservative* exchange models, which preserve the total wealth in each individual trade,

$$v^* + w^* = v + w. \tag{5.21}$$

In its most basic version, the microscopic interaction is determined by one single parameter $\lambda \in (0,1)$, which is the global *saving propensity.* In interactions, each agent keeps the corresponding fraction of his pre-trade wealth, while the rest $(1-\lambda)(v+w)$ is equally shared among the two trade partners,

$$v^* = \lambda v + \frac{1}{2}(1-\lambda)(v+w), \quad w^* = \lambda w + \frac{1}{2}(1-\lambda)(v+w). \tag{5.22}$$

As a result, all agents become equally rich eventually. Indeed, the variance of $f(w,t)$ satisfies

$$\frac{d}{dt}\int_{\mathbb{R}_+} (w-m)^2 f(w,t)\,dw = -\frac{1}{2}(1-\lambda^2)\int_{\mathbb{R}_+} (w-m)^2 f(w,t)\,dw. \tag{5.23}$$

The steady state $f_\infty(w) = \delta_0(w-m)$ is a Dirac delta concentrated at the mean wealth, and is approached at the exponential rate $(1-\lambda^2)/2$.

More interesting, non-deterministic variants of the model have been proposed, where the amount $(1-\lambda)(v+w)$ is not equally shared, but in a stochastic way:

$$v^* = \lambda v + \epsilon(1-\lambda)(v+w), \quad w^* = \lambda w + (1-\epsilon)(1-\lambda)(v+w), \tag{5.24}$$

with a random variable $\epsilon \in (0,1)$. Independently of the particular choice of ϵ, the characteristic function

$$\mathbf{S}(s) = \frac{1}{2}\Big(\langle[\lambda+\epsilon(1-\lambda)]^s\rangle + \langle[1-\epsilon(1-\lambda)]^s\rangle + [\langle\epsilon^s\rangle + \langle(1-\epsilon)^s\rangle](1-\lambda)^s\Big) - 1 \tag{5.25}$$

is negative for all $s > 1$, hence case (ST) applies. Though the steady state f_∞ is no longer explicit – for approximations see [259,270] – one concludes that its tail is slim. In conclusion, no matter how sophisticated the trade mechanism chosen, one-dimensional, strictly conservative trades always lead to narrow, *socialistic* distributions of wealth.

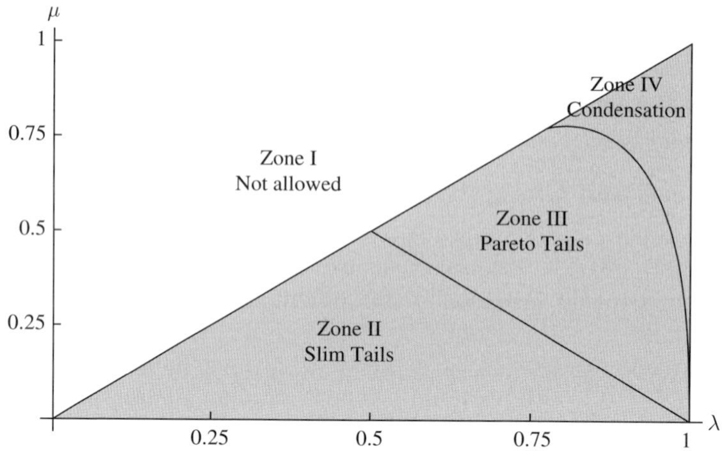

Fig. 5.3 The CPT model: regimes for the formation of Pareto tails.

The Cordier, Pareschi and Toscani model

In 2005, Cordier et al. [105] introduced the CPT model, which breaks with the paradigm of strict conservation. The idea is that wealth changes hands for a specific reason: one agent intends to *invest* his wealth in some asset, property, etc. in possession of his trade partner. Typically, such investments bear some risk, and either provide the buyer with some additional wealth, or lead to the loss of wealth in a non-deterministic way. An easy realization of this idea [229] consists in coupling the previously discussed rules (5.22) with some *risky investment* that yields an immediate gain or loss proportional to the current wealth of the investing agent,

$$v^* = \Big(\frac{1+\lambda}{2} + \eta_1\Big)v + \frac{1-\lambda}{2}w, \quad w^* = \Big(\frac{1+\lambda}{2} + \eta_2\Big)w + \frac{1-\lambda}{2}v. \tag{5.26}$$

The coefficients η_1, η_2 are random parameters, which are independent of v and w, and distributed so that $v^*, w^* \geq 0$, i.e. $\eta_1, \eta_2 \geq -\lambda$. Unless these random variables are centred, i.e. $\langle \eta_1 \rangle = \langle \eta_2 \rangle = 0$, it is immediately seen that the mean wealth is not preserved, but it increases or decreases exponentially (see the computations in [105]). For centred η_i,

$$\langle v^* + w^* \rangle = (1 + \langle \eta_1 \rangle)v + (1 + \langle \eta_2 \rangle)w = v + w, \tag{5.27}$$

implying conservation of the average wealth. Various specific choices for the η_i have been discussed [229]. The easiest one leading to interesting results is $\eta_i = \pm\mu$, where each sign comes with probability 1/2. The factor $\mu \in (0, \lambda)$ should be understood as the *intrinsic risk* of the market: it quantifies the fraction of wealth agents are willing to gamble on.

Figure 5.3 displays the various regimes for the steady state f_∞ depending on λ and μ, which follow from numerical evaluation of

$$\mathbf{S}(s) = \frac{1}{2}\Big[\Big(\frac{1+\lambda}{2} - \mu\Big)^s + \Big(\frac{1+\lambda}{2} + \mu\Big)^s\Big] + \Big(\frac{1-\lambda}{2}\Big)^s - 1. \tag{5.28}$$

Zone I is forbidden by the constraint $\mu < \lambda$. In Zone II, corresponding to low market risk, the wealth distribution again shows *socialistic* behaviour with slim tails. Increasing the risk, one falls into *capitalistic* Zone III, where the wealth distribution displays the desired Pareto tail. A minimum of saving ($\lambda > 1/2$) is necessary for this passage; this is expected since if wealth is spent too quickly after earning, agents cannot accumulate enough to become rich. Inside Zone III, the Pareto index α decreases from $+\infty$ at the border with Zone II to unity at the Zone IV border. Finally, in Zone IV, the steady wealth distribution is a delta in zero. Both risk and saving propensity are so high that a marginal number of individuals manage to monopolize all of the society's wealth. In the long-time limit, these few agents become infinitely rich, leaving all other agents truly poor.

Example 5.1 (Winner takes all). An illustration of the solution's behaviour in the (DD) case is provided by the winner takes all dynamics, described by (5.15). In this exceptional case, where $p_1 = q_1 = 1$ and $p_2 = q_2 = 0$, the Fourier-transformed equation (5.12) takes the simple form

$$\frac{\partial \widehat{f}(\xi,t)}{\partial t} + \widehat{f}(\xi,t) = \frac{1}{2}\left(\widehat{f}^2(\xi,t) + 1\right), \tag{5.29}$$

which can be equivalently written as

$$\frac{\partial \widehat{f}(\xi,t)}{\partial t} = \frac{1}{2}\left(\widehat{f}(\xi,t) - 1\right)^2. \tag{5.30}$$

We do not discuss uniqueness in detail here; this can be proven with an argument similar to Section 2.2.2. It is enough to remark that, taking $s = 1$, the right-hand side of equation (2.14) relative to the difference between two solutions of equation (5.29), namely

$$\frac{1}{2}\frac{\hat{f}_1(\xi)^2 - \hat{f}_2(\xi)^2}{|\xi|},$$

is such that

$$\frac{1}{2}\frac{|\hat{f}_1(\xi)^2 - \hat{f}_2(\xi)^2|}{|\xi|} \le \frac{1}{2}|\hat{f}_1(\xi) + \hat{f}_2(\xi)|\frac{|\hat{f}_1(\xi) - \hat{f}_2(\xi)|}{|\xi|} \le \sup_{\xi} \frac{|\hat{f}_1(\xi) - \hat{f}_2(\xi)|}{|\xi|}.$$

This implies that inequality (2.16) still holds, to give

$$d_1(f_1(t), f_2(t)) \le d_1(f_1(t=0), f_2(t=0)),$$

which implies uniqueness.

For any given ξ, let us introduce the function $\varrho(t) = \widehat{f}(\xi,t) - 1$. Then $\varrho(t)$ satisfies the differential equation

$$\frac{d\varrho}{dt} = \frac{1}{2}\varrho^2,$$

which has a solution given by

$$\varrho(t) = \left(\frac{1}{\varrho(0)} - \frac{t}{2}\right)^{-1}.$$

Therefore, the unique solution to equation (5.29) is explicitly given by

$$\hat{f}(\xi, t) = \frac{2\hat{f}_0(\xi) + (1 - \hat{f}_0(\xi))t}{2 + (1 - \hat{f}_0(\xi))t}. \tag{5.31}$$

Note that the solution (5.31) converges towards the (universal) steady state $\hat{f}_\infty(\xi) = 1$ with a rate of convergence which is linear in t. As noticed at the end of Section 2.5, while the steady state is explicitly known in the physical space (in this case $f_\infty(v) = \delta_0(v)$), knowledge of the explicit solution (5.31) at any time $t > 0$ in the Fourier space does not automatically imply knowledge of the solution $f(v, t)$ in the physical space. For a particular choice of the initial density, however, the solution can also be made explicit in the physical space. Consider as initial condition the exponential density, $f(w, t = 0) = \exp(-w)$. Then, since

$$\hat{f}_0(\xi) = \frac{1}{1 + i\xi}, \tag{5.32}$$

the solution (5.31) corresponding to (5.32) is

$$\hat{f}(\xi, t) = \frac{2 + i\xi t}{2 + i\xi(t + 2)}.$$

On the other hand, the numerator in the previous expression can be written as

$$2 + i\xi t = 2\left(\frac{2}{t+2} + \frac{t}{t+2}\right) + i\xi t = \frac{4}{t+2} + \frac{t}{t+2}\left(2 + i\xi(t+2)\right),$$

and

$$\hat{f}(\xi, t) = \frac{2}{t+2}\left(1 + i\xi\frac{t+2}{2}\right)^{-1} + \frac{t}{t+2}. \tag{5.33}$$

It remains to observe that, for any given constant $\alpha > 0$, the function

$$\hat{g}(\xi) = \frac{1}{1 + i\alpha\xi}$$

is the Fourier transform of the exponential density function of mean α,

$$g(w) = \frac{1}{\alpha}\exp\left\{-\frac{w}{\alpha}\right\},$$

to conclude with the explicit formula

$$f(w, t) = \left(\frac{2}{2+t}\right)^2 \exp\left(-\frac{2}{2+t}w\right) + \frac{t}{2+t}\,\delta_0(w), \tag{5.34}$$

which shows that the solution at any time $t > 0$ is the convex combination, with precise weights, of an exponential distribution of mean $(t+2)/2$ and a Dirac delta distribution concentrated at zero. Note that the average wealth is conserved at all finite times $t \geq 0$, so that the limit as $t \to \infty$ of $m_1(t)$ is equal to $m_1(0)$, but $f_\infty = \delta_0$ has vanishing average wealth. As predicted by the Fourier expression (5.31), at time $t > 0$ there is a mass of size $t/(t+2)$ concentrated at $w = 0$. ∎

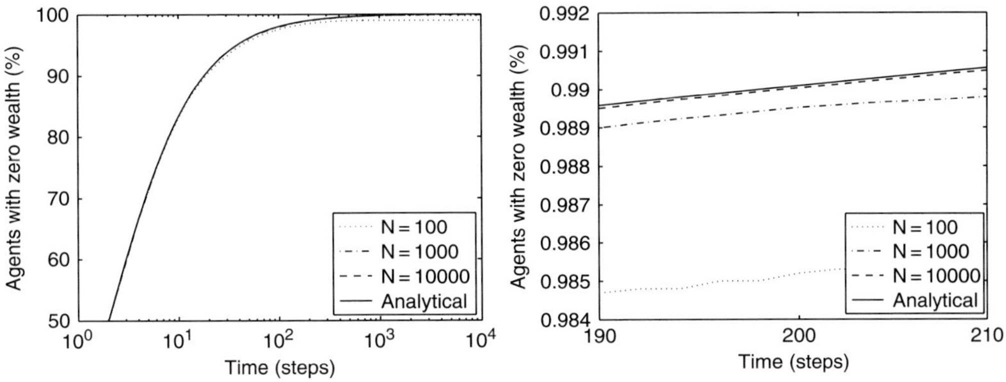

Fig. 5.4 Winner takes all model: Evolution of the fraction of agents with zero wealth (left) and blow up (right). Here, N is the total number of agents used in the Monte Carlo simulation.

5.3.2 Mathematical analysis of conservative models

This technical section deals with the mathematical analysis of the kinetic equation (5.13), with the principal objective to describe the large-time behaviour of the solution, its convergence to a stationary state, as well as its main properties. This analysis is complementary to the results of Section 5.5, where the simplicity of the binary interaction allows us to explicitly obtain steady states in many cases.

For general binary interactions it is indeed prohibitive to describe in a precise way the stationary solution, and the only possibility is to use mathematical instruments to investigate the accessible properties, like the description of its behaviour at infinity, as well as its regularity. It is interesting to remark that smoothness is a typical property of steady states which are not concentrated (like the one studied in Example 5.1).

The reader who is not interested in the details of the various proofs can simply look at the results by reading the statements of the forthcoming theorems and corollaries.

Following the analysis of Chapter 2, we resort to the Fourier transforms. The Fourier representation is indeed particularly adapted to the use of the Fourier metrics we introduced in Section 2.2.2. An auxiliary tool is the study of the evolution of moments.

The existence and uniqueness of solutions to equation (5.13) can easily be derived by resorting to the arguments of Chapter 2. Let f_1 and f_2 be two solutions of the kinetic equation (5.13), corresponding to initial values $f_{1,0}$ and $f_{2,0}$ in $\mathcal{P}_s$, with $s > 1$, and denote by $\widehat{f}_1$, $\widehat{f}_2$ their Fourier transforms. Let $s \ge 1$ be such that $d_s(f_{1,0}, f_{2,0})$ is finite. Then we obtain

$$\frac{\partial}{\partial t}\frac{\left(\widehat{f}_1(\xi) - \widehat{f}_2(\xi)\right)}{|\xi|^s} + \frac{\widehat{f}_1(\xi) - \widehat{f}_2(\xi)}{|\xi|^s} = \frac{\left\langle \widehat{f}_1(p_i\xi)\widehat{f}_1(q_i\xi) - \widehat{f}_2(p_i\xi)\widehat{f}_2(q_i\xi)\right\rangle_+}{|\xi|^s}. \tag{5.35}$$

Now, since $f_1(t)$ and $f_2(t)$ are probability densities, so that $|\widehat{f}_1(\xi,t)| \le 1$ and $|\widehat{f}_2(\xi,t)| \le 1$ for all $\xi \in \mathbb{R}$, we get

$$\left| \frac{\langle \widehat{f}_1(p_i\xi)\widehat{f}_1(q_i\xi) - \widehat{f}_2(p_i\xi)\widehat{f}_2(q_i\xi)\rangle_+}{|\xi|^s} \right|$$

$$\leq \left\langle |\widehat{f}_1(p_i\xi)| \left| \frac{\widehat{f}_1(q_i\xi) - \widehat{f}_2(q_i\xi)}{|q_i\xi|^s} \right| q_i^s \right\rangle_+ + \left\langle |\widehat{f}_2(q_i\xi)| \left| \frac{\widehat{f}_1(p_i\xi) - \widehat{f}_2(p_i\xi)}{|p_i\xi|^s} \right| p_i^s \right\rangle_+$$

$$\leq \sup_\xi \left| \frac{\widehat{f}_1(\xi) - \widehat{f}_2(\xi)}{|\xi|^s} \right| \langle p_i^s + q_i^s \rangle_+.$$

In terms of the auxiliary quantity

$$h(\xi, t) = \frac{\widehat{f}_1(\xi) - \widehat{f}_2(\xi)}{|\xi|^s},$$

the preceding computation shows that

$$\left| \frac{\partial h}{\partial t} + h \right| \leq \langle p_i^s + q_i^s \rangle_+ \|h\|_\infty.$$

Gronwall's lemma yields at once that

$$\|h(t)\|_\infty \leq \exp\left\{ (\langle p_i^s + q_i^s \rangle_+ - 1)t \right\} \|h_0\|_\infty. \tag{5.36}$$

This introduces into the game the quantity $\mathbf{S}(s)$ we defined in (5.14). Since $\|h(t)\|_\infty = d_s[f_1(t), f_2(t)]$, we obtain from (5.36)

$$d_s[f_1(t), f_2(t)] \leq \exp\left\{ \mathbf{S}(s) \cdot t \right\} d_s[f_{1,0}, f_{2,0}]. \tag{5.37}$$

In particular, if $\mathbf{S}(s)$ is negative, then the d_s distance of f_1 and f_2 decays exponentially in time.

Theorem 5.1 *Let $f_1(t)$ and $f_2(t)$ be two solutions of the Boltzmann equation (5.7), corresponding to initial values $f_{1,0}$ and $f_{2,0}$ in $\mathcal{P}_r$, $r > 1$. Let $s \geq 1$ be such that $d_s[f_{1,0}, f_{2,0}]$ is finite. Then, for all times $t \geq 0$, (5.37) holds.*

In particular, if $\mathbf{S}(s)$ is negative, then the Fourier-based distances of f_1 and f_2 decay exponentially in time.

Putting $f_{1,0} = f_{2,0} = f_0$ in (5.37), and using $s = 1$ yields:

Corollary 5.1 *If f_0 is a non-negative density in $\mathcal{P}_r$, $r > 1$. then there exists a unique weak solution $f(t)$ of the Boltzmann equation with $f(0) = f_0$.*

Evolution of moments

In Theorem 5.1 about the large-time behaviour of solutions to (5.7), the essential quantity $\mathbf{S}$ was introduced in a natural way by looking at the evolution in time of the Fourier-based metric. Below, we prove that the values $\mathbf{S}(s)$ also control the asymptotic behaviour of moments. In fact, if $\mathbf{S}(s)$ is negative for some $s > 0$, then the s-th moment of the solution,

$$m_s(t) = \int_{\mathbb{R}_+} v^s f(v,t)\,dv,$$

remains bounded for all times. On the other hand, if $\mathbf{S}(s)$ is positive for some $s > 1$, then $m_s(t)$ diverges exponentially fast as $t \to \infty$. We exploit this information to prove decay properties to the steady state.

To start with, we note that conservation of the total wealth allows us to conclude that at least all moments of order $s \le 1$ remain uniformly bounded. In fact, by Hölder's inequality,

$$\int_{\mathbb{R}_+} v^s f(v)\,dv \le \left(\int_{\mathbb{R}_+} v f(v)\,dv\right)^s \cdot \left(\int_{\mathbb{R}_+} f(v)\,dv\right)^{1-s} = m_1^s, \quad 0 < s < 1.$$

Now, let $s > 1$ and suppose that the initial density $f_0(v)$ satisfies

$$m_s(0) = \int_{\mathbb{R}_+} v^s f_0(v)\,dv < \infty. \tag{5.38}$$

Then, putting $\varphi(v) = v^s$ in the weak form (5.8), we obtain

$$\frac{d}{dt}\int_{\mathbb{R}_+} v^s f(v,t)\,dv = \frac{1}{2}\left\langle \int_{\mathbb{R}_+^2} \left(\sum_{i=1}^{2}(p_i v + q_i w)^s - v^s - w^s\right) f(v)f(w)\,dv\,dw \right\rangle_+. \tag{5.39}$$

In the following, we establish upper and lower bounds for the right-hand side of (5.39). These bounds rely on the following elementary inequality.

Lemma 5.1 *For arbitrary non-negative real numbers a and b, and $s > 1$,*

$$a^s + b^s + \theta_s(a^{s-1}b + ab^{s-1}) \le (a+b)^s \le a^s + b^s + \Theta_s(a^{s-1}b + ab^{s-1}), \tag{5.40}$$

$$\textit{with}\quad \theta_s = \begin{cases} s & (s > 3) \\ 2^{s-3}s & (2 \le s \le 3) \\ 0 & (1 < s < 2) \end{cases} \quad \textit{and}\quad \Theta_s = \begin{cases} s & (2 \le s \le 3) \\ 2^{s-3}s & (\textit{otherwise}). \end{cases}$$

Remark 5.1 *An investigation of the limit behaviour as $a \to 0$ and $b > 0$ makes clear that $\theta_s = 0$ for $1 < s < 2$ cannot be improved in general.*

Proof. By homogeneity, it suffices to prove the inequality for $a + b = 1$. Define, for $s > 1$,

$$\phi(s) = a^s + b^s + sab.$$

A calculation yields $\phi(2) = \phi(3) = 1$, independently of a and $b = 1 - a$. Furthermore, ϕ is convex in s since

$$\phi''(s) = a^s \log^2 a + b^s \log^2 b \geq 0.$$

Hence, $\phi(s) \leq 1$ if and only if $2 \leq s \leq 3$. Observe that $a^{s-2} + b^{s-2}$ is concave with respect to $a = 1 - b \in (0, 1)$ for $2 \leq s \leq 3$, and convex for all other $s > 1$; the expression attains its extremal value 2^{3-s} at $a = b = \frac{1}{2}$. Hence,

$$a^{s-1}b + ab^{s-1} = ab(a^{s-2} + b^{s-2}) \begin{cases} \leq 2^{3-s}ab \ (2 \leq s \leq 3) \\ \geq 2^{3-s}ab \ (otherwise). \end{cases}$$

Thus we obtain, for $2 \leq s \leq 3$,

$$a^s + b^s + 2^{s-3}s(a^{s-1}b + ab^{s-1}) \leq \phi(s) \leq 1 = (a+b)^s,$$

and with reversed inequalities for $1 < s < 2$ or $s > 3$.

Now let $s > 1$ be fixed and consider for $a \in [0, 1]$ the quantity

$$K_s(a) = a^s + b^s + s(a^{s-1}b + ab^{s-1}),$$

with $b = 1 - a$. Observe that $K_s(0) = K_s(1) = 1$, and furthermore

$$K_s'(a) = s(s-1)(a^{s-2}b - ab^{s-2}),$$

so that f_s has $\bar{a} = \frac{1}{2}$ as the only critical point in $(0, 1)$. For $s < 3$, one has $K_s(\bar{a}) = (s+1)2^{1-s} > 1$, so $\bar{a}$ is a maximum point and hence $K_s(a) \geq 1$ for $a \in [0, 1]$; for $s > 3$, it is a minimum point and $K_s(a) \leq 1$. Consequently, for $s \leq 3$,

$$a^s + b^s + s(a^{s-1}b + ab^{s-1}) = K_s(a) \geq 1 = (a+b)^s.$$

The reversed inequality holds for $s \geq 3$. ∎

Using the upper bound in (5.40), estimate

$$(p_i v + q_i w)^s \leq p_i^s v^s + q_i^s w^s + \Theta_s(p_i^{s-1}q_i v^{s-1}w + p_i q_i^{s-1} v w^{s-1})$$

under the integral in (5.39), leading to:

$$\begin{aligned} \frac{d}{dt}m_s(t) \leq \frac{1}{2}\Big\langle \int_{\mathbb{R}_+^2} &\big((p_1^s + p_2^s - 1)v^s + (q_1^s + q_2^s - 1)w^s\big) f(v)f(w)\,dv\,dw \Big\rangle \\ + \frac{\Theta_s}{2}\Big\langle \int_{\mathbb{R}_+^2} &\big((p_1^{s-1}q_1 + p_2^{s-1}q_2)v^{s-1}w \\ &+ (p_1 q_1^{s-1} + p_2 q_2^{s-1})vw^{s-1}\big) f(v)f(w)\,dv\,dw \Big\rangle \\ = \mathbf{S}(s)m_s(t) + \Theta_s\langle p_i^{s-1}q_i &+ p_i q_i^{s-1}\rangle_+ \cdot m_1(t) \cdot \int_{\mathbb{R}_+} w^{s-1} f(w)\,dw. \end{aligned} \tag{5.41}$$

Recall that the total wealth $\int v f(v)\,dv = m_1$ is conserved in time. Further, by Hölder's inequality, it follows that

$$\int_{\mathbb{R}_+} w^{s-1} f(w)\,dw \le \left(\int_{\mathbb{R}_+} w^s f(w)\,dw\right)^{1-\frac{1}{s}}.$$

Hence, we obtain

$$\frac{d}{dt} m_s(t) \le \mathbf{S}(s) m_s(t) + \Theta_s \mathfrak{K}(s) m_1 \cdot \left(\int_{\mathbb{R}_+} v^s f(v)\,dv\right)^{1-\frac{1}{s}}, \tag{5.42}$$

where

$$\mathfrak{K}(s) = \langle p_i^{s-1} q_i + p_i q_i^{s-1} \rangle_+ \le \langle p_i^s + q_i^s \rangle_+ = \mathbf{S}(s) + 1. \tag{5.43}$$

In particular, if $\mathbf{S}(s)$ is a finite number, then so is $\mathfrak{K}(s)$. In this case, the s-th moment grows at most exponentially, with rate $\mathbf{S}(s)$, if it was finite initially. Moreover, if $\mathbf{S}(s) < 0$, then the s-th moment remains uniformly bounded for all times. In fact, an upper bound on the s-th moment is determined by the solution of the associated ordinary differential equation

$$\dot{y} = \mathbf{S} \cdot y + \Theta_s \mathfrak{K}(s) m_1 y^{1-\frac{1}{s}}$$

with initial condition $y(0) = m_s(0)$. The solution is explicitly given by

$$y(t) = \left[\left(m_s(0) \exp\left\{t\mathbf{S}(s)\right\}\right)^{1/s} + \frac{\Theta_s \mathfrak{K}(s) m_1}{\mathbf{S}(s)} \left(\exp\left\{\frac{t}{s}\mathbf{S}(s)\right\} - 1\right)\right]^s.$$

Notice that the first term in the square bracket vanishes for $t \to \infty$ if $\mathbf{S}(s) < 0$, so that the limiting value depends on the initial condition f_0 only through the total wealth m_1.

By the same reasoning as above, we construct a bound *from below* on the time derivative of the integral. For this, we use the lower bound given in the elementary inequality (5.40). Replacing the respective expressions under the integral, we obtain

$$\begin{aligned}
\frac{d}{dt} m_s(t) \ge{}& \frac{1}{2} \int_{\mathbb{R}_+^2} \left(\langle p_1^s + p_2^s - 1\rangle v^s + \langle q_1^s + q_2^s - 1\rangle w^s\right) f(v) f(w)\,dv\,dw \\
&+ \frac{\theta_s}{2} \Big\langle \int_{\mathbb{R}_+^2} \Big((p_1^{s-1} q_1 + p_2^{s-1} q_2) v^{s-1} w \\
&\qquad + (p_1 q_1^{s-1} + p_2 q_2^{s-1}) v w^{s-1} \Big) f(v) f(w)\,dv\,dw \Big\rangle \\
={}& \mathbf{S}(s) m_s(t) + \theta_s \mathfrak{K}(s) m_1 \cdot \int_{\mathbb{R}_+} w^{s-1} f(w)\,dw.
\end{aligned}$$

We use Hölder's inequality to estimate

$$\int_{\mathbb{R}_+} w^{s-1} f(w)\, dw \geq \left(\int_{\mathbb{R}_+} w f(w)\, dw \right)^{s-1} = m_1^{s-1}.$$

By Gronwall's inequality, a lower bound is given by

$$m_s(t) \geq m_s(0) \cdot \exp\{t \cdot \mathbf{S}(s)\} + \frac{\theta_s \cdot \mathfrak{K}(s) \cdot m_1^s}{\mathbf{S}(s)} \left(\exp\{t \cdot \mathbf{S}(s)\} - 1 \right).$$

We conclude that if $\mathbf{S}(s) > 0$, then the moment m_s diverges exponentially in time.

In the special case that $\mathbf{S}(s) = 0$, similar but simpler arguments give that the corresponding moment remains either bounded (iff $\mathcal{K}(s) = 0$) or diverges, but only at the algebraic rate t^s. Finally, if $\mathbf{S}(s) = +\infty$, an easy argument shows that the s-th moment of $f(v,t)$ must be infinite for all positive times.

Theorem 5.2 **[229]** *Consider a solution f to the Boltzmann equation (5.7). Let $s > 1$ be such that $M_s(0) = \int_{\mathbb{R}_+} v^s f_0(v)\, dv < \infty$.*

1. *If $0 < \mathbf{S}(s) < +\infty$, then the s-th moment diverges exponentially fast as $t \to \infty$:*

$$m_s(0) + \frac{\theta_s \mathfrak{K}(s)}{\mathbf{S}(s)} \cdot m_1^s + o(1) \leq \frac{M_s(t)}{\exp\{t \cdot \mathbf{S}(s)\}} \leq \left[m_s^{1/s}(0) + \frac{\Theta_s \mathfrak{K}(s)}{\mathbf{S}(s)} \cdot m_1 \right]^s + o(1). \quad (5.44)$$

2. *If $\mathbf{S}(s) < 0$, then the s-th moment remains uniformly bounded as $t \to \infty$:*

$$\frac{\theta_s \mathfrak{K}(s)}{|\mathbf{S}(s)|} \cdot m_1^s + o(1) \leq M_s(t) \leq \left(\frac{\Theta_s \mathfrak{K}(s)}{|\mathbf{S}(s)|} \right)^s \cdot m_1^s + o(1). \quad (5.45)$$

3. *If $\mathbf{S}(s) = 0$, then the s-th moment either remains bounded or diverges at an algebraic rate:*

$$\left(\frac{\theta_s \mathfrak{K}(s)}{s} \right)^s \cdot m_1^s + o(1) \leq t^{-s} \cdot M_s(t) \leq \left(\frac{\Theta_s \mathfrak{K}(s)}{s} \right)^s \cdot m_1^s + o(1). \quad (5.46)$$

4. *Finally, if $\mathbf{S}(s) = +\infty$, then the s-th moment is infinite for all $t > 0$.*

Existence and tails of the steady state

The previous analysis shows that the long-time behaviour of solutions is essentially determined by the quantity $\mathbf{S}$. For this reason, let us investigate this function in more detail.

First, recall that, for an arbitrary non-negative number p, the exponential $s \mapsto p^s$ is convex in $s > 0$. Hence $\mathbf{S}(s)$, which is the average of convex functions, is convex on its domain. By the dominated convergence theorem, $\mathbf{S}(s)$ is well defined at least for $0 < s \leq 1$, but possibly $\mathbf{S}(s) = +\infty$ for all $s > s_\infty > 1$. Since $\mathbf{S}(1) = 0$, convexity leaves only three possibilities for the behaviour of $\mathbf{S}$:

1. $\mathbf{S}(s)$ is non-negative for all $s > 0$.

2. $\mathbf{S}(s)$ is negative for some $s \in (0,1)$, and positive for all $s > 1$.
3. $\mathbf{S}(s)$ is negative for all $1 < s < \bar{s}$, and positive for all $s < 1$ and all $s > \bar{s}$; here, either $\bar{s} = +\infty$, or $\mathbf{S}(\bar{s}) = 0$, or $\mathbf{S}(s) = +\infty$ for $s > \bar{s}$.

If $\mathbf{S}$ is differentiable at $s = 1$, then the first case corresponds to $\mathbf{S}'(1) = 0$, the second to $\mathbf{S}'(1) > 0$ and the last to $\mathbf{S}'(1) < 0$. These cases are now discussed in detail.

In the **first case**, no information about the existence of a long-time limit can be extracted.

In the **second case**, $r = -\mathbf{S}(s) > 0$ for some $s \in (0,1)$. Observe that $f_* = \delta_0$, corresponding to a mass concentrated at $v = 0$, trivially constitutes a stationary solution of the Boltzmann equation. Recall that the initial condition f_0 is a probability density of finite first moment. Since $s < 1$, it follows that $K = d_s(f_0, f_*) < \infty$. By the contraction estimate, $d_s(f(t), f_*) \leq K \exp(-rt) \to 0$ as $t \to \infty$. Thus, the solution $f(t)$ converges weakly to f_*. It is worthwhile to observe that, by Theorem 5.1, all moments $m_s(t)$ with $s > 1$ diverge as $t \to \infty$. So, although each $f(t)$ for $t > 0$ has the same (positive and finite) first moment as f_0, one *cannot* invoke Prokhorov's theorem [269] to conclude that the weak limit f_* also has a positive first moment.

The **third case** is the most interesting one. Choose some $s \in (1,2)$ with $s < \bar{s}$; then $r = -\mathbf{S}(s) > 0$. Assume that the initial datum of f possesses a moment of order $S > s$. In view of the completeness of $(\mathcal{P}_s, d_s)$ when $s \in (1,2)$ [22], the existence of the long-time limit can be concluded directly from the contractivity of the kinetic equation in d_s norm (cf. Theorem 5.1).

The same result can be achieved by means of the metric d_s. In fact, $f(v,t)$ has the Cauchy property in d_s; notice that $d_s[f(t), f_0]$ is always finite since $s < 2$ and the first moment (mean wealth) is conserved under evolution. Moreover, as we required $s < \bar{s}$, there exists s' with $s < s' < \min(\bar{s}, S)$; by Theorem 5.2, the moment of order s' remains uniformly bounded. It follows that $f(t)$ converges in d_s to a limit distribution $f_\infty(v)$, which is normalized and has the same first moment as $f(t)$.

This convergence implies that f_∞ is a steady state for the kinetic equation (5.7). Indeed, denote by $f_\infty(t)$ the solution to (5.7) with initial datum f_∞, then Theorem 5.1 gives

$$\begin{aligned} d_s[f_\infty(t), f_\infty] &\leq d_s[f_\infty(t), f(t+T)] + d_s[f(t+T), f_\infty] \\ &\leq e^{-rt} d_s[f_\infty, f(T)] + d_s[f(t+T), f_\infty]. \end{aligned}$$

The last expression can be made arbitrarily small by choosing T large enough, so that $f_\infty(t) = f_\infty$ for all $t \geq 0$. In fact, f_∞ is the *only* steady state with the respective value of the first moment; for if f'_∞ is another steady state with the same first moment, then $d_s[f_\infty, f'_\infty]$ is finite, and so, invoking Theorem 5.1 again,

$$d_s[f_\infty, f'_\infty] \leq e^{-r} d_s[f_\infty, f'_\infty],$$

which forces $f_\infty = f'_\infty$. Finally, consider a solution f which has arbitrary moments bounded initially. Theorem 5.2 gives a time-uniform bound for moments of order less than $\bar{s}$. Since convergence $f(t) \to f_\infty$ in d_s implies weak* convergence of the associated measures, it follows that f_∞ has finite moments of all orders less than $\bar{s}$. On the other hand, no moment of order larger than $\bar{s}$ can be finite. To see this, simply apply Theorem 5.2 to the steady-state solution $f_\infty(t) \equiv f_\infty$ to derive a contradiction.

We summarize the results of this section.

Theorem 5.3 [229] *Let $f(v,t)$ be the (unique) weak solution of the Boltzmann equation (5.7), which has initially finite moments up to order $S > 0$. Further assume that $\mathbf{S}(s) < 0$ for some $s \in (0, S)$. Then f converges exponentially fast in d_s to a steady state f_∞,*

$$d_s[f(t), f_\infty] \le d_s[f_0, f_\infty] \exp\{-|\mathbf{S}(s)|t\}. \tag{5.47}$$

If $s < 1$, then f_∞ is a Dirac distribution centred at $v = 0$, and there are no other steady states. If $s > 1$, then f_∞ has mean wealth equal to M, and it is the only steady state with this mean wealth. Moreover, if $\mathbf{S}(s') < 0$ exactly for $1 < s' < \bar{s}$, possibly with $\bar{s} = +\infty$, then f_∞ has finite moments of all orders less than $\bar{s}$, while moments of order larger than $\bar{s}$ are infinite.

Regularity of the steady wealth distribution

Even if the most important property of the steady-state distribution is its behaviour for large values of the wealth variable, which clarifies the eventual formation of Pareto tails, other characteristics can be extracted by a direct investigation of the collision integral. In particular, it is surprising that in most cases one obtains that the steady state is smooth.

Theorem 5.4 [229] *Assume that there are positive numbers r and δ such that*

$$p_i^r + q_i^r \ge 1 + \delta^r \quad a.s. \tag{5.48}$$

Assume further that $\mathbf{S}(s) < 0$ for some $s > 1$, so that a non-trivial steady state f_∞ to the kinetic equation (5.7) exists. Then either f_∞ is a Dirac distribution centred at $v = M$, or f_∞ is a smooth function and belongs to the r-th Gevrey class, i.e.

$$\big|\hat{f}_\infty(\xi)\big| \le \exp\big(-\mu|\xi|^r\big) \quad for\ |\xi| \ge \rho,$$

with suitable positive numbers ρ and μ [276].

Remark 5.2 *Since the mean wealth is preserved, i.e. $\langle p_i + q_i\rangle_+ = 1$, it follows that $r < 1$ in (5.48). Notice that condition (5.48) imposes no serious restrictions on the distribution of the interaction coefficients. For instance, it allows any probability measure which is compactly supported in the interior of the first quadrant of the p_i-q_i plane.*

The proof of Theorem 5.4 is done at the level of Fourier transforms, and is divided into a series of Lemmas. The first is concerned with the behaviour of $\hat{f}_\infty$ near the origin:

Lemma 5.2 *Under the hypotheses of Theorem 5.4, either f_∞ is a Dirac distribution, i.e. $|\hat{f}| \equiv 1$, or there exist positive numbers ρ and κ s.t.*

$$|\hat{f}_\infty(\xi)| \le \exp(-\kappa\xi^2) \quad for\ |\xi| < \rho. \tag{5.49}$$

Proof. The following proof uses no properties of f_∞ other than the finiteness of the first moment. First assume that f_∞ also has a finite second moment. Then $\hat{f}_\infty$ is twice differentiable at $\xi = 0$, and

$$\hat{f}_\infty(\xi) = 1 - iM\xi - \frac{1}{2}Q\xi^2 + o(\xi^2),$$

where M and Q denote the first and second moment of f_∞, respectively. This means that

$$|\hat{f}_\infty(\xi)|^2 = 1 - (Q - M^2)\xi^2 + o(\xi^2).$$

Now either $Q = M^2$, which implies that f_∞ is concentrated at $\xi = M$, or $\kappa = \frac{1}{4}(Q - M^2) > 0$ makes (5.49) true for some small $\rho > 0$.

In the case that the second moment of $\hat{f}_\infty$ is infinite, we write

$$\hat{f}_\infty(\xi) = \int_{\mathbb{R}} f_\infty(v)e^{-i\xi v}\,dv = 1 - \int_{\mathbb{R}} \big(1 - \cos(\xi v)\big)df_\infty(v) - i\int_{\mathbb{R}} \sin(\xi v)df_\infty(v),$$

so that, with the obvious meaning of H_1 and H_2,

$$|\hat{f}_\infty(\xi)|^2 = 1 - 2H_1(\xi) + H_1(\xi)^2 + H_2(\xi)^2.$$

The elementary inequalities $1 - \cos x \le |x|$ and $|\sin x| \le |x|$ for $x \in \mathbb{R}$ imply

$$H_1(\xi) \le |\xi| \int_{\mathbb{R}} v\,df_\infty(v) = M|\xi|, \qquad |H_2(\xi)| \le |\xi| \int_{\mathbb{R}} v\,df_\infty(v) = M|\xi|.$$

On the other hand, with the elementary inequalities

$$1 - \cos x \ge \frac{1}{3}x^2 \quad \text{for } |x| \le 1, \qquad 1 - \cos x \ge 0 \quad \text{for } x \in \mathbb{R},$$

one obtains, for arbitrary $\xi \in \mathbb{R}$,

$$H_1 \ge \frac{\xi^2}{3} \int_{|\xi v|<1} v^2 df_\infty(v) = A(\xi)\xi^2.$$

Here, $A(\xi)$ is a positive, non-increasing function with $\lim_{\xi\to 0} A(\xi) = +\infty$, since the second moment of f_∞ diverges. Choosing $\rho > 0$ so small that $A(\rho) \ge 2M^2$, it follows that

$$|\hat{f}_\infty(\xi)|^2 \le 1 - 2M^2\xi^2 \quad \text{for } |\xi| < \rho.$$

Hence (5.49) holds with $\kappa = \frac{1}{2}M^2$, possibly after diminishing ρ suitably. ∎

From now on, we will make use of a method first introduced by Bobylev and Cercignani in [45]. Gevrey regularity for the elastic Boltzmann equation for Maxwell molecules recently has been shown in [125]. We introduce the fixed point operator

$$\mathcal{R}[\psi](\xi) = \begin{cases} \hat{f}_\infty(\xi) & \text{if } |\xi| < \rho \\ \langle \psi(p_i\xi)\psi(q_i\xi)\rangle_+ & \text{if } |\xi| \geq \rho \end{cases}$$

on bounded functions $\psi : \mathbb{R} \to \mathbb{C}$. Notice that $\mathcal{R}$ is closely related to the Fourier transform of the collision kernel. Furthermore, for $\mu \geq 0$, define K_μ as the set of functions ψ with $\psi(\xi) = \hat{f}_\infty(\xi)$ for $|\xi| < \rho$, satisfying the estimate

$$|\psi(\xi)| \leq \exp(-\mu|\xi|^r) \quad \text{for } |\xi| \geq \rho. \tag{5.50}$$

Here, ρ comes from the constant in Lemma 5.2 above. Eventually, we wish to prove that $\hat{f}_\infty \in K_\mu$ for a suitable $\mu > 0$.

Lemma 5.3 *Under the hypotheses of Theorem 5.4, and assuming that f_∞ is not a Dirac distribution, there exists some $\mu > 0$, depending only on ρ and κ from* (5.49), *such that $\mathcal{R}$ maps K_μ into itself.*

Proof. Let $\psi \in K_\mu$ and $\xi \in \mathbb{R}$ with $|\xi| \geq \rho$ be arbitrary. Lemma 5.2 assures that the estimate (5.49) holds with ψ in place of $\hat{f}_\infty$. We show that

$$e^{\mu|\xi|^r} \cdot \big|\mathcal{R}[\psi](\xi)\big| \leq \langle e^{\mu|\xi|^r} \cdot |\psi(p_i\xi)| \cdot |\psi(q_i\xi)|\rangle_+ \leq 1. \tag{5.51}$$

To this end, we prove that the term inside the expectation value is less than or equal to 1 a.s. We distinguish four cases:

1. If both $p_i|\xi| \geq \rho$ and $q_i|\xi| \geq \rho$, then the hypothesis (5.48) yields

$$e^{\mu|\xi|^r} \cdot |\psi(p_i\xi)| \cdot |\psi(q_i\xi)| \leq e^{\mu|\xi|^r - \mu p_i^r|\xi|^r - \mu q_i^r|\xi|^r} \leq e^{-\mu|\xi|^r\delta^r} \leq 1.$$

2. If both $p_i|\xi| < \rho$ and $q_i|\xi| < \rho$, then we conclude from (5.48) that $p_i^2 + q_i^2 \geq 2^{1-2/r}$, so

$$e^{\mu|\xi|^r} \cdot |\psi(p_i\xi)| \cdot |\psi(q_i\xi)| \leq e^{\mu|\xi|^r - \kappa p_i^2|\xi|^2 - \kappa q_i^2|\xi|^2} \leq e^{-|\xi|^r(\mu - \kappa\cdot 2^{1-2/r}\rho^{2-r})} \leq 1,$$

   provided that $\mu \leq 2^{1-2/r}\rho^{2-r} \cdot \kappa$.

3. Now assume that $p_i|\xi| < \rho$ while $q_i|\xi| \geq \rho$. Using (5.48) once again, one finds

$$e^{\mu|\xi|^r} \cdot |\psi(p_i\xi)| \cdot |\psi(q_i\xi)| \leq e^{\mu|\xi|^r - \kappa p_i^2|\xi|^2 - \mu q_i^r|\xi|^r} \leq e^{\mu(p_i^r - \delta^r)|\xi|^r - \kappa p_i^2|\xi|^2}.$$

   Obviously, the last expression is less than 1 if $p_i < \delta$. Assume now that $p_i \geq \delta$. Since $|\xi| \geq \rho$, one has $p_i|\xi| \geq \delta\rho$ and hence the exponent can be estimated by

$$\mu p_i^r|\xi|^r - \kappa p_i^2|\xi|^2 \leq p_i^r|\xi|^r(\mu - \kappa\delta^{2-r}\rho^{2-r}).$$

   The last expression is negative provided that $\mu \leq (\delta\rho)^{2-r} \cdot \kappa$.

4. The last case, $p_i|\xi| \geq \rho$ and $q_i|\xi| < \rho$, is treated in the same way and yields the same condition on μ.

In summary, (5.51) holds if μ is the minimum of $2^{1-2/r}\rho^{2-r} \cdot \kappa$ and $(\delta\rho)^{2-r} \cdot \kappa$. ■

Lemma 5.4 *Under the hypotheses of Theorem 5.4, and assuming that f_∞ is not a Dirac distribution, $\hat{f}_\infty \in K_\mu$ with the μ from Lemma 5.3 above.*

Proof. Let $\psi_0(\xi) = \hat{f}_\infty(\xi)$ for $|\xi| < \rho$, and $\psi_0(\xi) = 0$ for $|\xi| \geq \rho$. Define inductively the sequence $\psi_{n+1} = \mathcal{R}[\psi_n]$. By Lemma 5.3, each ψ_n belongs to $K_\mu \subset K_0$. Moreover, by the same estimates that led to Theorem 5.1, one proves that $\mathcal{R}$ acts as a contraction on K_0 in the following sense. Clearly, $\mathcal{R}$ maps K_0 into itself, and any two functions ψ, $\varphi \in K_0$ satisfy

$$\frac{\left|\mathcal{R}[\psi](\xi) - \mathcal{R}[\varphi](\xi)\right|}{|\xi|^s} \leq \sup_{\xi' \neq 0} \left(\frac{|\psi(\xi') - \varphi(\xi')|}{|\xi'|^s}\right) \cdot \langle p_i^s + q_i^s \rangle_+. \tag{5.52}$$

The supremum on the right-hand side is obviously finite since $\psi(\xi') = \varphi(\xi') = \hat{f}_\infty(\xi')$ for $|\xi'| < \rho$. By hypothesis, $\mathbf{S}(s) < 0$, so $0 < \langle p_i^s + q_i^s \rangle_+ < 1$, so (5.52) is a genuine contraction estimate and thus implies local uniform convergence of the sequence ψ_n to a limit function ψ_∞. Local uniform convergence in combination with the pointwise estimates (5.50) is more than enough to invoke Lebesgue's dominated convergence theorem and conclude

$$\psi_\infty(\xi) = \langle \psi_\infty(p_i\xi)\psi_\infty(q_i\xi)\rangle_+, \qquad \text{for } |\xi| \geq \rho, \tag{5.53}$$

i.e. ψ_∞ is a fixed point of the operator $\mathcal{R}$. It is, in fact, the *unique* fixed point of $\mathcal{R}$ on K_0 because of its contractivity (5.52). But, clearly, $\hat{f}_\infty \in K_0$ is *one* fixed point. So $\hat{f}_\infty = \psi_\infty \in K_\mu$. ■

By definition of K_μ, Theorem 5.4 now follows immediately from Lemma 5.4.

5.4 Non-conservative kinetic market models

A crucial assumption made for the models considered in Section 5.3.2 is the conservation (at least in a statistical sense) of the average wealth per agent, i.e. the first moment of the wealth distribution, over time. Wealth conservation sounds plausible on a microscopic level, whereas on a macroscopic level it is arguable that the apparent conservation is, in reality, a mixture of two effects. On one hand, *wealth is created* through the production of goods, interests on savings, etc. On the other hand, (monetary) *wealth is lost* through inflation.

Kinetic models which take these two effects into account were proposed in [105,287], and were further developed in [254]. In order to incorporate the *creation* of wealth, the respective trade rules are designed to reward agents for trading activity. In the CPT model (5.26), this can be achieved by assuming that the market risk satisfies

$\langle \eta_i \rangle = \varepsilon > 0$. In other words, the risky investment is more likely to create additional wealth than to destroy existing wealth. *This is a genuine motivation for agents to engage in trades!* The effect of inflation is modelled by a time-dependent rescaling $f(w,t) \rightsquigarrow g(v,t)$ of the wealth distribution,

$$g(v,t) = e^{\varepsilon t} f(w,t), \quad w = e^{\varepsilon t} v, \tag{5.54}$$

chosen so that the mean wealth of $g(v,t)$ is kept constant. *The monetary unit is adapted in such a way that people stay equally wealthy on the average.* As discussed in Section 1.5, the kinetic equation (5.7) is respectively modified by an additional drift term,

$$\frac{\partial}{\partial t} g = Q_+(g,g) - g + \varepsilon \frac{\partial}{\partial v}(v\, g). \tag{5.55}$$

Solutions to this equation have been proven [254] to converge to a steady state g_∞, which may or may not have a Pareto tail. Again, the evolution of moments can be analyzed, and leads to a classification of the tail size in terms of properties of the η_i.

Slanina's model

Motivated by the analogy with a dissipative Maxwell gas, F. Slanina [287] introduced an increasing wealth model where

$$p_1 = \lambda, \quad q_1 = 1 - \lambda + \varepsilon; \qquad p_2 = 1 - \lambda + \varepsilon, \quad q_2 = \lambda. \tag{5.56}$$

In (5.56), the growth parameter ε is a fixed positive constant, which implies that the total wealth has increased after the trade,

$$v^* + w^* = (1+\varepsilon)(v+w). \tag{5.57}$$

The aforementioned model is a particular case of the general mixing model introduced in [30], where the collision rules are of the type (1.83).

Similarly to the p_i and q_i in (5.6), the interaction parameters p and q determine the agents' behaviour in binary trades. It suffices to consider *constant* p and q instead of random coefficients. Choosing $\varphi(v) = v$ in (5.8) shows that

$$m(t) = \int_{\mathbb{R}+} v f(v,t)\, dv = m(0) \exp\left\{(p+q-1)t\right\}. \tag{5.58}$$

Therefore, unless $p+q=1$, which implies that the model is pointwise conservative, the mean wealth is increasing if $p+q>1$, or decreasing if $p+q<1$, exponentially in time. In both cases, however, stationary solutions do not exist. However, the large–time behaviour of the model can be described by *self-similar* solutions.

The standard way to look for self–similarity is to scale the solution according to the rule indicated in (5.54). More precisely, define the rescaled solution g by

$$g(v,t) = m(t) f\big(m(t)v, t\big), \tag{5.59}$$

which implies that $\int v g(v,t)\, dv = 1$ for all $t \geq 0$. Moreover, $g = g(v,t)$ satisfies the equation

$$\frac{d}{dt}\int_{\mathbb{R}_+} \varphi(v)g(v,t)\,dv - (p+q-1)\int_{\mathbb{R}_+} \varphi(v)\frac{\partial}{\partial v}(vg)\,dv$$
$$= \int_{\mathbb{R}_+^2} g(v)g(w)(\varphi(v^*) - \varphi(v))\,dv\,dw. \tag{5.60}$$

Choosing $\varphi(v) = e^{-sv}$, with $s \in \mathbb{R}_+$, gives the representation of equation (5.60) in terms of the Laplace transform $\tilde{g}$ of g,

$$\frac{\partial \tilde{g}}{\partial t} + s(p+q-1)\frac{\partial \tilde{g}}{\partial s} = \tilde{g}(ps)\tilde{g}(qs) - \tilde{g}(s). \tag{5.61}$$

Steady solutions to equation (5.61) satisfy

$$s(p+q-1)\frac{\partial \tilde{g}}{\partial s} = \tilde{g}(ps)\tilde{g}(qs) - \tilde{g}(s). \tag{5.62}$$

Suppose $\sqrt{p} + \sqrt{q} = 1$. Then, since

$$p + q - 1 = -\sqrt{2p}\sqrt{2q},$$

direct computations show that the function

$$\tilde{g}_\infty(s) = \left(1 + \sqrt{2s}\right) e^{-\sqrt{2s}} \tag{5.63}$$

solves (5.62).

The steady solution (5.63) is independent of the values of p and q, provided that $\sqrt{p}+\sqrt{q} = 1$. This gives us the possibility of obtaining its analytic expression without resorting to the inverse Laplace transform.

5.4.1 Fokker–Planck equations

Let us use a second-order Taylor expansion of $\varphi(v^*)$ around v,

$$\varphi(v^*) - \varphi(v) = \big((p-1)v + qw\big)\varphi'(v) + \frac{1}{2}\big((p-1)v + qw\big)^2\varphi''(\tilde{v}),$$

where, for some $0 \le \theta \le 1$,

$$\tilde{v} = \theta v^* + (1-\theta)v.$$

Inserting this expansion in the collision operator, we obtain the equality

$$\frac{d}{dt}\int_{\mathbb{R}_+} \varphi(v)g(v,t)\,dv + q\int_{\mathbb{R}_+} \varphi'(v)(v-1)g(v)\,dv$$
$$= \frac{1}{2}\int_{\mathbb{R}_+} g(v)\left((p-1)^2v^2 + q^2w^2 + 2(p-1)qvw\right)\varphi''(v)\,dv + R(p,q),$$

where

$$R(p,q) = \frac{1}{2}\int_{\mathbb{R}_+^2} \big((p-1)v + qw\big)^2 \big(\varphi''(\tilde{v}) - \varphi''(v)\big)g(v)g(w)\,dv\,dw. \tag{5.64}$$

For small values of the parameter q, let us set

$$\tau = qt, \quad h(v,\tau) = g(v,t), \tag{5.65}$$

which implies that $h(v,\tau)$ satisfies the equation

$$\begin{aligned} &\frac{d}{d\tau}\int_{\mathbb{R}_+} \varphi(v)h(v,\tau)\,dv + \int_{\mathbb{R}_+} \varphi'(v)(v-1)h(v)\,dv \\ &\quad = \frac{1}{2}\int_{\mathbb{R}_+} h(v)\frac{(p-1)^2}{q}v^2\varphi''(v)\,dv + R_1(p,q), \end{aligned} \tag{5.66}$$

where the remainder R_1 is given by

$$R_1(p,q) = \frac{1}{2}\int_{\mathbb{R}_+} \left(qw^2 + 2(p-1)vw\right)\varphi''(v)\,dv + \frac{1}{q}R(p,q).$$

Since equation (5.66) coincides with the original Boltzmann equation, if p and q satisfy the condition $\sqrt{p}+\sqrt{q} = 1$, the function (5.63) remains a stationary solution. On the other hand, under the same conditions on the parameters, $p = (1-\sqrt{q})^2$ implies that

$$\lim_{q\to 0}\frac{(p-1)^2}{q} = 4. \tag{5.67}$$

Then, equation (5.66) is approximated well by

$$\frac{d}{d\tau}\int_{\mathbb{R}_+} \varphi(v)h(v,\tau)\,dv + \int_{\mathbb{R}_+} \varphi'(v)(v-1)h(v)\,dv = 2\int_{\mathbb{R}_+} h(v)v^2\varphi''(v)\,dv. \tag{5.68}$$

Equation (5.68) is nothing but the weak form of the Fokker–Planck equation

$$\frac{\partial h}{\partial \tau} = 2\frac{\partial^2}{\partial v^2}\left(v^2 h\right) + \frac{\partial}{\partial v}\left((v-1)h\right), \tag{5.69}$$

which admits a unique stationary state of unit mass, given by the Γ-distribution [55, 105]

$$P_\infty(v) = \frac{1}{\sqrt{2\pi}}\frac{\exp\left(-\frac{1}{2v}\right)}{v^{5/2}}. \tag{5.70}$$

This stationary distribution exhibits a Pareto power-law tail for large v's. From the asymptotic equivalence of the Boltzmann equation (5.60) with the Fokker–Planck equation (5.68), and from the invariance of the steady state, we can conclude that $P_\infty(v)$ has Laplace transform $\tilde{g}(s)$, and for this reason is a steady state of the Boltzmann equation in the case $\sqrt{p}+\sqrt{q} = 1$. In this case, however, the quantity $p+q-1$, which can produce exponential growth of wealth (when positive), or exponential dissipation of wealth (when negative), is negative. This is quite remarkable since it shows that this uneven distribution of wealth which characterizes most western economies may not only be produced as the effect of a growing economy but also under critical economical circumstances. This simple example, in addition, shows that the old intuition of Amoroso [7] about the shape of wealth distribution was exact.

Apart from an investigation of moments, the Boltzmann equations (5.7) or (5.55) are hard to analyze, even in the stationary regime. The method of the previous section provides a useful way to generate from the kinetic equation a *Fokker–Planck equation*, i.e. a parabolic differential equation of second order, which is more accessible.

This asymptotic procedure, which has been used in Section 5.4 in the case of non-conservative economies, still applies to conservative (in the mean) models. Consider the CPT model (5.26) with saving propensity $\lambda = 1 - \beta^2$ and market risks $\beta\eta_i$,

$$v^* = (1-\beta^2)v + \beta\eta_1 v + \beta^2 w, \quad w^* = (1-\beta^2)w + \beta\eta_2 w + \beta^2 v, \tag{5.71}$$

where $\beta > 0$ is a small parameter, and η_1 and η_2 are two equally distributed, centred random variables with $\sigma^2 = \langle\eta_i^2\rangle$. Expanding the collisional operator in terms of β,

$$\begin{aligned} &Q_+(f,f)[\varphi] - \int_{\mathbb{R}_+} \varphi(v)\,dv \\ &\quad = \int_{\mathbb{R}_+^2} \Big(\varphi'(v)[\beta^2(w-v) + \beta\langle\eta\rangle v] \\ &\qquad\qquad + \frac{1}{2}\varphi''(v)[\beta^2(w-v) + \beta\langle\eta\rangle v]^2 + o(\beta^4)\Big) f(v)f(w)\,dv\,dw \\ &\quad = \beta^2 \int_{\mathbb{R}_+} \varphi(v)\Big(-\frac{\partial}{\partial v}[(m-v)f(v)] + \frac{\sigma^2}{2}\frac{\partial^2}{\partial v^2}[v^2 f(v)]\Big)\,dv + o(\beta^4). \end{aligned}$$

Finally, increase the collision frequency by rescaling $t \rightsquigarrow t/\beta^2$. In the limit $\beta \to 0$, the Boltzmann equation turns into the Fokker–Planck equation

$$\frac{\partial}{\partial t} f = \frac{\sigma^2}{2}\frac{\partial^2}{\partial v^2}[v^2 f] + \frac{\partial}{\partial v}[(v-m)f], \tag{5.72}$$

which possesses an explicit stationary solution of Amoroso type [7],

$$f_\infty(v) = C_{\sigma,m} \exp\left(-\frac{2m}{\sigma^2 v}\right) v^{-(2+2/\sigma^2)}. \tag{5.73}$$

For instance, in agreement with the results for the CPT model, more risky trades (larger σ) induce fatter Pareto tails (decreasing index $\alpha = 1 + 2/\sigma^2$). As shown in Figure 5.5 the solution f_∞ constitutes a good approximation of the steady state of the respective (kinetic) CPT model for sufficiently small $\beta > 0$ [105].

5.4.2 Mathematical analysis of non-conservative models

The results of Sections 5.3.2 can easily be extended to equation (5.7), even in the case of non-conservative trades. In particular, Theorem 5.1 and Corollary 5.1 remain valid, and insure the existence of a unique solution to the Boltzmann equation for the density f. The real problem, on the other hand, is related to the scaled density $g(w,t)$, which satisfies equation (5.60).

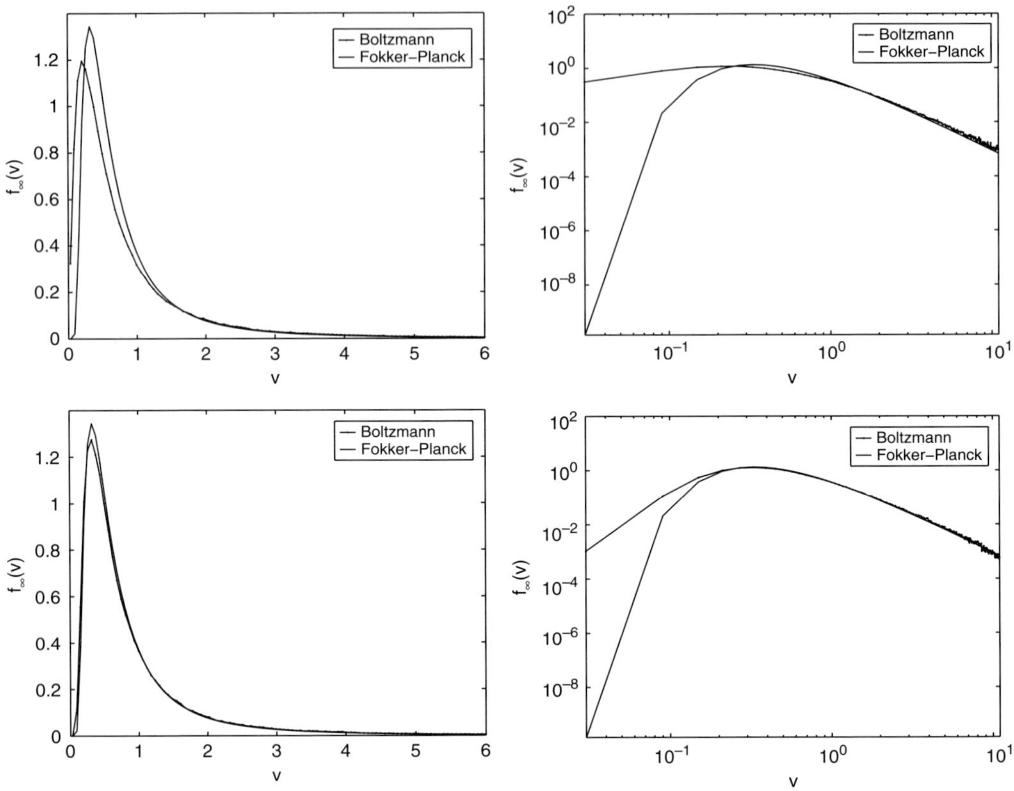

Fig. 5.5 CPT kinetic market model. Asymptotic behaviour of the Fokker–Planck model and the Boltzmann model for $\alpha = 2.0$ and $\beta^2 = 0.1$ (top), $\beta^2 = 0.01$ (bottom). The figures on the right are in log-log scale. The Boltzmann solution has been computed with a Monte Carlo simulation using $N = 10000$ agents [105].

Convergence to self-similar solutions

As a consequence of the scaling property ii) of Proposition 2.1, whenever g is defined by (5.59),

$$\widehat{g}(\xi) = \widehat{f}\Big(\frac{\xi}{m(t)}\Big),$$

so that (5.37) implies that the solution to equation (5.60) satisfies the bound

$$d_s[g_1(t), g_2(t)] = \sup_{\xi \in \mathbb{R}} \frac{|\widehat{g}_1(\xi, t) - \widehat{g}_2(\xi, t)|}{|\xi|^s} = \Big(\frac{1}{m(t)}\Big)^s d_s[f_1(t), f_2(t)]. \tag{5.74}$$

Consequently, if $g_1(t)$ and $g_2(t)$ are two solutions of the scaled Boltzmann equation (5.60) corresponding to initial values $f_{1,0}$ and $f_{2,0}$ in $\mathcal{P}_s$, for some $1 \le s \le 2$, for all times $t \ge 0$,

$$d_s[g_1(t), g_2(t)] \le \exp\{[(p^s + q^s - 1) - s(p + q - 1)]\, t\}\, d_s[f_{1,0}, f_{2,0}]. \tag{5.75}$$

Let us define, for $s \geq 1$,

$$\mathbf{R}(s) = p^s + q^s - 1 - s\,(p + q - 1)\,. \tag{5.76}$$

Then the sign of $\mathbf{R}$ determines the asymptotic behaviour of the distance $d_s[g_1(t), g_2(t)]$. We give below the main result which characterizes the sign of the function (5.76).

Lemma 5.5 *There exists some $\bar{s} \in (0, +\infty]$ such that $\mathbf{R}(s) < 0$ for all $1 < s < \bar{s}$. If $p + q \geq 1$, but both $p, q \leq 1$, then $\bar{s} = +\infty$, while $\bar{s}$ is finite in all other cases.*

The main consequence of Lemma 5.5 is the following theorem.

Theorem 5.5 [**254**] *Let $g_1(t)$ and $g_2(t)$ be two solutions of the Boltzmann equation (5.60), corresponding to initial values $f_{1,0}$ and $f_{2,0}$ in $\mathcal{P}_s$ for some $1 \leq s \leq 2$. There exists a constant $\bar{s} > 1$ such that, if $1 < s < \bar{s}$, then*

$$d_s[g_1(t), g_2(t)] \leq \exp\{-C_s t\}\, d_s[f_{1,0}, f_{2,0}] \tag{5.77}$$

for all times $t \geq 0$. The constant $C_s = -\mathbf{R}(s)$ is strictly positive, so that the distance d_s is contracting exponentially in time.

The existence and uniqueness of the stationary solution to equation (5.60) follow along the same lines as Section 5.3.2, now using $\mathbf{R}$ as a convex function instead of $\mathbf{S}$. The main result is stated in the following theorem.

Theorem 5.6 [**254**] *Let $s > 1$ be such that $\mathbf{R}(s) < 0$, and let $g_\infty(v)$ be the unique stationary solution (of given mean wealth $m > 0$) to the Boltzmann equation (5.60). Let $g(v,t)$ be the weak solution of equation (5.60) corresponding to the initial density $f_0 \in \mathcal{P}_s$. Then*

$$\int_{\mathbb{R}_+} v^s g(v,t)\, dv \leq c_{\bar{s}} < \infty,$$

with a time-uniform constant $c_{\bar{s}}$ depending only on p and q. Moreover, $g(v,t)$ converges exponentially fast in Fourier metric towards stationarity,

$$d_s[g(t), g_\infty] \leq d_s[f_0, g_\infty] \exp\{-|\mathbf{R}(s)|t\}\,, \tag{5.78}$$

where $\mathbf{R}(s)$ is defined in (5.76).

Depending on the values of the mixing parameters p and q, the stationary solution g_∞ may have overpopulated tails. The Fourier transform of g_∞ satisfies

$$-(p + q - 1)\xi \frac{\partial \widehat{g}_\infty}{\partial \xi} + \widehat{g}_\infty(\xi) = \widehat{g}_\infty(p\xi)\widehat{g}_\infty(q\xi). \tag{5.79}$$

In the following, let $\bar{s} \in (1, 2)$. We would like to decide if the stationary solution g_∞ has a fat tail of Pareto index $\bar{s}$. More restrictively, we ask if g_∞ belongs to every $\mathcal{P}_s$ for $s < \bar{s}$, but not to $\mathcal{M}_{M,\bar{s}}$. If this is the case, then its Fourier transform satisfies

$$\widehat{g}_\infty(\xi) = 1 - i\xi + A|\xi|^{\bar{s}} + o(|\xi|^{\bar{s}}). \tag{5.80}$$

Substitution of the expansion (5.80) into equation (5.79) shows that the coefficient of the power $|\xi|^{\bar{s}}$ is $A\mathbf{R}(\bar{s})$. Thus, the term $A|\xi|^{\bar{s}}$ can appear in the expansion of $\widehat{g}(\xi)$ as soon as $\mathbf{R}(\bar{s}) = 0$. As before, tails in the stationary distributions are present in all cases in which there exists a $\bar{s} \in (1,2)$ with $\mathbf{R}(\bar{s}) = 0$. With slight modifications, the argument also works for $\bar{s} \geq 1$. Thus, the question of existence of a fat-tailed steady state is answered by Lemma 5.5.

To finish the discussion, we mention that the passage to the Fokker–Planck equation can be made rigorous.

Theorem 5.7 [254] *Let an initial condition $f_0 \in \mathcal{P}_s$ be given, with $m > 0$ and $s > 1$. Consider the corresponding family of solutions $h^{(q)}(t)$ to the Boltzmann equation (5.66) for the scaled densities $h^{(q)}(v,\tau) = g^{(q)}(v,t)$, with $\tau = qt$. The super-index (q) refers to the mixing parameter $q > 0$, and*

$$(p-1)^2 = \lambda q$$

for some fixed $\lambda > 0$. Then, as $q \to 0$, the solutions $h^{(q)}(v,\tau)$ converge weakly to a limit $h(v,\tau)$. The latter constitutes a weak solution of the Fokker–Planck equation

$$\frac{\partial h}{\partial \tau} = \frac{\lambda}{2}\frac{\partial^2}{\partial v^2}\left(v^2 h\right) + \frac{\partial}{\partial v}\left((v-1)h\right). \tag{5.81}$$

Regularity of self-similar solutions

The Fourier transform of the steady solution g_∞ of the scaled Boltzmann equation (5.60) solves (5.79). The regularity of g_∞ can be recovered from the results of Section 5.3.2 by rewriting equation (5.79) in an interesting way, as proposed by Bobylev and Cercignani [45]. With the definitions

$$r = 1/(p+q-1) \quad \text{and} \quad F[\psi](\xi) = \psi(p\xi)\psi(q\xi),$$

equation (5.79) takes the form

$$r^{-1}\xi\widehat{g}'_\infty(\xi) + F[\widehat{g}_\infty](\xi) - \widehat{g}_\infty(\xi) = 0. \tag{5.82}$$

Equation (5.82) can be rewritten as

$$\frac{d}{d\xi}\left(\xi^{-r}\widehat{g}_\infty(\xi)\right) = -\frac{r}{\xi^{r+1}}F[\widehat{g}_\infty](\xi). \tag{5.83}$$

In analogy to Section 5.3.2, we shall represent $\widehat{g}_\infty$ as solution to a fixed-point equation,

$$\widehat{g}_\infty = \mathcal{R}[\widehat{g}_\infty], \tag{5.84}$$

where the integral operator $\mathcal{R}$ is now defined by

$$\mathcal{R}[\psi](\xi) = r\xi^r \int_\xi^\infty \frac{F[\psi](\eta)}{\eta^{r+1}}\,d\eta = r\int_1^\infty \frac{F[\psi](\tau\xi)}{\tau^{r+1}}\,d\tau.$$

The function θ with $\theta(\tau) = r/(\tau^{r+1})$ for $\tau \geq 1$, and $\theta(\tau) = 0$ for $\tau < 1$ is a probability density function,

$$\int_{\mathbb{R}+} \theta(\tau)\, d\tau = \int_1^\infty \frac{r}{\tau^{r+1}}\, d\tau = 1.$$

Therefore, recalling the definition of F, the fixed-point equation (5.84) takes the form

$$\widehat{g}_\infty(\xi) = \langle \widehat{g}_\infty(\tilde{p}\xi)\widehat{g}_\infty(\tilde{q}\xi)\rangle\,, \tag{5.85}$$

where now $\tilde{p} = \Theta p$ and $\tilde{q} = \Theta q$ are random variables and Θ is distributed on $\mathbb{R}_+$ with density $\theta(\tau)$. Thus, the results of Section 5.3.2 apply. In particular, since $\Theta \geq 1$, condition (5.48) is always satisfied for some $r > 0$.

Theorem 5.8 *Let $r > 0$ be such that the mixing parameters satisfy condition* (5.48)*. Denote by g_∞ the non-trivial steady state to the kinetic equation* (5.55) *of given mean $M > 0$. Then g_∞ is a smooth function and belongs to the r-th Gevrey class, i.e.*

$$\left|g_\infty(\xi)\right| \leq \exp\left(-\mu|\xi|^r\right) \quad \textit{for } |\xi| \geq \rho,$$

with suitable positive numbers ρ and μ.

5.5 Exact solutions

A shortcoming of most of the kinetic models for wealth distribution is the unpleasant fact that steady states are not explicitly known, even if many properties, such as regularity and tails, can be recovered by various technical means. In other words, contrary to what happens in classical kinetic theory of rarefied gases, where the equilibrium is explicitly known, in kinetic models for wealth distribution the general structure of binary trades does not allow exact solutions.

Despite the high number of studies devoted to the subject, well documented by various recent review papers [87, 167, 260, 321, 322], analytical solutions or explicit steady states for wealth distribution densities are, in fact, rarely present in the literature. For continuous models, there are few exceptions.

The first of these was discussed in Section 5.3, Example 5.1, where the simplified model relative to the *winner takes all* trade was shown to lead, for some well-chosen initial distribution, to an explicit solution for all times.

The second exception is the self-similar solution with Pareto tails found by Slanina [287] for the kinetic model of a non-conservative economy, as discussed in Section 5.4.

Last, we quote the steady-state solution (the exponential distribution) which appears both in the Chackraborti and Chackrabarti model [91] when the saving propensity is assumed to be zero [259, 270], and in a linear kinetic equation modelling taxation and uniform redistribution [309]. For discrete models, explicit expressions for steady states have been found in [282]. In a suitable scaling limit, some of these results can be used to re-obtain the exponential distribution as equilibrium for the underlying continuous model.

A fully solvable model, which allows us to recover and describe the properties of the equilibrium distributions was considered in [24]. This model is obtained from model [91], with binary interactions (5.22), and zero saving propensity (pure gambling trade interactions). With respect to previous studies on the same model [259, 270], in [24] the random fraction of wealth which governs the microscopic interaction is assumed to be different from the uniform distribution.

Due to its relative simplicity, and by taking advantage as usual from the possibility of making use of the Laplace transform version of the kinetic equation, two different situations, which correspond to requiring that the binary trade is pointwise conservative or not [132, 229], can be analyzed in detail.

Among the various cases which are explicitly solvable, and lead to an analytic expression of the steady profile, the most important ones emerge by choosing the random fraction to be a symmetric Beta distribution. In this case one obtains as equilibrium a Gamma distribution with a parameter which depends on the parameter of the Beta distribution. A special case emerges here, since, as previously observed [259, 270], the exponential distribution emerges for a random fraction uniformly distributed. Also, the case of the *winner takes all* game discussed in Section 5.3, Remark 5.1 can be studied in detail as a limit of $Beta(\varepsilon, \varepsilon)$, when the parameter ε tends to zero. In all these cases, however, the equilibria possess moments of any order.

Interestingly enough, the case of the *conservative in the mean* gambling trade [105, 229] can be treated likewise, by choosing the random fractions to be an inverse Beta distribution. In this case, the steady-state distribution is shown to be an inverse Gamma distribution with fat tails, which contains as a particular case the distribution found by Slanina [287] in a different context (cf. Section 5.4).

Through a simple example, this finding clarifies the role of the *social* use of the wealth present in the market. The formation of the rich class is consequent to the possibility to access the wealth available in the common *reservoir*.

5.5.1 The gambling process

In a pure gambling process, the entire sum of wealths of two agents is up for gambling. Some random fraction of this sum is shared by one agent and the rest goes to the other. The randomness is introduced into the model through a parameter θ which is a random number drawn from a probability distribution in $[0, 1]$. In general it is assumed that θ is independent of a pair of agents, so that a pair of agents is not likely to share the same fraction of aggregate wealth the same way when they interact repeatedly. The interaction rule reads:

$$\begin{aligned} w_i(\tau+1) &= \theta[w_i(\tau) + w_j(\tau)], \\ w_j(\tau+1) &= (1-\theta)[w_i(\tau) + w_j(\tau)], \end{aligned} \tag{5.86}$$

where the pair of agents (indicated by i and j) are chosen randomly. The amount of wealth that is exchanged is now $\Delta w = \theta[w_i(\tau) + w_j(\tau)] - w_i(\tau)$.

The interaction rule (5.86) is a particular case of the model introduced by Chackraborti and Chackrabarti [91] and detailed in (5.22), given by

$$\begin{aligned} w_i(\tau+1) &= \lambda w_i(\tau) + \theta(1-\lambda)[w_i(\tau) + w_j(\tau)], \\ w_j(\tau+1) &= \lambda w_j(\tau) + (1-\theta)(1-\lambda)[w_i(\tau) + w_j(\tau)], \end{aligned}$$

and corresponds to the choice $\lambda = 0$. By bearing in mind that λ measures the saving propensity of the agents, it appears natural to refer to this model as a *pure gambling* model. Both numerical experiments [167] and theoretical studies [259, 270] show that, if θ is a random number drawn from a uniform distribution in $[0, 1]$ and $\lambda = 0$, the individual wealth distribution at equilibrium emerges as the exponential distribution. The same studies (cf. [92] for an exhaustive review), however, show that, in the case $\lambda > 0$, the previous model produces at equilibrium only Gamma-like distributions, while the form of the exact solution is still an open question.

The main feature of trade (5.86) is that the amount of wealth which is restituted in a single trade coincides with the amount of wealth the two agents use for gambling (*pointwise conservative trade*). In this gambling market there is no rule of the (conserved) total amount of wealth initially in the hands of the agents. In other words, agents do not take advantage from the amount of wealth available in the market.

The idea of using this wealth as a *reservoir* for trades can be easily implemented, allowing agents to trade with random profit:

$$\begin{aligned} w_i(\tau+1) &= \theta_1[w_i(\tau) + w_j(\tau)], \\ w_j(\tau+1) &= \theta_2[w_i(\tau) + w_j(\tau)], \end{aligned} \tag{5.87}$$

where the parameters (θ_1, θ_2) are now random numbers drawn from a joint probability distribution such that

$$\langle \theta_1 + \theta_2 \rangle = 1, \tag{5.88}$$

where $\langle \cdot \rangle$ denotes, as usual, the mathematical expectation. Within this picture, $w_i(\tau + 1) + w_j(\tau+1)$ can be strictly less than $w_i(\tau) + w_j(\tau)$, and in this case the lost wealth is achieved by the market, or the reverse situation is verified, and the additional wealth is taken from the market. Condition (5.88), however, guarantees that in the mean the wealth present in the market is left unchanged. Note that if one assumes that $\theta_i \geq \gamma > 0$, $i = 1, 2$, then agents are prevented from losing all their wealth in a single trade. Note also that, choosing $(\theta_1, \theta_2) = (\theta, 1 - \theta)$, the collision rule (5.87) becomes (5.86).

In order to produce a fair game, it will be assumed that the random numbers θ_1 and θ_2 are identically distributed. For trades of type (5.86), this assumption simply forces the random fraction θ to be symmetric with respect to the value 1/2.

The time evolution of the wealth distribution obeys a kinetic equation of type (1.79):

$$\begin{aligned} &\frac{d}{dt} \int_{\mathbb{R}_+} f(v,t)\phi(v)\,dv \\ &\quad = \frac{1}{2} \Big\langle \int_{\mathbb{R}_+^2} \big(\phi(v^*) + \phi(w^*) - \phi(v) - \phi(w)\big) f(v,t) f(w,t)\,dv\,dw \Big\rangle, \end{aligned} \tag{5.89}$$

where the post-trade wealths (v^*, w^*) obey the rule (5.87),

$$v^* = \theta_1(v + w), \qquad w^* = \theta_2(v + w),$$

which preserves both mass density and the total mean. Differing from the Fourier description (5.13) used before, we will resort here to the Laplace transformed version.

This is obtained by setting $\phi(w) = \exp\{-\xi w\}$ in (5.89). The kinetic equation for the Laplace transform $\widetilde{f}$ of f, where

$$\widetilde{f}(\xi, t) = \int_{\mathbb{R}+} e^{-\xi w} f(w, t)\, dw,$$

is then

$$\frac{\partial \widetilde{f}(\xi, t)}{\partial t} + \widetilde{f}(\xi, t) = \frac{1}{2}\left\langle \widetilde{f}(\xi\theta_1, t)^2 + \widetilde{f}(\xi\theta_2, t)^2 \right\rangle.$$

As proven in Section 5.3.2, by resorting to the Fourier transform version of the kinetic equation, explicitly computable conditions on the gambling variables θ_i, $i = 1, 2$, guarantee that the distribution $f(w, t)$ weakly converges to a *universal* probability distribution $f_\infty(w)$. In point of fact, the results of Section 5.3.2 can easily be adapted to show that if the condition

$$\mathbf{S}(s)'(1) < 0 \tag{5.90}$$

holds, the Laplace transform of $f_\infty(w)$ is a solution of

$$\widetilde{f}_\infty(\xi) = \frac{1}{2}\left\langle \widetilde{f}_\infty(\xi\theta_1)^2 + \widetilde{f}_\infty(\xi\theta_2)^2 \right\rangle.$$

Moreover, this solution is unique in the class of all probability functions with unit mean, a property that can be equivalently written as

$$\widetilde{f}_\infty'(\xi)|_{\xi=0} = -1.$$

By further requiring that the random variables θ_1 and θ_2 are distributed with the same law, the equilibrium solution $\widetilde{f}_\infty$ is found to be the unique solution to

$$\widetilde{f}_\infty(\xi) = \left\langle \widetilde{f}_\infty(\xi\theta)^2 \right\rangle, \tag{5.91}$$

where the random variable θ is distributed according to the common law of θ_1 and θ_2. Equation (5.91) can be better understood by saying that, if Z is a random variable with law f_∞, the law of Z, defined in (5.91), is a distributional fixed point of the equation [200]

$$Z \stackrel{d}{=} \theta(Z_1 + Z_2), \tag{5.92}$$

where $\stackrel{d}{=}$ means identity in distribution (in law) and one assumes that the random variables Z_1, Z_2 and Z have the same probability law, while the variables Z_1, Z_2 and θ are assumed to be stochastically independent. Equations of type (5.92) are well known and extensively studied, see e.g. [137, 212].

5.5.2 Exact solutions for the pure gambling model

Let us start from the pure gambling case in which $(\theta_1, \theta_2) = (\theta, 1 - \theta)$. Having in mind that numerical experiments are usually done with a random number drawn from a uniform distribution in $[0, 1]$, which leads the discrete market to an exponential distribution at equilibrium [167], we fix the random number θ to be a symmetric $Beta(a, a)$ random variable, with $(a > 0)$ (cf. Appendix B).

The case $a = 1$, where θ is a random number uniformly distributed on $(0,1)$, confirms the numerical outcome. In this case, in fact, the nonlinear equation (5.91) which characterizes the stationary distributions becomes

$$\widetilde{f}_\infty(\xi) = \int_0^1 \widetilde{f}^2_\infty(\xi x)\, dx.$$

It is easy to see that the function

$$\widetilde{f}_\infty(\xi) = (1+\xi)^{-1}$$

is a solution of (5.91), such that $\widetilde{f}'_\infty(\xi)|_{\xi=0} = -1$. Since $(1+\xi)^{-1}$ is the Laplace transform of the exponential distribution of unit mean,

$$f_\infty(w) = e^{-w} \qquad (w \geq 0), \tag{5.93}$$

the exponential distribution is an analytical steady solution to the pure gambling trade market, in case when θ is a uniform random number in $(0,1)$. In this case, since $0 < \theta < 1$ with probability 1,

$$\mathbf{S}(s)'(1) = \langle \theta \log \theta \rangle < 0,$$

and condition (5.90) is verified. Consequently, on the basis of the results of [25, 229], reported in Section 5.3.2, the exponential distribution is the unique solution to (5.91) with unit mean.

To treat the more general case it suffices to recall that, if $\alpha_1, \alpha_2, \xi > 0$, then formula 3.197.4 in [164] gives

$$\frac{\Gamma(\alpha_1+\alpha_2)}{\Gamma(\alpha_1)\Gamma(\alpha_2)} \int_0^1 \frac{x^{\alpha_1-1}(1-x)^{\alpha_2-1}}{(1+\xi x)^{\alpha_1+\alpha_2}} dx = (1+\xi)^{-\alpha_1}. \tag{5.94}$$

Using identity (5.94) with $\alpha_1 = \alpha_2 = a$, one obtains that, for a general $a > 0$, the function

$$\widetilde{f}_\infty(\xi) = \left(1 + \xi \frac{1}{a}\right)^{-a}$$

solves equation (5.91). Hence, the equilibrium solution is a Gamma distribution of unit mean, with shape parameter a and scale parameter $1/a$, the *Gamma*$(a, 1/a)$ distribution for short,

$$f_\infty(w) = \frac{a^a w^{a-1} e^{-aw}}{\Gamma(a)} \qquad (w > 0). \tag{5.95}$$

Again, since θ is a symmetric *Beta*(a,a) random variable, so that $0 < \theta < 1$ with probability 1, the function

$$\mathbf{S}(s) = 2\langle \theta^s \rangle - 1$$

satisfies condition (5.90), and (5.95) is the unique equilibrium solution with unit mean.

Remark 5.3 *The uniform distribution, which corresponds to the choice $a = 1$, determines a natural separation between two different behaviours of the equilibrium solutions. In case $a < 1$, the Gamma$(a, 1/a)$ densities (5.95) are monotonically decreasing, starting from $f_\infty(0) = +\infty$. In the opposite case $a > 1$, a peak appears* around *the unit mean value. The average wealth is unchanged for every a but, when $a > 1$, the number of agents with wealth closer to the average value increases or, in other words, the wealth distribution becomes more fair for larger a. Analogously, for $a < 1$ the distribution gives more weight to agents with wealth close to zero.*

As a consequence, measures of the inequality of the wealth distribution, such as the Gini coefficient, increase for decreasing a and tend to zero for $a \to +\infty$. Some insight can be gained by the simple computation of the variance of a random variable X_∞ distributed according to f_∞. It holds

$$Var(X_\infty) = \int_{\mathbb{R}+} w^2 f_\infty(w)\, dw - 1 = \frac{1}{a}.$$

Hence, the variance (the spreading) decreases as a increases.

Remark 5.4 *There are two interesting limiting cases. The first one is obtained by letting $a \to +\infty$. In this case ϵ converges in distribution to the constant value $1/2$ and one can immediately see that the steady state is a degenerate distribution concentrated on the value 1 of the mean wealth. This corresponds to a perfectly fair distribution, in which all agents end up with the same wealth. The other limit case is the* winner takes all *game treated in Section 5.3, which can be obtained by letting $a \to 0$. In this case, the steady state is concentrating on 0, while its variance is blowing up. This corresponds to the discrete situation in which a finite number of agents end up with no wealth, except one which* takes all. *In the continuous case, the value $a = 0$ can not be assumed directly, since the exchange of the limits $a \to 0$ and $t \to \infty$ is not allowed by the lack of regularity of the equilibrium solution for $a = 0$.*

5.5.3 Analytic equilibria with heavy tails

In agreement with the previous section, we assume that

$$\theta_i = \frac{1}{4\bar{\theta}_i}, \qquad i = 1, 2, \tag{5.96}$$

where, for a given $a > 1$, $\bar{\theta}_i$ $(i = 1, 2)$ is a $Beta(a + 1/2, a - 1/2)$ random variable. It is clear immediately that the random variables θ_i are such that $\langle \theta_i \rangle = 1/2$, $i = 1, 2$. Consequently, the model is conservative in the mean. We can invoke again [25, 229] to prove that there is convergence to a unique equilibrium solution such that its Laplace transform satisfies

$$\widetilde{f}_\infty(\xi) = \int_0^1 \widetilde{f}_\infty(\xi/(4x))^2 \beta_{a+1/2,a-1/2}(x)dx, \tag{5.97}$$

with $\widetilde{f}_\infty^{\,\prime}(\xi)|_{\xi=0} = -1$. In this case one needs to show that condition (5.90) holds true for

$$\begin{aligned}\mathbf{S}(s) &= \langle \epsilon_1^s + \epsilon_2^s \rangle - 1 \\ &= 2\int_0^1 \frac{1}{(4x)^s}\beta_{a+1/2,a-1/2}(x)dx - 1 \\ &= \frac{2^{1-2s}\Gamma(2a)\Gamma(a-s+\frac{1}{2})}{\Gamma(2a-s)\Gamma(a+\frac{1}{2})} - 1.\end{aligned} \tag{5.98}$$

The proof of this condition is not direct. In order to follow directly the rest of the proof, we postpone the computations to the end of this section.

Now, we shall prove that the solution of (5.97) is obtained by taking the Laplace transform of the so-called Inverse-Gamma distribution [295] of shape parameter a and scale parameter $a-1$, that is

$$f_\infty(v) = \frac{(a-1)^a}{\Gamma(a)} \frac{e^{-\frac{(a-1)}{v}}}{v^{a+1}}, \tag{5.99}$$

which is peaked around the mean value 1 and has heavy tails, in that it decays at infinity like $v^{-(a+1)}$. An analytical solution of type (5.99) with a polynomial decay corresponding to $a = 3/2$ has been discovered in [287] as a self-similar solution of a kinetic model of a non-conservative economy (cf. Section 5.4). As briefly discussed there, Slanina considered a non-conservative collision rule given by

$$\begin{aligned}&w' = pw + qw_*, \quad w'_* = qw + pw_*, \\ &p \geq q > 0, \quad \sqrt{p} + \sqrt{q} = 1.\end{aligned} \tag{5.100}$$

Note that, within condition (5.100) on the mixing parameters p and q,

$$w' + w'_* = (1 - 2\sqrt{pq})\,(w + w_*) < w + w_*,$$

which implies that the mean value $m(t)$ at time t decays exponentially to zero at the rate $2\sqrt{pq}$. Define the scaled solution g by

$$g(w,t) = m(t) f\big(m(t)w, t\big),$$

which implies that $\int wg(w,t)\,dw = 1$ for all $t \geq 0$.

In terms of the Laplace transform $\tilde{g}$ of g, it is found that the equation satisfied by $\tilde{g}$ reads [134]

$$\frac{\partial \tilde{g}}{\partial t} + \xi(p+q-1)\frac{\partial \tilde{g}}{\partial \xi} = \tilde{g}(p\xi)\tilde{g}(q\xi) - \tilde{g}(\xi). \tag{5.101}$$

Steady solutions to equation (5.101) satisfy

$$\xi(p+q-1)\frac{\partial \tilde{g}}{\partial \xi} = \tilde{g}(p\xi)\tilde{g}(q\xi) - \tilde{g}(\xi). \tag{5.102}$$

Direct computations then show that the function

$$\tilde{g}_\infty(\xi) = \left(1 + \sqrt{2\xi}\right) e^{-\sqrt{2\xi}} \tag{5.103}$$

solves (5.102) for all values of p and q satisfying the constraint $\sqrt{p}+\sqrt{q}=1$. Note that (5.103) is the explicit Laplace transform of

$$g_\infty(w) = \frac{(1/2)^{3/2}}{\Gamma(3/2)} \frac{e^{-\frac{1}{2w}}}{w^{5/2}}. \tag{5.104}$$

Let us set $p=q=1/4$ in (5.102). Then the steady solution (5.103) satisfies

$$-\frac{\xi}{2}\frac{\partial \tilde{g}}{\partial \xi} + \tilde{g}(\xi) = \tilde{g}\left(\frac{\xi}{4}\right)^2. \tag{5.105}$$

Following [50], Sect. 6, (5.105) can be written equivalently in integral form as

$$\tilde{g}(\xi) = \int_0^1 \tilde{g}\left(\frac{\xi}{4}\rho^{-1/2}\right)^2 d\rho,$$

or, setting $\rho^{1/2}=x$,

$$\tilde{g}(\xi) = \int_0^1 2\tilde{g}\left(\frac{\xi}{4x}\right)^2 x\,dx,$$

which is nothing but (5.97) with $a=3/2$. Consequently, the distribution (5.104) solves (5.97) with $a=3/2$. This argument establishes a connection between the present problem and the non-conservative one introduced by Slanina [287], which leads to explicit computations.

In order to prove that the Laplace transform of (5.99) is the solution of (5.97), since a direct proof does not seem as straightforward as in the previous case, we recast the problem in a more probabilistic way. First of all, let us note that an Inverse-Gamma random variable Y of parameter $(a, a-1)$ can be obtained by taking $Y=1/X$, where X is a $Gamma(a, 1/(a-1))$ random variable. Recall that X is a $Gamma(a, 1/(a-1))$ random variable if its density is (cf. Appendix B)

$$\frac{(a-1)^a w^{a-1} e^{(a-1)w}}{\Gamma(a)}.$$

Recall also that its Laplace transform reads

$$(1+\xi/(a-1))^{-a}. \tag{5.106}$$

Equation (5.97) can be rewritten in an equivalent way as

$$Y \stackrel{d}{=} \frac{1}{4\theta}[Y_1+Y_2], \tag{5.107}$$

where Y_1, Y_2, θ are independent random variables, Y, Y_1, Y_2 have density f_∞, while θ has density $\beta_{a+1/2, a-1/2}$. Recall that the symbol $\stackrel{d}{=}$ means identity in distribution. Hence, to prove (5.107) it suffices to show that $Y^{-1} \stackrel{d}{=} 4\theta[Y_1+Y_2]^{-1}$ or, equivalently,

$$X \stackrel{d}{=} \frac{4\theta X_1 X_2}{X_1+X_2}, \tag{5.108}$$

where X, X_1, X_2 are independent $Gamma(a, 1/(a-1))$ random variables. Summarizing, the result follows if one is able to show that

$$\frac{4\theta X_1 X_2}{X_1 + X_2}, \tag{5.109}$$

has Laplace transform (5.106). Setting $G = X_1 + X_2$ and $B = X_1/(X_1 + X_2)$, one rewrites (5.109) as

$$4\theta GB(1 - B).$$

It is a classical result of probability theory that, given X_1 and X_2 which are independent $Gamma(a, 1/(a-1))$ distributed, the random variables G and B are stochastically independent and, moreover, that G has $Gamma(2a, 1/(a-1))$ distribution, while B has $Beta(a, a)$ distribution. For the proof of this property, we refer, for instance, to Chapter 10.4 in [151]. Moreover, using relation (5.94), one can reckon that θG has $Gamma(a + 1/2, 1/(a-1))$ distribution simply by computing its Laplace transform. Now using the fact that θG and B are mutually independent, one can write the Laplace transform of $4\theta GB(1 - B)$ in the point ξ as

$$\begin{aligned} L(\xi) &= \int_0^1 \frac{\beta_{a,a}(x)}{\left[1 + 4(a-1)^{-1}\xi x(1-x)\right]^{a+\frac{1}{2}}}\, dx \\ &= 2\int_0^{\frac{1}{2}} \frac{\beta_{a,a}(x)}{\left[1 + 4(a-1)^{-1}\xi x(1-x)\right]^{a+\frac{1}{2}}}\, dx. \end{aligned}$$

At this stage, a simple change of variable shows that

$$\begin{aligned} L(\xi) &= \frac{\Gamma(2a)}{2^{2a-1}\Gamma(a)^2} \int_0^1 \frac{z^{a-1}(1-z)^{\frac{1}{2}-1}}{(1 + \xi z/(a-1))^{a+\frac{1}{2}}}\, dz \\ &= \int_0^1 \frac{1}{(1 + \xi z/(a-1))^{a+\frac{1}{2}}} \beta_{a,1/2}(z)\, dz. \end{aligned}$$

The last identity follows by the duplication formula $\Gamma(2a) = 2^{2a-1}\Gamma(a)\Gamma(a + \frac{1}{2})/\Gamma(\frac{1}{2})$ (see 8.335.1 in [164]). Using relation (5.94) once again, we get $L(\xi) = (1 + \xi/(a-1))^{-a}$ and (5.108) is proved.

Within the choice (5.96), the conservative in the mean trade (5.87) is such that the two agents maintain at least 1/4 of the total wealth used to trade. Thus, trade (5.87) is, in a sense, less risky than the conservative trade (5.86), where one of the two agents can emerge from the trade with almost no wealth. In addition, it follows that the number of moments of the explicit equilibrium state which are finite increase with a. On the other hand, when a increases, the area described by the distribution of the random fraction θ on the interval $[1, +\infty)$ decreases, and the probability of using the wealth of the society is also decreasing. Hence, a fat Pareto tail is obtained through significant use of the common wealth.

Remark 5.5 *Due to its simplicity, modelling redistribution of wealth in a simple market economy in which trades are described by a standard gambling game leads to a large* *variety of explicit equilibria. In particular, analytical solutions can be obtained in the case in which the post-trade wealths depend on the pre-trade ones through random variables which are Beta distributed. Previously known analytical solutions are here shown to appear for particular values of the underlying parameters of this model. The results enlighten the role of the interaction in producing Pareto tails. The lesson we can draw from this elementary market model is that Pareto tails cannot be produced through binary trade interactions in which agents only redistribute their wealths, while Pareto tails are produced through binary trades in which agents can take advantage of the amount of wealth available in the market. As discussed from a mathematical point of view in Section 5.3.2, this behaviour is common to more general binary trades, in which exchange of money is not pointwise conservative, thus allowing the creation of a common reservoir.*

Proof that $\mathbf{S}'(1) < 0$

We conclude the analysis of the pure gambling model by proving that $\mathbf{S}'(1) < 0$ when $\mathbf{S}$ is defined as in (5.98). The proof is very technical, and it requires knowledge of a certain number of properties of some special functions, which can be found in the classic book of Gradshteyn and Ryzhik [164]. Starting from (5.98), differentiation shows that

$$\mathbf{S}'(s) = -2^{1-2s}\frac{\Gamma(2a)\Gamma(a-s+\frac{1}{2})}{\Gamma(2a-s)\Gamma(a+\frac{1}{2})}\cdot \\ \cdot\Big\{2\log(2)+\psi(a-s+\frac{1}{2})-\psi(2a-s)\Big\},$$

where $\psi(x) = \Gamma'(x)/\Gamma(x)$ is the Digamma function. Using the duplication formula $2\psi(2x) = \psi(x) + \psi(x+1/2) + 2\log(2)$, see 8.365.6 in [164], one obtains

$$\mathbf{S}'(1) = -\frac{1}{4}\frac{\Gamma(2a)\Gamma(a-s+\frac{1}{2})}{\Gamma(2a-s)\Gamma(a+\frac{1}{2})}Q(a),$$

where $Q(a) = 2\log(2) + \psi(a-1/2) + \psi(a)$.

Now, $\psi(1/2) = -\gamma - 2\log 2$ and $\psi(1) = -\gamma$ (γ being the Eulero–Mascheroni constant), see 8.366.1/2 in [164], and then $Q(1) = 0$.

The classical expansion formula (see 8.363.8 in [164])

$$\psi'(x) = \sum_{k\geq 0}\frac{1}{(x+k)^2}$$

allows us to conclude that, for every $a > 1$,

$$Q'(a) = \sum_{k\geq 0}\frac{1}{(a-1/2+k)^2} - \sum_{k\geq 0}\frac{1}{(a+k)^2} > 0.$$

Hence, $Q(a)$ is strictly monotone in $[1,+\infty)$ and, since $Q(1) = 0$, it follows that $Q(a) > 0$ for every $a > 1$. This shows that $\mathbf{S}'(1) < 0$.

5.6 Modelling heterogeneous traders

Among the possible generalizations of the models presented in the previous sections, one of them is related to the statistical description of agent-based models constituted by agents from n different countries or social groups of individuals which can trade with each other. These groups shall be identified with countries, or social classes inside a country. The main outcome one expects from these types of models is to reach stationary profiles for wealth distribution able to capture phenomena which are present in the recent history of economies, but impossible to obtain on the basis of binary exchanges in a homogeneous group.

5.6.1 Interacting groups of traders

In general, to simplify the models without affecting the possibility of a general outcome, we adopt the hypothesis that all agents belonging to one group share a common saving rate parameter. This hypothesis can be further relaxed by assuming that the saving rate is a random quantity, with a statistical mean which is different for different social groups. Here we describe the model proposed in [136] which is based on the CPT conservative model. A related problem, based on increasing wealth, has recently been introduced and numerically studied in [101].

This can be seen as an analogue of the physical problem of a mixture of gases, where the molecules of the different gases exchange momentum in collisions [46]. When two agents from the *same country* with pre-trade wealths v and w interact – a domestic trade event – then their post-trade wealths v^* and w^* are supposed to be given by (5.26) with a common saving rate parameter which is characteristic for this country. On the other hand, in the case of an international trade, i.e. when two agents of *different countries* interact, we assume that each agent uses the transaction parameter which is characteristic for his country. Hence, when two agents, one from country i $(i = 1, 2, \dots, n)$ with pre-trade wealth v and the other from country j $(j = 1, 2, \dots, n)$ with pre-trade wealth w interact, their post-trade wealths v^* and w^* are given by

$$v^* = (1 - \gamma_i\gamma)v + \gamma_j\gamma w + \eta_{ij}v, \tag{5.110a}$$

$$w^* = (1 - \gamma_j\gamma)w + \gamma_i\gamma v + \eta_{ji}w. \tag{5.110b}$$

In (5.110), the trade depends on the transaction parameters γ and γ_i $(i = 1, \dots, n)$, while the risks of the market are described by η_{ij} $(i, j = 1, \dots, n)$, which are equally distributed random variables with zero mean and variance σ_{ij}^2. The different variances for domestic trades in each country and for international trades reflect different risk structures in these trades. For example, investments and trades inside different countries or markets may be subject to different types and quantities of risk, and international trading may face additional risks compared to domestic trades.

The trading rule (5.110) preserves, as in the original conservative CPT model, the total wealth in the statistical mean,

$$\langle v^* + w^* \rangle = \left(1 + \langle \eta_{ij} \rangle\right) v + \left(1 + \langle \eta_{ji} \rangle\right) w = v + w. \tag{5.111}$$

In this setting, we are led to study the evolution of the distribution function for each country as a function depending on the wealth $w \in \mathbb{R}_+$ and time $t \in \mathbb{R}_+$, $f_i = f_i(w, t)$.

By analogy with the classical kinetic theory of mixtures of rarefied gases, the time evolution of the distributions will obey a system of n Boltzmann-like equations, given by

$$\frac{\partial}{\partial t} f_i(w,t) = \sum_{j=1}^{n} \frac{1}{\tau_{ij}} \mathcal{Q}(f_i, f_j)(w), \quad i = 1, \dots, n. \tag{5.112}$$

Herein, τ_{ij} are suitable relaxation times, which depend on the velocity of money circulation [324]. The Boltzmann-like collision operators are derived by the standard methods presented in Chapter 1. The $\mathcal{Q}$ operator now reads

$$\mathcal{Q}(f_i, f_j)(w) = \left\langle \int_{\mathbb{R}_+} \left(\frac{1}{J_{ij}} f_i(v_*) f_j(w_*) - f_i(v) f_j(w) \right) dv \right\rangle. \tag{5.113}$$

In (5.113), (v_*, w_*) denote the pre-trade pair that produces the post-trade pair (v, w), following rules like (5.110), while J_{ij} denotes the Jacobian of the transformation of (v, w) into (v^*, w^*). As before, we can fruitfully consider the weak form

$$\int_{\mathbb{R}_+} \mathcal{Q}(f_i, f_j)(w) \phi(w)\, dw = \left\langle \int_{\mathbb{R}_+^2} (\phi(v^*) - \phi(v)) f_i(v) f_j(w)\, dv\, dw \right\rangle. \tag{5.114}$$

A related system of Fokker–Planck equations

As briefly remarked in Section 5.4.1, it is rather difficult to describe analytically the behaviour of the solution to the kinetic system (5.112). By adopting the asymptotic procedure sketched in Section 5.4.1, we can reduce our system to a system of Fokker–Planck type equations. By means of this approach it is easier to identify steady states while retaining important information on the microscopic interaction at a macroscopic level. In the present case, this asymptotic procedure corresponds to consider the joint limits $\gamma \to 0$, $\sigma_{ij}^2 \to 0$ and $\sigma_{ij}^2/\gamma \to \mu_{ij}$.

The weak form of (5.112) is given by $(i = 1, \dots, n)$

$$\frac{d}{dt} \int_{\mathbb{R}_+} f_i(w,t) \phi(w)\, dw = \int_{\mathbb{R}_+} \sum_{j=1}^{n} \frac{1}{\tau_{ij}} \mathcal{Q}(f_i, f_j)(w) \phi(w)\, dw, \tag{5.115}$$

where the terms on the right-hand side are given by (5.114). To study the situation for large times, i.e. close to the steady state, we introduce for $\gamma \ll 1$ the transformation

$$\tau = \gamma t, \quad g_i(w, \tau) = f_i(w, t), \quad i = 1, \dots, n. \tag{5.116}$$

This implies $f_{i,0} = g_{i,0}$ and the evolution of the scaled densities $g_i(w, \tau)$ follows $(i = 1, \dots, n)$

$$\frac{d}{d\tau} \int_{\mathbb{R}_+} g_i(w, \tau) \phi(w)\, dw = \frac{1}{\gamma} \int_{\mathbb{R}_+} \sum_{j=1}^{n} \frac{1}{\tau_{ij}} \mathcal{Q}(g_i, g_j)(w) \phi(w)\, dw. \tag{5.117}$$

By the trading rule (5.110), it holds that

$$v^* - v = \gamma(\gamma_j w - \gamma_i v) + \eta_{ij} v. \tag{5.118}$$

Using a second-order Taylor expansion of ϕ around v, we obtain

$$\phi(v^*) - \phi(v) = \phi'(v)[\gamma(\gamma_j w - \gamma_i v) + \eta_{ij} v] + \tfrac{1}{2}\phi''(\tilde{v})[\gamma(\gamma_j w - \gamma_i v) + \eta_{ij} v]^2, \tag{5.119}$$

with $\tilde{v} = \theta v' + (1-\theta)v$ for some $0 \leq \theta \leq 1$.

Inserting this expansion into the collision operators yields

$$\begin{aligned}
&\frac{d}{d\tau}\int_{\mathbb{R}_+} g_i(w,\tau)\phi(w)\,dw \\
&\quad = \frac{1}{\gamma}\sum_{j=1}^{n}\frac{1}{\tau_{ij}}\left\langle \int_{\mathbb{R}_+^2}\Big(\phi'(v)[\gamma(\gamma_j w - \gamma_i v) + \eta_{ij} v] + \tfrac{1}{2}\phi''(v)[\gamma(\gamma_j w - \gamma_i v) + \eta_{ij} v]^2\Big)\times\right. \\
&\qquad\qquad \left. \times\, g_i(v,\tau) g_j(w,\tau)\,dv\,dw \right\rangle + R(\gamma,\sigma_{ij}),
\end{aligned}$$

where the remainder $R(\gamma,\sigma_{ij})$ converges to zero as $\gamma, \sigma_{ij} \to 0$ in such a way that $\sigma_{ij}^2 \to \mu_{ij}\gamma$, provided further hypotheses on the random variables η_{ij} and on ϕ hold (cf. [105]).

Recalling that $\langle \eta_{ij}\rangle = 0$ and $\langle \eta_{ij}^2\rangle = \sigma_{ij}^2$, in the same limit process we obtain

$$\begin{aligned}
&\lim_{\gamma\to 0}\frac{1}{\gamma}\sum_{j=1}^{n}\frac{1}{\tau_{ij}}\left\langle \int_{\mathbb{R}_+^2}\Big(\phi'(v)[\gamma(\gamma_j w - \gamma_i v) + \eta_{ij} v] + \tfrac{1}{2}\phi''(v)[\gamma(\gamma_j w - \gamma_i v) + \eta_{ij} v]^2\Big)\times\right. \\
&\qquad\qquad \left. \times\, g_i(v,\tau) g_j(w,\tau)\,dv\,dw \right\rangle \\
&\quad = \sum_{j=1}^{n}\frac{1}{\tau_{ij}}\int_{\mathbb{R}_+}[\phi'(v)(\gamma_j m_j - \gamma_i v\rho_j) + \tfrac{\mu_{ij}}{2} v^2 \rho_j \phi''(v)] g_i(v,\tau)\,dv,
\end{aligned} \tag{5.120}$$

where

$$\rho_j(\tau) = \int_{\mathbb{R}_+} g_j(w,\tau)\,dw, \quad m_j(\tau) = \int_{\mathbb{R}_+} w g_j(w,\tau)\,dw \tag{5.121}$$

denote the total population and the total wealth of the j-th country, respectively. This expression is nothing but the right-hand side of the weak form of the system of Fokker–Planck equations ($i = 1, \ldots, n$)

$$\frac{\partial g_i}{\partial \tau} = \sum_{j=1}^{n}\Big[\frac{\mu_{ij}}{2\tau_{ij}}\frac{\partial^2}{\partial v^2}\big(v^2\rho_j g_i\big) + \frac{1}{\tau_{ij}}\frac{\partial}{\partial v}\big((\gamma_i v \rho_j - \gamma_j m_j) g_i\big)\Big]. \tag{5.122}$$

Using the same arguments as in Section 5.4.1, we obtain in this case the following:

Theorem 5.9 **[136]** *Let initial probability densities of finite mean value such that* $f_{0,i} \in \mathcal{P}_p$, *where* $p > 2$, *be given, and assume that the symmetric random variables*

η_{ij} have a density in $\mathcal{P}_s$ with $s > p$. Consider the family – parameterized by $\gamma > 0$ – of weak solutions $g_i^{(\gamma)}$ of the Boltzmann system (5.112) for the scaled densities $g_i^{(\gamma)}(w,\tau) = f_i^{(\gamma)}(w,t)$ with $\tau = \gamma t$.

Then, as $\gamma \to 0$ and $\sigma_{ij} \to 0$ in such a way that $\sigma_{ij} = \mu_{ij}\gamma$, the weak solutions converge, up to extraction of a subsequence, to probability densities $g_i(w,\tau)$. These densities constitute weak solutions of the Fokker–Planck system (5.122).

The (relatively) easy structure of the Fokker–Planck system (5.122) allows us to study in some simple cases the evolution of the mean wealth in each community, as well as the steady-state solution. For the sake of simplicity, we will limit ourselves to the case of two populations, in which $\mu_{ij} = \tau_{ij} = 1$ for $i = 1,2$. From the equations (5.122) it follows that the masses $\rho_1(\tau)$ and $\rho_2(\tau)$ do not vary with time, while the total wealths $m_1(\tau)$ and $m_2(\tau)$ satisfy the system of ordinary differential equations

$$\frac{dm_1}{d\tau} = -\left(\gamma_1\rho_2 m_1 - \gamma_2\rho_1 m_2\right) \tag{5.123a}$$

$$\frac{dm_2}{d\tau} = +\left(\gamma_1\rho_2 m_1 - \gamma_2\rho_1 m_2\right). \tag{5.123b}$$

System (5.123) can be solved exactly, to give

$$m_1(\tau) = m_1(0)e^{-(\gamma_1\rho_2+\gamma_2\rho_1)\tau} + \frac{\gamma_2\rho_1}{\gamma_1\rho_2+\gamma_2\rho_1}\left(1 - e^{-(\gamma_1\rho_2+\gamma_2\rho_1)\tau}\right) \tag{5.124a}$$

$$m_2(\tau) = m_2(0)e^{-(\gamma_1\rho_2+\gamma_2\rho_1)\tau} + \frac{\gamma_1\rho_2}{\gamma_1\rho_2+\gamma_2\rho_1}\left(1 - e^{-(\gamma_1\rho_2+\gamma_2\rho_1)\tau}\right). \tag{5.124b}$$

Choosing equally large populations, $\rho_1(0) = \rho_2(0)$, it follows from formulas (5.124) that the total wealths $m_1(\tau)$ and $m_2(\tau)$ converge at an exponential rate towards steady values $m_{1,\infty}$ and $m_{2,\infty}$. The latter are inversely proportional to the γ's (a bigger γ produces a smaller mean wealth),

$$m_{1,\infty} = \frac{\gamma_2}{\gamma_1+\gamma_2}; \qquad m_{2,\infty} = \frac{\gamma_1}{\gamma_1+\gamma_2}. \tag{5.125}$$

This effect can be directly deduced by examining the analytic expressions of the steady states, given by

$$g_{1,\infty}(v) = \frac{c_1}{\rho}\frac{1}{v^{2+2\gamma_1/\lambda}}\exp\left\{-\frac{2(\gamma_1 m_1+\gamma_2 m_2)}{\lambda\rho v}\right\} \tag{5.126a}$$

$$g_{2,\infty}(v) = \frac{c_2}{\rho}\frac{1}{v^{2+2\gamma_2/\lambda}}\exp\left\{-\frac{2(\gamma_1 m_1+\gamma_2 m_2)}{\lambda\rho v}\right\}. \tag{5.126b}$$

In (5.126) the constants $c_1, i = 1,2$, are chosen to have masses ρ_1 (respectively ρ_2) for the steady states. Note that here the size of the tail of $g_{i,\infty}(v)$ is proportional to γ_i. Hence, the smaller the γ is, the smaller the number of bounded moments of the steady state. Taking the sum of the densities in (5.126) gives expression for the total density,

$$g_\infty(v) = \frac{1}{\rho}\left(\frac{c_1}{v^{2+2\gamma_1/\lambda}} + \frac{c_2}{v^{2+2\gamma_2/\lambda}}\right)\exp\left\{-\frac{2(\gamma_1 m_1+\gamma_2 m_2)}{\lambda\rho v}\right\}. \tag{5.127}$$

Analysis of the steady density (5.127) reveals, depending on the values of the various parameters involved, the formation of a bimodal distribution. In fact, the extremals of the (non-negative) function

$$\psi(v) = \left(\frac{a}{v^p} + \frac{b}{v^q}\right) \exp\{-c/v\}, \tag{5.128}$$

where a, b, c are positive constants and $q > p$, are located in the points solutions of the equation

$$\Phi(v) = -pav^{q-p+1} - qbv + cav^{q-p} + cb = 0. \tag{5.129}$$

On the other hand, since $\Phi(0) > 0$, while $\Phi(+\infty) = -\infty$, the curve $y = \Phi(v)$ crosses the axis $y = 0$ either at a single point or at three points. In this last case, we have two maxima and one minimum outside $v = 0$, and consequently a bimodal distribution.

5.6.2 A numerical example

In a first example, we consider two groups with $N_1 = N_2 = 5000$ agents. We investigate the relaxation behaviour when the random variables η_{ij}, $i, j \in \{1, 2\}$, attain values $\pm\mu$ with probability 1/2 each. We set the coefficient $\gamma = 1$. We set $\mu = 0.15$ and $\tau_{ij} = 1$ for $i, j \in \{1, 2\}$. If we choose $\gamma_1 = \gamma_2 = 0.875$ and $\gamma_1 = \gamma_2 = 0.99$, respectively, the system reduces to the standard CPT model. The probability density for both cases is plotted in Figure 5.6. The cumulative distribution functions show a Pareto tail; see Figure 5.7. The Pareto index α of the tail is determined by the non-trivial root of (5.28) – strictly speaking this holds for the limit $N_{1,2} \to \infty$ – which is given by 28.068 and 1.875, respectively. These tail indices are indicated in Figure 5.7 by a thick line.

Now we choose $\gamma_1 = 0.875$ and $\gamma_2 = 0.99$ and keep $\mu = 0.15$ and $\tau_{ij} = 1$ for $i, j \in 1, 2$. The probability density for the whole population is plotted in Figure 5.8 (left plot). It shows a bimodal shape. The distance between the two peaks in the distribution decreases with decreasing difference between γ_1 and γ_2. Such bimodal distributions (and a polymodal distribution, in general) are also reported with real data for income distributions in Argentina [147, 167]. This distribution features transport of wealth from one group to the other, which makes it different from the probability distribution for the union of two groups with the same parameters which do not interact, see Figure 5.8 (right plot).

The associated cumulative distribution functions are shown in Figure 5.9.

Remark 5.6 *The features typically incorporated in kinetic trade models are saving effects and randomness. As explained, saving means that each agent is guaranteed to retain at least a certain minimal fraction of his initial wealth at the end of the trade. This concept was introduced in [91], where a fixed saving rate for all agents was proposed, and generalized in [96] by introducing individual saving rate. The distributed saving gives rise to an additional interesting feature when a special case is considered where the saving parameter is assumed to take only two fixed values, preferably widely separated. In this case, the steady distribution of wealth can result in a bimodal distribution [167]. The numerical output evolves towards a robust and distinct two-peak distribution as the difference in the two saving parameters is increased systematically. A population can be imagined to have two distinctly different kinds of people: some*

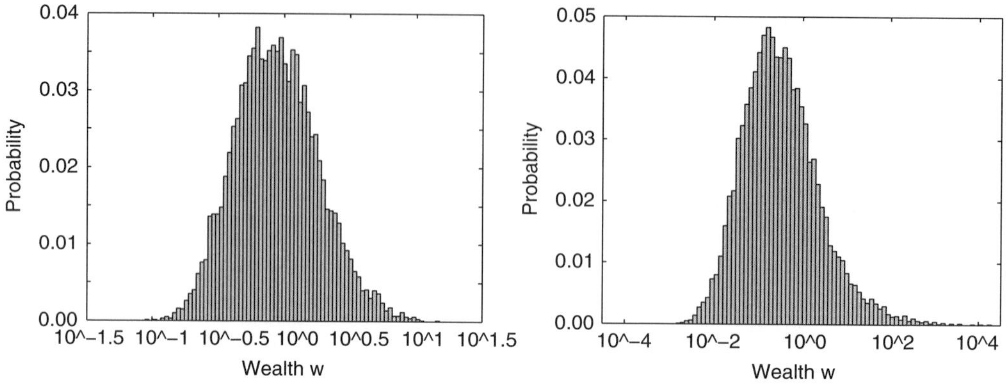

Fig. 5.6 Histogram of steady-state distribution for $\gamma_1 = \gamma_2 = 0.875$ (left) and for $\gamma_1 = \gamma_2 = 0.99$ (right).

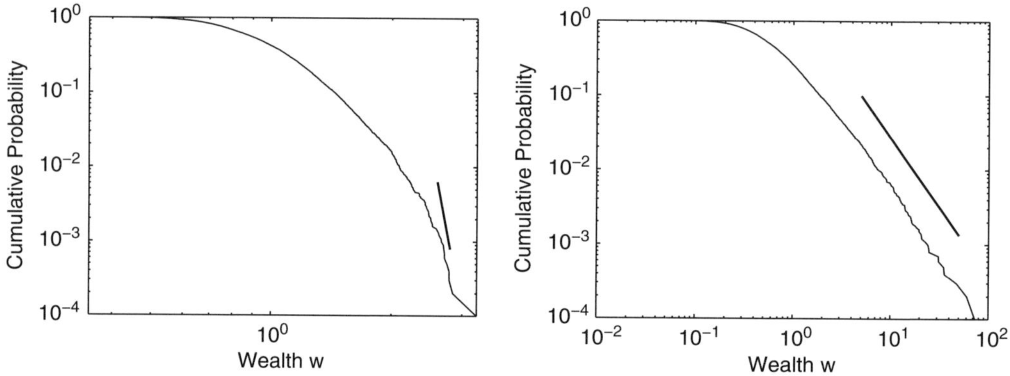

Fig. 5.7 Cumulative wealth distribution for $\gamma_1 = \gamma_2 = 0.875$ (left) and for $\gamma_1 = \gamma_2 = 0.99$ (right).

of them tend to save a very large fraction (fixed) of their wealth and the others tend to save a relatively small fraction (fixed) of their wealth. Bimodal distributions (and a polymodal distribution, in general) are, for example, reported with real data for the income distributions in Argentina [147], and are the consequence of a deep crisis in which the mean class decreased in a substantial way.

Bimodal distributions have recently been found in a different context, by studying a suitable modification of the non-conservative trade model of Slanina [287] to account for different outcomes which depend on the individual wealth [101]. Numerical experiments on the model show that the (normalized) wealth distribution tends to develop a bimodal behaviour with a power-law profile for large wealths. Despite the fact that the model in [221] is non-conservative, while the analysis of Gupta [167] refers to a conservative one, there is a common feature which produces the bimodal effect, which can easily be recognized in a separation of rules in trades between poor and rich people.

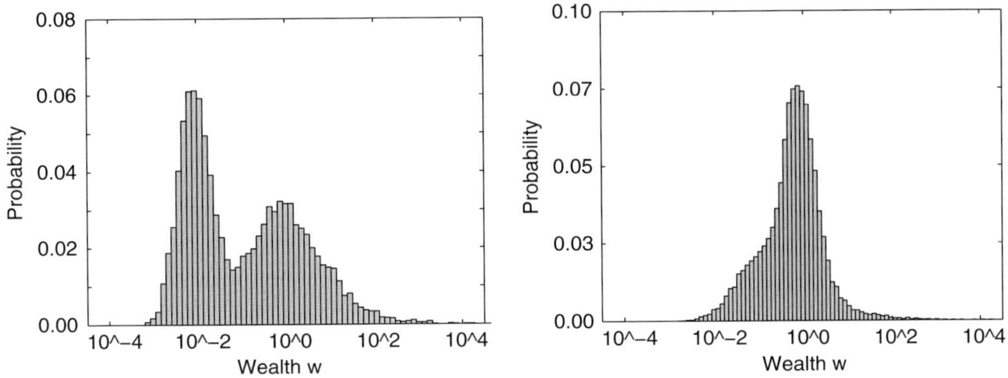

Fig. 5.8 Histogram of steady-state distribution for $\gamma_1 = 0.875$ and $\gamma_2 = 0.99$ (left) in comparison with the histogram for the union of two disjunct populations with the same parameters (right).

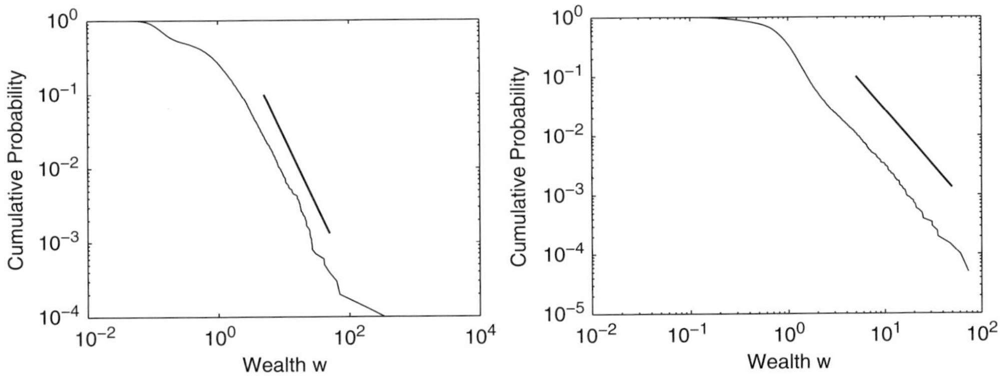

Fig. 5.9 Cumulative wealth distribution for $\gamma_1 = 0.875$ and $\gamma_2 = 0.99$ (left) in comparison with the cumulative wealth distribution for the union of two disjunct populations with the same parameters (right).

5.7 Individual preferences

The *Chatterjee–Chakrabarti–Manna* (CCM) model introduces into Angle's original trade a notable novelty. Arguing that agents are not indistinguishable in reality, but have personal trading preferences, Chatterjee et al. [96] introduced the concept of *quenched saving propensity.* Now λ is not a global quantity, but characterizes the agents. The current state of an agent is consequently described by two numbers, his wealth $w > 0$ *and* his personal saving propensity $\lambda \in (0, 1)$. We shall only discuss the case where λ does not change with time. Trade rules which allow the agents to adapt their saving strategy in time (annealed saving) have been investigated [94, 96], but seemingly do not exhibit genuinely novel effects.

The configuration of the kinetic system is described by the *extended density* function $f(\lambda, w, t)$. The wealth distribution $h(w, t)$ is recovered from the density $f(\lambda, w, t)$ as a marginal,

$$h(w,t) = \int_0^1 f(\lambda, w, t)\, d\lambda, \tag{5.130}$$

but is no longer sufficient to characterize the configuration completely. The other marginal yields the time-independent density of saving propensities,

$$\chi(\lambda) = \int_{\mathbb{R}_+} f(\lambda, w, t)\, dw. \tag{5.131}$$

Clearly, $\chi(\lambda)$ is determined by the initial condition $f(\lambda, w, t = 0)$, and should be considered as the defining parameter of the model. The collision rules are the same as originally (5.24), but take into account the individual characteristics: two agents with pre-trade wealth v, w and saving propensities λ, μ, respectively, exchange wealth according to

$$v^* = \lambda v + \epsilon[(1-\lambda)v + (1-\mu)w], \tag{5.132}$$

$$w^* = \mu w + (1-\epsilon)[(1-\lambda)v + (1-\mu)w]. \tag{5.133}$$

Clearly, wealth is strictly conserved, $v^* + w^* = v + w$, so the mean wealth m is constant in time. The Boltzmann equation (5.7) is now posed on a two-dimensional domain, $(\lambda, w) \in (0,1) \times (0, \infty)$. The collisional gain operator Q_+ satisfies

$$Q_+(f,f)[\varphi](\lambda) = \int_{\mathbb{R}_+^2} \int_0^1 \langle \varphi(v^*) \rangle f(\lambda, v) f(\mu, w)\, dv\, dw\, d\mu \tag{5.134}$$

after integration against a regular test function $\varphi(w)$. For simplicity, we assume that ϵ is symmetric around 1/2.

Pareto tail of the wealth distribution

Due to its two-dimensionality, the CCM model behaves very differently from the strictly conservative model (5.24). In particular, $h_\infty(w)$ may possess a Pareto tail.

By analogy with $\mathbf{S}(s)$ from (5.14), define the function

$$\mathbf{Q}(r) = \int_0^1 \frac{\chi(\lambda)}{(1-\lambda)^r}\, d\lambda, \tag{5.135}$$

which determines the properties of the steady wealth distribution $h_\infty(w)$ as follows [230]:

(PT′) if $\mathbf{Q}(1) < +\infty$, and $\alpha \in [1, +\infty)$ is the infimum of r for which $\mathbf{Q}(r) = +\infty$, then $h_\infty(w)$ has a Pareto tail of index α;
(ST′) if $\mathbf{Q}(r) < +\infty$ for all $r \geq 1$, then $h_\infty(w)$ has a slim tail;
(DD′) if $\mathbf{Q}(1) = +\infty$, then $h_\infty(w) = \delta_0(w)$.

To derive these results, it is useful to think of the global wealth distribution $h_\infty(w)$ as a superposition of λ-specific steady wealth distributions $f_\infty(\lambda, w)/\chi(\lambda)$, i.e., the wealth distributions of all agents with a certain personal saving propensity λ. The individual λ-specific distributions are conjectured [96,258] to resemble the wealth distributions associated with the one-dimensional model (5.24), but their features are, so far, unknown. However, they are conveniently analyzed in terms of the λ-specific moments

$$\hat{m}_s^*(\lambda) = \frac{1}{\chi(\lambda)} \int_{\mathbb{R}_+} w^s f_\infty(\lambda, w)\, dw. \tag{5.136}$$

Integration of the stationary Boltzmann equation

$$f_\infty(\lambda, w) = Q_+(f_\infty, f_\infty) \tag{5.137}$$

against $\varphi(w) = w^s$ for a *non-negative integer* s gives

$$\hat{m}_s^*(\lambda) = \frac{1}{\chi(\lambda)} \int_{\mathbb{R}_+^2} \int_0^1 \left\langle \left([\lambda + \epsilon(1-\lambda)]v + \epsilon(1-\mu)w \right)^s \right\rangle f_\infty(\lambda, v) f_\infty(\mu, w)\, d\mu\, dv\, dw.$$

After simplifications,

$$\begin{aligned} &(1-\lambda)\phi_s(\lambda)\hat{m}_s^*(\lambda) \\ &\quad = \sum_{k=0}^{s-1} \binom{s}{k} \langle \epsilon^{s-k}[\lambda + \epsilon(1-\lambda)]^k \rangle \hat{m}_s^*(\lambda) \int_0^1 (1-\mu)^{s-k} \hat{m}_{s-k}^*(\mu)\chi(\mu)\, d\mu, \end{aligned} \tag{5.138}$$

where $\phi_s(\lambda)$ is a polynomial with no roots in $[0,1]$. *The λ-specific steady wealth distributions have slim tails, and moments of arbitrary order can be calculated recursively from (5.138).* From

$$\hat{m}_0^*(\lambda) \equiv 1, \quad \hat{m}_1^*(\lambda) = \frac{m}{\mathbf{Q}(1)}(1-\lambda)^{-1}, \tag{5.139}$$

it follows inductively that

$$\hat{m}_s^*(\lambda) = r_s(\lambda)(1-\lambda)^{-s}, \tag{5.140}$$

and $r_s(\lambda)$ is a continuous, strictly positive function for $0 \le \lambda \le 1$. By Jensen's inequality, formula (5.140) extends from integers s to all real numbers $s \ge 1$. In conclusion, the total momentum

$$m_s^* = \int_0^1 \hat{m}_s^*(\lambda)\, d\lambda \propto \int_0^1 \frac{\chi(\lambda)}{(1-\lambda)^s}\, d\lambda \tag{5.141}$$

is finite exactly if $\mathbf{Q}(s)$ is finite.

$\mathbf{Q}(1) = +\infty$ would imply infinite average wealth per agent in the steady wealth distribution by formula (5.141). This clearly contradicts the conservation of the mean wealth at finite times. In reality, the first moment vanishes, and h_∞ is a Dirac distribution; see the discussion about the function $\mathbf{S}(s)$ in Section 5.3.

We emphasize this fact since a notable number of theoretical and numerical studies have been devoted to the calculation of h_∞ for *uniformly distributed* λ, i.e. $\chi(\lambda) \equiv 1$, where clearly $\mathbf{Q}(1) = +\infty$. In the corresponding experiments [94, 96, 97, 258, 261] with finite ensembles of N agents, an almost perfect Pareto tail $h_\infty(w) = C_N w^{-2}$ of index $\alpha = 1$ has been observed over a wide range $w_N < w < W_N$. However, the true tail of $h_\infty(w)$, for $w \gg W_N$, is slim. As the system's size N increases, $W_N \propto N$ also increases and $C_N \propto 1/\log N \to 0$. In fact, weak convergence of $h_\infty(w)$ to $\delta_0(w)$ in the thermodynamic limit $N \to \infty$ has been proved [230].

Rates of relaxation: Pareto tail

The discussion of relaxation is more involved than in one dimension, and we restrict our attention to the deterministic CCM model, $\epsilon \equiv 1/2$, in the case (PT′) of Pareto tails of index $\alpha > 1$. In fact, it is believed [97] that the randomness introduced by ϵ has little effect on the large-time behaviour of the kinetic system.

The stationary state of the deterministic CCM model is characterized by the *complete stop of wealth exchange.* This is very different from the steady states for the one-dimensional models, where the macroscopic wealth distribution is stationary despite the fact that wealth *is* exchanged on the microscopic level. Stationarity in (5.132) and (5.133) is achieved precisely if $v(1-\lambda) = w(1-\mu)$ for arbitrary agents with wealth v, w and saving propensities λ, μ, respectively. Correspondingly, the particle density concentrates in the plane on the curve

$$K_\infty = \{\,(\lambda, w) \,|\, (1-\lambda)w = m/\mathbf{Q}(1)\}, \tag{5.142}$$

and the steady wealth distribution is explicitly given by Mohanty's formula [237],

$$h_\infty(w) = \frac{m}{w^2}\chi\Big(1 - \frac{m}{w}\Big), \tag{5.143}$$

with the convention that $\chi(\lambda) = 0$ for $\lambda < 0$.

The conjectured [95, 261] time scale for relaxation of solutions is $t^{-(\alpha-1)}$. It has been proven [132] that relaxation is *at most* $\alpha - 1$, and cannot occur on a faster time scale. The complete statement, however, has been, so far, made rigorous only for $1 < \alpha < 2$ [230].

The key tool for the analysis is the equation for the λ-specific mean wealth,

$$\frac{d}{dt}\hat{m}_1(\lambda, t) = -\frac{1-\lambda}{2}\hat{m}_1(\lambda, t) + \int_0^1 \frac{1-\mu}{2}\hat{m}_1(t;\mu)\chi(\mu)\,d\mu. \tag{5.144}$$

Intuitively, the slow algebraic relaxation is explained by the temporal behaviour of the richest agents. By (5.144), the λ-specific average wealth $\hat{m}_1(\lambda, t)$ grows at most linearly in time,

$$\hat{m}_1(\lambda, t) \le t + \hat{m}_1(0;\lambda). \tag{5.145}$$

Thus, the tail of the wealth curve $h(w, t)$ becomes slim for $w \gg t$. The cost to fill up the fat tail $h_\infty(w) \propto w^{-(\alpha+1)}$ is approximately given by

$$\int_t^\infty w\, h_\infty(w)\,dw \propto \int_t^\infty w^{-\alpha}\,dw \propto t^{-(\alpha-1)}. \tag{5.146}$$

That equilibration works *no slower* than this (at least for $1 < \alpha < 2$) follows from a detailed analysis of the relaxation process. In [230], it has been proven that

$$\int_0^1 \left| \hat{m}_1(\lambda, t) - \frac{m}{\lambda \mathbf{Q}(1)} \right| \chi(\lambda)\, d\lambda \propto t^{-(\alpha-1)} \tag{5.147}$$

by relating (5.144) to the radiative transfer equation [154]. Moreover, the λ-specific variance

$$\hat{V}(\lambda, t) = \hat{m}_2(\lambda, t) - \hat{m}_1(\lambda, t)^2 \tag{5.148}$$

was shown to satisfy

$$\int_0^1 (1-\lambda)^2 \hat{V}(\lambda, t) \chi(\lambda)\, d\lambda \propto t^{-\alpha}, \tag{5.149}$$

provided $1 < \alpha < 2$.

Moreover, relaxation may be decomposed into two processes. The first is concentration of agents at the λ-specific mean wealth $\hat{m}_1(\lambda, t)$; i.e., all agents with the same saving propensity become approximately equally rich. According to (5.149), this process happens on a time scale $t^{-\alpha/2}$. Second, the localized mean values tend towards their respective terminal values $m/\lambda \mathbf{Q}(1)$. Thus, agents of the same saving propensity simultaneously adjust their wealth. By (5.147), the respective time scale is $t^{-(\alpha-1)}$, which is indeed slower than the first provided $\alpha < 2$.

Rates of relaxation: Dirac delta

Lastly, let us consider the deterministic CCM model with a density $\chi(\lambda)$ that satisfies $\lim_{\lambda \to 0} \chi(\lambda) > 0$, e.g. $\chi \equiv 1$ on $[0, 1]$. Clearly, $\mathbf{Q}(1) = +\infty$. An analysis of (5.144) provides for $\lambda < 1$ the estimate [230]

$$\frac{c}{1-\lambda} \le \log t \cdot \hat{m}_1(\lambda, t) \le \frac{C}{1-\lambda} \quad (t > T_\lambda), \tag{5.150}$$

with $0 < c < C < +\infty$, and $T_\lambda \to +\infty$ as $\lambda \to 1$. Convergence of $h(w, t)$ to a delta in $w = 0$ is a direct consequence, since $\hat{m}_1(\lambda, t)$ tends to zero for each $0 \le \lambda < 1$ as $t \to \infty$.

Estimate (5.150) has a direct interpretation. Agents of very high saving propensity $\lambda \approx 1$ drain all wealth out of the remaining society as follows. At intermediate times $t \gg 1$, agents equilibrate in microscopic trades so that the product $(1-\lambda)w$ becomes approximately a global constant $m(t)$. Agents with low saving propensity $\lambda < 1 - m(t)/t$ indeed satisfy $w \approx m(t)/(1-\lambda)$. Agents with higher saving propensity, however, are in general far from this (apparent) equilibrium; their target wealth $m(t)/(1-\lambda)$ is very large, whereas their actual wealth is bounded by t on the average. Correspondingly, a Pareto region of shape $h(w, t) \approx \chi(1) m(t) w^{-2}$ forms over a range $1 \ll w \le t$, whereas the tail of $h(w, t)$ for $w \gg t$ is slim. The average wealth per agent contained in the Pareto region amounts to

$$\int_1^t w h(w, t)\, dw \approx \chi(1) m(t) \log t. \tag{5.151}$$

By conservation of the average wealth, the global constant $m(t)$ tends to zero logarithmically in t and gives rise to (5.150).

A Fokker–Planck equation for distributed trading rate

The two-dimensional model of Chatterjee, Chakrabarti and Manna, namely a kinetic model of Boltzmann type, in which the saving propensity represents an additional variable, can lead, in the same spirit, to a two-dimensional model of Fokker–Planck type [136]. The idea is to generalize the Fokker–Planck system (5.122) to the case in which the trading rate is randomly distributed on the interval $(0, 1)$, with distribution $\Gamma(\lambda)$, where

$$\Gamma(\lambda) = \int_{\mathbb{R}_+} g_0(\lambda, v)\, dv, \quad 0 < \lambda < 1,$$

is the λ-marginal of the initial density of wealth. In this case, the unknown densities $g_i(v, \tau)$ $(i = 1, \ldots, n)$ are substituted by $g(\tau; \lambda, v)$, while $\sum_{j=1}^{n} g_i$ becomes

$$\langle g(\lambda, v)\rangle_\lambda = \int_0^1 g(\lambda, v)\, d\lambda.$$

Note that here and below we will denote

$$\langle \Phi(\lambda)\rangle_\lambda = \int_0^1 \Phi(\lambda)\, d\lambda.$$

If we assume $\mu_{ij} = \mu$ and $\tau_{ij} = 1$ $(i, j = 1, \ldots, n)$ in (5.122), the corresponding Fokker–Planck equation with a continuous varying trading rate reads

$$\begin{aligned} \frac{\partial g(\lambda, v)}{\partial \tau} &= \frac{\mu}{2}\frac{\partial^2}{\partial v^2}\left(v^2 \langle \rho(\tau; \lambda)\rangle_\lambda\, g(\lambda, v)\right) \\ &\quad + \frac{\partial}{\partial v}\left(\left(\lambda v \langle \rho(\tau; \lambda)\rangle_\lambda - \langle \lambda M(\tau; \lambda)\rangle_\lambda\right) g(\lambda, v)\right). \end{aligned} \tag{5.152}$$

Taking into account that the total mass is preserved, $\langle \rho(\lambda)\rangle_\lambda \equiv 1$, while the distribution

$$\Gamma(\tau; \lambda) = \int_{\mathbb{R}_+} g(\tau; \lambda, v)\, dv$$

does not depend on time, i.e. $\Gamma(\tau; \lambda) = \Gamma(\lambda)$, equation (5.152) simplifies to

$$\frac{\partial g(\lambda, v)}{\partial \tau} = \frac{\mu}{2}\frac{\partial^2}{\partial v^2}\left(v^2 g(\lambda, v)\right) + \frac{\partial}{\partial v}\left(\left(\lambda v - \langle \lambda M(\lambda)\rangle_\lambda\right) g(\lambda, v)\right). \tag{5.153}$$

The analytical study of the behaviour of the solution to equation (5.152) would certainly deserve attention. The main difference between the CCM model and this one is related to the fact that the Fokker–Planck equation (5.153) is obtained from the CPT model with risky components.

5.8 Taxation and wealth redistribution

The exact description of the steady states of the pure gambling model presented in Section 5.5 emphasizes that a substantial difference in the final behaviour of the model (the presence or not of tailed steady states) can be observed depending on the fact that binary trades are pointwise conservative, or conservative in the mean. In all the aforementioned kinetic models, however, the asymptotic distribution of wealth depends completely on the microscopic structure of binary trades, and all the other effects which in real economies drive the wealth distribution towards a fixed objective are missing. Recently, the pertinent research started to develop in at least two new directions. On the one hand, new kinetic models have recently been proposed, which, while keeping the kinetic description, introduce more sophisticated rules for trading. On the other hand, in a number of papers [42, 158, 166, 263], there have been efforts to include non-microscopic effects like global taxation (and subsequent redistribution). While postponing a short description of some of these new efforts to Chapter 7, we dedicate this section to the description of a possible model for taxation and redistribution, by resorting mainly to the recent contribution [42].

The kinetic model in [42] consists of a nonlinear kinetic equation, similar to the ones introduced in this chapter. The novelty was to introduce a simple taxation mechanism at the level of the single binary trade to generate a portion of the mean wealth of the society, totally redistributed to agents, to keep the total wealth constant. The mechanism of redistribution was assumed sufficiently flexible to be able to redistribute to agents a constant amount of wealth independently of the wealth itself, or to redistribute proportionally (or inversely proportionally) to their wealth. As happens in the kinetic description of simple general binary trades, the nonlinearity of the model, however, does not allow explicit descriptions of the resulting steady distributions, nor to relate the structure of these states to the taxation and redistribution mechanism.

In order to best represent the role of taxation and redistribution in the steady wealth distribution, in [309] it appeared convenient to rely on a simple linear kinetic model, reminiscent of the taxation and redistribution mechanism introduced in [42], in which taxation and redistribution depend on the action of an external (fixed in time) subject. This subject has the function of the fixed background of field particles in kinetic theory, widely discussed in Chapter 1, suitably adapted to the present situation.

As discussed in Chapter 1, Sections 1.3.2 and 1.3.3, the wealth distribution of the market is driven here by the *collisions* undergone with a background $M(w)$, and the wealth distribution function will consequently be governed by a single linear integro-differential kinetic equation. The interaction with the background is built up in such a way that not all wealth which enters into the trade is instantaneously restituted to the agents. The part of the wealth which is not restituted in a single trade with the background is subsequently uniformly redistributed among agents. Consequently, the total mean wealth is kept constant in time, and the wealth distribution is shown to converge towards a steady profile, which depends both on the background density and the redistribution operator.

The mathematical description takes advantage of the linearity of the model, which in some particular cases leads to an explicit expression of the steady profile. In particular, a total taxation plus a uniform redistribution is responsible for convergence towards the exponential distribution. In all the other cases, the steady states are shown to be probability distributions of convergent series of a special structure, called perpetuities. Steady states of this structure are known to occur in the realm of insurance and finance as a sum of discounted payment streams (cf. [6, 159] and the references therein).

5.8.1 A simple taxation model

We aim to describe the time evolution of the distribution of wealth in a population of agents which are subject to *microscopic* trades with a fixed background. In this picture each trade is interpreted as an interaction where a fraction of the money moves to the background, which at the same time can restitute a fraction of its money, according to some fixed strategy. In order to mimic taxation, the balance of the mean wealth is negative and, in the absence of redistribution, the mean wealth of the population decreases. Hence, following the idea of [42], a uniform redistribution operator is introduced into the model to keep the mean wealth constant. As a consequence of these trade rules, we expect to obtain for large time a stationary wealth distribution $f_\infty(v)$.

Let $M(w)$, $w \geq 0$, denote the probability distribution of the wealth of the (fixed) background. We will assume, moreover, that $M(w)$ has a bounded mean wealth, so that

$$\int_{\mathbb{R}_+} M(w)\,dw = 1, \quad \int_{\mathbb{R}_+} wM(w)\,dw = m_B. \tag{5.154}$$

The interaction with the background will be described by a trade in which the agent with wealth v comes out with wealth

$$v^* = (1-\varepsilon)v + \delta w. \tag{5.155}$$

In (5.155), $0 < \varepsilon \leq 1$ and $\delta \leq \varepsilon$ are positive constants. As a consequence of the interaction (5.155), each agent will move a proportion εv of his wealth v to the external background, which on the other hand will restitute part of its wealth w to the agent. It is clear that the structure of the background will induce a certain taxation policy.

For example, if we assume that the background is a random variable uniformly distributed on the interval $(0, a)$, where $a > 0$ is a fixed constant, if the wealth v of the agent is bigger than a, the agent will come out of the interaction with a wealth v^* less than v. Hence, interactions with the background are favourable to agents with small wealth.

Following the construction of Chapter 1, the effect of collisions of type (5.155) can be quantitatively described by a linear kinetic equation for the wealth distribution $f(v,t)$, which in weak form reads

$$\frac{d}{dt}\int_{\mathbb{R}_+} \phi(v) f(v,t)\,dv = \int_{\mathbb{R}_+^2} \big(\phi(v^*) - \phi(v)\big) f(v,t) M(w)\,dw\,dv. \tag{5.156}$$

Clearly, the mean wealth $m(t)$ is not preserved, since

$$\frac{dm(t)}{dt} = -\varepsilon m(t) + \delta m_B. \tag{5.157}$$

Note that equation (5.157) can be explicitly solvable to give

$$m(t) = m_o e^{-\varepsilon t} + \frac{\delta}{\varepsilon} m_B \left(1 - e^{-\varepsilon t}\right).$$

Hence, the mean wealth would relax towards the constant value $\bar{m} = \delta m_B/\epsilon$.

In order to keep the mean wealth constant, and to mimic (uniform) redistribution of wealth among agents, we introduce a transport operator. This transport operator is given by the formula

$$T^{\varepsilon,\delta}(f) = -\left(\varepsilon m(t) - \delta m_B\right)\frac{\partial f}{\partial v}. \tag{5.158}$$

Coupling taxation and redistribution leads to the kinetic equation

$$\begin{aligned}\frac{d}{dt}\int_{\mathbb{R}_+}\phi(v)f(v,t)\,dv &= \int_{\mathbb{R}_+^2}\left(\phi(v^*)-\phi(v)\right)f(v,t)M(w)\,dw\,dv\\ &\quad +(\varepsilon\, m(t)-\delta\, m_B)\int_{\mathbb{R}_+}\phi'(v)f(v,t)\,dv.\end{aligned} \tag{5.159}$$

Note that (5.159) implies that $f(v,t)$ remains a probability density if it is so initially,

$$\int_{\mathbb{R}_+} f(v,t)\,dv = \int_{\mathbb{R}_+} f_0(v)\,dv = 1. \tag{5.160}$$

Moreover, on the basis of (5.155), choosing $\phi(v) = v$ shows that the simultaneous presence in (5.159) of collisional and transport terms is such that the total mean wealth is also preserved in time:

$$m(t) = \int_{\mathbb{R}_+} vf(v,t)\,dv = \int_{\mathbb{R}_+} vf_0(v)\,dv = m(0). \tag{5.161}$$

We remark that only in the case in which $m(0) = \bar{m}$ does the coefficient in front of the transport term vanish, and the model reduces to the linear kinetic equation (5.156).

5.8.2 Details on the redistribution operator

The redistribution operator defined in (5.158) is composed of two parts, which can be analyzed separately to better understand their effect on the modification of the wealth distribution. The effect of the constant transport is easily understood. Indeed, the partial differential equation (pure transport)

$$\frac{\partial f}{\partial t} = \delta m_B \frac{\partial f}{\partial v} \tag{5.162}$$

is exactly solvable. By direct inspection it can easily be verified that, starting from an initial wealth distribution $f_0(v)$, the function

$$f(v,t) = f_0(v + \delta m_B t)$$

solves (5.162). The physical meaning of this solution is clear. The operator moves wealth from rich to poor people, and this action is linear with respect to time. As time goes on, the wealth located far away from the origin (at a distance $\delta m_B t$), is moved left towards the origin. Thus, the process will end up with all wealth located at $v = 0$.

The second part of the redistribution operator is a particular case of a more general one, studied in [42], of the form

$$R_\chi^\varepsilon(f)(v,t) = \varepsilon \frac{\partial}{\partial v}\Big[\left(\chi v - (\chi+1)m(t)\right) f(v,t)\Big]. \tag{5.163}$$

Here, $m(t)$ denotes the first moment of f, which, in general, makes the operator R_χ^ε nonlinear. The weight factor multiplying the distribution function inside the square brackets in (5.163) has been taken to be linear in v for simplicity, also in order to involve in the mechanism only the most meaningful moments, those of order zero and one. Such a weight function contains only one disposable real parameter χ, a constant that characterizes the type of redistribution, and that determines the slope of the straight line as well as the value of v, whether physical or non-physical, at which the weight itself vanishes. The other parameter has been determined by the constraint that the redistribution operator preserves the number of agents and actually redistributes the total amount of money that is being collected by taxation. In fact, note that we have, whatever the constant χ,

$$\int_{\mathbb{R}_+} v R_\chi^\varepsilon(f)(v,t)\, dv = \varepsilon\, m(t), \tag{5.164}$$

and, provided $f(v,t)$ also satisfies the boundary condition $f(0,t) = 0$,

$$\int_{\mathbb{R}_+} R_\chi^\varepsilon(f)(v,t)\, dv = 0. \tag{5.165}$$

The additional boundary condition would not be needed in the special case $\chi = -1$, which reproduces the standard anti-drift operator considered by Slanina [287].

A further insight into the effect of the operator (5.163) comes from the analytic expression of the solution to the equation

$$\frac{\partial}{\partial t} f(v,t) = R_\chi^\varepsilon(f)(v,t) \tag{5.166}$$

with initial density $f_0(v)$ and corresponding initial average wealth $m(0)$. First, it is immediately realized that the process actually creates wealth, with $m(t) = m(0)\, \exp(\varepsilon\, t)$. This renders the kinetic equation (5.166) linear, and it can be verified by direct inspection that the solution reads

$$f(v,t) = e^{\varepsilon\chi t} f_0 \left[e^{\varepsilon\chi t} v - m(0) \left(e^{\varepsilon(\chi+1)t} - 1 \right) \right], \quad t > 0. \tag{5.167}$$

Note that, if $f_0(v) = 0$ for $v < 0$, the solution at time $t > 0$ vanishes for

$$v < m(0)\, \frac{e^{\varepsilon(\chi+1)t} - 1}{e^{\varepsilon\chi t}}, \tag{5.168}$$

which implies that no agents with wealth below the right-hand side of (5.168) are present at time t. For $\chi > -1$ the threshold increases in time and diverges as $O(e^{\varepsilon t})$ for $t \to +\infty$.

It is not difficult to check that the actual redistribution function among the population of agents (namely the normalized fraction $\psi(v)$ of the total collected money that is supplied to each agent of wealth v) can be deduced from (5.163):

$$\psi(v,t)\frac{\partial f}{\partial v} = -\left[\frac{\chi}{m(t)}v - (\chi+1)\right]\frac{\partial f}{\partial v} - \frac{\chi}{m(t)}f(v,t). \tag{5.169}$$

Clearly, ψ depends on f in a complicated way. However, its qualitative analysis versus v is sufficient for an estimate of the actual redistribution to agents with different wealths when the characteristic parameter χ of the operator R_χ^ϵ is varied. The case in which the total yield from taxation is equally distributed among all agents corresponds to the option $\psi(v) = 1$, and is achieved in (5.169) by the special option $\chi = 0$. In all other cases the redistribution is selective, and may correspond to some partition strategy. It is not guaranteed that ψ should be positive for any v, namely the redistribution might take away further money (beyond taxation) from a part of the population, and reinforce another part. On the other hand, in this model the processes described by the two distinct operators introduced so far proceed simultaneously, and gain or loss is determined by the overall balance only. It is easy to see that, for the most reasonable trends for the wealth profiles f, the redistribution function ψ is negative or positive in a right neighbourhood of $v = 0$ according to whether $\chi < -1$ or $\chi > -1$, respectively. Analogously, when v is large enough, ψ turns out to be positive or negative for $\chi < 0$ or $\chi > 0$, respectively. In a sense, for positive values of the parameter χ, money is redistributed to agents with little wealth, whereas agents of large wealth are taxed once more. When $\chi < -1$, we would have the opposite situation in which the poorest part of the population supplies additional resources to the richest part, and for this reason we will exclude this range of parameter values from the analysis below. Finally, in the intermediate case $-1 < \chi < 0$, both the richest and poorest agents are favoured by the taxation/redistribution mechanism, at the expense of the middle class.

5.8.3 Details on the linear model

Having explained the action of the redistribution operator, we now proceed to study equation (5.159). We assign both to the background and to the initial density a unit mean

$$\int_{\mathbb{R}_+} wM(w)\,dw = \int_{\mathbb{R}_+} vf_0(v)\,dv = 1. \tag{5.170}$$

In addition, we assume further that the background density has some bounded moment bigger than one, typically the second:

$$\int_{\mathbb{R}_+} w^2 M(w)\,dw = \sigma^2 < +\infty. \tag{5.171}$$

By (5.170), the Boltzmann equation assumes the simpler form

$$\begin{aligned}\frac{d}{dt}\int_{\mathbb{R}_+}\phi(v)f(v,t)\,dv &= \int_{\mathbb{R}_+^2}\big(\phi(v^*) - \phi(v)\big)f(v,t)M(w)\,dw\,dv \\ &\quad + (\varepsilon - \delta)\int_{\mathbb{R}_+} f(v,t)\phi'(v)\,dv.\end{aligned} \tag{5.172}$$

Note that, since, by construction, $\delta \le \varepsilon$, part of the money that comes out of the trades with the background is redistributed uniformly. Thanks to (5.171), if the initial density f_0 has a bounded second moment, it holds that

$$\int v^2 f(v,t)\,dv \le \max\left\{\int v^2 f_0(v);\ \frac{2\varepsilon(1-\delta)+\delta^2\sigma^2}{2\varepsilon-\varepsilon^2}\right\}. \tag{5.173}$$

The two extremal cases, corresponding to $\delta = 0$ ($\delta = \varepsilon$ respectively) give rise to a notable simplification. If $\delta = 0$, the outgoing wealth does not depend on the background M, and the Boltzmann equation simplifies to

$$\frac{d}{dt}\int \phi(v)f(v,t)\,dv = \int \big(\phi(v^*) - \phi(v)\big) f(v,t)\,dv + \varepsilon \int \phi' f(v,t)\,dv. \tag{5.174}$$

In this case, (5.174) does not depend on the background, and the post-interaction wealth is simply

$$v^* = (1-\varepsilon)v. \tag{5.175}$$

The Boltzmann equation (5.174) in this case represents the evolution of a wealth density in which a constant part of the wealth is taken away in each interaction (taxation) and the same portion of wealth is redistributed uniformly among agents.

If, on the contrary, $\delta = \varepsilon$, the redistribution operator in (5.172) disappears, and the Boltzmann equation simplifies to

$$\frac{d}{dt}\int_{\mathbb{R}_+} f(v,t)\phi(v)\,dv = \int_{\mathbb{R}_+^2} \big(\phi(v^*) - \phi(v)\big) f(v,t)M(w)\,dw\,dv. \tag{5.176}$$

In (5.176) the post-interaction wealth is now

$$v^* = (1-\varepsilon)v + \varepsilon w. \tag{5.177}$$

Following the methodology of Chapter 2, most of the analytical properties of the solution are obtained by means of Fourier analysis. In this case, the Fourier-transformed equation reads

$$\frac{\partial \widehat{f}(\xi,t)}{\partial t} = \hat{f}((1-\varepsilon)\xi,t)\hat{M}(\delta\xi) - \hat{f}(\xi,t) - i(\varepsilon-\delta)\xi\hat{f}(\xi,t). \tag{5.178}$$

In (5.178) we used the fact that $M(v)$ is a probability density function, so that $\hat{M}(0) = 1$. The initial conditions (5.170) turn into

$$\widehat{f_0}(0) = 1 \quad \text{and} \quad \widehat{f_0}'(0) = -\,i. \tag{5.179}$$

Existence and uniqueness of the solution follows along the same lines as Section 2.2.2, owing to the Fourier-based metric (2.17). In this simple linear case one can prove that the operator on the right-hand side of (5.178) is Lipschitz continuous in this metric.

In order to show Lipschitz continuity, let us define

$$\hat{g}(\xi,t) = \hat{f}(\xi,t)\exp\{i(\varepsilon-\delta)\xi t\}. \tag{5.180}$$

Then it is easily shown that, whenever $\hat{f}$ satisfies (5.172), $\hat{g}$ satisfies

$$\frac{\partial \hat{g}(\xi,t)}{\partial t} + \hat{g}(\xi,t) = \hat{g}((1-\varepsilon)\xi,t)\hat{M}(\delta\xi)\exp\{i(\varepsilon-\delta)\varepsilon\xi t\}. \tag{5.181}$$

Therefore, if $\hat{g}_1(\xi,t)$ and $\hat{g}_2(\xi,t)$ are two solutions to equation (5.181) corresponding to the initial data $\hat{g}_{0,1}(\xi)$ and $\hat{g}_{0,2}(\xi)$, it holds that

$$\begin{aligned}\frac{\partial}{\partial t}\frac{\hat{g}_1(\xi,t)-\hat{g}_2(\xi,t)}{|\xi|^s} &+ \frac{\hat{g}_1(\xi,t)-\hat{g}_2(\xi,t)}{|\xi|^s} \\ &= \frac{\hat{g}_1((1-\varepsilon)\xi,t)-\hat{g}_2((1-\varepsilon)\xi,t)}{|\xi|^s}\hat{M}(\delta\xi)\exp\{i(\varepsilon-\delta)\varepsilon\xi t\}.\end{aligned} \tag{5.182}$$

On the other hand, since $|\hat{M}(\xi| \le 1$ the right-hand side of (5.182) is bounded above by

$$(1-\varepsilon)^s d_s(g_1(t),g_2(t)). \tag{5.183}$$

Application of Gronwall's inequality to (5.182) gives

$$d_s(g_1(t),g_2(t)) \le d_s(g_{0,1},g_{0,2})\exp\{((1-\varepsilon)^s-1))t\}. \tag{5.184}$$

Considering that, by definition,

$$d_s(g_1(t),g_2(t)) = d_s(f_1(t),f_2(t)), \tag{5.185}$$

while $g(v,t=0) = f(v,t=0) = f_0(v)$, we conclude with the estimate

$$d_s(f_1(t),f_2(t)) \le \exp\{((1-\varepsilon)^s-1))t\}\,d_s(f_{1,0},f_{2,0}), \tag{5.186}$$

which is valid any time that the initial values $f_{1,0}$ and $f_{2,0}$ satisfying conditions (5.170) are such that $d_s(f_{1,0},f_{2,0})$ is finite.

We remark that in the present case, since the solution to equation (5.172) conserves both mass and momentum, the natural setting is to choose initial data with bounded first moment, and consequently to set $s = 1+\beta$, where $0 \le \beta < 1$.

Also, it can be proven that the weak solution of the kinetic equation (5.172), which initially has finite moments up to order 2, converges exponentially fast in d_s to a steady state f_∞,

$$d_s(f(t),f_\infty) \le d_s(f_0,f_\infty)\exp\{((1-\varepsilon)^s-1))t\}. \tag{5.187}$$

In addition, f_∞ has mean wealth equal to 1, and it is the only steady state with this mean wealth. To prove this claim, one can proceed step by step as in Section 5.3.2.

5.8.4 Steady states as perpetuities

Once we know the existence and uniqueness of a stationary profile for the kinetic equation (5.172), we can try to extract information on it. This profile satisfies the equation

$$\hat{f}_\infty((1-\varepsilon)\xi)\hat{M}(\delta\xi) - \hat{f}_\infty(\xi) - i(\varepsilon-\delta)\xi\hat{f}_\infty(\xi) = 0. \qquad (5.188)$$

Equation (5.188) can be rewritten as

$$\hat{f}_\infty(\xi) = \frac{\hat{M}(\delta\xi)}{1+i(\varepsilon-\delta)\xi}\hat{f}_\infty((1-\varepsilon)\xi), \qquad (5.189)$$

which, due to the fact that $f_\infty(0) = 1$, recursively gives

$$\hat{f}_\infty(\xi) = \prod_{k=0}^{\infty} \frac{\hat{M}\left(\delta(1-\varepsilon)^k\xi\right)}{1+i(\varepsilon-\delta)(1-\varepsilon)^k\xi}. \qquad (5.190)$$

This formula completely identifies the (unique) steady state to equation (5.172).

Let us describe in some detail both the structure and the properties of $f_\infty(v)$. By direct inspection it can easily be verified that, for a given constant $\alpha > 0$, the function

$$\hat{G}_\alpha(\xi) = \frac{1}{1+i\alpha\xi}, \qquad \xi \in \mathbb{R} \qquad (5.191)$$

is the Fourier transform of the exponential density ($G_1 = G$) of mean α,

$$G_\alpha(v) = \frac{1}{\alpha}\exp\left\{-\frac{v}{\alpha}\right\}, \quad v \geq 0; \qquad G_\alpha(v) = 0 \quad v < 0. \qquad (5.192)$$

Consequently, if Z is a random variable with law f_∞, the law of Z, defined in (5.189), is a distributional fixed point of the equation

$$Z \stackrel{d}{=} \delta X + (\varepsilon-\delta)Y + (1-\varepsilon)Z, \qquad (5.193)$$

where we recall that $\stackrel{d}{=}$ means that the random variables on both sides have the same probability law, and as usual the variables X, Y and Z are assumed to be independent of each other. In (5.193) the random variable X is distributed according to $M(v)$, while Y is distributed according to the exponential distribution G. The law of Z is often called perpetuity due to its occurrence in the realm of insurance and finance as a sum of discounted payment streams (cf. [6, 159] and the references therein). As discussed in [6], it is quite difficult to extract properties of perpetuities in general. What is proven is that such distributions are of pure type, namely degenerate, absolutely continuous, or continuously singular. Moreover, necessary and sufficient criteria are presented in [6] for the finiteness of polynomial as well as exponential moments.

In our case, due to the fact that we know the laws of the random variables X and Y, one can extract more information about regularity and behaviour at the point $v = 0$.

The simplest case corresponds to the extremal choice $\delta = 0$, $\varepsilon = 1$, which leads to equation (5.174) with $\varepsilon = 1$. This case corresponds to total taxation (all the wealth put into the trade moves to the background), and uniform redistribution. Setting $\delta = 0$ and $\varepsilon = 1$ in (5.189) gives

$$\hat{f}_\infty(\xi) = \frac{1}{1 + i\xi}, \tag{5.194}$$

and the steady state is an exponential distribution of unit mean

$$f_\infty(v) = G(v). \tag{5.195}$$

Choosing $\delta = 0$, $\varepsilon < 1$, which leads to the equation (5.174), we obtain from (5.190) the expression

$$\hat{f}_\infty(\xi) = \prod_{k=0}^{\infty} \frac{1}{1 + i\varepsilon(1-\varepsilon)^k \xi}, \tag{5.196}$$

namely an infinite convolution of independent exponential distributions of means $m_k = \varepsilon(1-\varepsilon)^k$. In this case, regularity of the steady state follows by showing that it belongs to the Sobolev spaces of high degree [1],

$$\|g\|^2_{\dot{H}^\eta(\mathbb{R})} = \int_{\mathbb{R}} |\xi|^{2\eta} |\hat{g}(\xi)|^2 \, d\xi, \tag{5.197}$$

with $\eta > 0$. In fact, since

$$|\hat{G}_\alpha(\xi)| = \left| \frac{1}{1 + i\alpha\xi} \right| = \left(\frac{1}{1 + \alpha^2 |\xi|^2} \right)^{1/2}, \tag{5.198}$$

it follows that, for a given $N \in \mathbb{N}$,

$$\begin{aligned} \left| \prod_{k=0}^{N} \frac{1}{1 + i\varepsilon(1-\varepsilon)^k \xi} \right| &= \left(\prod_{k=0}^{N} \frac{1}{1 + \varepsilon^2 (1-\varepsilon)^{2k} |\xi|^2} \right)^{1/2} \\ &\le \left(\frac{1}{1 + c_{N,\varepsilon} |\xi|^{2N+2}} \right)^{1/2}. \end{aligned} \tag{5.199}$$

In (5.199), the positive constant $c_{N,\varepsilon}$ is given by

$$c_{N,\varepsilon} = \prod_{k=0}^{N} \varepsilon^2 (1-\varepsilon)^{2k}. \tag{5.200}$$

Consequently, for any given $N \in \mathbb{N}$,

$$|\hat{f}_\infty(\xi)|^2 \le \frac{1}{1 + c_{N,\varepsilon} |\xi|^{2N+2}}, \tag{5.201}$$

and this shows that f_∞ belongs to the Sobolev space $\dot{H}^N(\mathbb{R})$ [1]. This result guarantees that $f_\infty(v = 0) = 0$. In fact, since $f_\infty(v) \in \dot{H}^N$, with $N > 1/2$, by the Sobolev imbedding [1], this regularity is enough to guarantee that the steady state is a bounded and continuous function. Since the trade mechanism (5.155) is such that the post-trade wealth is nonnegative, while the (transport) redistribution operator moves the wealth density on the right, a solution density which is initially distributed on $\mathbb{R}_+$ remains distributed on $\mathbb{R}_+$ at any subsequent time, so that $f(v,t) = 0$ if $v < 0$. The same property thus holds for the steady state f_∞. On the other hand, since $f_\infty(v)$ is continuous in $v = 0$, the condition $f_\infty(v = 0) = 0$ follows.

The same conclusion can be drawn in the case in which $\delta > 0$, due to the fact that inequality (5.199) still holds with a different constant,

$$c_{N,\varepsilon,\delta} = \prod_{k=0}^{N} (\varepsilon - \delta)^2 (1 - \varepsilon)^{2k}. \tag{5.202}$$

The case $\delta = \varepsilon$ has to be treated separately. In this case, regularity of the steady state follows provided the background density satisfies

$$|\hat{M}(\xi)| \le \left(\frac{1}{1 + \mu|\xi|^2} \right)^\eta, \tag{5.203}$$

for some positive constants μ and η.

These arguments can be put in the form of a theorem, which characterizes various properties of the steady distribution.

Theorem 5.10 *Let $f_\infty(v)$ be the (unique) steady solution of the Boltzmann equation (5.172). Then, for all $0 < \varepsilon < 1$, $\delta < \varepsilon$, f_∞ belongs to $\dot{H}^N(\mathbb{R})$, for all $N > 0$, and $f_\infty(v = 0) = 0$. The same result holds if $\delta = \varepsilon$, and the background density satisfies condition (5.203). Finally, if $\varepsilon = 1$ while $\delta = 0$, $f_\infty(v)$ equals the exponential distribution of unit mean.*

Lastly, we examine the boundedness of moments of the steady distribution. The computations leading to (5.173) can easily be extended to show that the number of moments which remain bounded are linked to the number of bounded moments of the background. In fact, if

$$\int_{\mathbb{R}_+} v^{\bar{s}} M(v)\, dv = m_{\bar{s}} < +\infty, \qquad \bar{s} > 2 \tag{5.204}$$

we obtain, for $s \le \bar{s}$

$$\begin{aligned} \frac{d}{dt} \int_{\mathbb{R}_+} v^s f(v,t)\, dv &= (\varepsilon - \delta) s \int_{\mathbb{R}_+} v^{s-1} f(v,t)\, dv \\ &\quad + \int_{\mathbb{R}_+^2} \left[((1-\varepsilon)v + \delta w)^s - v^s \right] f(v,t) M(w)\, dv\, dw. \end{aligned} \tag{5.205}$$

Let us denote by $[s]$ the entire part of s, so that $s = [s] + \nu$, where $0 \le \nu < 1$. Given any two constants a and b, the following inequality holds:

$$(|a| + |b|)^\nu \le |a|^\nu + |b|^\nu. \tag{5.206}$$

Therefore, for any two constants α and δ smaller than 1,

$$(\alpha v + \delta w)^s \le (\alpha v + \delta w)^{[s]} \left((\alpha v)^\nu + (\delta w)^\nu \right), \tag{5.207}$$

so that

$$(\alpha v + \delta w)^s - v^s \le -\left(1 - \alpha^s\right) v^s + \delta^s w^s + P_{[s]}(v, w), \tag{5.208}$$

and the polynomial function $P_{[s]}$ contains only powers of v and w of order less than or equal to $[s]$. Substituting formula (5.208) into (5.205), and considering that the coefficient of the s-th moment of f is negative, we obtain

$$\begin{aligned} \frac{d}{dt} \int_{\mathbb{R}_+} v^s f(v,t)\, dv \le &-\left(1 - (1-\varepsilon)^s\right) \int_{\mathbb{R}_+} v^s f(v,t)\, dv \\ &+ \delta^s \int_{\mathbb{R}_+} w^s M(v)\, dw + M_{[s]}. \end{aligned} \tag{5.209}$$

Since the last term $M_{[s]}$ contains only moments of order less than or equal to $[s]$, while the coefficient of the s-th moment of $f(v,t)$ is negative, the boundedness of the moment of order $[s]$ implies the uniform boundedness of the moment of order s. Starting from $s = 1$ recursively concludes the analysis. The previous analysis shows that the steady state and the background have the same number of moments. Hence, independently of the value of the taxation parameter, a background with Pareto tails produces a steady state with the same tails.

The case $\delta = 0$ leads to a completely different situation. In this case, in fact, we can prove that the steady state has bounded exponential moments. For a given ε, let α be such that $\varepsilon\alpha < 1$. Then

$$\begin{aligned} \frac{d}{dt} \int_{\mathbb{R}_+} e^{\alpha v} f(v,t)\, dv &= \int_{\mathbb{R}_+} \left(e^{\alpha(1-\varepsilon)v} - e^{\alpha v} \right) f(v,t)\, dv + \varepsilon\alpha \int_{\mathbb{R}_+} e^{\alpha v} f(v,t)\, dv \\ &= \int_{\mathbb{R}_+} e^{\alpha v} \left(e^{-\varepsilon\alpha v} - 1 + \varepsilon\alpha \right) f(v,t)\, dv. \end{aligned} \tag{5.210}$$

Since

$$v \ge \bar{v} = \frac{1}{\varepsilon\alpha} \log \frac{2}{1 - \varepsilon\alpha} \tag{5.211}$$

implies

$$e^{-\varepsilon\alpha v} \le \frac{1 - \varepsilon\alpha}{2}, \tag{5.212}$$

we have

$$\int_{\mathbb{R}_+} e^{\alpha v} \left(e^{-\varepsilon\alpha v} - 1 + \varepsilon\alpha \right) f(v,t)\, dv$$

$$\le \int_{v \le \bar{v}} e^{\alpha v} \varepsilon\alpha f(v,t)\, dv + \int_{v > \bar{v}} e^{\alpha v} \left(e^{-\varepsilon\alpha v} - 1 + \varepsilon\alpha \right) f(v,t)\, dv \tag{5.213}$$

$$\le \varepsilon\alpha \left(\frac{2}{1 - \varepsilon\alpha} \right)^{1/\varepsilon} - \frac{1 - \varepsilon\alpha}{2} \int_{\mathbb{R}_+} e^{\alpha v} f(v,t)\, dv. \tag{5.214}$$

From (5.214), the uniform boundedness of the exponential moments follows. Hence we have:

Theorem 5.11 *Let $f_\infty(v)$ be the (unique) steady solution of the Boltzmann equation (5.172). Then, for all $0 < \varepsilon < 1$, $0 < \delta < \varepsilon$, f_∞ has the same moments as the background. If $\delta = 0$, f_∞ has exponential moments of order α, for all $\alpha < 1/\varepsilon$, that is*

$$\int_{\mathbb{R}_+} e^{\alpha v} f(v,t)\,dv < +\infty, \qquad \alpha < 1/\varepsilon. \tag{5.215}$$

Remark 5.7 *Theorem 5.11, coupled with the decay formula (5.186), provides an interesting and somewhat counterintuitive result for the case described by the Boltzmann equation (5.172). By Theorem 5.11, the steady profile is more peaked around the mean wealth, and consequently closer to a Dirac delta function when the taxation is small. On the other hand, the relaxation time predicted by formula (5.186) is slower if ε is small. Therefore, if we assume that the ideal steady distribution is obtained when all agents reach the same wealth, we need to assume a very small taxation parameter. Of course, this will imply a very slow rate of convergence, and the steady profile cannot be obtained in reasonable times. These considerations are in favour of the choice of intermediate coefficients of taxation, as is usually done in most western economies, to balance an unequal distribution of wealth with the time required to see the effects of the redistribution.*

5.8.5 A result about optimal taxation

The results of Sections 5.5 and 5.8 show that, for some simplified models of wealth distribution, it is possible to recover the exact analytic expression of the stationary solution. In particular, the exponential distribution (5.192) appears as a steady solution of the kinetic equation for the pure gambling model (5.89) when the random parameter that characterizes the binary trade (5.86) is uniformly distributed on the interval $[0,1]$. Surprisingly enough, the same exponential distribution appears as a steady state of the linear kinetic model with taxation introduced in Section 5.8.1, as soon as one assumes total taxation.

These results suggest that a suitable candidate for an elementary kinetic model for wealth redistribution and taxation is the pure gambling model (5.89) with a uniform random parameter, coupled with the redistribution operator (5.162). In this case, we assume that part of the post-trade wealths in (5.89) is taken away, so that

$$v_\varepsilon^* = (1-\varepsilon)\theta(v+w), \qquad w_\varepsilon^* = (1-\varepsilon)(1-\theta)(v+w). \tag{5.216}$$

In (5.216), θ is a uniform distribution on $[0,1]$, while $0 < \varepsilon \le 1$. Therefore

$$v_\varepsilon^* + w_\varepsilon^* = (1-\varepsilon)(v+w) < v+w.$$

The proportion $\varepsilon(v+w)$ of wealth which is taken away from the trade is restituted among all agents through the redistribution operator (5.162). Thus, the nonlinear kinetic model with taxation and redistribution reads

$$\frac{d}{dt}\int_{\mathbb{R}_+} f(v,t)\phi(v)\,dv = \left\langle \int_{\mathbb{R}_+^2} (\phi(v_\varepsilon^*) - \phi(v))f(v,t)f(w,t)\,dv\,dw \right\rangle + \varepsilon m(t)\int_{\mathbb{R}_+} \phi'(v)f(v,t). \tag{5.217}$$

By construction, the kinetic equation (5.217) is such that both the mass and the mean value of the solution remain constant in time. Thus, $m(t)$ is equal to the initial mean value, say m. The interesting feature of the kinetic equation (5.217) is that its solution converges towards the exponential distribution $G_m(v)$ defined in (5.192) in the two extremal cases $\varepsilon = 0, 1$, which correspond to absence of taxation (respectively total taxation).

A main question now arises. If we assume as highly reasonable that the main scope of taxation and redistribution is to eliminate as much as possible the differences in wealth among the agents, can we reckon from equation (5.217) an optimal value for taxation? The simplest way to calculate an explicit (optimal) value for ε is to look for the variance of the solution. Indeed, the variance is a measure of the spreading of the wealth of the population around its mean value, and there are no doubts that it can be assumed as a (weak) measure of the efficacy of a taxation policy.

Let us take $\phi(v) = v^2$ in (5.217). Then, the second moment

$$m_2(t) = \int_{\mathbb{R}_+} v^2\, f(v,t)\,dv$$

obeys the equation

$$\begin{aligned}\frac{dm_2(t)}{dt} &= \left\langle \int_{\mathbb{R}_+^2} ((v_\varepsilon^*)^2 - v^2)f(v,t)f(w,t)\,dv\,dw \right\rangle + 2\varepsilon m\int_{\mathbb{R}_+} vf(v,t)\,dv \\ &= \int_{\mathbb{R}_+^2} ((1-\varepsilon)^2\langle\theta^2\rangle(v+w)^2 - v^2)f(v,t)f(w,t)\,dv\,dw \Big\rangle + 2\varepsilon m^2 \\ &= \frac{1}{3}(1-\varepsilon)^2\left(2\,m_2(t) + 2m^2\right) - m_2(t) + 2\varepsilon m^2.\end{aligned} \tag{5.218}$$

Hence, $m_2(t)$ converges towards the value

$$\bar{m}_2 = \frac{2(1-\varepsilon)^2 - 6\varepsilon}{3 - 2(1-\varepsilon)^2}\, m^2,$$

so that the variance tends to the value

$$Var(f_\infty) = \left(\frac{2(1-\varepsilon)^2 - 6\varepsilon}{3 - 2(1-\varepsilon)^2} - 1\right) m^2.$$

It is clear that the variance attains its minimum value when $\varepsilon = \bar{\varepsilon}$, where

$$\bar{\varepsilon} = \frac{5 - 3\sqrt{2}}{2},$$

which is approximately equal to 38%. This optimal (with respect to the variance) percentage of taxation depends on the choice of the random variable θ which characterizes the returns in the binary trade. A different choice, which changes the value of its second moment $\langle \theta^2 \rangle$ in (5.218), leads to a different optimal ε. However, independently of the choice of the outcome of the binary trade, there exists an optimal value for ε which characterizes the solution with minimal variance.

Despite its essential simplicity, the previous computations illuminate the possibility of using kinetic modelling to understand the large-time behaviour of essential quantities, of paramount importance in the description of real economies. As a matter of fact, it is noticeable that this toy model provides a prediction of optimal taxation of about 38%, which is quite close to the total taxation percentage of most western countries.

6

Opinion modelling and consensus formation

> The deterministic conception of nature holds in its very being a real motive of weakness because irremediably contradicts the most evident data of our conscience. G. Sorel tried to compose this dysfunction by distinguishing between artificial nature and natural nature (this later being a-causal), although in this way he denied the unity of Science. On the other hand, the formal analogy between the statistical laws of Physics and those of Social Sciences supports the opinion that also human actions were submitted to a rigid determinism. It is important then, that the principles of Quantum Mechanics have led to a recognition (as well as a certain absence of objectivity in the description of phenomena) of the statistical character of the ultimate laws of elemental processes. This conclusion has made substantial the analogy between Physics and Social Sciences, and has produced between them an identity of value and method.
>
> E. Maiorana, The value of statistical laws in physics and the social sciences, *Scientia*, 1942.

6.1 Opinion formation

Microscopic models of both social and political phenomena describing collective behaviour and self-organization in a society have recently been introduced and analyzed by several authors [157, 210, 244, 292, 289, 298, 326].

The modelling of opinion formation has attracted the interest of increasing number of researchers (cf. [116, 117, 210, 244, 298] and the references therein). The starting point for a large part of these models is represented by cellular automata, where the lattice points are the agents, and where any of the agents of a community are initially associated with a random distribution of numbers, one of which is the opinion. Hence, society is modelled as a graph, where each agent interacts with his neighbours in an iterative way.

Very recently, other attempts have been successfully applied [31, 288], with the aim of describing the formation of opinion by means of mean field equations. These models are, in general, described by systems of ordinary differential equations or partial differential equations of diffusive type that can, in some cases, be treated analytically to give explicit steady states.

In [31], the attention has been focused on two aspects of opinion formation, which in principle could be responsible for the formation of coherent structures. The first one is the remarkably simple compromise process, in which pairs of agents reach a fair compromise after exchanging opinions [34, 35, 79, 116, 117, 175, 294, 327]. The bounded confidence model introduced by Defuant et al. [116] is a popular model for such scenarios.

The second one is the diffusion process, which allows individual agents to change their opinions in a random diffusive fashion. While the compromise process has its basis in the human tendency to settle conflicts, diffusion accounts for the possibility that people may change opinion through global access to information. At the present time, this aspect is gaining importance due to the emergence of new forms of communication (among them electronic mail and web navigation [271]).

Following this line of thought, a wide class of kinetic models of opinion formation, based on two-body interactions involving both compromise and diffusion properties in exchanges between individuals, has been introduced in [308]. These models are sufficiently general to take into account a large variety of human behaviours, and to reproduce in many cases explicit steady profiles from which one can easily elaborate information on the behaviour of opinion. This type of modelling has subsequently been applied to various situations in [56, 57].

It is essential to outline that the binary interactions which express the microscopic change of opinion differ in many aspects from the binary interactions considered in the previous chapters. The first difference is that, in the pertinent literature, opinion is identified with a number which can take values in a bounded interval. If we identify this interval with the subset $[-1, 1]$ we clearly intend that the values ± 1 represent the opposite extremal opinions. Even if, in principle, one could assume opinions to vary in an unbounded interval, this choice would generate several difficulties in the modelling process. Indeed, in most of the models it is assumed that a compromise process can be realized only if the difference between the opinions of the two agents belongs to a bounded interval. This hypothesis reflects the reasonable idea that any agent can influence people in his neighbourhood only if the respective opinions are not too far from each other.

Secondly, the post-interaction opinions are not a linear transformation of the pre-interaction ones. Indeed, it is realistic to assume that people with a neutral opinion are more likely to change it, while the opposite phenomenon happens with people who have extreme opinions. Hence, the mixing coefficients in general depend themselves on the opinion variable. This introduces a substantial difference from the models of wealth distribution considered in Chapter 5, since an essential part of the mathematical techniques we introduced in Chapter 2 are no longer applicable. This is true in particular for Fourier techniques, which need linear binary collisions to be used. Also, nonlinearity introduces additional difficulties in the study of macroscopic quantities of essential importance for the evolution, like moments and other observable quantities.

On the other hand, the asymptotic techniques introduced in Chapter 1, Section 1.4, whose goal it is to reduce the complexity of the kinetic model, remain applicable. The main idea in this context is to make dominant interactions which produce small changes of opinions. This asymptotic limit (hereafter called the quasi-invariant opinion limit) leads to partial differential equations of Fokker–Planck type for the distribution of opinion among individuals.

As explained in in Chapter 1, Section 1.4, this approach is an alternative to other techniques. Indeed, similar diffusion equations can be obtained by resorting to the mean field approximation. The interested reader can fruitfully learn about this approach by taking a look at the paper [288], which, starting from discrete models for opinion formation like the well-known Sznajd model [298], illustrates the mean field passage to partial differential equations for the evolution of the opinion density.

Unlike the binary kinetic model, the equilibrium state of the Fokker–Planck equation can be computed explicitly and reveals the formation of peaks in correspondence to the points where diffusion is missing.

The rest of this chapter is organized as follows. In the first part we deal with kinetic models of opinion formation and introduce the basic notions and properties of such models. Then we study their asymptotic behaviour by means of the quasi-invariant limit. Next we consider the situation when agents are influenced by external sources which drive them towards a finite set of possible options, like the case of media in a voting dynamic. The formation of steady profiles is investigated both analytically and numerically. Finally, we discuss the presence of leaders, which modifies the homogeneity of the set of interacting agents. Some kinetic models are introduced and numerical experiments reported.

6.2 Kinetic models of opinion formation

In opinion dynamics, the goal of a kinetic model is to describe the evolution of the distribution of opinions in a society by means of *microscopic* interactions among agents or individuals which exchange information.

As discussed in Section 6.1, in agreement with the usual assumptions of the pertinent literature, we associate opinion with a variable w which varies continuously from -1 to 1, where -1 and 1 denote two (extreme) opposite opinions.

The unknown of our model is the density (or distribution function) $f = f(w,t)$, where $w \in I = [-1,1]$ and the time $t \geq 0$, whose time evolution is described, as shown later, by a kinetic equation of Boltzmann type.

The precise meaning of the density f is the following. Given the population to study, if the opinions are defined on a subdomain $D \subset I$, the integral

$$\int_D f(w,t)\,dw$$

represents the "number" of individuals with opinion included in D at time $t > 0$. It is assumed that the density function is normalized to 1, that is

$$\int_I f(w,t)\,dw = 1.$$

As always happens when dealing with a kinetic problem in which the variable belongs to a bounded domain, this choice introduces supplementary mathematical difficulties in the correct definition of binary interactions. In fact, it is essential to consider only interactions that do not produce opinions outside the allowed interval, which corresponds to imposing that the extreme opinions cannot be crossed. This crucial limitation emphasizes the difference between the present *social* interactions, where not all outcomes are permitted, and the classical interactions between molecules, or, more generally, the wealth trades we considered in Chapter 5, where the only limitation for trades was to insure that the post-collision wealths had to be non-negative.

In order to build a possibly realistic model, this severe limitation has to be coupled with a reasonable physical interpretation of the process of opinion forming. In other words, the impossibility of crossing the boundaries has to be a by-product of good modelling of binary interactions.

From a microscopic viewpoint, we can describe the binary interaction by the rules

$$\begin{aligned} w' &= w - \gamma P(|w|)(w - w_*) + \Theta D(|w|), \\ w'_* &= w_* - \gamma P(|w_*|)(w_* - w) + \Theta_* D(|w_*|). \end{aligned} \tag{6.1}$$

In (6.1), the pair (w, w_*), with $w, w_* \in I$, denotes the opinions of two arbitrary individuals before the interaction, and (w', w'_*) their opinions after exchanging information between each other and with the exterior. The coefficient $\gamma \in (0, 1/2)$ is a given constant, while Θ and Θ_* are random variables with the same distribution, with zero mean and variance σ^2, taking values on a set $\mathcal{B} \subseteq \mathbb{R}$. The constant γ and the variance σ^2 measure respectively the compromise propensity and the degree of spreading of opinion due to diffusion, which describes possible changes of opinion due to personal access to information. Finally, the functions $P(\cdot)$ and $D(\cdot)$ take into account the local relevance of compromise and diffusion for a given opinion.

Remark 6.1 *We remark that, in this chapter, in order to emphasize the different kind of binary interactions, we denote with primes the post-interaction opinion pair, like* (w', w'_*)*, in contrast with the asterisk superscript, like* (v^*, w^*)*, used in the previous chapters to describe the wealth of the agents.*

Let us describe in detail the interaction on the right-hand side of (6.1). The first part is related to the compromise propensity of the agents, and the last contains the diffusion effects due to individual deviations from the average behaviour. Note that the pre-interaction opinion w increases (getting closer to w_*) when $w_* > w$ and decreases in the opposite situation. The presence of both the functions $P(\cdot)$ and $D(\cdot)$ is linked to the hypothesis that openness to change of opinion is linked to the opinion itself, and decreases as one gets closer to extremal opinions. This corresponds to the natural idea that extreme opinions are more difficult to change. We will present later on various realizations of these functions. In all cases, however, we assume that both $P(|w|)$ and $D(|w|)$ are non-increasing with respect to $|w|$, and in addition $0 \le P(|w|) \le 1$, $0 \le D(|w|) \le 1$.

In the absence of the diffusion contribution ($\Theta, \Theta_* \equiv 0$), (6.1) implies

$$\begin{aligned} w' + w'_* &= w + w_* + \gamma(w - w_*)\left(P(|w|) - P(|w_*|)\right), \\ w' - w'_* &= \left(1 - \gamma(P(|w|) + P(|w_*|))\right)(w - w_*). \end{aligned} \tag{6.2}$$

Thus, unless the function $P(\cdot)$ is assumed constant, $P = 1$, the total *momentum* is not conserved and it can increase or decrease depending on the opinions before the interaction. If $P(\cdot)$ is assumed constant, the conservation law is reminiscent of analogous conservations which take place in kinetic theory. In such a situation, thanks to the upper bound on the coefficient γ, equations (6.1) correspond to a granular-gas-like interaction (or to a traffic flow model [194]) where the stationary state is a Dirac delta centred on the average opinion (usually referred to as synchronized traffic state in traffic flow modelling). This behaviour is a consequence of the fact that, in a single interaction, the compromise propensity implies that the difference of opinion is diminishing, with $|w' - w'_*| = (1 - 2\gamma)|w - w_*|$. Thus, all agents in the society will end up with exactly the same opinion. Note that in this elementary case a constant part of the relative opinion is restituted after the interaction. This property does not remain true if the function P depends on the opinion variable. In this case,

$$|w' - w'_*| = (1 - \gamma(P(|w|) + P(|w_*|)))\,|w - w_*|.$$

In fact, since $\gamma \in (0, 1/2)$ and $0 \le P(|w|) \le 1$, $0 \le D(|w|) \le 1$,

$$0 \le \varepsilon(w, w_*) = 1 - \gamma(P(|w|) + P(|w_*|)) \le 1.$$

Hence, the general case corresponds to a granular gas interaction with a variable coefficient of restitution $\varepsilon(w, w_*)$ [307].

We remark, moreover, that, in the absence of diffusion, the lateral bounds are not violated, since

$$\begin{aligned} w' &= (1 - \gamma P(|w|))w + \gamma P(|w|)w_*, \\ w'_* &= (1 - \gamma P(|w_*|))w_* + \gamma P(|w_*|)w, \end{aligned} \tag{6.3}$$

imply

$$\max\{|w'|, |w'_*|\} \le \max\{|w|, |w_*|\}.$$

Let $f(w,t)$ denote the distribution of opinion $w \in I$ at time $t \ge 0$. Along the same lines as Chapter 1, we recover the time evolution of f as a balance between bilinear gain and loss of opinion terms, described in weak form by the integro-differential equation of Boltzmann type

$$\begin{aligned} &\frac{d}{dt}\int_I \varphi(w) f(w,t)\,dw = (Q(f,f),\varphi) = \\ &\left\langle \int_{I^2} \beta_{int} f(w) f(w_*)(\varphi(w') + \varphi(w'_*) - \varphi(w) - \varphi(w_*))dw_* dw \right\rangle, \end{aligned} \tag{6.4}$$

where (w', w'_*) are the post-interaction opinions generated by the pair (w, w_*) in (6.1). In (6.4), the kernel β_{int} is related to the details of the binary interaction.

Among other possible choices, one can assume that the transition rate is of the form

$$\beta_{int} = \beta_{int}(w, w_*, \Theta, \Theta_*) = \chi(|w'| \le 1)\chi(|w'_*| \le 1), \tag{6.5}$$

where $\chi(\cdot)$ is the indicator function. Hence, the presence of the rate function β_{int} ensures that the post-interaction opinions belong to the correct interval I. We remark that in general the random variables Θ and Θ_* take values on a subset of I to prevent diffusion generating a complete change of opinion. This property can be weakened by assuming, for example, diffusion as a random variable normally distributed but well concentrated on zero.

For a general probability density $\Theta(\cdot)$, the rate function β_{int} depends on the opinion variables (w, w_*) through the indicator function χ. This fact recalls a similar property of the classical Boltzmann equation [82, 121], where the rate function depends on the relative velocity.

The simplified situation occurs when a suitable choice of the function $D(\cdot)$ in (6.1) coupled with a small support $\mathcal{B}$ of random variables implies that both $|w'| \leq 1$ and $|w'_*| \leq 1$, and the kernel β_{int} does not depend on the opinion variables (w, w_*). In this case the kinetic equation (6.4) corresponds to the classical Boltzmann equation for Maxwell molecules, like the ones studied in Chapters 1 and 5.

6.2.1 Bounded confidence models

We note that in this model the term $\gamma(w - w_*)$ always contributes to a change of opinion, independent of the size of the relative opinion $|w - w_*|$. Therefore, this model assumes that a certain compromise is always possible, independent of the distance between the opinions of the agents. On the other hand, the presence of the function $P(|w|)$, which is decreasing in $|w|$, renders the compromise effect smaller for agents with an extremal opinion. The size of the relative opinion can be easily taken into account. There are two possible approaches, the first one based on modifying the microscopic binary interaction rule, whereas the second one is based on the introduction of an interaction kernel dependent on the relative opinion.

In the first case we can consider the modified interaction

$$\begin{aligned} w' &= w - \gamma P(|w - w_*|)(w - w_*) + \Theta D(|w|), \\ w'_* &= w_* - \gamma P(|w - w_*|)(w_* - w) + \Theta_* D(|w_*|). \end{aligned} \tag{6.6}$$

For example, we can choose

$$P(|w - w_*|) = (\Delta - |w - w_*|)_+, \tag{6.7}$$

where $\Delta \leq 2$ is a positive constant, while h_+ denotes as usual the positive part of h, that is $h_+ = h$ if $h > 0$, while $h_+ = 0$ if $h \leq 0$. With this choice, the opinion can be modified by compromise if and only if the difference between the opinions is less than a fixed value Δ. In (6.7) there is a linear decay in $|w - w_*|$, but one can easily modify the expressions to have a standard bounded confidence selection simply by replacing the positive part function with the indicator function

$$P(|w - w_*|) = \chi(|w - w_*| \leq \Delta). \tag{6.8}$$

The resulting kinetic model is similar to the one introduced in [31], which assumes a compromise effect on agents only in a fixed and bounded interval around their opinion. Note that from the modelling viewpoint the above dynamic again produces a Boltzmann equation of type (6.4) with the interacting velocities (6.6) and with a kernel β_{int} that, for a suitable choice of the diffusion function $D(\cdot)$, does not depend on the opinion variables (w, w_*).

A different approach to the bounded confidence modelling, similar to [116], consists in allowing, through the effect of the collision kernel, the standard binary interaction (6.1) only when the relative opinion is inside a given confidence interval $|w - w_*| \leq \Delta$. The resulting kinetic model is now reminiscent of the Boltzmann equation for hard spheres and has the general weak form (6.4) with the interacting velocities (6.1), but where now the kernel depends on the relative opinion and is given by

$$\beta_{int} = \beta_{int}(w, w_*, \Theta, \Theta_*) = \chi(|w - w_*| \leq \Delta). \tag{6.9}$$

Similarly, an analogue of (6.7) may consist in choosing

$$\beta_{int} = \beta_{int}(w, w_*, \Theta, \Theta_*) = (\Delta - |w - w_*|)_+, \tag{6.10}$$

and, more generally, the choice $\beta_{int} = P(|w - w_*|)$ yields analogous compromise dynamics in the two approaches. Note, however, that the resulting kinetic models are considerably different. For example, the action of the diffusion term in models (6.7)–(6.8) is not limited by the relative opinion as happens in models (6.9)–(6.10).

Remark 6.2 *In the absence of diffusion, the bounded confidence interaction model (6.4) characterized by (6.1) with $P(|w|) = 1$ and (6.9) can easily be expressed in strong form using the results from Section 4.3. In this way, for $\gamma \neq 1/2$, we recover the kinetic model derived in [161]:*

$$\begin{aligned}\frac{d}{dt} f(w,t) = \frac{1}{J} \int_{w-J\Delta}^{w+J\Delta} & f\left(\frac{(1-\gamma)w - \gamma w_*}{J}\right) f\left(\frac{(1-\gamma)w_* - \gamma w}{J}\right) dw_* \\ & - f(w) \int_{w-\Delta}^{w+\Delta} f(w_*)\, dw_*,\end{aligned} \tag{6.11}$$

where $J = 2\gamma - 1$.

6.2.2 Simplifications and main properties

Given a system of agents with a given initial distribution of opinions, and a precise interaction rule for exchange of opinion, the main problem in opinion dynamics is concerned with the study of the consequent evolution of the opinion density towards a stationary profile which contains all the information about the final distribution of opinion in the system. In the kinetic picture this corresponds to an investigation of the large-time behaviour of the density of opinion $f(w,t)$, which obeys the Boltzmann-type equation (6.4).

By reason of the nonlinear structure of the interactions among agents, in general it is quite difficult to extract from the kinetic equation the explicit evolution of observable quantities.

To simplify computations, at first we restrict our analysis to the cases in which the kernel β does not depend on the opinion variables (the Maxwellian case). To this end let us briefly discuss the importance of the support $\mathcal{B}$ of the probability density function Θ in connection with the possible simplification of the kernel. To clarify the point, let us set

$$D(|w|) = 1 - |w|.$$

This function satisfies all the requirements listed in Section 6.2. Then, since

$$(1-\gamma P(|w|))w + \gamma P(|w|)w_* + (1-|w|)\Theta \le (1-\gamma P(|w|))w + \gamma P(|w|) + (1-|w|)\Theta,$$

the condition $|w'| \le 1$ holds true if

$$(1-\gamma P(|w|))w + \gamma P(|w|) + (1-|w|)\Theta \le 1,$$

or, equivalently,

$$(1-|w|)\Theta \le (1-\gamma P(|w|))(1-w). \tag{6.12}$$

Since $P(|w|) \le 1$, bound (6.12) is verified for all $w \ge 0$ as soon as $\Theta \le 1-\gamma$. An analogous result holds if $w \le 0$. Hence, if $D(|w|) = 1-|w|$ and $\mathcal{B} = (-(1-\gamma), 1-\gamma)$, both w' and w'_* belong to the right interval.

Any choice of $D(|w|)$ and $\mathcal{B}$ which is suitable for preserving the lateral bounds of extreme opinions allows us to study in detail the dynamics of the model with a significant simplification. In these cases, in fact, the kernel β_{int} is constant.

From (6.4), conservation of the total opinion is obtained for $\varphi(w) = 1$, which represents in general the only conservation property satisfied by the system. The choice $\varphi(w) = w$ is of particular interest since it gives the time evolution of the average opinion. We have

$$\begin{aligned}\frac{d}{dt}\int_I wf(w,t)\,dw = &\left\langle \int_{I^2} f(w)f(w_*)\gamma(P(|w|)w_* - P(|w|)w)dw_*dw \right\rangle \\ &+ \left\langle \int_{I^2} f(w)f(w_*)\Theta D(|w|)dw_*dw \right\rangle.\end{aligned}$$

The first integral on the right-hand side represents the contribution of the exchange of information to the variation of momentum. In the case $P(|w|) = 1$ this contribution disappears, since, by symmetry,

$$\int_{I^2} f(w)f(w_*)\gamma(w_* - w)dw_*dw = 0.$$

In this case, since the mean value of the random variable Θ is zero,

$$\frac{d}{dt}\int_I wf(w,t)\,dw = \int_{I^2} D(|w|)f(w)f(w_*)dw_*dw = 0. \tag{6.13}$$

This shows that P constant implies conservation of the average opinion. The situation changes if we have a bounded confidence interval or when P is not constant. In the latter case, the time evolution of the average opinion is given by

$$\frac{d}{dt}\int_I wf(w,t)\,dw = \gamma\int_I P(|w|)f(w)\,dw \int_I wf(w)\,dw - \gamma\int_I wP(|w|)f(w)\,dw. \tag{6.14}$$

Note that equation (6.14) is not closed, since the right-hand side depends on higher moments of the density.

Let us now fix $\varphi(w) = w^2$. We have

$$\frac{d}{dt}\int_I w^2 f(w,t)\,dw = \frac{1}{2}\left\langle \int_{I^2} f(w)f(w_*)(w'^2 + w_*'^2 - w^2 - w_*^2)dw_*dw \right\rangle. \quad (6.15)$$

Bearing in mind that Θ has zero mean and variance σ^2, by easy computations one shows that

$$\begin{aligned}
&\frac{1}{2}\left\langle \int_{I^2} f(w)f(w_*)(w'^2 + w_*'^2 - w^2 - w_*^2)dw_*dw \right\rangle = \\
&\quad \gamma^2 \int_{I^2} P(|w|)^2 (w - w_*)^2 f(w)f(w_*)\,dwdw_* \\
&\quad -2\gamma \int_{I^2} P(|w|)w(w - w_*)f(w)f(w_*)\,dwdw_* + \sigma^2 \int_I D(|w|)^2 f(w)dw.
\end{aligned}$$

The choice $P(|w|) = 1$ leads to the simpler evolution equation

$$\begin{aligned}
\frac{d}{dt}\int_I w^2 f(w,t)\,dw &= -2\gamma(1-\gamma)\left[\int_I w^2 f(w)\,dw - m^2\right] \\
&\quad + \sigma^2 \int_I D(|w|)^2 f(w)dw,
\end{aligned} \quad (6.16)$$

where m is the constant value of the average opinion,

$$m = \int_I w f(w,t)dw.$$

Let us fix

$$D(|w|) = \sqrt{1 - w^2}.$$

Then $D(|w|)$ satisfies all the properties listed in Section 6.2, and equation (6.16) takes the form

$$\frac{d}{dt}\int_I w^2 f(w,t)\,dw = -\left[2\gamma(1-\gamma) + \sigma^2\right]\int_I w^2 f(w)\,dw + 2\gamma(1-\gamma)\,m^2 + \sigma^2. \quad (6.17)$$

Equation (6.17) for the second moment is explicitly solvable. If

$$r = r(\gamma,\sigma) = \sigma^2 + 2\gamma(1-\gamma),$$

we obtain

$$\int_I w^2 f(w,t)\,dw = \int_I w^2 f_0(w)\,dw\, e^{-rt} + \frac{\sigma^2 + 2\gamma(1-\gamma)\,m^2}{\sigma^2 + 2\gamma(1-\gamma)}\left(1 - e^{-rt}\right). \quad (6.18)$$

Even if the possibility of these exact computations is linked to a very particular choice of the functions $P(|w|)$ and $D(|w|)$, nevertheless the evolution of the second moment of the opinion density in (6.18) contains some interesting features which should be noted.

Remark 6.3 *The first remark is concerned with the rate of relaxation of the second moment towards its final value. This rate, expressed by the constant $r(\gamma,\sigma)$, for a given value of the variance σ^2, attains its maximum value in correspondence to $\gamma = 1/2$. Since we fixed $P(|w|) = 1$, the case $\gamma = 1/2$, in the absence of diffusion, corresponds to a total compromise, where the two agents end up with exactly the same opinion. Hence, as expected, a higher value of the compromise intensity increases the rate of convergence. Notably, since the rate r is the sum of the effects due to compromise and diffusion, the presence of diffusion always speeds up the rate of convergence to the final value.*

Since the mean opinion m is always strictly less than 1, the final value of the second moment, which is a measure of the spreading of opinion, is an increasing function of σ^2 and a decreasing function of γ. In particular, the minimal variance of the opinion density is attained at $\gamma = 1/2$ and takes the value

$$V_{min} = \frac{4\sigma^2}{4\sigma^2+1}(1-m^2).$$

These computations show that, except in some trivial cases, the evolution of moments does not provide useful information on the properties of the (possible) steady opinion. On the other hand, since $|w| \leq 1$, all moments of the density function $f(w,t)$ are bounded. This implies that, in all cases, we can draw conclusions on the large-time convergence of the class of probability densities $\{f(w,t)\}_{t\geq 0}$. By virtue of Prokhorov's theorem (cf. [200]) the existence of a uniform bound on moments implies, in fact, that this class is tight, so that any sequence $\{f(w,t_n)\}_{n\geq 0}$ contains an infinite subsequence which converges weakly to some probability measure f_∞.

6.2.3 The quasi-invariant opinion limit

In Chapter 1, Section 1.4, we highlighted that a particular asymptotic of the kinetic equation results in simplified models (generally of Fokker–Planck type), for which it is relatively easier to find steady states and to prove their stability. This idea is particularly useful in the case we are treating here, in which the interactions are given by nonlinear relations.

In order to give a physical basis to this asymptotic approach, let us discuss the interaction rule (6.1) from a slightly different point of view. Leaving for the moment the general case, we will assume $P(|w|) = 1$, so that, as proven in Section 6.2.2, conservation both of mass and momentum holds. Then the following properties follow from (6.1):

$$\langle w' + w'_*\rangle = w + w_*, \qquad \langle w' - w'_*\rangle = (1-2\gamma)(w - w_*). \tag{6.19}$$

The first equality in (6.19) describes the property of mean conservation of opinion. The second refers to the compromise propensity, which favours a decrease (in mean) of the distance of opinions after the interaction. This tendency is a universal consequence of the rule (6.1), in that it holds whatever distribution one assigns to Θ, namely to the random variable which accounts for the effects of the external world in opinion formation.

The second property in (6.19) is analogous to the similar one that holds in a collision between molecules in a granular gas. There, the quantity $e = 2\gamma$ is called the *coefficient of restitution*, and describes the peculiar fact that energy is dissipated [307].

We now consider the situation in which most of the interactions produce a very small exchange of opinion ($\gamma \to 0$), while at the same time both properties (6.19) remain bounded away from zero at a macroscopic level. This corresponds to require that, while $\gamma \to 0$,

$$\int_{I^2} (w + w_*) f(w) f(w_*) dw dw_* = 2 \int_I w f(w) dw = 2m(t) \tag{6.20}$$

remains constant, and

$$\frac{1}{2} \int_{I^2} (w - w_*)^2 f(w) f(w_*) dw dw_* = \int_I w^2 f(w)\, dw - m_0^2 = C_f(t) \tag{6.21}$$

varies with time, and decays to zero when the diffusion is not present (i.e. $\sigma = 0$).

Since in our case the kernel β does not depend on the opinion variables, (6.13) implies that $m(t) = m_0$ independently of the value of γ. Moreover, using the computations of Section 6.2.2, one obtains that $C_f(t)$ varies with law

$$\frac{dC_f(t)}{dt} = -2\gamma(1-\gamma)C_f(t) + \sigma^2 \int_I D(|w|)^2 f(w) dw. \tag{6.22}$$

Hence, if we set

$$\tau = \gamma t, \quad g(w, \tau) = f(w, t), \tag{6.23}$$

which implies $f_0(w) = g_0(w)$, it follows that

$$\frac{dC_g(\tau)}{d\tau} = -2\,(1-\gamma)\,C_g(\tau) + \frac{\sigma^2}{\gamma} \int_I D(|w|)^2 f(w) dw. \tag{6.24}$$

Now letting both $\gamma \to 0$ and $\sigma \to 0$ in such a way that $\sigma^2/\gamma = \lambda$, (6.24) becomes, in the limit,

$$\frac{dC_g(\tau)}{d\tau} = -2C_g(\tau) + \lambda \int_I D(|w|)^2 f(w) dw. \tag{6.25}$$

This argument shows that the value of the ratio σ^2/γ is of paramount importance in obtaining asymptotics which maintain a memory of the microscopic interactions. It is remarkable that, by virtue of (6.23), $t = \tau/\gamma$, so that the limit $\gamma \to 0$ describes the large-time behaviour of $f(v, t)$. On the other hand, since $f(w, t) = g(w, \tau)$, the large-time behaviour of $f(w, t)$ is close to the large-time behaviour of $g(w, \tau)$.

Remark 6.4 *The balance $\gamma \to 0$ and $\sigma \to 0$ in such a way that $\sigma^2/\gamma = \lambda$ allows us to recover in the limit the contributions due both to compromise propensity and diffusion. Other limits can be considered, like diffusion dominated ($\sigma^2/\gamma \to \infty$) or compromise dominated ($\sigma^2/\gamma \to 0$). As we shall present later, however, the formation of an asymptotic profile for opinion is linked to the first balance [31].*

In the rest of this section we shall present, starting from the Boltzmann equation for the opinion density $g(w,\tau)$, the derivation of the Fokker–Planck model when both $\gamma \to 0$ and $\sigma \to 0$ in such a way that $\sigma^2/\gamma \to \lambda$. For the sake of simplicity, we will assume that $P(|w|) = 1$. This derivation is close to the one of Chapter 1, Section 1.4.

The scaled density $g(v,\tau) = f(v,t)$ satisfies the equation (in weak form)

$$\frac{d}{d\tau}\int_I g(w)\varphi(w)\,dw = \frac{1}{\gamma}\left\langle \int_{I^2} g(w)g(w_*)(\varphi(w') - \varphi(w))dw_*dw \right\rangle. \tag{6.26}$$

Some definitions are needed at this point. Let $\mathcal{F}_s(I)$ be the class of all real functions h on I such that $h(\pm 1) = h'(\pm 1) = 0$, and $h^{(m)}(v)$ is Hölder continuous of order δ,

$$\|h^{(m)}\|_\delta = \sup_{v \neq w} \frac{|h^{(m)}(v) - h^{(m)}(w)|}{|v-w|^\delta} < \infty. \tag{6.27}$$

The integer m and the number $0 < \delta \leq 1$ are such that $m + \delta = s$, and $h^{(m)}$ denotes the m-th derivative of h.

Recall that the probability densities Θ and Θ_* belong to $\mathcal{P}_{2+\alpha}$, for some $\alpha > 0$. To show the rate of decay of moments, we assume that these densities are obtained from a given random variable Y, with zero mean and unit variance, that belongs to $\mathcal{P}_{2+\alpha}$. Thus, Θ of variance σ^2 is the density of σY. By this assumption, we can easily obtain the dependence on σ of the moments of Θ. In fact, for any $p > 0$ such that the p-th moment of Y exists,

$$\langle |\Theta|^p \rangle = \langle |\sigma Y|^p \rangle = \sigma^p \langle |Y|^p \rangle.$$

Given $0 < \delta \leq \alpha$, let us set $\varphi \in \mathcal{F}_{2+\delta}(I)$.

By (6.1),

$$w' - w = \gamma(w_* - w) + \Theta D(|w|).$$

Then, if we use a second-order Taylor expansion of φ around w,

$$\varphi(w') - \varphi(w) = (\gamma(w_* - w) + \Theta D(|w|))\varphi'(w) + \frac{1}{2}\left(\gamma(w_* - w) + \Theta D(|w|)\right)^2 \varphi''(\tilde{w}),$$

where, for some $0 \leq \theta \leq 1$,

$$\tilde{w} = \theta w' + (1-\theta)w.$$

Inserting this expansion in the collision operator, we get

$$\begin{aligned}\frac{d}{d\tau}\int_I g(w)\varphi(w)\,dw = \frac{1}{\gamma}\Bigg\langle \int_{I^2} &[(\gamma(w_* - w) + \Theta D(|w|))\varphi'(w) \\ &+\frac{1}{2}(\gamma(w_* - w) + \Theta D(|w|))^2\varphi''(w)] \\ &g(w)g(w_*)dw_*dw \Bigg\rangle + R(\gamma,\sigma),\end{aligned} \tag{6.28}$$

where the remainder $R = R(\gamma,\sigma)$ is

$$R = \frac{1}{2\gamma}\left\langle \int_{I^2} (\gamma(w_* - w) + \Theta D(|w|))^2 \left(\varphi''(\tilde{w}) - \varphi''(w)\right) g(w)g(w_*)dw_*dw \right\rangle. \tag{6.29}$$

Since $\varphi \in \mathcal{F}_{2+\delta}(I)$, and $|\tilde{w} - w| = \theta|w' - w|$,

$$|\varphi''(\tilde{w}) - \varphi''(w)| \le \|\varphi''\|_\delta |\tilde{w} - w|^\delta \le \|\varphi''\|_\delta |w' - w|^\delta. \tag{6.30}$$

Hence,

$$|R(\gamma,\sigma)| \le \frac{\|\varphi''\|_\delta}{2\gamma} \left\langle \int_{I^2} |\gamma(w_* - w) + \Theta D(|w|)|^{2+\delta} g(w)g(w_*)dw_*\, dw \right\rangle.$$

By virtue of the inequality

$$\begin{aligned}|\gamma(w_* - w) + \Theta D(|w|)|^{2+\delta} &\le 2^{1+\delta}\left(|\gamma(w_* - w)|^{2+\delta} + |\Theta D(|w|)|^{2+\delta}\right)\\ &\le 2^{2+\delta}\gamma^{1+\delta} + 2^{1+\delta}|\Theta|^{2+\delta},\end{aligned}$$

we finally obtain the bound

$$|R(\gamma,\sigma)| \le 2^{1+\delta}\|\varphi''\|_\delta \left(\gamma^{1+\delta} + \frac{1}{2\gamma}\langle|\Theta|^{2+\delta}\rangle\right). \tag{6.31}$$

Since Θ is a probability density with zero mean and $\lambda\gamma$ variance, and Θ belongs to $\mathcal{M}_{2+\alpha}$, for $\alpha > \delta$,

$$\langle|\Theta|^{2+\delta}\rangle = \langle|\lambda\gamma Y|^{2+\delta}\rangle = (\lambda\gamma)^{1+\delta/2}\langle|Y|^{2+\delta}\rangle,$$

and $\langle|Y|^{2+\delta}\rangle$ is bounded. Using this equality in (6.31) shows that $R(\gamma,\sigma)$ converges to zero as both γ and σ converge to zero in such a way that $\sigma^2 = \lambda\gamma$. Within the same scaling,

$$\begin{aligned}&\lim_{\gamma\to 0}\frac{1}{\gamma}\left\langle \int_{I^2} [(\gamma(w_* - w) + \Theta D(|w|))\varphi'(w)\right.\\ &\qquad \left. + \frac{1}{2}(\gamma(w_* - w) + \Theta D(|w|))^2\varphi''(w)] g(w)g(w_*)dw_*\, dw \right\rangle \\ &\quad = \int_I \left[(m - w)\varphi'(w) + \frac{\lambda}{2}D(|w|)^2\varphi''(w)\right] g(w)dw.\end{aligned} \tag{6.32}$$

Considering that $\varphi \in \mathcal{F}_s(I)$, we can integrate back by parts. This shows that the right-hand side of (6.32) coincides with the weak form of the Fokker–Planck equation

$$\frac{\partial g}{\partial \tau} = \frac{\lambda}{2}\frac{\partial^2}{\partial w^2}\left(D(|w|)^2 g\right) + \frac{\partial}{\partial w}\left((w - m)g\right). \tag{6.33}$$

Last, since the solution to the kinetic model conserves mass and momentum, while the second moment is uniformly bounded in time, conservation of both mass and momentum pass to the limit. We note that these conservations are difficult to prove directly on the Fokker–Planck equation, due to the fact that $\varphi(v) = v$ does not belong to $\mathcal{F}_s(I)$. Hence, we have proved:

Theorem 6.1 *Let the probability density $f_0 \in \mathcal{P}(I)$, and let the random variable Y which characterizes the kernel have a density in $\mathcal{P}_{2+\alpha}$, with $\alpha > \delta$. Then, as $\gamma \to 0$, $\sigma \to 0$ in such a way that $\sigma^2 = \lambda\gamma$, the weak solution to the Boltzmann equation for the scaled density $g_\gamma(v,\tau) = f(v,t)$, with $\tau = \gamma t$, converges, up to the extraction of a subsequence, to a probability density $g(w,\tau)$. This density is a weak solution of the Fokker–Planck equation (6.33) such that the average opinion is conserved.*

Remark 6.5 *In the case of a bounded confidence compromise (6.6) with a constant kernel $\beta_{int} = 1$, the same computations formally lead to the nonlinear Fokker–Planck equation*

$$\frac{\partial g}{\partial \tau} = \frac{\lambda}{2}\frac{\partial^2}{\partial w^2}\left(D(|w|)^2 g(w)\right) + \frac{\partial}{\partial w}\left(g(w)\mathcal{K}[g](w)\right), \tag{6.34}$$

where

$$\mathcal{K}[g](w) = \int_I P(|w - w_*|)(w - w_*)g(w_*)dw_*. \tag{6.35}$$

It is possible to show that, for a given choice of the bounded confidence model, both approaches introduced in Section 6.2.1, in the quasi-invariant limit and in the absence of diffusion, originate the same nonlinear drift equation:

$$\frac{\partial g}{\partial \tau} = \frac{\partial}{\partial w}\left(g(w)\mathcal{K}[g](w)\right). \tag{6.36}$$

In contrast, the presence of the diffusion term in the case of an interaction kernel dependent on the relative opinion $\beta_{int} = P(|w-w_*|)$ gives rise to the nonlinear Fokker–Planck equation

$$\frac{\partial g}{\partial \tau} = \frac{\lambda}{2}\frac{\partial^2}{\partial w^2}\left(D(|w|)^2 g(w)\mathcal{R}[g](w)\right) + \frac{\partial}{\partial w}\left(g(w)\mathcal{K}[g](w)\right), \tag{6.37}$$

with

$$\mathcal{R}[g](w) = \int_I P(|w - w_*|)g(w_*)dw_*. \tag{6.38}$$

We will consider Fokker–Planck systems for bounded confidence models in the presence of leaders in Section 6.5.2.

6.3 Other Fokker–Planck models of opinion formation

Theorem 6.1 can be generalized in many ways. Always remaining within the simplification of a constant kernel, we can choose a general function $P(|w|)$ in the interaction rule (6.1). The main difference with respect to the proof of Theorem 6.1 is the evaluation of the first-order term in (6.28), which, using the fact that the mean value of Θ is zero, reads

$$\int_{I^2} P(|w|)(w_* - w)\varphi'(w)g(w)g(w_*)dw_*\,dw = \int_I P(|w|)(m(\tau) - w)\varphi'(w)g(w)\,dw,$$

where $m(\tau)$ is the value of the average opinion at time $\tau \geq 0$,

$$m(\tau) = \int_I wg(w,\tau)dw.$$

Note that, by (6.23):

$$m(\tau) = \int_I wg(w,\tau)dw = \int_I wf(w,t)dw. \tag{6.39}$$

Hence, by (6.14), the evolution of $m(\tau)$ obeys the law

$$\frac{dm(\tau)}{d\tau} = m(\tau)\int_I P(|w|)g(w,\tau)\,dw - \int_I wP(|w|)g(w,\tau)\,dw. \tag{6.40}$$

Finally, in the limit $\gamma \to 0$ we obtain that $g(w,\tau)$ satisfies the Fokker–Planck equation

$$\frac{\partial g}{\partial \tau} = \frac{\lambda}{2}\frac{\partial^2}{\partial w^2}\left(D(|w|)^2 g\right) + \frac{\partial}{\partial w}\left(P(|w|)(w - m(t))g\right). \tag{6.41}$$

The presence of a general propensity function $P(|w|)$ introduces nonlinearity into the drift term of the Fokker–Planck equations which is difficult to treat. The nonlinearity is due to the fact that the average opinion is no longer constant, and the evaluation of the drift term requires the evaluation of (6.40).

The Fokker–Planck equation (6.41) was first derived in [308]. Related pure diffusion and drift equations were recently introduced in [288]. These equations, in our picture, refer to diffusion-dominated ($\sigma^2/\gamma \to \infty$) or compromise-dominated ($\sigma^2/\gamma \to 0$) limits. Looking at the proof of Theorem 6.1, it is almost immediate to conclude that the diffusion-dominated limit takes into account only the second-order term in the Taylor expansion. To verify this, suppose that

$$\frac{\sigma^2}{\gamma^\alpha} \to \lambda, \quad \alpha < 1.$$

Then we can set

$$\tau = \gamma^\alpha t, \quad g(w,\tau) = f(w,t), \tag{6.42}$$

where now $g(w,\tau)$ satisfies

$$\begin{aligned}\frac{d}{d\tau}\int_I g(w)\varphi(w)\,dw = \frac{1}{\gamma^\alpha}\Bigg\langle\int_{I^2} &[(\gamma(w_* - w) + \Theta D(|w|))\varphi'(w)\\ &+\frac{1}{2}(\gamma(w_* - w) + \Theta D(|w|))^2\varphi''(w)]\\ &g(w)g(w_*)dw_*dw\Bigg\rangle + R(\gamma,\sigma),\end{aligned} \tag{6.43}$$

with the obvious meaning of the remainder. Since $\alpha < 1$, the first-order term in the Taylor expansion vanishes in the limit, and g satisfies the diffusion equation

$$\frac{\partial g}{\partial \tau} = \frac{\lambda}{2}\frac{\partial^2}{\partial w^2}\left(D(|w|)^2 g\right). \tag{6.44}$$

The choice

$$D(|w|) = \sqrt{1-w^2}, \quad \lambda = 2,$$

leads to the diffusion equation

$$\frac{\partial g}{\partial \tau} = \frac{\partial^2}{\partial w^2}\left[(1-w^2)g\right]. \tag{6.45}$$

The same diffusion equation has been derived as a mean field approximation [288] of the well-known Sznajd model [298] in the simplified version proposed by Ochrombel [244].

Likewise, the compromise-dominated ($\sigma^2/\gamma \to 0$) limit can be considered. In this case,

$$\frac{\sigma^2}{\gamma^\alpha} \to \lambda, \quad \alpha > 1,$$

and we can set

$$\tau = \gamma t, \quad g(w,\tau) = f(w,t). \tag{6.46}$$

The diffusion part disappears in the limit and we obtain the pure drift equation

$$\frac{\partial g}{\partial \tau} = \frac{\partial}{\partial w}\left(P(|w|)(w-m(t))g\right). \tag{6.47}$$

Note that, due to (6.46), the evolution of the mean opinion $m(t)$ obeys the law (6.39). The choice

$$P(|w|) = 1 - w^2$$

was considered in [5]. In this case,

$$\frac{\partial g}{\partial \tau} = \frac{\partial}{\partial w}\left((1-w^2)(w-m(t))g\right), \tag{6.48}$$

where

$$\frac{dm(\tau)}{d\tau} = -m(\tau)\int_I w^2 g(w,\tau)\,dw + \int_I w^3 g(w,\tau)\,dw. \tag{6.49}$$

Thus, equation (6.48) differs from the pure drift in magnetization obtained in [288] as the mean field limit of the Sznajd model [298] in the case of two opinions. There, the first-order partial differential equation reads

$$\frac{\partial g}{\partial \tau} = -\frac{\partial}{\partial w}\left((1-w^2)wg\right). \tag{6.50}$$

Note that the sign in front of the drift (6.50) is now opposite to the sign in (6.48), and the equation is now linear in g, even if the evolution of the mean opinion is not closed. Equation (6.48) was extensively studied in [5], as a particular case of the equation

$$\frac{\partial g}{\partial \tau} = \nu\frac{\partial}{\partial w}\left((1-w^2)(w-m(t)),g\right),$$

where the parameter $\nu = 1$. In [5], both the cases $\nu = 1$ and $\nu = -1$ were considered. While, in the latter case, which corresponds to a nonlinear concentration, it is possible to obtain a precise picture of the limiting behaviour, in the former case there is no clear idea of the way in which the initial opinion splits between the two poles $w = -1$ and $w = 1$.

6.3.1 Stationary solutions of the Fokker–Planck opinion model

In this section we analyze in some detail various cases of the interaction dynamics in the Boltzmann equation from which it is possible to derive a Fokker–Planck equation with an explicitly computable steady state. The structure of the steady state then represents the formation of opinion consequent to the choice of the interaction dynamics. In most cases we are forced to suppose $P(|w|) = 1$, which implies conservation of the average opinion, and the Fokker–Planck (6.33) as the underlying quasi-invariant opinion limit. For any of these choices, we briefly discuss the link between $D(|w|)$ and the maximal support $\mathcal{B}$ of the diffusion variable.

The first model of the diffusion dependence on opinion that we propose is

$$D(|w|) = 1 - w^2.$$

Since $P(|w|) = 1$, the interaction rules are

$$w' = w - \gamma(w - w_*) + \Theta(1 - w^2),$$

$$w'_* = w_* - \gamma(w_* - w) + \Theta_*(1 - w_*^2).$$

In this case, $|\Theta|(1+|w|) \leq 1-\gamma$ implies $|w'| \leq 1$. Hence, for a given opinion we obtain $\mathcal{B} = (-(1-\gamma)/(1+|w|), (1-\gamma)/(1+|w|))$, which shows that when the opinion w is close to zero and γ is small, the effects of global access to information can move opinion towards extremals. This possibility reduces as soon as $|w|$ increases. The steady-state distribution of opinion is a solution to

$$\frac{\lambda}{2}\frac{\partial}{\partial w}\left((1-w^2)^2 g\right) + (w - m)g = 0, \tag{6.51}$$

where m is a given constant (the average initial opinion), $-1 < m < 1$. The solution to (6.51) is easily found as

$$g_\infty(w) = c_{m,\lambda}\,(1+w)^{-2+m/(2\lambda)}\,(1-w)^{-2-m/(2\lambda)}\exp\left\{-\frac{1-mw}{\lambda(1-w^2)}\right\}, \tag{6.52}$$

where the constant $c_{m,\lambda}$ is such that the mass of g_∞ is equal to one. Note that the presence of the exponential assures that $g_\infty(\pm 1) = 0$. The solution is regular, but not symmetric unless $m = 0$. Hence, the initial opinion distribution reflects on the steady state through the mean opinion. In any case, the stationary distribution has two peaks (on the right and on the left of zero) with intensity depending on λ.

A similar result is expected from the choice

$$D(|w|) = 1 - |w|.$$

As discussed in Section 6.2, $|\Theta| \leq 1 - \gamma$ implies $|w'| \leq 1$, and $\mathcal{B} = (-(1-\gamma), 1-\gamma)$. Once more, the support of the random variable, for γ small, covers the whole domain of opinions. The steady-state distribution of opinion is a solution to

$$\frac{\lambda}{2}\frac{\partial}{\partial w}\left((1-|w|)^2 g\right) + (w - m)g = 0, \tag{6.53}$$

where m, the average initial opinion, is a given constant such that $-1 < m < 1$. The solution to (6.51) is easily found as

$$g_\infty(w) = c_{m,\lambda} (1 - |w|)^{-2-2/\lambda} \exp\left\{ -\frac{1 - mw/|w|}{2\lambda(1-|w|)} \right\}, \tag{6.54}$$

where, as usual, the constant $c_{m,\lambda}$ is such that the mass of g_∞ is equal to one. We note that the low regularity of $D(|w|)$ reflects on the steady solution, which has a jump at $w = 0$. The jump disappears only when the mean $m = 0$, and only in this case do we have a symmetric distribution. As in the first case, the presence of the exponential assures that $g_\infty(\pm 1) = 0$, and the initial opinion distribution reflects on the steady state through the mean opinion.

Last, we consider

$$D(|w|) = \sqrt{1 - w^2},$$

in the Fokker–Planck model (6.33). The steady-state distribution of opinion solves

$$\frac{\lambda}{2} \frac{\partial}{\partial w} \left((1 - w^2) g \right) + (w - m) g = 0, \tag{6.55}$$

and equals

$$g_\infty(w) = c_{m,\lambda} \left(\frac{1}{1+w} \right)^{1-(1+m)/\lambda} \left(\frac{1}{1-w} \right)^{1-(1-m)/\lambda}. \tag{6.56}$$

As before, the constant $c_{m,\lambda}$ is such that the mass of g_∞ is equal to one. Since $-1 < m < 1$, g_∞ is integrable on I. Differently from the previous cases, however, $g_\infty(w)$ tends to infinity as $w \to \pm 1$, and it has no peaks inside the interval I. The explanation comes from a deep insight into the connection between $D(|w|)$ and the support $\mathcal{B}$ in this case.

It can immediately be verified that, in order to satisfy the constraint in (6.1), one cannot directly choose $D(|w|) = \sqrt{1 - w^2}$. With this choice, in fact, choosing $w_* = 1$, the first equality in (6.1) gives for Θ the upper bound

$$\Theta \le (1 - \gamma) \frac{(1 - w)}{\sqrt{1 - w^2}}.$$

Since the right-hand side converges to zero as $w \to 1$, it follows that there is no way to satisfy the constraint. A different choice which gives, in the limit, the Fokker–Planck equation with $D(|w|) = \sqrt{1 - w^2}$, is the following. For any given γ, we set

$$D(|w|) = \sqrt{(1 - (1 + \gamma^p) w^2)_+}, \quad p > 0,$$

where f_+ denotes as usual the positive part of f, that is $f_+ = f$ if $f > 0$, while $f_+ = 0$ if $f \le 0$. In this case, one can show that, to respect the constraint on post-interaction opinions, it is sufficient that Θ satisfies the condition

$$|\Theta| \le a_\gamma = \frac{1 - \gamma}{\sqrt{1 + \gamma^p}} \gamma^{p/2}. \tag{6.57}$$

To give an example, let us assume that Θ is uniformly distributed on the interval $(-a_\gamma, a_\gamma)$. Then σ^2 behaves like $\gamma^{(3p)/2}$, and it is enough to set $p = 2/3$ to obtain

$\lambda = 1$. The previous discussion shows that the choice $D(|w|) = \sqrt{1-w^2}$ in the Fokker–Planck equation (6.33) corresponds to a kinetic interaction in which diffusion is of the order of γ^p, where p is taken so that σ^2/γ tends to a finite limit λ as $\gamma \to 0$. In this case, the *smallness* of the interval of diffusion produces the peaks on $w = \pm 1$.

The interesting feature of the Fokker–Planck equation

$$\frac{\partial g}{\partial \tau} = \frac{\lambda}{2} \frac{\partial^2}{\partial w^2} \left((1-w^2)g\right) + \frac{\partial}{\partial w}\left((w-m)g\right) \tag{6.58}$$

is that, as happens for the Boltzmann-type equation (6.4) corresponding to the same choice of the functions $P(|w|)$ and $D(|w|)$ studied in Section 6.2.2, it leads to a closed evolution of moments. We remark that equation (6.58) is to be studied with the conservation of both mass and momentum.

6.3.2 Further remarks on opinion modelling

In the previous sections we introduced and discussed kinetic models of opinion formation based on binary interactions involving both compromise and diffusion properties in exchanges between individuals. Then, a suitable scaling of compromise and diffusion allows us to derive, from the Boltzmann-type kinetic model, related Fokker–Planck equations for which it is easy to recover the stationary distribution of opinion. Among these Fokker–Planck equations, one has emerged (6.58) that takes the role of the analogous one obtained in [55, 105] for the evolution of wealth (cf. Chapter 5). The main feature of this equation is that moments can be evaluated in closed form.

The starting point for the construction of the kinetic model was the definition of the binary interaction between agents, expressed by (6.1). In normal life, however, the formation of a personal opinion results from a continuous exchange of information with others, and, contrary to what happens to molecules in a dilute gas, it is difficult to split this change into a sequence of binary interactions which obey identical rules.

Consequently, the kinetic approach seems to reveal an objective weakness, and one can question the validity of this approach and the meaning of the stationary distributions of opinion we introduced in Section 6.3.1.

We can account for the continuous exchange of opinions by assuming that, instead of a single interaction, at any time $t > 0$ any agent compares his opinion with a certain number of other agents, say $N \gg 1$, following the binary rules we fixed in (6.1). Then, his new opinion moves to the mean of these values. By doing this, we substitute the opinion w_* at time t in (6.1) with

$$\bar{w}' = \frac{1}{N}\sum_{i=1}^{N}(w^i)' = w + P(|w|)\left(\frac{1}{N}\sum_{i=1}^{N} w_*^i - w\right) + \Theta D(|w|). \tag{6.59}$$

In (6.59) we denoted with w_*^i the opinion of the i-th agent who, at time $t > 0$, enters into contact with the selected one. If the number of contacts is sufficiently high,

$$\frac{1}{N}\sum_{i=1}^{N} w_*^i \cong m(t), \tag{6.60}$$

where

$$m(t) = \int_I w f(w,t) dw, \tag{6.61}$$

is the mean opinion at time t. Hence, it seems natural in this case to define the change of opinion at time $t > 0$ as

$$\bar{w}'(t) = w + P(|w|)\,(m(t) - w) + \Theta D(|w|), \tag{6.62}$$

where $m(t)$ is given by (6.61). If we assume this interaction, then the consequent kinetic equation, in weak form, reads

$$\frac{d}{dt}\int_I \varphi(w) f(w,t)\, dw = \left\langle \int_I (\varphi(\bar{w}'(t) - w)\, f(w) \right\rangle. \tag{6.63}$$

In (6.63), nonlinearity enters into the post-interaction opinion $\bar{w}'(t)$.

Applying to equation (6.63) the quasi-invariant limit procedure of Section 6.2.3, one obtains in the limit the same Fokker–Planck equation (6.41). This shows that this limit equation is suitable for describing a large variety of opinion dynamics, which at the kinetic level can differ in a sensible way.

6.4 Choice formation and influence of external factors

In the first part of this chapter we introduced a dynamical model for the formation of opinion in a multiagent society in which people exchanged ideas without any particular preference. In many aspects of life, however, as happens during political elections, agents choose from a finite set of possible options and the media try to push them in a specific direction. In this case, the goal of a kinetic model is to describe the evolution of the distribution of opinions in a closed society around a well-defined statement. If we maintain the fundamental rules of the process of opinion formation, the formation of choice still depends on *microscopic* interactions among agents or individuals which can change their viewpoints and which can be influenced by the others in a random fashion. This process has recently been studied in [103] (see also [58]) as regards a finite number of questions, the simplest being given by the choice between two options. Many important situations (e.g. a referendum) require, in fact, that, at the end of the process of opinion formation, the individuals should be close to accept, or reject, a very precise statement.

In addition to the evolving distribution of opinions $f(w,t)$ discussed in Section 6.2, a crucial role is now played by the fixed distribution of choices $M(w)$, which is nothing but a fixed background of field particles (cf. Chapter 1) adapted to the present situation. Indeed, the choice formation of the population is driven by the *collisions* undergone with the background $M(w)$, and the opinion distribution function will consequently be governed by a single linear kinetic equation. If we only consider opinions concerning a finite set of choices, it is reasonable to assume that the background $M(w)$ is given by a mass-concentrated distribution

$$M(w) = \sum_{i=0}^{N} \alpha_i \delta(w - \bar{w}_i), \quad \sum_{i=0}^{N} \alpha_i = 1. \tag{6.64}$$

In (6.64), $\delta(w-\bar{w})$ denotes, as usual, a Dirac delta function concentrated at the opinion point $w = \bar{w} \in I$.

Within this representation of equilibria, one can identify two parameters. The first is the constant $0 < \alpha_i < 1$, which represents the probability that one can perform the i-th choice. The second one is the numerical value $0 < |\bar{w}_i| < 1$, which represents the *stability* of the i-th choice. This parameter, which assumes a particular importance in our model, reflects the social fact that extremal opinions are more difficult to change.

The background distribution can, of course, be modified in many ways. The form (6.64), however, corresponds to the reasonable hypothesis that the possible choices are well defined and localized in the interval I, and at the same time allows for a simple treatment of the underlying equation. What is important here, for example in a problem of formation of opinion between two options, is the mean value of the resulting steady state, which indicates which of the two possible choices, ends up with the prevalence of the consensus. For example, in the case of only two possible choices, the reader can easily recognize that a background distribution is represented by the distribution of a two-valued random variable, where the two Dirac delta functions are localized at two points w_- and w_+ of the interval I, and $w_- < 0$ (respectively $w_+ > 0$). Then, the symmetric case $\alpha_1 = \alpha_2 = 1/2$, $w_+ = -w_-$, corresponds to the distribution of the random experiment of tossing a coin. In view of Remark 6.29, the choice is between two alternatives which are equally probable and stable. A non-symmetric case (with zero mean) can be produced by setting $\alpha_1 = 1/3$, $\alpha_2 = 2/3$ while $w_- = -2/3$, $w_+ = 1/3$. In this case, the less probable choice is more stable than the other.

A reasonable and quite general model for the binary interaction with the fixed background, which is based on the same assumptions (6.1), is given by

$$w' = w - \gamma P(|w|)(w - w_*) + \Theta D(|w|). \tag{6.65}$$

As in (6.1), the pair (w, w'), with $w, w' \in I$, denotes the opinions of an arbitrary individual respectively before and after interaction with the background $M(w_*)$. We refer to the discussion of Section 6.2 for the description of the various quantities involved in (6.65).

Let $\varphi = \varphi(w)$ be a suitably regular test function. As usual, we define the weak form of the interaction operator induced by (6.65) as

$$(Q(f, M), \varphi) = \left\langle \int_{I^2} (\varphi(w') - \varphi(w)) \, f(w,t) M(w_*) dw \, dw_* \right\rangle. \tag{6.66}$$

In (6.66), $f(w,t)$ denotes the distribution of opinion $w \in I$ at time $t \geq 0$, while $M(w_*)$ represents the fixed background equilibrium distribution, which we assume to take the form (6.64). Then the Boltzmann equation in weak form reads

$$\frac{d}{dt} \int_I \varphi(w) f(w,t) \, dw = (Q(f, M), \varphi) \,. \tag{6.67}$$

Let us proceed as in Section 6.2.2 to evaluate the principal moments of the solution. Conservation of the total opinion is obtained for $\varphi(w) = 1$, which represents the only conservation property satisfied by equation (6.67). Therefore, it is enough to assume that $f(w,t=0)$ is a probability density for the solution to remain a probability density at any subsequent time.

The choice $\varphi(w) = w$ gives the time evolution of the average opinion. Considering that $\langle\Theta\rangle = 0$, we have

$$\begin{aligned}\frac{d}{dt}\int_I wf(w,t)\,dw &= \left\langle \int_{I^2} (w'-w)f(w,t)M(w_*)dw\,dw_* \right\rangle \\ &= -\gamma\int_{I^2} P(|w|)(w-w_*)f(w,t)M(w_*)dw\,dw_* \\ &= -\gamma\int_I \mathcal{K}[M](w)f(w,t)dw.\end{aligned} \tag{6.68}$$

In (6.19), $\mathcal{K}[M]$ is obtained from the convolution between the background and the non-local compromise function,

$$\mathcal{K}[M](w) = \int_I P(|w|)(w-w_*)M(w_*)dw_*. \tag{6.69}$$

If $P(|w|)$ is assumed constant, $P(|w|) = 1$, $\mathcal{K}[M](w)$ is linear,

$$\mathcal{K}[M](w) = w - \int_I w_* M(w_*)dw_* = w - \langle M\rangle.$$

Substituting this value into (6.68) shows that in this case the mean value of the solution converges exponentially to the mean value of the background. The situation is not explicitly solvable for a general P. Let us fix now $\varphi(w) = w^2$. Considering that $\langle\Theta\rangle = 0$, and $\langle\Theta^2\rangle = \sigma^2$, we obtain

$$\begin{aligned}\frac{d}{dt}\int_I w^2 f(w,t)\,dw &= \left\langle \int_{I^2} \left((w')^2 - w^2\right) f(w,t)M(w_*)dw\,dw_* \right\rangle \\ &= \gamma^2\int_{I^2} P(|w|)^2(w-w_*)^2 f(w)M(w_*)\,dwdw_* \\ &\quad -2\gamma\int_{I^2} P(|w|)w(w-w_*)f(w)M(w_*)dwdw_* \\ &\quad +\sigma^2\int_I D(|w|)^2 f(w)dw \\ &= \gamma^2\int_I \mathcal{J}[M](w)f(w)\,dw - 2\gamma\int_I \mathcal{K}[M](w)wf(w)\,dw \\ &\quad +\sigma^2\int_I D(|w|)^2 f(w)dw.\end{aligned} \tag{6.70}$$

In (6.70) we defined

$$\mathcal{J}[M](w) = \int_I P(|w|)^2(w-w_*)^2 M(w_*)dw_*. \tag{6.71}$$

In this second case, as in (6.17), the choice $P(|w|) = 1$ and $D(|w|)^2 = 1 - w^2$ leads to a closed evolution equation for the second moment:

$$\begin{aligned} \frac{d}{dt}\int_I w^2 f(w,t)\,dw = &-(2\gamma + \sigma^2 - \gamma^2)\int_I w^2 f(w)\,dw \\ &+2(\gamma - \gamma^2)\langle M\rangle \int_I w f(w)\,dw + \gamma^2\langle M^2\rangle + \sigma^2. \end{aligned} \tag{6.72}$$

Since for $P(|w|) = 1$ we showed that the mean value of the solution converges to the mean value of the background $\langle M\rangle$, equation (6.72) implies that the second moment of the solution converges to the limit value

$$m_2 = \frac{2\gamma - \gamma^2}{2\gamma - \gamma^2 + \sigma^2}\langle M^2\rangle + \frac{\sigma^2}{2\gamma - \gamma^2 + \sigma^2}.$$

We note that, since the value of $\langle M^2\rangle < 1$, the value of m_2 is always lower than the second moment of the background.

Remark 6.6 *Despite the fact that the case $P(|w|) = 1$ leads to explicit computations for the moments of the solution, one immediately comes to the conclusion that this case is unrealistic. In fact, since the mean opinion is converging to the mean opinion of the equilibrium, if we start with a Maxwellian of mean equal to zero, the final opinion will have zero mean. Thus, this mechanism of exchange will not produce a choice, independently of the diffusion rule. The situation is clearly different for a non-constant $P(|w|)$.*

6.4.1 A Fokker–Planck equation for choice formation

The strategy of Section 6.2.3 can also be applied in this framework, to obtain from the kinetic model a Fokker–Planck equation as the limit of the quasi-invariant opinion. In the case in which the quantities involved at the limit satisfy the hypotheses of Theorem 6.1, that is as $\gamma \to 0$, $\sigma \to 0$ in such a way that $\sigma^2 = \lambda\gamma$, for any $\varphi \in \mathcal{F}_{2+\delta}(\mathbb{R})$ with $\delta < \alpha$, the weak solution to the equation (6.67) for the scaled density $g_\gamma(w,\tau) = f(w,t)$ with $\tau = \gamma t$, converges, up to the extraction of a subsequence, to a probability density $g(w,\tau)$, a weak solution of the Fokker–Planck equation

$$\frac{\partial g}{\partial \tau} = \frac{\lambda}{2}\frac{\partial^2}{\partial w^2}\left(D(|w|)^2 g\right) + \frac{\partial}{\partial w}\left(\mathcal{K}[M](w)g\right). \tag{6.73}$$

As described before, the Fokker–Planck equation (6.73) describes the large-time behaviour of the solution of the Boltzmann equation (6.67) when the parameters $\gamma \ll 1$, $\sigma \ll 1$, while $\sigma^2 = \lambda\gamma$. In this range of the parameters, the interaction dynamics in the Boltzmann equation lead to a Fokker–Planck equation in which the random change of opinion enters into the coefficient of diffusion, while the compromise is related to the drift.

In equation (6.73) the drift depends on the convolution operator $\mathcal{K}[M]$, and in general it is not explicitly solvable. Consequently, the steady profile g_∞, which depends on the choice of the two functions $P(\cdot)$ and $D(\cdot)$, can in general be computed only from a numerical point of view.

6.4.2 The steady opinion profile

Steady profiles, say $g_\infty(w)$, of the Fokker–Planck model (6.73) presented in the previous section are solutions of the first-order differential equation

$$\frac{d}{dw}\left(D(|w|)^2 g\right) = -\frac{2}{\lambda}\mathcal{K}[M](w) g \tag{6.74}$$

subject to the mass constraint

$$\int_{\mathbb{R}} g_\infty(w)\, dw = 1. \tag{6.75}$$

The solutions g_∞ represent the final distribution of opinions around the possible choices. It is extremely important to check how these solutions behave on the lateral boundaries. We will exclude solutions that do not satisfy the constraints

$$g_\infty(-1) = g_\infty(1) = 0. \tag{6.76}$$

Indeed, solutions for which the extremal opinions are so attracting that mass is accumulated at the boundaries, so that conditions (6.76) are not satisfied, can be considered unrealistic. Examples of interactions for which this phenomenon can happen were discussed in [308].

The steady profile g_∞ of equation (6.74) depends on the choice of the two functions $P(\cdot)$ and $D(\cdot)$, the first one being directly connected with the compromise propensity of the underlying population of agents, the latter with the diffusion behaviour of opinions.

In addition, one has to deal with the characteristic parameter λ, which represents the quotient between the compromise and diffusion parameters ($\lambda = \gamma/\sigma^2$). The choice of the value of λ, as numerical simulation shows, is a very delicate question, related to the description of the choice formation in a compromise (respectively diffusion) dominated society. A bad choice of λ will, in fact, produce a steady profile that does not satisfy conditions (6.76).

However, it is remarkable that the value of λ has no effect on the final mean opinion, since the diffusion operator, in the absence of boundary terms, has no role in the evolution of either the mass or momentum. Hence, the choice of λ is only connected with the shape of the steady profile, which is more peaked for small values of the parameter.

In the forthcoming simulations, we will assume that the diffusion function $D(\cdot)$ is of the form

$$D(w^2) = (1 - w^2)^\alpha. \tag{6.77}$$

This function satisfies all the assumptions made in Section 6.2, including the realistic idea that extremal opinions are more difficult to change. In fact, $D(\cdot)$ is a decreasing function of w^2, with $D(1) = 0$. Moreover, the choice of α can be related to the choice of the zone (around the neutral choice point $w = 0$), where the diffusion of opinions is more pronounced.

Lastly, we assume that the function $P(\cdot)$, which is related both to the compromise propensity and to the influence zone of the Maxwellian opinion (6.64), could assume one of the two profiles

$$P(w) = e^{-r|w-\bar{w}|}, \tag{6.78}$$

or

$$P(w) = \chi\left(|w - \bar{w}| \leq \rho\right). \tag{6.79}$$

In (6.78) and (6.79), $\bar{w}$ is one of the points where the Maxwellian (6.64) is localized, r and ρ are positive constants. As previously, $\chi(\cdot)$ is the indicator function. Note that, within the first choice of P, the option localized at a point $\bar{w}$ can attract the agent of opinion w, independently of the distance between the two opinions. In the second case, the domain of attraction of the option localized at $\bar{w}$ is limited to opinions which belong to a limited interval centred at $\bar{w}$. In this last case, agents modify their opinion in favour of a certain choice only if their opinion is not too far from this choice. The functions (6.78) and (6.79) appear in various contexts, and are widely used to model parameters of biological aggregation [305, 306]. They are, in fact, suitable for modelling both long and short range interactions between agents. The intensity of this interaction is described by the positive parameters r and ρ. Of course, other functions can be introduced to model both long and short interactions. As numerical simulations indicate, however, no essential modifications to the shape of the steady distribution of opinions can be introduced in this way. For the sake of simplicity, and to better test the role of the various parameters, we will limit simulations to the case of the formation of choice between two opinions. Hence, we will refer to the Maxwellian

$$M_2(w) = p\delta(w - w_-) + q\delta(w - w_+), \qquad p + q = 1, \qquad w_- < 0, w_+ > 0. \tag{6.80}$$

Clearly, the case of multiple choices can also be simulated.

A simple example, in which the steady profile can be evaluated directly, clarifies the role of the various parameters. Let us set $\alpha = 1/2$ in (6.77), and $r = 0$ in (6.78). These choices imply $D(w^2)^2 = 1 - w^2$ and $P(w) = 1$. As described in Section 6.4, $\mathcal{K}[M](w) = w - \langle M \rangle$ and the steady profile g_∞ is a solution to

$$\frac{d}{dw}\left((1 - w^2)g\right) = -\frac{2}{\lambda}(w - m)g \tag{6.81}$$

where m denotes the mean value of the Maxwellian $m = \langle M \rangle = pw_- + qw_+$. Note that, unless $w_- = -1$ and $w_+ = 1$, $-1 < m < 1$. Equation (6.81) can easily be integrated to give

$$g_\infty(w) = C\,(1 + w)^{\frac{1+m}{\lambda} - 1}\,(1 + w)^{\frac{1-m}{\lambda} - 1}. \tag{6.82}$$

Thanks to the constraints on m, it follows that the function (6.82) is integrable on the interval I. This implies that the mass constraint (6.75) can always be satisfied by a suitable choice of the constant C. On the other hand, the constraints (6.76) are satisfied only if

$$\min\left\{\frac{1 + m}{\lambda} - 1, \frac{1 - m}{\lambda} - 1\right\} > 0. \tag{6.83}$$

Note that condition (6.83) implies an upper bound on the values assumed by the constant λ.

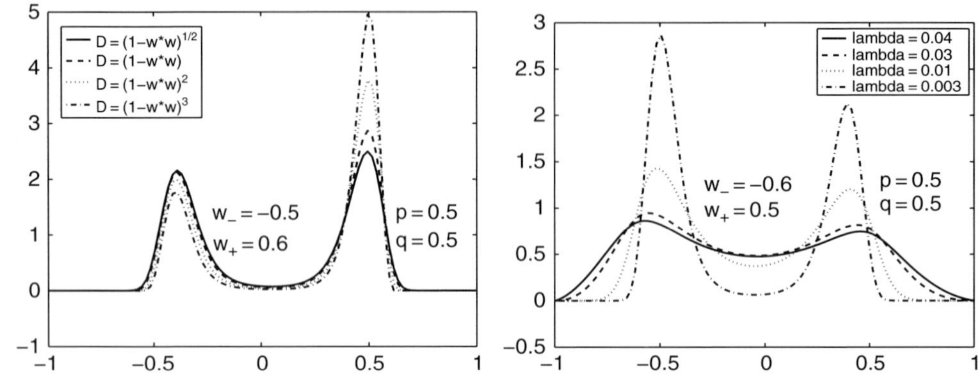

Fig. 6.1 Left: Various realizations of the diffusion function. Right: Effect of diffusion parameter λ.

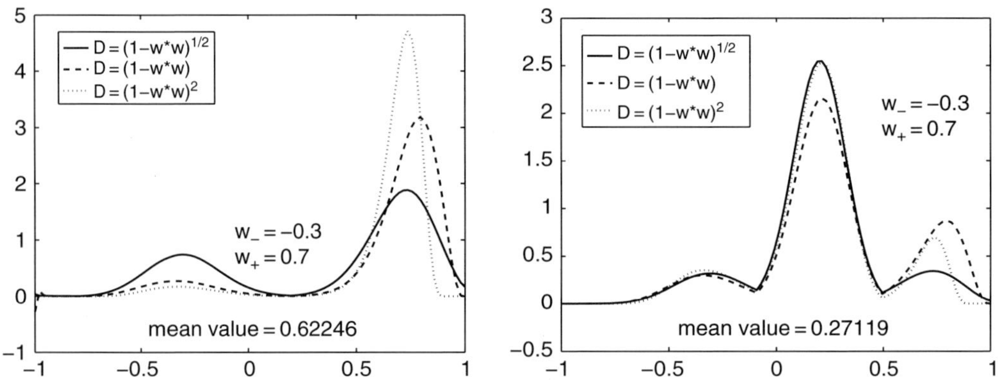

Fig. 6.2 $P(w) = \chi(|w - \bar{w}| \leq \rho)$ with $\rho = 0.5$ (left) and $\rho = 0.8$ (right), $\lambda = 0.003$ and $p = q = 0.5$.

We illustrate the influence of the various parameters in the formation of the steady profile through a series of numerical experiments. The first experiment (cf. Figure 6.1 on the left) examines the role of the diffusion function $D(w^2)$ by varying the values of the exponent α in (6.77).

The interaction parameter is the short-range interaction (6.79), with $\rho = 0.5$. It is evident that a larger value of the parameter α results in a more peaked steady state. The more stable option ($w_+ = 0.6$) becomes more attractive in the presence of a more peaked D. Also, the pivotal influence of the diffusion parameter λ can be observed in Figure 6.1 on the right. The results refer to the choice $\alpha = 1/2$ in (6.77). Small variations of the parameter λ induce an evident variation of the shape of the steady profile. In the presence of a *relatively* big diffusion coefficient, the steady profile tends to lose its dependence on the Maxwellian equilibrium. In other words, a too-pronounced self-thinking destroys the different opinions.

The influence of the short-range interaction (6.79) has been plotted in Figure 6.2,

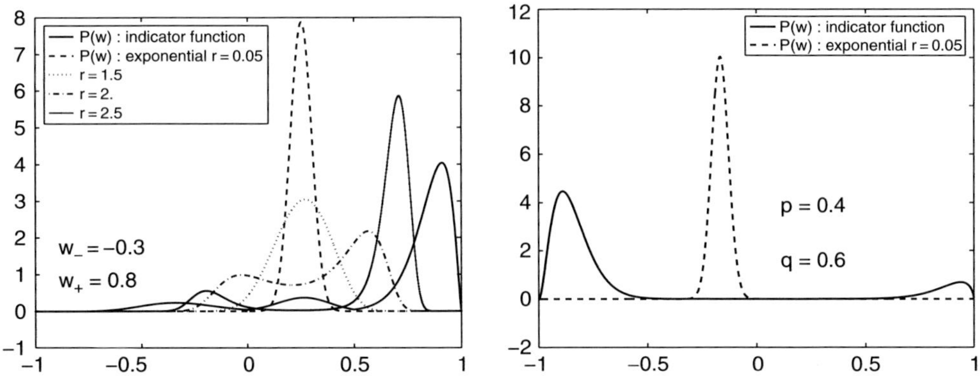

Fig. 6.3 Steady profile g corresponding to the choices of Maxwellian localization $w_- = -0.8$ and $w_+ = 0.8$. $p = q = 0.5$ (left) and $p = 0.6, q = 0.4$ (right).

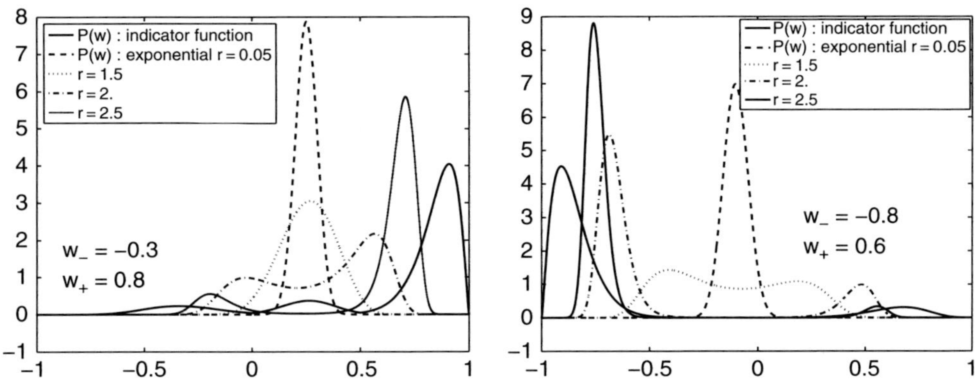

Fig. 6.4 Steady profile g for $p = q = 0.5$, corresponding to different choices of Maxwellian localization $w_- = -0.3$ and $w_+ = 0.8$ (left), $w_- = -0.8$ and $w_+ = 0.6$ (right).

for different values of the interaction interval ρ, and in the presence of an asymmetric situation for the two possible choices. Here, a small value of ρ forces the opinion to stay around the more stable choice. The influence of the range of interaction is still more evident in Figure 6.3, where the short and long range interaction functions (6.78) and (6.79) are compared. In the presence of short-range interactions the two options remain well separated, and the final profile is peaked around them in case of long-range interactions, the steady profile is dramatically changed, and the steady profile has a unique maximum which is located approximately on the neutral opinion.

The difference is still more evident when an asymmetric distribution of opinion is assumed for the Maxwellian equilibrium. Figure 6.4 shows the influence of both short and long range interaction in this situation. Various choices of the parameter r in (6.78) reveal that the final difference of opinions around the possible choices is heavily dependent on a rapid decay of the long-range interactions.

6.5 Opinion formation in the presence of leaders

The models presented so far are founded on the hypothesis that all agents in the society play the same role with respect to the exchange of information, which is at the basis of the formation of opinion, and its statistical distribution among people. There is no doubt, however, that nowadays distinguished agents in society play an essential role in the formation of opinion. Due to their importance in political elections, these agents are usually known as leaders.

Opinion leadership is, among several sociological models, one of the most convincing in trying to explain the formation of opinions in a society. It is a concept which is present in the literature from the middle of the last century [203]. In the course of their study of the presidential elections in the USA in 1940, Lazarsfeld, Berelson and Gaudet [203] found interpersonal communication to be much more influential than direct media effects. They formulated a theory of a two-step flow of communication where the so-called opinion leaders who are active media users are selecting, interpreting, modifying, facilitating and finally transmitting information from the media to the less active parts of the population.

Fig. 6.5 Paul Felix Lazarsfeld[1]

Later on, sociologists obtained a new view on opinion leader characteristics, by developing the notion of public individuation. Public individuation describes how people feel the urge to differentiate themselves and act differently from other people [228]. This is a necessity for an opinion leader, because she or he must be willing to set herself or himself apart from the ordinary people. Certain, typical personal characteristics are supposed to characterize opinion leaders: high confidence, high self-esteem, a strong need to be unique and the ability to withstand criticism. An opinion leader is socially active, highly connected and held in high esteem by those accepting his or her opinions.

Although different from the others, opinion leaders are still related to their followers and not always easy to distinguish from them. This is because opinion leadership is specific to a subject and can change over time. Someone who is a strong opinion leader in one field may be a follower in another.

As already discussed in Section 6.1, in the last decade, new communication forms like email, web navigation, blogs and instant messaging have globally changed the way information is disseminated and opinions are formed in the society [271]. Nevertheless, opinion leadership continues to play a critical role in all these processes, independently of the underlying technology.

Opinion leadership appears in many aspects of our daily life. Examples range from the advertisement of commercial products, where people are convinced by product reviewers in the media who have a deeper knowledge and background than average

[1]**Paul Felix Lazarsfeld**: born on 13 February 1901 in Vienna (Austria), died on 30 August 1976 in New York (USA). He earned a doctoral degree in mathematics from the University of Vienna in 1925. Lazarsfeld is considered among the founders of modern empirical sociology and mathematical sociology. Opinion leadership comes from the theory of two-step flow of communication propounded by Paul Lazarsfeld and Elihu Katz.

consumers, to the dissemination of new technologies, where *early adopters* play an important role, either as lighthouse customers who assist in the development or as individuals who recommend a new product to others.

6.5.1 A kinetic model with opinion leaders

The idea of taking the kinetic model for opinion formation of Section 6.1 as a building block for more general situations, including the presence of leaders, was developed by Düring, Markowich, Pietschmann and Wolfram in [131].

The model of [131] describes a society of agents constituted by two different groups of individuals which interact with each other. The leading idea is that human societies typically contain a set of individuals who can influence opinion through their strong personalities, financial means, access to media or others. The kinetic modelling of their effect on public opinion is based on the hypothesis that the opinions of these *distinguished* agents do not change through interactions with *normal* agents in the society.

This easily leads to two groups, one identified with such *strong opinion leaders* and the other with their followers, the *ordinary people.* According to the bounded confidence interaction law (6.6), all agents belonging to one group share a common compromise parameter. This hypothesis could be further relaxed by assuming that the compromise parameter is a random quantity, with a statistical mean which is different for the two groups. If two individuals from the same group meet, the interaction is given by ($i = 1, 2$)

$$
\begin{aligned}
w' &= w - \gamma_i P_i(|w - w_*|)(w - w_*) + \Theta_{i,1} D_i(w), \\
w'_* &= w_* - \gamma_i P_i(|w_* - w|)(w_* - w) + \Theta_{i,2} D_i(w_*).
\end{aligned}
\tag{6.84}
$$

If one individual from the group of ordinary people with opinion w meets a strong opinion leader with opinion w_*, their post-interaction opinions are given by

$$
\begin{aligned}
w' &= w - \gamma_3 P_3(|w - w_*|)(w - w_*) + \Theta_{1,1} D_1(w), \\
w'_* &= w_*.
\end{aligned}
\tag{6.85}
$$

Again, the functions $P_i(\cdot)$, $i = 1, 2, 3$, and $D_i(\cdot)$, $i = 1, 2$, take into account the local relevance of the compromise and diffusion for a given opinion. Finally, $\gamma_k \in (0, \frac{1}{2})$ ($k = 1, 2, 3$) are constant *compromise parameters*, which control the speed of attraction of two different opinions. For simplicity, it is assumed that all individuals in the society share a common compromise parameter γ. The quantities $\Theta_{i,j}$, $i, j = 1, 2$, are random variables assuming values on a set $\mathcal{B} \subset \mathbb{R}$, with zero mean and variance σ_{ij}^2.

According to the discussion of Section 6.1, the functions $D_j(\cdot)$ are decreasing functions of v^2 with $D_j(1) = 0$. In (6.85), (6.84) it is important that only interactions that guarantee $w', w'_* \in I$ should be allowed. To this end, one has to assume additionally that

$$
0 \le P_i(|w - w_*|) \le 1, \quad 0 \le D_j(w) \le 1.
$$

The choice of the set $\mathcal{B}$ depends on the particular choice for $D_j(\cdot)$. Let us consider the upper bound at $w_* = 1$ first. To ensure that individuals' opinions do not leave I, we need

$$w' = w - \gamma_i P_i(|w - w_*|)(w - w_*) + \Theta_{ik} D_j(w) \leq 1.$$

Obviously, the worst case is $w_* = 1$, where we have to ensure

$$\Theta_{ik} D_j(w) \leq 1 - w + \gamma_i(w - 1) = (1 - w)(1 - \gamma_i).$$

Hence, if $D_j(w)/(1 - w) \leq K_+$ it suffices to have $|\Theta_{ik}| \leq (1 - \gamma_i)/K_+$. A similar computation for the lower boundary shows that if $D_j(w)/(1 + w) \leq K_-$ it suffices to have $|\Theta_{ik}| \leq (1 - \gamma_i)/K_-$.

With these interaction rules, we are led to study the evolution of the distribution function for each group as a function depending on the opinion $w \in I$ and time $t \in \mathbb{R}_+$, $f_i = f_i(w, t)$. By analogy with the classical kinetic theory of mixtures of rarefied gases, the time evolution of the distributions will obey a system of two Boltzmann-like equations, given by

$$\frac{\partial}{\partial t} f_1(w, t) = \frac{1}{\tau_{11}} \mathcal{Q}_{11}(f_1, f_1)(w) + \frac{1}{\tau_{12}} \mathcal{Q}_{12}(f_1, f_2)(w), \tag{6.86}$$

$$\frac{\partial}{\partial t} f_2(w, t) = \frac{1}{\tau_{22}} \mathcal{Q}_{22}(f_2, f_2)(w). \tag{6.87}$$

Herein, τ_{ij} are suitable relaxation times which allow control of the interaction frequencies of opinion leaders and followers. The Boltzmann-like collision operators $\mathcal{Q}_{11}$ and $\mathcal{Q}_{22}$ relate to the microscopic interaction (6.84), whereas $\mathcal{Q}_{12}$ relates to (6.85).

As always, the collision operators are easily written in weak form

$$\begin{aligned} &\int_I \mathcal{Q}_{ij}(f_i, f_j)(w)\varphi(w)\, dw \\ &\quad = \frac{1}{2} \left\langle \int_{I^2} \left(\varphi(w_*') + \varphi(w') - \varphi(w) - \varphi(w)\right) f_i(w) f_j(w_*)\, dw\, dw_* \right\rangle. \end{aligned} \tag{6.88}$$

Remark 6.7 *With respect to the Boltzmann equation (6.4), equation (6.86) contains as additional parameters the relaxation times τ_{ij}. This is an important issue, which is typical of kinetic equations for mixtures. Indeed, the presence of these relaxation times allows us to understand and therefore control the role of the frequencies between groups. A similar situation was encountered in Chapter 5, Section 5.6, where the distribution of wealth among groups of traders was studied.*

6.5.2 Fokker–Planck description

Following step by step the computations of Section 6.2.3, we can derive the quasi-invariant opinion limit of the Boltzmann system (6.86) by allowing $\gamma\sigma_{ij} \to 0$ and

$\sigma_{ij}^2/\gamma = \lambda_{ij}$ (for simplicity we set here $\gamma_i = \gamma$, $i = 1, 2, 3$). To study the situation for large times, i.e. close to the steady state, we introduce for $\gamma \ll 1$ the transformation

$$\tau = \gamma t, \quad g_i(w, \tau) = f_i(w, t), \quad i = 1, 2,$$

which implies $f_i(w, 0) = g_i(w, 0)$. Denote by $M_i = \int g_i \, dv$ ($i = 1, 2$) the masses of the opinion leaders and followers, respectively. Then we obtain the following system of Fokker–Planck limit equations:

$$\begin{aligned}
\frac{\partial}{\partial \tau} g_1(w,\tau) =& \frac{\partial}{\partial w}\left(\left(\frac{1}{\tau_{11}}\mathcal{K}_1(w,\tau) + \frac{1}{2\tau_{12}}\mathcal{K}_3(w,\tau)\right) g_1(w,\tau)\right) \\
&+ \left(\frac{\lambda_{11}M_1}{2\tau_{11}} + \frac{\lambda_{12}M_2}{4\tau_{12}}\right)\frac{\partial^2}{\partial w^2}\left(D_1^2(w) g_1(w,\tau)\right), \\
\frac{\partial}{\partial \tau} g_2(w,\tau) =& \frac{\partial}{\partial w}\left(\frac{1}{\tau_{22}}\mathcal{K}_2(w,\tau) g_2(w,\tau)\right) + \frac{\lambda_{22}M_2}{2\tau_{22}}\frac{\partial^2}{\partial w^2}\left(D_2^2(w) g_2(w,\tau)\right),
\end{aligned} \tag{6.89}$$

subject to the following no flux boundary conditions (which result from the integration by parts):

$$\begin{aligned}
&\left(\frac{1}{\tau_{11}}\mathcal{K}_1(w,\tau) + \frac{1}{2\tau_{12}}\mathcal{K}_3(w,\tau)\right) g_1(w,\tau) \\
&\quad + \left(\frac{\lambda_{11}M_1}{2\tau_{11}} + \frac{\lambda_{12}M_2}{4\tau_{12}}\right)\frac{\partial}{\partial w}\left(D_1^2(w) g_1(w,\tau)\right) = 0 \qquad \text{on } w = \pm 1, \\
&\frac{1}{\tau_{22}}\mathcal{K}_2(w,\tau) g_2(w,\tau) + \frac{\lambda_{22}M_2}{2\tau_{22}}\frac{\partial}{\partial w}\left(D_2^2(w) g_2(w,\tau)\right) = 0 \qquad \text{on } w = \pm 1, \\
&\qquad D_1^2(w) g_1(w) = D_2^2(w) g_2(w) = 0 \qquad \text{on } w = \pm 1.
\end{aligned} \tag{6.90}$$

Note that, if the solutions g_1 and g_2 are sufficiently regular, the third condition in (6.90) holds automatically since $D_1(w) = D_2(w) = 0$ for $w = \pm 1$. The operators appearing in the drift term are defined as

$$\mathcal{K}_i(w,\tau) = \int_I P_i(|w - w_*|)(w - w_*) g_i(w_*,\tau)\, dw_*, \quad \text{for } i = 1, 2, \tag{6.91}$$

$$\mathcal{K}_3(w,\tau) = \int_I P_3(|w - v|)(w - w_*) g_2(w_*,\tau)\, dw_*. \tag{6.92}$$

6.5.3 Stationary profiles

As often happens when dealing with Fokker–Planck equations, system (6.90) has explicitly computable stationary solutions. These steady states are of paramount importance, in that they describe the final distribution of opinion which follows from the microscopic interactions between groups. To compute explicit solutions, as in Section 6.3.1, we consider the special case $P_i(|w - w_*|) = 1$ ($i = 1, 2, 3$). This choice leads

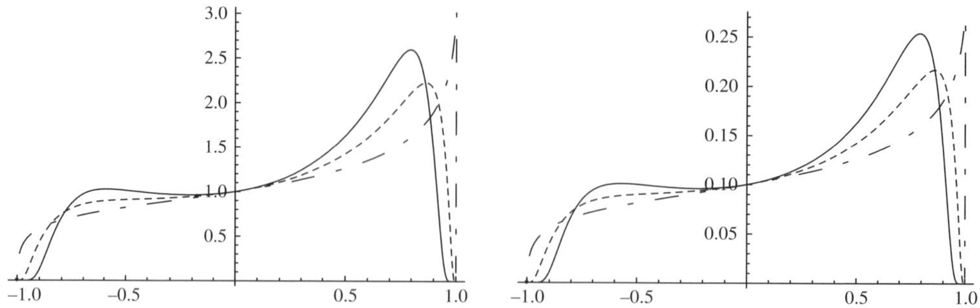

Fig. 6.6 Illustration of stationary solutions (6.96) for different values of α and parameters given by (6.97). The solid line corresponds to $\alpha = 1$, the dashed one to $\alpha = 0.75$ and the dashed-dotted one to $\alpha = 0.5025$.

to a marked simplification, implying conservation of the average opinion and of the first momentum for the second equation in (6.89). In addition, we consider

$$D_1(w) = D_2(w) = D(w) = \left(1 - w^2\right)^{\alpha}, \tag{6.93}$$

with $\alpha > \frac{1}{2}$ as a model for the diffusion, which is consistent with the requirement that post-collisional opinions have to be in I, at least when the ranges of the random variables Θ_{ij} are sufficiently small.

The steady state of (6.89) is a solution of

$$\begin{aligned} &\left(\frac{wM_1 - m_1}{\tau_{11}} + \frac{wM_2 - m_2}{2\tau_{12}}\right) g_{1,\infty}(w) \\ &+ \left(\frac{\lambda_{11} M_1}{2\tau_{11}} + \frac{\lambda_{12} M_2}{4\tau_{12}}\right) \left(D^2(w) g_{1,\infty}(w)\right)_w = 0, \\ &\frac{wM_2 - m_2}{\tau_{22}} g_{2,\infty} + \frac{\lambda_{22} M_2}{2\tau_{22}} \left(D^2(w) g_{2,\infty}\right)_w = 0. \end{aligned} \tag{6.94}$$

We denote the masses of the opinion leaders and followers by $M_i = \int g_{i,\infty}\, dv$ with $i = 1, 2$ and their first-order moments by $m_i = \int w_* g_{i,\infty}\, dw_*$, $i = 1, 2$.

The last equation in (6.94) can be written as

$$-\frac{wM_2 - m_2}{D^2(w)} f_2 = \frac{\lambda_{22} M_2}{2} \frac{d}{dw} f_2,$$

with $f_2 = D^2(w) g_{2,\infty}$. Therefore,

$$f_2 = c_2 e^{-\frac{2}{\lambda_{22} M_2} \int_0^w \frac{w_* M_2 - m_2}{(1 - w_*^2)^{2\alpha}}\, dw_*},$$

and hence

$$g_{2,\infty} = \frac{c_2}{(1 - w^2)^{2\alpha}} \exp\left\{-\frac{2}{\lambda_{22} M_2} \int_0^w \frac{w_* M_2 - m_2}{(1 - w_*^2)^{2\alpha}}\, dw_*\right\}. \tag{6.95}$$

In equation (6.95), c_2 is chosen such that the mass of $g_{2,\infty}$ is equal to M_2. Note that since $|m_2| < M_2$ and $\alpha > \frac{1}{2}$, $g_{2,\infty}(\pm 1) = 0$.

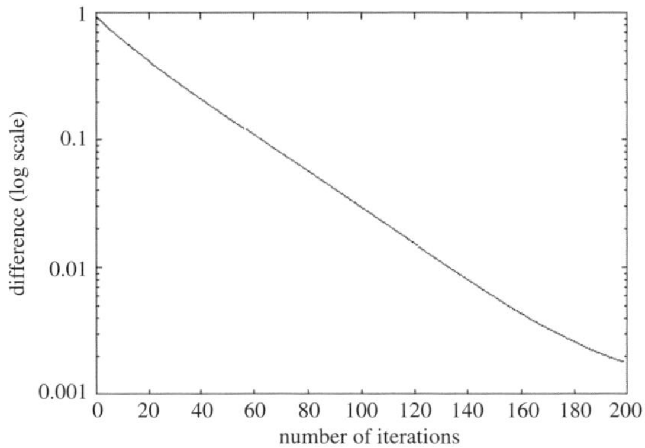

Fig. 6.7 Difference between approximate solution g_2 and exact solution $g_{2,\infty}$ in L^1-Norm.

Then the solution of the first equation in (6.94) can be calculated using the same arguments:

$$g_{1,\infty} = \frac{c_1}{(1-w^2)^{2\alpha}} \exp\left\{-k\int_0^w \left[\frac{w_* M_1 - m_1}{\tau_{11}\,(1-w_*^2)^{2\alpha}} + \frac{w_* M_2 - m_2}{2\tau_{12}(1-w_*^2)^{2\alpha}}\right] dw_*\right\}, \tag{6.96}$$

with

$$k = \frac{4\tau_{11}\tau_{12}}{2\lambda_{11}M_1\tau_{12} + \lambda_{12}M_2\tau_{11}}.$$

Integration of (6.94) leads to

$$M_2 m_1 - m_2 M_1 = 0.$$

Therefore, we can fix m_1 and rewrite (6.96) as

$$g_{1,\infty} = \frac{c_1}{(1-w^2)^{2\alpha}} \exp\left\{-k\left(\frac{M_1}{\tau_{11}} + \frac{M_2}{2\tau_{12}}\right)\int_0^w \frac{w_*}{(1-w_*^2)^{2\alpha}}\,dw_*\right\}$$
$$\cdot \exp\left\{k m_2\left(\frac{1}{2\tau_{12}} + \frac{M_1}{\tau_{11}M_2}\right)\int_0^w \frac{1}{(1-w_*^2)^{2\alpha}}\,dw_*\right\}.$$

From $|m_2| < M_2$ we conclude that if $\alpha > \frac{1}{2}$ then c_1 can be determined such that the mass of $g_{1,\infty}$ equals M_1. Figure 6.6 illustrates the behaviour of the stationary solutions for different values of α.

Here, we chose the following parameters:

$$\begin{gathered} M_1 = 1, \quad M_2 = 0.05, \quad m_2 = 0.01, \quad m_1 = \tfrac{m_2 M_1}{M_2} = 0.2, \\ \tau_{11} = 1, \quad \tau_{12} = 10, \quad \lambda_{11} = \lambda_{12} = \lambda_{22} = 1. \end{gathered} \tag{6.97}$$

Note that the stationary solutions are symmetric with respect to the y-axis if the first-order moments vanish.

Numerical simulations provide strong evidence that solutions converge exponentially fast to their steady state, see Figure 6.7 (here, all parameters are set to one, except $m_1 = m_2 = 0$). We refer the interested reader to [131] for an example application of the model in a real scenario, the 2004 and 2009 state elections in Carinthia, the southernmost state of Austria.

7
A further insight into economics and social sciences

> Every individual is continually exerting himself to find out the most advantageous employment for whatever capital he can command. It is his own advantage, indeed, and not that of the society which he has in view. But the study of his own advantage naturally, or rather necessarily, leads him to prefer that employment which is most advantageous to society. . . . He intends only his own gain, and he is in this, as in many other cases, led by an invisible hand to promote an end which was not part of his intention.
>
> A. Smith, *An Inquiry into the Nature and Causes of the Wealth of Nations*, 1776.

7.1 Towards more realistic models

In Chapter 5 we illustrated a variety of elementary models for wealth distribution, in which the evolution of the density is dependent on binary trades. In general, the richness of the steady states for these models is the main remarkable difference from the theory of the classical Boltzmann equation, for which the Maxwell distribution (1.2) is the universal steady profile. This richness of steady states makes both the theoretical and numerical study of these models highly interesting. Indeed, it corresponds mathematically to identifying the limit distribution in a problem which has many analogies with a generalized central limit theorem [165, 200]. Also, it allows us to clarify the types of binary interactions which produce fat tails in the distribution at infinity (Pareto tails), which are shown to form in the profiles of the distribution of wealth in western societies [321].

The objective weakness of models based on linear binary trades of type (5.6) is mainly linked to the fact that people trade according to rules which are in general difficult to justify from a microeconomic point of view. The common ingredients which appear in these trades are the saving propensity concept [91] (agents usually do not trade with all their wealth) and the risk concept [105] (there is no certainty of gain). While highly reasonable, these types of interaction do not clarify why agents decide to trade.

On the other hand, the real socio-economic behaviour of a multiagent market is an extremely complex phenomenon, which involves contributions from various fields. As a matter of fact, agents are pushed to trade by many reasons, which include, in addition to the desire to increase their wealth, a personal perception of risk, and the pressure to follow what others are doing. For a better understanding of some of the main outcomes of markets in western societies, in recent years the kinetic community started to improve the economic modelling of agent systems by including more realistic assumptions, paying close attention to a fair balance between the introduction of these new effects and the possibility of having a not-too-complicated model to handle. Some recent papers in particular fit into this line of thinking [61, 104, 135, 202, 222, 286].

One of the fundamental assumptions of price theory in economics is that people trade to improve its utility. One way of looking at this is contained in [104]. However, there the concept of utility is not introduced at the level of a single binary trade. The coupling of binary trades with this idea of utility was developed in [61].

To describe binary interactions which are motivated by an increase in the individual utility, the basic ingredient in [61] is furnished by the Edgeworth box [138], which is frequently used in general equilibrium theory, and can aid in representing the competitive equilibrium of a simple system or a range of such outcomes that satisfy economic efficiency. The Edgeworth box can fruitfully be applied in the presence of an agent-based system in which agents possess a finite number of goods of $n \geq 2$ different types. Inspired by this mechanism of increasing utility and competitive equilibrium, [61] introduced a kinetic equation of the Boltzmann type for the evolution of the distribution density of the quantities of two goods in a system of agents. The incomplete knowledge of the game in the binary interaction is taken into account by introducing randomness in the Edgeworth box outcome. There, the balance between utility and risk is shown to concentrate the density along the price line, and at the same time to distribute goods along this line with the possible formation of a tailed density.

It is remarkable that the basic problem of exchanging goods has previously been treated within the methods of statistical mechanics only in [286]. There, an exchange market consisting of many agents and two goods was introduced by allowing agents to have stochastic preferences between them. However, the model considered in [286] has steady-state solutions which are described by Gamma distributions, and Pareto tails are not present.

As a second example, in [104] a kinetic description of the behaviour of a simple financial market was derived where each agent can create their own portfolio between two investment alternatives: a stock and a bond. In this case, the variation of density is not based on linear binary collisions like (5.6), but is derived starting from the Levy–Levy–Solomon microscopic model for price formation [207, 208]. The model in [104] consists of a linear Boltzmann equation for the wealth distribution of the agents coupled with an equation for the price of the stock. From this model, by applying an asymptotic procedure similar to that applied many times in this book, the behaviour of the density was shown to obey a Fokker–Planck equation. Notably, this Fokker–Planck equation admits a self-similar lognormal behaviour. For the first time, the kinetic model in [104] attempts to join to simple financial rules a kinetic equation of the Boltzmann type, able to describe a complex behaviour that could then mimic the market and explain the macroscopic consequences of the microscopic price formation mechanism.

Next, a third interesting example couples the economic models with the consensus modelling described in Chapter 6 (see [222]).

In [222], a relatively simple kinetic model is introduced for a financial market characterized by a single stock and an interplay between two different trader populations, chartists and fundamentalists, which, as a consequence of their different behaviours, determine the price dynamics of the stock. The model is inspired by the microscopic Lux–Marchesi model [215–217]. The main novelty here is to couple the financial rules with the opinion of traders through a kinetic model of opinion formation, close to the one introduced in Chapter 6. Moreover, some psychological and behavioural components of the agents, like the way they interact with each other and perceive risk, which may produce non-rational behaviours, are taken into account. This is done by means of a suitable *value function* in agreement with the Prospect Theory by Kahneman and Tversky [188, 189]. As they demonstrate, people systematically overreacting produce substantial instabilities in the stock market.

Fig. 7.1 Daniel Kahneman[1]

Let us also mention a related approach in [202] where the authors introduce a simple mean-field model for the dynamical formation of a price based on a population of agents consisting of two groups namely one group of buyers of a certain good and one group of vendors of the same good. In such model, however, no psychological or behavioural components of the agents are considered.

Finally, as a last example we consider a general kinetic model for conservative economies and the passage to suitable hydrodynamic equations [135]. In this model the density of wealth depends on additional parameters, like the propensity to invest. As in kinetic theory of rarefied gases, the closure depends on the knowledge of the homogeneous steady wealth distribution of the underlying kinetic model. The interaction operator used here is the FokkerPlanck operator introduced in [55] and derived as a suitable asymptotic limit of a Boltzmann-like model in Section 5.4.1.

In the rest of this chapter we will describe briefly these models, paying particular attention to the underlying economic and social arguments which are at the basis of the consequent interactions.

7.2 A kinetic model for trading goods

7.2.1 The Edgeworth box

As discussed in the introduction, most of the existing kinetic models for wealth distribution are based on rigid assumptions which, although they are shared, are not deeply related to economic principles, like price theory [150]. The aim of this section

[1]**Daniel Kahneman**: born on 5 March 1934 in Tel Aviv (Israel). In their 1979 paper on Prospect Theory, Kahneman and Tversky examined a simple problem of economic risk and, rather than stating the optimal, rational answer, they quantified how most real people consistently make a less-rational choice. Kahneman was awarded the Nobel Prize in Economics in 2002 for his work with Tversky, who died before the award was bestowed.

is to introduce a new framework for trades, which is derived directly from the basic principles of economics.

Individuals exchange goods. The benefits they receive depend on how much they exchange and on what terms. Price theory tries to give an answer to this fundamental question. In the case of a binary trade, there may be many different exchanges, each of which would be beneficial to both parties; some exchanges will be preferred by one person, some by the other. There are then two different questions to be settled. One is how to squeeze as much total gain as possible out of the opportunities for trade; the other is how that gain is to be divided. The two individuals who are trading have a common interest in getting as much total gain as possible but are likely to disagree about the division.

Let us suppose in the following that agents in the system possess the same two types of goods, and there is no production, so that the total amount of goods remains unchanged. If (x_A, y_A) ((x_B, y_B) respectively) denote the quantity of goods of two agents A and B, the quantities

$$p_A = \frac{x_A}{x_A + x_B}, \qquad q_A = \frac{y_A}{y_A + y_B}, \tag{7.1}$$

are the proportions of goods of the first agent. Note that the point (p_A, q_A) belongs to the square $\mathcal{S} = [0,1] \times [0,1]$. In order to give a meaning to the reasons for trading, it is classically assumed that each agent's behaviour is driven by a utility function. One of the most popular of these functions is the Cobb–Douglas utility function

$$U(p,q) = p^\alpha q^\beta, \quad \alpha + \beta = 1. \tag{7.2}$$

Each agent will tend to maximize its utility by trading. The values α and β are linked to the preferences that the agent assigns to the two goods. If $\alpha > \beta$, the agent prefers to possess goods of the first type (numbered by x). The choice $\alpha = \beta = 1/2$ clearly means that the two goods are equally important for A. Given the proportion point (p_A, q_A) of the agent A, the curve

$$U(p,q) = U(p_A, q_A),$$

denotes the *indifference curve* for agent A. Indeed, any point on the indifference curve for A has the same utility. Note that the indifference curve for A belongs entirely to $\mathcal{S}$, and splits the square into the two regions

$$U_A^- = \{(p,q) : U(p,q) < U(p_A, q_A)\}, \quad U_A^+ = \{(p,q) : U(p,q) > U(p_A, q_A)\}.$$

Clearly, any trade which will move the proportions of agent A into the region U_A^+ will increase his utility function, and will be acceptable for the agent.

In the case of a binary trade, agent B also has a Cobb–Douglas utility function, with parameters of preference which in general are different from the parameters of agent A. Likewise, there will be an indifference curve for B, and a good region in which the utility of B is increased after trading. Ultimately, a trade will be acceptable for both agents if their proportions after the trade belong to the regions in which their utilities are increased.

Fig. 7.2 Francis Ysidro Edgeworth[2]

An ingenious way of looking at such a situation is the Edgeworth box, named after Francis Y. Edgeworth [138]. The Edgeworth box is built up by rotating the square $\mathcal{S}$ which contains the indifference curve of agent B 180° around the centre of the square $(1/2, 1/2)$, and considering the indifference curve of A and the rotated indifference curve of B together on the same square.

Any point inside the box represents a possible division of the total quantity of A and B, with the share of A measured from the lower left-hand corner, and the share of B from the upper right-hand corner. Any possible trade is represented by a movement from one point in the box to another.

By construction, since the Cobb–Douglas utility function is jointly convex, the utility of A increases if the point moves up and to the right; the utility of B increases if the point moves down and to the left. The region in which this happens is exactly the region common to U_A^+ and to U_B^+ rotated through 180°.

The view of trading furnished by the Edgeworth box has interesting consequences. If the proportions of A and B are such that there is a possible region for trading, then once the trade has been done the point has moved inside the region, and the new region for trading is smaller. If the two agents were so smart as to choose a point inside the region at which the two indifference curves are tangent to each other, since they curve in opposite directions, this means that starting from this point, any point that is on a higher indifference curve for the first agent must be on a lower curve for the second; any trade that makes one better off makes the other worse off. The point at which the two curves are tangent to each other is not unique. The set of all points from which no further mutually beneficial trading is possible (Pareto optimal points) is called the *contract curve.*

The contract curve is related to prices. At that tangency of the two indifference curves, in fact, the slope of the tangent represents the relative prices for the two goods. Hence, there are relative prices that will be consistent with the Pareto optimum, and these prices will maximize the possible budget for both agents.

This short discussion on utility functions, trading and prices is the basis of the forthcoming construction of the binary trade. In what follows, we suppose that all agents in the system have the same utility function (7.2). If agent A has proportions p and q in his Edgeworth box, we consider as possible trades the movements of the point (p, q) into $(p^*, q^*) \in \mathcal{S}$, where

$$\begin{aligned} p^* &= p + \lambda\beta(q - p) + \eta(q - p), \\ q^* &= q + \lambda\alpha(p - q) + \tilde{\eta}(p - q). \end{aligned} \tag{7.3}$$

[2]**Francis Ysidro Edgeworth**: born on 8 February 1845 in Edgeworthstown (Ireland), died on 13 February 1926 in Oxford (United Kingdom). Edgeworth was the first to apply certain formal mathematical techniques to individual decision making in economics. In *Mathematical Psychics: An Essay on the Application of Mathematics to the Moral Sciences*, published in 1881, he developed utility theory, introducing the indifference curve and the famous 'Edgeworth box', which have become standards in economic theory.

In (7.3), $0 < \lambda \le 1$, and η and $\tilde{\eta}$ are random variables with zero mean and finite variance. We will assume, moreover, that

$$0 < \lambda\beta + \eta < 1, \quad 0 < \lambda\alpha + \tilde{\eta} \le 1. \tag{7.4}$$

Under this condition, the post-trade point (p^*, q^*) belongs to $\mathcal{S}$, and it is an admissible point for the Edgeworth box.

The dynamics consequent to (7.3) are such that

$$|p^* - q^*| = |1 - \lambda - (\eta + \tilde{\eta})|\, |p - q|. \tag{7.5}$$

Hence, if the random variables η and $\tilde{\eta}$ satisfy

$$-\lambda < \eta + \tilde{\eta} < 2 - \lambda, \tag{7.6}$$

the difference between p and q decreases in time and the system moves asymptotically towards the contract curve. In particular, this behaviour holds true in the absence of randomness.

Moreover, unless $p = q$, and in the absence of randomness, that is if

$$\begin{aligned} p^* &= p + \lambda\beta(q - p), \\ q^* &= q + \lambda\alpha(p - q), \end{aligned} \tag{7.7}$$

the trade increases the Cobb–Douglas utility function of agent A. Indeed,

$$\frac{d}{d\lambda} U(p^*, q^*) = \alpha\beta(1 - \lambda)(p - q)^2 p^{\alpha-1} q^{\beta-1} > 0,$$

which, coupled with the convexity of the utility function, gives

$$\begin{aligned} U(p^*, q^*) &= U(p, q) + \alpha\beta(p - q)^2 p^{\alpha-1} q^{\beta-1}\lambda + \frac{1}{2}\frac{d^2}{d\lambda^2} U(\tilde{p}^*, \tilde{q}^*)\lambda^2 \\ &\ge U(p, q) + \alpha\beta(p - q)^2 p^{\alpha-1} q^{\beta-1}\lambda. \end{aligned}$$

Note, moreover, that the trade (7.7) is such that $p^* \ne q^*$, unless $p = q$. Concerning agent B, since it starts from the point $(1 - p, 1 - q) \in \mathcal{S}$, according to (7.7) it moves to the point

$$\begin{aligned} \bar{p} &= 1 - p + \lambda\beta(1 - q - (1 - p)), \\ \bar{q} &= 1 - q + \lambda\alpha(1 - p - (1 - q)), \end{aligned} \tag{7.8}$$

which, proceeding as before, implies

$$U(\bar{p}, \bar{q}) > U(1 - p, 1 - q).$$

Hence, in the absence of randomness, the trade (7.7) is such that the Cobb–Douglas utility functions of both agents are increased, locating the new point in $\mathcal{S}$ in the interior of the allowed area for trades predicted by the Edgeworth box. Clearly, this trade refers to the purely theoretical situation in which the traders A and B know all about the trade and its result. In general, this is not realistic, and all that traders can hope for, after a certain number of trades, is that the majority of their trades had the result of increasing their utility. This can be expressed by saying that a reasonable result for traders is to obtain, in a single trade, an increase in their utility only in the mean. If one agrees with this, then trades of type (7.3) satisfy all the constraints of the Edgeworth box, but a single trade can move the point to a region which is convenient only for one of the two traders, or to a region which is not convenient for either.

Once the trade has been done, according to the rules of price theory, the quantities of goods of agents A and B have been changed accordingly. Making use of the constraints $x_A + x_B = x_A^* + x_B^*$ and $y_A + y_B = y_A^* + y_B^*$, (7.3) implies for the agents A and B the post-trade quantities of goods:

$$\begin{aligned}
x_A^* &= x_A + (\lambda\beta + \eta)\left(\frac{x_A + x_B}{y_A + y_B} y_A - x_A\right), \\
y_A^* &= y_A + (\lambda\alpha + \tilde{\eta})\left(\frac{y_A + y_B}{x_A + x_B} x_A - y_A\right), \\
x_B^* &= x_B + (\lambda\beta + \eta)\left(\frac{x_A + x_B}{y_A + y_B} y_B - x_B\right), \\
y_B^* &= y_B + (\lambda\alpha + \tilde{\eta})\left(\frac{y_A + y_B}{x_A + x_B} x_B - y_B\right),
\end{aligned} \tag{7.9}$$

We note that, contrary to what is usually assumed in wealth trades of type (5.6), the post-trade quantities in (7.9) are related to the pre-trade quantities by nonlinear relations. It can be verified, however, that the constraints about the conservation of the total quantities of goods in the trade are verified, and

$$x_A^* + x_B^* = x_A + x_B, \quad y_A^* + y_B^* = y_A + y_B.$$

7.2.2 A linear Boltzmann equation for trading goods

Let $f(x, y, t)$ denote the density of agents with quantities x and y of the two goods at time $t \geq 0$. Without loss of generality, we will assume in the following that (x, y) are non-negative real numbers. A Boltzmann-like equation of Maxwell type for the time evolution of $f(x, y, t)$ can be written by observing that the *collision rules* here are given by (7.9). A useful way of writing the Boltzmann equation was recalled in the introduction, where the Boltzmann equation (5.7) was rewritten resorting to the weak form (5.8). This corresponds to writing, for any given smooth function φ, the spatially homogeneous Boltzmann equation in the form

$$\frac{d}{dt}\int_{\mathbb{R}_+^2} \varphi(x, y) f(x, y, t)\, dx\, dy = \int_{\mathbb{R}_+^2} \varphi(x, y) Q(f, f)(x, y)\, dx dy. \tag{7.10}$$

The right-hand side of equation (7.10) describes the change of density due to collisions of type (7.9). Given two agents with quantities of goods given by (x, y) and (x_1, y_1), one obtains from (7.9) the post-trade quantities

$$\begin{aligned}
x^* &= x + (\lambda\beta + \eta)\left(\frac{x + x_1}{y + y_1} y - x\right), \\
y^* &= y + (\lambda\alpha + \tilde{\eta})\left(\frac{y + y_1}{x + x_1} x - y\right).
\end{aligned} \tag{7.11}$$

Hence,

$$\int_{\mathbb{R}^2_+} \varphi(x,y)Q(f,f)(x,y)\,dxdy = \\ \left\langle \int_{\mathbb{R}^4_+} (\varphi(x^*,y^*) - \varphi(x,y)) f(x,y,t) f(x_1,y_1,t)\,dx\,dy\,dx_1\,dy_1 \right\rangle. \tag{7.12}$$

Alternatively, one can use the identity

$$\int_{\mathbb{R}^2_+} \varphi(x,y)Q(f,f)\,dxdy = \frac{1}{2}\left\langle \int_{\mathbb{R}^4_+} (\varphi(x^*,y^* + \varphi(x_1^*,y_1^*) \right. \\ \left. -\varphi(x,y) - \varphi(x_1,y_1)) f(x,y,t) f(x_1,y_1,t)\,dx\,dy\,dx_1\,dy_1 \right\rangle. \tag{7.13}$$

By choosing $\varphi(x,y) = x$ (respectively $\varphi(x,y) = y$), one verifies that the mean values

$$m_x(t) = \int_{\mathbb{R}^2_+} x f(x,y,t)\,dx\,dy, \\ m_y(t) = \int_{\mathbb{R}^2_+} y f(x,y,t)\,dx\,dy, \tag{7.14}$$

remain constant in time. $m_x(t) = m_x(0)$ (respectively $m_y(t) = m_y(0)$). These conservations are a consequence of the constraints of conservation of the quantities of goods in the trades. Since the exchanges of goods of type (7.11) are nonlinear, the study of the properties of the solution to the Boltzmann equation (7.10) is difficult. In particular, it is cumbersome to find the evolution of the higher moments of the solution in a closed form.

In order to derive a more treatable model, which maintains most of the features of the nonlinear one described by equation (7.10), let us study in detail the exchange rule driven by the Edgeworth box, assuming that, instead of single binary trades, any agent is allowed to exchange goods with part of the agent market, composed of a certain number of agents. We note that the exchange rule (7.9) for the quantity of the first type of goods can be fruitfully rewritten as

$$x_A^* = x_A + (\lambda\beta + \eta)\left(\frac{x_A + x_B}{y_A + y_B} y_A - x_A\right) = x_A + (\lambda\beta + \eta)\left(\frac{\frac{x_A+x_B}{2}}{\frac{y_A+y_B}{2}} y_A - x_A\right). \tag{7.15}$$

This clarifies that, in the post-trade quantity of goods in (7.15), the coefficient of y_A is the ratio between the mean values of the quantities x and y of the two agents exchanging goods.

If one agent exchanges goods, according to the Edgeworth box strategy, with N agents, the quantity of the first type of goods changes according to

$$x_A^* = x_A + (\lambda\beta + \eta)\left(\frac{x_A + x_1 + \cdots + x_N}{y_A + y_1 + \cdots + y_N} y_A - x_A\right),$$

and, if N is sufficiently large,

$$\frac{x_A + x_1 + \cdots + x_N}{y_A + y_1 + \cdots + y_N} = \frac{\frac{x_A + x_1 + \cdots + x_N}{N+1}}{\frac{y_A + y_1 + \cdots + y_N}{N+1}} \cong \frac{m_x}{m_y}. \tag{7.16}$$

Since the ratio between the mean values is a conserved quantity, by using approximation (7.16) we obtain a consistent linear version of the trade of goods, clearly valid under the assumptions specified above. Therefore, by allowing agents to interact with part of the market according to the rules of the Edgeworth box, the result of the trade can be well approximated by the linear exchange rules

$$\begin{aligned} x^* &= x + (\lambda\beta + \eta)\left(\frac{m_x}{m_y} y - x\right), \\ y^* &= y + (\lambda\alpha + \tilde{\eta})\left(\frac{m_y}{m_x} x - y\right), \end{aligned} \tag{7.17}$$

in which the action of the market appears only through the mean values m_x and m_y. Consequently, if $f(x, y, t)$ denotes the density of agents with quantities x and y of the two goods at time $t \geq 0$, $f(x, y, t)$ now satisfies the linear Boltzmann equation

$$\frac{d}{dt} \int_{\mathbb{R}_+^2} \varphi(x, y) f(t) dx dy = \left\langle \int_{\mathbb{R}_+^2} (\varphi(x^*, y^*) - \varphi(x, y)) f(t) dx dy \right\rangle. \tag{7.18}$$

Also, it can be easily verified for equation (7.18) that the mean values (7.14) remain constant in time.

Due to its linearity, the Boltzmann equation (7.18) can be studied by resorting to Fourier transforms, by standard methods available for the Boltzmann equation for Maxwell molecules, and their extensions to kinetic models of wealth distribution [133, 229]. Among other properties, the analysis of [229] in terms of Fourier-based metrics allows us to prove the existence (and uniqueness) of a steady distribution $f_\infty(x, y)$, under a few reasonable conditions on the initial distribution of goods, like existence of a certain number of principal moments.

A quite useful expression for the Fourier-transformed equation follows by writing the linear interaction (7.17) as

$$\begin{aligned} m_y x^* &= m_y x + (\lambda\beta + \eta)(m_x y - m_y x), \\ m_x y^* &= m_x y + (\lambda\alpha + \tilde{\eta})(m_y x - m_x y). \end{aligned} \tag{7.19}$$

The trade (7.19) corresponds to a dissipative interaction of particles with a random addition proportional to the relative velocity, similar to the one studied in [74]. Then the pair $(x' = m_y x, y' = m_x y)$ can be fruitfully used as new set of variables, and $f(x', y', t)$ still satisfies the Boltzmann equation (7.18). By choosing

$$\varphi(x, y) = \exp\{-i(x\xi + y\zeta)\},$$

and using (7.19), one directly obtains from (7.18) the equation for the Fourier transform $\hat{f}$ of f, that takes the form

$$\frac{\partial \hat{f}(\xi, \zeta)}{\partial t} = \langle \hat{f}(\xi^*, \zeta^*) \rangle - \hat{f}(\xi, \zeta). \tag{7.20}$$

In (7.20), the post-interaction variables are given by

$$\begin{aligned} \xi^* &= (1 - (\lambda\beta + \eta))\xi + (\lambda\alpha + \tilde{\eta})\zeta, \\ \zeta^* &= (\lambda\beta + \eta)\xi + (1 - (\lambda\alpha + \tilde{\eta}))\zeta. \end{aligned} \tag{7.21}$$

Note that, by conditions (7.4), the random coefficients in (7.21) are non-negative. It follows immediately that the post-trade Fourier variables satisfy the constraint

$$\xi^* + \zeta^* = \xi + \zeta. \tag{7.22}$$

This implies that the eventual stationary solutions are of the form

$$\hat{f}_\infty(\xi, \zeta) = \hat{f}_\infty(\xi + \zeta). \tag{7.23}$$

Hence, the solution to the Boltzmann equation (7.18) will concentrate for large times along the line $m_x y = m_y x$. Indeed, any generalized function on the plane x, y concentrated on the line $x = y$ can be written in the form

$$F(x, y) = \phi(x + y)\delta_0(x - y),$$

where, as usual, $\delta(x)$ represents a Dirac delta function concentrated at $x = 0$, and therefore has a Fourier transform equal to $\frac{1}{2}\widehat{\phi}\left(\frac{1}{2}(\xi + \zeta)\right)$.

To verify that this behaviour holds true, one resorts to the standard Fourier-based norm, introduced in Chapter 2 in dimension one of the Fourier space by (2.17):

$$d_s(f, g) = \sup_{\xi, \zeta \neq 0} \frac{|\hat{f}(\xi, \zeta) - \hat{g}(\xi, \zeta)|}{(|\xi|^2 + |\zeta|^2)^{s/2}}, \qquad s > 0. \tag{7.24}$$

Uniqueness follows by an argument similar to that of Chapter 2, Section 2.2.2. Then, for any fixed time $T > 0$ one obtains

$$d_s(f_1(t), f_2(t)) \le d_s(f_{0,1}, f_{0,2}) \exp\left\{(C_s - 1)\, t\right\}, \tag{7.25}$$

where

$$C_s = \sup_{\xi, \zeta \neq 0} \left\langle \frac{(|\xi^*|^2 + |\zeta^*|^2)^{s/2}}{(|\xi|^2 + |\zeta|^2)^{s/2}} \right\rangle < +\infty. \tag{7.26}$$

The case in which $C_s < 1$ would lead to better decay. In fact, if this is the case, for all times $t \ge 0$ the d_s distance of f_1 and f_2 decays exponentially in time, and, proceeding as in Chapter 5, Section 5.3.2, it can be proven that, if $f(x, y, t)$ is a weak solution of the Boltzmann equation (7.18), which has initially finite moments up to order 2, then f converges exponentially fast in d_s to a steady state f_∞. In addition, f_∞ has the same mean values, given by (7.14), as the initial datum, and it is the only steady state with these averages.

Verifying that the constant C_s can be strictly less than one is not, in general, direct. However, the following argument shows that we can easily prove that the d_s norm is at least non-increasing. Let us set

$$A = \lambda\beta + \eta, \qquad B = \lambda\alpha + \tilde{\eta}.$$

From condition (7.4), both A and B are strictly positive. It is clear that

$$A\xi^* - B\zeta^* = (1 - A - B)(A\xi - B\zeta). \tag{7.27}$$

Therefore, if the random variables η and $\tilde{\eta}$, in addition to (7.4), satisfy the additional constraint

$$\langle |1 - A - B|^s \rangle = \langle |1 - \lambda - \eta - \tilde{\eta}|^s \rangle < 1, \tag{7.28}$$

for some positive s, it holds that

$$C_s = \left\langle \frac{(|\xi^* + \zeta^*|^2 + |A\xi^* - B\zeta^*|^2)^s}{(|\xi + \zeta|^2 + |A\xi - B\zeta|^2)^s} \right\rangle \le 1.$$

Hence, by changing to the new variables

$$\xi_1 = \xi + \zeta, \qquad \zeta_1 = A\xi - B\zeta,$$

one can prove that, with respect to this set of variables, $d_s(f, g)$ is not increasing for all $s > 0$. This argument implies the global existence of a unique solution to the Boltzmann equation (7.18), but it is not conclusive with respect to the convergence of the global solution towards a unique (in terms of the initial density) steady state. Also, it is not clear if, along the line in which the measure steady solution is concentrated, there is a behaviour at infinity with fat tails. This is related to the possibility that, under the constraint of conservation of the total number of goods in a binary exchange, there is the possibility of forming a class of rich agents.

Instead of working directly on the Boltzmann equation (7.18), in the next section we will deal with the derivation of a Fokker–Planck equation which, while keeping the main properties of the Boltzmann equation, will be easier to treat in connection with the aforementioned question about tails.

7.2.3 A Fokker–Planck model

As illustrated many times in this book, suitable asymptotics of the kinetic equation of Boltzmann type allows us to recover Fokker–Planck-type equations, which are relatively easier to study. Starting from (7.19), let us define

$$v = m_y x + m_x y, \qquad w = m_y x - m_x y. \tag{7.29}$$

Note that $v \in \mathbb{R}_+$, while $w \in \mathbb{R}$. In addition, by construction $|w| \le v$. With respect to the new variables (v, w), the trade induced by the Edgeworth box reads

$$\begin{aligned} v^* &= v + [\lambda(\alpha - \beta) + \tilde{\eta} - \eta]w, \\ w^* &= (1 - \lambda + \tilde{\eta} + \eta)\, w. \end{aligned} \tag{7.30}$$

These new variables better enlighten the outcome of the exchange and the consequences we described by Fourier transform methods in Section 7.2.2. In the absence of randomness, the interaction is dissipative in the w variable, since

$$|w^*| = (1-\lambda)|w|.$$

The same property remains true if the random variables η and $\tilde{\eta}$ are such that, for a given λ:

$$\left\langle (1-\lambda+\tilde{\eta}+\eta)^2 \right\rangle = (1-\lambda)^2 + \left\langle (\tilde{\eta}+\eta)^2 \right\rangle < 1. \tag{7.31}$$

Note that condition (7.31) is the analogue of condition (7.28), which relates the smallness of the support of the random variables to the size of λ. If (7.31) holds true, in view of the dissipation, the solution to the linear Boltzmann equation (7.18) will concentrate in time on the line $m_y x = m_x y$.

As outlined at the end of Section 7.2.2, the interesting question to answer is to describe the equilibrium profile in the v variable to understand if the action of interaction (7.30) could produce Pareto tails or not. Since the new variables (v, w) seem more suitable for describing the problem, let us set $g(v, w, t) = f(x, y, t)$. The condition $|w| \le v$ implies that the new density vanishes outside the allowed set of values, so that

$$g(v, w, t) = 0, \quad \text{if } |w| > v. \tag{7.32}$$

On the set $|w| \le v$, g satisfies the linear Boltzmann equation

$$\frac{d}{dt}\int \varphi(v, w) g(t)\, dv\, dw = \left\langle \int (\varphi(v^*, w^*) - \varphi(v, w)) g(t)\, dv\, dw \right\rangle. \tag{7.33}$$

In the rest of the section, let us assume that the random variables η and $\tilde{\eta}$ are identically distributed, with zero mean and variance σ^2. Thanks to conditions (7.4), the random variables also possess bounded moments of any order. Let us expand the smooth function $\varphi(v^*, w^*)$ in a Taylor series up to order two. We obtain

$$\begin{aligned}
&\langle \varphi(v^*, w^*) - \varphi(v, w) \rangle = \\
&\lambda \left[(\alpha-\beta) w \frac{\partial \varphi}{\partial v} - w \frac{\partial \varphi}{\partial w} \right] + \frac{1}{2}\sigma_1^2 w^2 \frac{\partial^2 \varphi}{\partial v^2} + \frac{1}{2}\sigma_2^2 w^2 \frac{\partial^2 \varphi}{\partial w^2} \\
&+ \frac{1}{2}\lambda^2 \left[(\alpha-\beta)^2 w^2 \frac{\partial^2 \varphi}{\partial v^2} + w^2 \frac{\partial^2 \varphi}{\partial w^2} - 2(\alpha-\beta) w^2 \frac{\partial^2 \varphi}{\partial v \partial w} \right] + R(v, w).
\end{aligned}$$

Clearly, $R(v, w)$ denotes the remainder of the Taylor expansion.

Let us set $\tau = \varepsilon t$, and let us consider the situation in which $\lambda \to \varepsilon\lambda$, while $\eta \to \sqrt{\varepsilon}\eta$ and $\tilde{\eta} \to \sqrt{\varepsilon}\tilde{\eta}$. Then $h_\varepsilon(v, w, \tau) = g(v, w, t)$ satisfies

$$\begin{aligned}
&\frac{d}{d\tau}\int \varphi(v, w) h_\varepsilon(v, w, \tau)\, dv\, dw = \\
&\int \left[\lambda(\alpha-\beta) w \frac{\partial \varphi}{\partial v} - \lambda w \frac{\partial \varphi}{\partial w} + \frac{1}{2}\sigma_1^2 w^2 \frac{\partial^2 \varphi}{\partial v^2} + \frac{1}{2}\sigma_2^2 w^2 \frac{\partial^2 \varphi}{\partial w^2} \right] h_\varepsilon(\tau)\, dv\, dw \\
&+ \varepsilon \frac{1}{2}\lambda^2 \int \left[(\alpha-\beta)^2 w^2 \frac{\partial^2 \varphi}{\partial v^2} + w^2 \frac{\partial^2 \varphi}{\partial w^2} - 2(\alpha-\beta) w^2 \frac{\partial^2 \varphi}{\partial v \partial w} \right] h_\varepsilon(\tau) + R_\varepsilon(v, w).
\end{aligned}$$

We remark that, by construction, the remainder $R_\varepsilon(v,w)$ depends in a multiplicative way on higher moments of the random variables $\sqrt{\varepsilon}\eta$ and $\sqrt{\varepsilon}\tilde{\eta}$, so that $R_\varepsilon(v,w)/\varepsilon \ll 1$ for $\varepsilon \ll 1$ (cf. the discussion in [105, 308], where similar computations have been done explicitly).

As $\varepsilon \to 0$, $h_\varepsilon(v,w,\tau) \to h(v,w,\tau)$ satisfying

$$\begin{aligned}\frac{d}{d\tau}\int \varphi(v,w)h(v,w,\tau)dvdw = \\ \int \left[\lambda(\alpha-\beta)w\frac{\partial\varphi}{\partial v} - \lambda w\frac{\partial\varphi}{\partial w} + \frac{1}{2}\sigma_1^2 w^2\frac{\partial^2\varphi}{\partial v^2} + \frac{1}{2}\sigma_2^2 w^2\frac{\partial^2\varphi}{\partial w^2}\right]h(\tau)dvdw.\end{aligned} \tag{7.34}$$

Equation (7.34) is the weak form of the Fokker–Planck equation

$$\frac{\partial h}{\partial\tau} = \frac{1}{2}\sigma_1^2 w^2\frac{\partial^2 h}{\partial v^2} + \frac{1}{2}\sigma_2^2\frac{\partial^2(w^2 h)}{\partial w^2} - \lambda(\alpha-\beta)w\frac{\partial h}{\partial v} + \lambda\frac{\partial(wh)}{\partial w}. \tag{7.35}$$

The coefficients in equation (7.35) are given by $\sigma_1^2 = \langle(\eta-\tilde{\eta})^2\rangle$, and $\sigma_2^2 = \langle(\eta+\tilde{\eta})^2\rangle$. Clearly, $\sigma_1^2 = \sigma_2^2 = 2\sigma^2$ if the errors are uncorrelated.

The formal derivation of the Fokker–Planck equation (7.35) can be made rigorous by repeating the analogous computations of [105, 308], which refer to one-dimensional models. The meaning of this derivation is that we allow for small changes in a single trade of goods. Then, in order that a macroscopic change be visible, the system needs to wait a sufficiently long time. We call this procedure the *quasi-invariant trade limit.* It is interesting to note that the balance $\sigma^2/\lambda = C$ is the right one to maintain in the limit equation both the effects of the exchange of goods in terms of the intensity λ, as well as the effects of the randomness (through the variance σ). As explained in [308] for the case of opinion formation, different balances give, in the limit, purely diffusive equations, or purely drift equations.

The Fokker–Planck equation (7.35) is reminiscent of all the parameters which contributed to the exchange mechanism. In particular, it contains the values α and β, with $\alpha+\beta = 1$, which are the exponent in the Cobb–Douglas utility function. It is remarkable that these parameters appear only in terms of their difference, so that the case $\alpha = \beta$ leads to a simplification of the equation. Also, the values of λ and σ^2 are linked to the differential terms of the equation, the former to the drift, and the latter to the diffusion.

Maybe the most important characteristic of the solution to the Fokker–Planck equation is that it is immediately recognizable that it develops fat tails. This property follows by evaluating moments relative to the w variable. Let us set $\varphi(v,w) = |w|^{1+r}$, with $r > 0$. Then, assuming that the solution is rapidly vanishing at infinity, one obtains

$$\frac{d}{d\tau}\int |w|^{1+r}h(v,w,\tau)dvdw = \left(\frac{1}{2}r\sigma_2^2 - \lambda\right)(1+r)\int |w|^{1+r}h(v,w,\tau)dvdw.$$

The sign of the right-hand side determines the large-time behaviour of the solution. If $r \geq 2\lambda/\sigma_2^2$, the moment $m_{1+r} = \int |w|^{1+r}h(v,w,t)dvdw$ blows up as time goes to infinity. On the other hand, in agreement with the behaviour of the Fourier-transformed

Boltzmann equation studied in Section 7.2.2, if $r < 2\lambda/\sigma_2^2$ the solution concentrates along the line $w = 0$, that is $m_y x = m_x y$. Moreover, since $h(v, w, t) = 0$ whenever $w > v$,

$$\int v^{1+r} h(v, w, \tau) dv dw \geq \int |w|^{1+r} h(v, w, \tau) dv dw. \tag{7.36}$$

Hence, if $r \geq 2\lambda/\sigma_2^2$, the principal moment of order r with respect to the v variable also blows up when time goes to infinity, and the stationary solution, which is concentrated along the line $w = 0$, has Pareto tails.

Remark 7.1 *Equation (7.35) is a kinetic equation for the evolution of the probability distribution of two goods among a huge population of agents. In the kinetic description, the leading idea was to describe the trading of these goods by means of some fundamental rules in prize theory, in particular by using Cobb–Douglas utility functions for the binary exchange, and the Edgeworth box for the description of the common exchange area in which utility is increasing for both agents. Also, to take into account the intrinsic risks of the market, there is randomness in the trade, which does not affect the microscopic conservations of the trade, namely the conservation of the total number of each type of goods in the binary exchange.*

Despite the microscopic constraints, sufficiently high moments of the solution blow up with time. This is in contrast with the behaviour of the simplified one-dimensional models for wealth distribution expressed by equation (5.7), consequent to trades of type (5.6), studied in Chapter 5, where the microscopic conservation of wealth in the binary exchange has been shown to imply an exponential decay of the stationary profile at infinity, thus excluding the possibility of Pareto tails.

7.3 Modelling speculative financial markets

One of the aims of kinetic modelling is motivated by the desire to have a more realistic description of the speculative dynamics in multiagent models. An interesting improvement in this direction was obtained in [104], which derived a kinetic description of the behaviour of a simple financial market where a population of homogeneous agents can create their own portfolio between two investment alternatives: a stock and a bond. The model is closely related to the Levy–Levy–Solomon (LLS) microscopic model in finance [207, 208].

In this non-stationary financial market model, the average wealth is not conserved and this produces price variations. Let us point out that, even if the model is linear since no binary interaction dynamics between agents is present, the study of the large-time behaviour is not immediate. In fact, despite conservation of the total number of agents, as happens in the opinion dynamics models studied in Chapter 6, there are no additional conservation equations, and the determination of an explicit form of the asymptotic wealth distribution of the kinetic equation remains difficult and requires the use of suitable numerical methods.

By resorting to a particular asymptotic which maintains the characteristics of the solution to the original problem for large times, one can also prove in this case that the Boltzmann model converges towards a Fokker–Planck-type equation for the distribution of wealth among individuals.

In this case, however, due to the variation in time of the average wealth, no steady solutions exist, and one can only show that the Fokker–Planck equation admits self-similar solutions that can be computed explicitly and which are lognormal distributions.

7.3.1 The Levy–Levy–Solomon model

Let us consider a set of financial agents $i = 1, \ldots, N$ who can create their own portfolio between two alternative investments: a stock and a bond. We denote by w_i the wealth of agent i and by n_i the number of stocks of the agent. Additionally, we use the notations S for the price of the stock and n for the total number of stocks.

The essence of the dynamics is the choice of the agent's portfolio. More precisely, at each time step each agent selects what fraction of wealth to invest in bonds and what fraction in stocks. We indicate with r the (constant) interest rate of bonds. The bond is assumed to be a riskless asset yielding a return at the end of each time period. The stock is a risky asset with overall returns rate x composed of two elements: a capital gain or loss and the distribution of dividends.

To simplify the notation, let us neglect for the moment the effects due to the stochastic nature of the process, the presence of dividends, and so on. Thus, if an agent has invested $\gamma_i w_i$ of its wealth in stocks and $(1-\gamma_i)w_i$ of its wealth in bonds, at the next time step in the dynamics he will achieve the new wealth value

$$w_i' = (1-\gamma_i)w_i(1+r) + \gamma_i w_i(1+x), \tag{7.37}$$

where the rate of return of the stock is given by

$$x = \frac{S'-S}{S}, \tag{7.38}$$

and S' is the new price of the stock.

Since we have the identity

$$\gamma_i w_i = n_i S, \tag{7.39}$$

we can also write

$$\begin{aligned} w_i' &= w_i + w_i(1-\gamma_i)r + w_i\gamma_i\left(\frac{S'-S}{S}\right) && (7.40)\\ &= w_i + (w_i - n_i S)r + n_i(S'-S). && (7.41)\end{aligned}$$

Note that, independently of the number of stocks of the agent at the next time level, it is only the price variation of the stock (which is unknown) that characterizes the gain or loss of the agent on the stock market at this stage.

The dynamics now is based on the agent's choice of the new fraction of wealth he wants to invest in stocks at the next stage. Each investor i is confronted with a decision where the outcome is uncertain: what is the new optimal fraction γ_i' of wealth to invest in stock? According to the standard theory of investment, each investor is characterized by a *utility function* (of its wealth) $U(w)$ that reflects the personal risk taking preference. The optimal γ_i' is the one that maximizes the expected value of $U(w)$.

In Section 7.2.1 we presented some utility functions (the Cobb–Douglas utility functions (7.2)) in the presence of two variables. In the present case, different models can be used for this (see [208, 320]). One can consider a von Neumann–Morgenstern utility function with a constant risk aversion of the type

$$U(w) = \frac{w^{1-\alpha}}{1-\alpha}, \tag{7.42}$$

where $0 < \alpha < 1$ is the risk aversion parameter, or a logarithmic utility function

$$U(w) = \log(w). \tag{7.43}$$

Fig. 7.3 Oskar Morgenstern[3]

As they don't know the future stock price S', the investors estimate the stock's next period return distribution and find an optimal mix of the stock and the bond that maximizes their expected utility $\langle U \rangle$. In practice, for any hypothetical price S^h, each investor finds the hypothetical optimal proportion $\gamma_i^h(S^h)$ which maximizes his/her expected utility evaluated at

$$w_i^h(S^h) = (1-\gamma_i^h)w_i'(1+r) + \gamma_i^h w_i'(1+x'(S^h)), \tag{7.44}$$

where $x'(S^h) = (S^h - S')/S'$ and S' is estimated in some way. For example, in [208] the investors' expectations for x' are based on extrapolating the past values.

Note that, if we assume that all investors have the same risk aversion α in (7.42), then they will have the same proportion of investment in stocks regardless of their wealth, thus $\gamma_i^h(S^h) = \gamma^h(S^h)$.

Once each investor decides on the hypothetical optimal proportion of wealth γ_i^h that he/she wishes to invest in stocks, one can derive the number of stocks $n_i^h(S^h)$ he/she wishes to hold corresponding to each hypothetical stock price S^h. Since the total number of shares in the market, n, is fixed there is a particular value of the price S' for which the sum of the $n_i^h(S^h)$ equals n. This value S' is the new market equilibrium price and the optimal proportion of wealth is $\gamma_i' = \gamma_i^h(S')$.

More precisely, following [208], each agent formulates a *demand curve*

$$n_i^h = n_i^h(S^h) = \frac{\gamma^h(S^h)w_i^h(S^h)}{S^h},$$

characterizing the desired number of stocks as a function of the hypothetical stock price S^h. This number of share demands is a monotonically decreasing function of the hypothetical price S^h. As the total number of stocks

[3]**Oskar Morgenstern**: born on 24 January 1902 in Görlitz (German Empire), died on 26 July 1977 in Princeton (New Jersey, U.S.). In collaboration with mathematician John von Neumann, he wrote *Theory of Games and Economic Behavior* which is recognized as the first book on game theory [239]. Morgenstern is also known for his skepticism about economic measurement (see [240]).

$$n = \sum_{i=1}^{N} n_i \tag{7.45}$$

is preserved, the new price of the stock at the next time level is given by the so-called *market clearance condition*. Thus, the new stock price S' is the unique price at which the total demand equals the supply:

$$\sum_{i=1}^{N} n_i^h(S') = n. \tag{7.46}$$

This will fix the value w' in (7.37) and the model can be advanced to the next time level. To make the model more realistic, typically a source of stochastic noise, which characterizes all factors causing the investor to deviate from his/her optimal portfolio, is introduced in the proportion of investments γ_i and in the rate of return of the stock x'.

7.3.2 Kinetic modelling

As usual, we define $f = f(w,t)$, $w \in \mathbb{R}_+$, $t > 0$, as the distribution of wealth w, which represents the probability of an agent having wealth w. We assume that at time t the proportion of wealth invested is of the form $\gamma(Z) = \mu(S) + Z$, where Z is a random variable in $[-z, z]$, and $z = \min\{-\mu(S), 1 - \mu(S)\}$ is distributed according to some probability density $\Lambda(\mu(S), Z)$ with zero mean and variance ς^2. This probability density characterizes the individual strategy of an agent around the optimal choice $\mu(S)$. We assume Λ to be independent of the wealth of the agent. Here, the optimal demand curve $\mu(\cdot)$ is assumed to be a given monotonically non-increasing function of the price $S \geq 0$ such that $0 < \mu(0) < 1$.

Note that, given $f(w,t)$, the actual stock price S satisfies the demand–supply relation

$$S = \frac{1}{n}\langle \gamma w \rangle, \tag{7.47}$$

and $f(w,t)$ has been normalized

$$\int_0^\infty f(w,t)dw = 1.$$

More precisely, since γ and w are independent, at each time t, the price $S(t)$ satisfies

$$S(t) = \frac{1}{n}\langle \gamma \rangle \langle w \rangle = \frac{1}{n}\mu(S(t))\bar{w}(t), \tag{7.48}$$

with

$$\bar{w}(t) = \langle w \rangle = \int_0^\infty f(w,t)w\,dw, \tag{7.49}$$

being the mean wealth and, by construction,

$$\mu(S) = \int \Lambda(\mu(S), Z)Z\,dZ.$$

At the next round in the market, the new wealth of the investor will depend on the future price S' and the percentage γ of wealth invested according to

$$w'(S', \gamma, \eta) = (1-\gamma)w(1+r) + \gamma w(1 + x(S', \eta)), \tag{7.50}$$

where the expected rate of return of stocks is given by

$$x(S', \eta) = \frac{S' - S + D + \eta}{S}. \tag{7.51}$$

In the above relation, $D \geq 0$ represents a constant dividend paid by the company and η is a random variable distributed according to $\Theta(\eta)$ with zero mean and variance σ^2, which takes into account fluctuations due to price uncertainty and dividends [208,176]. We assume η to take values in $[-d, d]$ with $0 < d \leq S' + D$ so that $w' \geq 0$ and thus negative wealths are not allowed in the model. Note that equation (7.51) requires estimation of the future price S', which is unknown.

The dynamics is then determined by the agent's new fraction of wealth invested in stocks, $\gamma'(Z') = \mu(S') + Z'$, where Z' is a random variable in $[-z', z']$ and $z' = \min\{\mu(S'), 1-\mu(S')\}$ is distributed according to $\Lambda(\mu(S'), Z')$. We have the demand–supply relation

$$S' = \frac{1}{n}\langle \gamma' w' \rangle, \tag{7.52}$$

which permits us to write the following equation for the future price:

$$S' = \frac{1}{n}\langle \gamma' \rangle \langle w' \rangle = \frac{1}{n}\mu(S')\langle w' \rangle. \tag{7.53}$$

Now

$$w'(S', \gamma, \eta) = w(1+r) + \gamma w(x(S', \eta) - r), \tag{7.54}$$

thus

$$\langle w' \rangle = \langle w \rangle (1+r) + \langle \gamma w \rangle (\langle x(S', \eta) \rangle - r) \tag{7.55}$$

$$= \bar{w}(t)(1+r) + \mu(S)\bar{w}(t)\left(\frac{S' - S + D}{S} - r\right). \tag{7.56}$$

This gives the identity

$$S' = \frac{1}{n}\mu(S')\bar{w}(t)\left[(1+r) + \mu(S)\left(\frac{S' - S + D}{S} - r\right)\right]. \tag{7.57}$$

Using equation (7.48) we can eliminate the dependence on the mean wealth and write

$$\begin{aligned} S' &= \frac{\mu(S')}{\mu(S)}\left[(1-\mu(S))S(1+r) + \mu(S)(S' + D)\right] \\ &= \frac{(1-\mu(S))\mu(S')}{(1-\mu(S'))\mu(S)}(1+r)S + \frac{\mu(S')}{1-\mu(S')}D. \end{aligned} \tag{7.58}$$

Equation (7.58) determines implicitly the future value of the stock price. Let us set

$$g(S) = \frac{1-\mu(S)}{\mu(S)}S. \tag{7.59}$$

Then the future price is given by the equation

$$g(S') = g(S)(1 + r) + D \tag{7.60}$$

for a given S. Note that

$$\frac{dg(S)}{dS} = -\frac{d\mu(S)}{dS}\frac{S}{\mu(S)^2} + \frac{1 - \mu(S)}{\mu(S)} > 0,$$

so the function $g(S)$ is strictly increasing with respect to S. This guarantees the existence of a unique solution

$$S' = g^{-1}\left(g(S)(1 + r) + D\right) > S. \tag{7.61}$$

Moreover, if $r = 0$ and $D = 0$, the unique solution is $S' = S$ and the price remains unchanged in time.

For the average stock return, we have

$$\bar{x}(S') - r = \frac{(\mu(S') - \mu(S))(1 + r)}{(1 - \mu(S'))\mu(S)} + \frac{\mu(S')D}{S(1 - \mu(S'))}, \tag{7.62}$$

where

$$\bar{x}(S') = E[x(S', \eta)] = \frac{S' - S + D}{S}. \tag{7.63}$$

Now, the right-hand side of (7.62) has non-constant sign since $\mu(S') \leq \mu(S)$. In particular, the average stock return is above the bonds rate r only if the (negative) rate of variation of the investments is above a certain threshold:

$$\frac{\mu(S') - \mu(S)}{\mu(S)\mu(S')}S \geq -\frac{D}{(1 + r)}.$$

In the constant investment case $\mu(\cdot) = C$, with $C \in (0, 1)$ constant, then we have $g(S) = (1 - C)S/C$ and

$$S' = (1 + r)S + \frac{C}{1 - C}D, \tag{7.64}$$

which corresponds to a dynamics of growth of the prices at rate r. As a consequence, the average stock return is always larger than the constant return of bonds:

$$\bar{x}(S') - r = \frac{D}{S(1 - C)} \geq 0.$$

The previous dynamics originate the linear kinetic equation for the evolution of the wealth distribution, which is fruitfully written in weak form:

$$\frac{d}{dt}\int \varphi(w)f(w, t)dw = \left\langle \int_{\mathbb{R}_+} \beta(w \to w')\left(\varphi(w') - \varphi(w)\right) f(w, t) \right\rangle. \tag{7.65}$$

The above equation takes into account all possible variations that can occur in the distribution of a given wealth w, according to (7.54) and equation (7.61) for the price. In (7.65), the kernel β takes the form

$$\beta(w \to w') = \Lambda(\mu(S), \xi)\Theta(\eta). \tag{7.66}$$

The distribution function $\Lambda(\mu(S), \xi)$, together with the function $\mu(\cdot)$, characterize the behaviour of the agents on the market (more precisely, they characterize the way the agents invest their wealth as a function of the actual price of the stock).

From (7.65) follows the conservation of the total number of investors if $\varphi(w) = 1$. If we set $\varphi(w) = w$ we obtain the time evolution of the average wealth which characterizes the price behaviour. The mean wealth is not conserved since we have

$$\frac{d}{dt}\int_0^\infty f(w,t)w\,dw = \left(r + \mu(S)\left(\frac{S' - S + D}{S} - r\right)\right)\int_0^\infty f(w,t)w\,dw. \tag{7.67}$$

Note that since the sign of the right-hand side is non-negative, the mean wealth is non-decreasing in time. In particular, we can rewrite the equation as

$$\frac{d}{dt}\bar{w}(t) = \left((1 - \mu(S))r + \mu(S)\bar{x}(S')\right)\bar{w}(t). \tag{7.68}$$

From this we get the equation for the price,

$$\frac{d}{dt}S(t) = \frac{\mu(S(t))}{\mu(S(t)) - \dot{\mu}(S(t))S(t)}\left((1 - \mu(S(t)))r + \mu(S(t))\bar{x}(S'(t))\right)S(t), \tag{7.69}$$

where S' is given by (7.58) and

$$\dot{\mu}(S) = \frac{d\mu(S)}{dS} \leq 0.$$

Now, since from (7.62) it follows by the monotonicity of μ that

$$\bar{x}(S') \leq M {=} r + \frac{D}{S(0)(1 - \mu(S(0)))},$$

using (7.68) we have the bound

$$\bar{w}(t) \leq \bar{w}(0)\exp{(Mt)}. \tag{7.70}$$

From (7.48) we immediately obtain

$$\frac{S(t)}{\mu(S(t))} \leq \frac{S(0)}{\mu(S(0))}\exp{(Mt)},$$

which gives

$$S(t) \leq S(0)\exp{(Mt)}. \tag{7.71}$$

For a constant $\mu(\cdot) = C$, $C \in (0, 1)$, we have the explicit expression for the growth of the wealth (and consequently of the price):

$$\bar{w}(t) = \bar{w}(0)\exp(rt) - (1 - \exp(rt))\frac{nD}{1 - C}. \tag{7.72}$$

7.3.3 Fokker–Planck asymptotics and wealth distribution

As usual, it is difficult to study in detail the large-time behaviour of the system. We can therefore apply a quasi-invariant limit technique described in the previous chapters to derive simplified models whose behaviour is easier to analyze. Here we consider the limit of large times in which the market originates a very small exchange of wealth (small rates of return r and x).

In order to study the asymptotic behaviour of the distribution function $f(w,t)$, we set

$$\tau = rt, \quad \tilde{f}(w,\tau) = f(w,t), \quad \tilde{S}(\tau) = S(t), \quad \tilde{\mu}(\tilde{S}) = \mu(S),$$

which implies that $\tilde{f}(w,\tau)$ satisfies the weak form of the kinetic equation,

$$\begin{aligned} &\frac{d}{d\tau}\int_0^\infty \tilde{f}(w,\tau)\phi(w)dw \\ &= \frac{1}{r}\int_0^\infty \int_{-d}^{d}\int_{-z}^{z} \Lambda(\tilde{\mu}(\tilde{S}),\xi)\Theta(\eta)\tilde{f}(w,\tau)(\phi(w')-\phi(w))d\xi\, d\eta\, dw, \end{aligned} \tag{7.73}$$

and consider a second-order Taylor expansion of ϕ around w,

$$\phi(w')-\phi(w) = w(r+\gamma(x(S',\eta)-r))\phi'(w) + \frac{1}{2}w^2(r+\gamma(x(S',\eta)-r))^2\phi''(\tilde{w}),$$

where, for some $0 \le \vartheta \le 1$,

$$\tilde{w} = \vartheta w' + (1-\vartheta)w.$$

Now we insert this expansion into the right-hand side, and compute the limit of very small values of the constant rate r. In order for such a limit to make sense and preserve the characteristics of the model, we must assume that

$$\lim_{r\to 0}\frac{\sigma^2}{r} = \nu, \quad \lim_{r\to 0}\frac{D}{r} = \lambda. \tag{7.74}$$

Note that the above limits in (7.61) immediately imply that

$$\lim_{r\to 0}\tilde{S}' = \tilde{S}. \tag{7.75}$$

Omitting the details of the computations, sending $r \to 0$ under these assumptions, we obtain the weak form

$$\begin{aligned} &\frac{d}{d\tau}\int_0^\infty \tilde{f}(w,\tau)\varphi(w)dw \\ &= \left(1+\tilde{\mu}(\tilde{S})\left((\kappa(\tilde{S})-1)+\frac{\tilde{\mu}(\tilde{S})(\kappa(\tilde{S})-1)+1}{1-\tilde{\mu}(\tilde{S})}\frac{\lambda}{\tilde{S}}\right)\right)\int_0^\infty \tilde{f}(w,\tau)w\varphi'(w)\,dw \\ &\quad +\frac{1}{2}\frac{(\tilde{\mu}(\tilde{S})^2+\zeta^2)}{\tilde{S}^2}\nu\int_0^\infty \tilde{f}(w,\tau)w^2\varphi''(w)\,dw, \end{aligned}$$

with

$$0 < \kappa(\tilde{S}) = \frac{\tilde{\mu}(\tilde{S})(1-\tilde{\mu}(\tilde{S}))}{\tilde{\mu}(\tilde{S})(1-\tilde{\mu}(\tilde{S}))-\tilde{S}\dot{\tilde{\mu}}(\tilde{S})} \le 1, \quad \dot{\tilde{\mu}}(\tilde{S}) = \frac{d\tilde{\mu}(\tilde{S})}{d\tilde{S}} \le 0. \tag{7.76}$$

This corresponds to the Fokker–Planck equation

$$\frac{\partial}{\partial \tau}\tilde{f} = \frac{\partial}{\partial w}\left[-A(\tau)w\tilde{f} + \frac{1}{2}B(\tau)\frac{\partial}{\partial w}w^2\tilde{f}\right], \tag{7.77}$$

with

$$A(\tau) = 1 + \tilde{\mu}(\tilde{S})\left((\kappa(\tilde{S}) - 1) + \frac{\tilde{\mu}(\tilde{S})(\kappa(\tilde{S}) - 1) + 1}{1 - \tilde{\mu}(\tilde{S})}\frac{\lambda}{\tilde{S}}\right) \tag{7.78}$$

$$B(\tau) = \frac{(\tilde{\mu}(\tilde{S})^2 + \zeta^2)}{\tilde{S}^2}\nu. \tag{7.79}$$

In order to search for self-similar solutions, we consider the scaling

$$\tilde{f}(w,\tau) = \frac{1}{w}\tilde{g}(\chi,\tau), \quad \chi = \log(w). \tag{7.80}$$

Simple computations show that $\tilde{g}(\chi,\tau)$ satisfies the linear convection–diffusion equation

$$\frac{\partial}{\partial \tau}\tilde{g}(\chi,\tau) = \left(\frac{B(\tau)}{2} - A(\tau)\right)\frac{\partial}{\partial \chi}\tilde{g}(\chi,\tau) + \frac{B(\tau)}{2}\frac{\partial^2}{\partial \chi^2}\tilde{g}(\chi,\tau),$$

which admits the self-similar solution

$$\tilde{g}(\chi,\tau) = \frac{1}{(2b(\tau)\pi)^{1/2}}\exp\left(-\frac{(\chi + b(\tau)/2 - a(\tau))^2}{2b(\tau)}\right), \tag{7.81}$$

where

$$a(\tau) = \int_0^\tau A(s)\,ds + C_1, \qquad b(\tau) = \int_0^\tau B(s)\,ds + C_2.$$

Reverting to the original variables, we obtain the lognormal asymptotic behaviour of the model,

$$\tilde{f}(w,\tau) = \frac{1}{w(2b(\tau)\pi)^{1/2}}\exp\left(-\frac{(\log(w) + b(\tau)/2 - a(\tau))^2}{2b(\tau)}\right). \tag{7.82}$$

The constants $C_1 = a(0)$ and $C_2 = b(0)$ can be determined from the initial data at $t = 0$. If we denote by $\bar{w}(0)$ and $\bar{e}(0)$ the initial values of the first two central moments, we get

$$C_1 = \log(\bar{w}(0)), \qquad C_2 = \log\left(\frac{\bar{e}(0)}{(\bar{w}(0))^2}\right).$$

7.3.4 A numerical example

In this section we report the results of a simple numerical simulation of the proposed kinetic equations. In the test case, we use $N = 1000$ agents and $n = 10000$ shares. Initially, each investor has a total wealth of 1000 composed of 10 shares, at a value of 50 per share, and 500 in bonds. The random variables ξ and η are assumed distributed according to truncated normal distributions so that negative wealth values are avoided

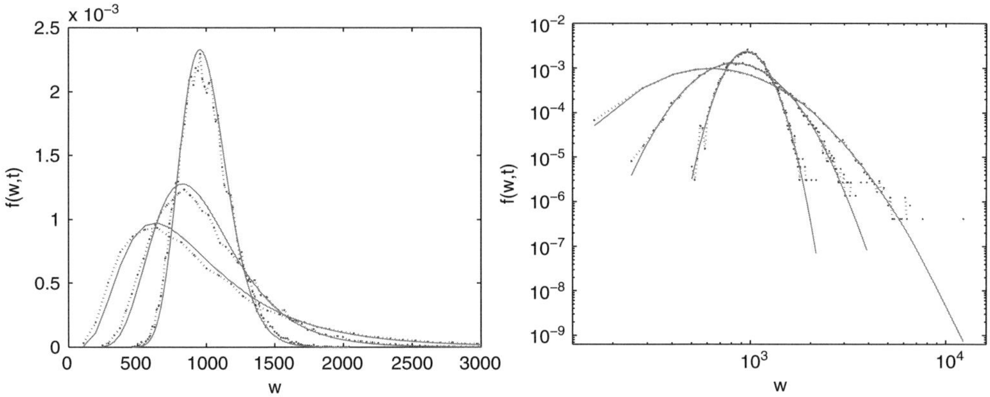

Fig. 7.4 Distribution function at $t = 50, 200, 500$. The continuous line is the lognormal Fokker–Planck solution. The right plot is in log-log scale.

(no borrowing and no short selling). In the test case we consider the time-averaged Monte Carlo asymptotic behaviour of the kinetic model and compare its numerical self-similar solution with the explicit one computed in the last section using the Fokker–Planck model.

To this end, we consider the self-similar scaling (7.80) and compute the solution for the values $r = 0.001$, $D = 0.0015$ with ξ and $\eta/S(0)$ distributed with standard deviation 0.05. We report the numerical solution for a constant value of $\mu = 0.5$ at different times $t = 50, 200, 500$ in Figure 7.4. A very good agreement between the Boltzmann and the lognormal Fokker–Planck solutions is observed, as expected from the results of the last section. We also compute the corresponding Lorentz curve $L(F(w,t))$ defined as

$$L(F(w,t)) = \frac{\displaystyle\int_0^w f(v,t)v\,dv}{\displaystyle\int_0^\infty f(v,t)v\,dv}, \qquad F(w,t) = \int_0^w f(v,t)\,dv,$$

and the Gini coefficient $G \in [0,1]$,

$$G = 1 - 2\int_0^1 L(F(w,t))\,dw.$$

The Gini coefficient is a measure of the inequality in the wealth distribution. A value of 0 corresponds to the line of perfect equality depicted in Figure 7.5 together with the different Lorentz curves. It is clear that inequalities grow in time due to the speculative dynamics.

7.4 A model for different groups of traders

The description of a market in terms of binary trades among agents constituted the basis of kinetic modelling introduced in Chapter 5. Some attempts at more realistic

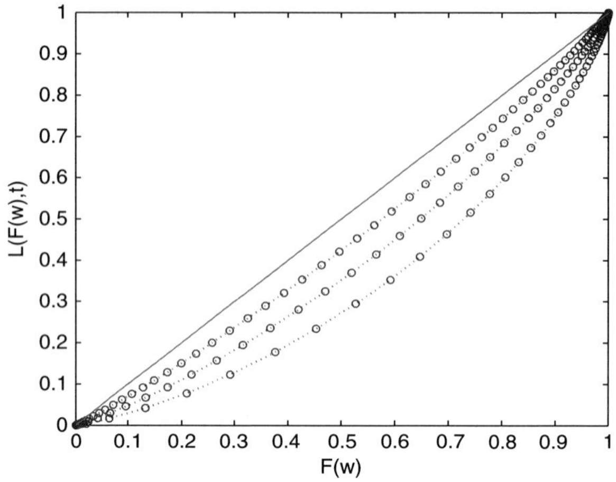

Fig. 7.5 Test 3. The corresponding Lorentz curves. The Gini coefficients are $G = 0.1$, $G = 0.2$ and $G = 0.3$ respectively.

descriptions of these phenomena have been presented in this chapter. With few exceptions, like the one discussed in Chapter 5, Section 5.6, the unifying aspect of the totality of models considered so far is that the population of agents has a common behaviour with respect to trading. On the other hand, it is clear that different people exhibit different reactions with respect to many facts of real life, including aspects related to wealth. Very recently, this aspect of the market has been considered from the kinetic modelling point of view in [222]. There, a Boltzmann-like model for a speculative market characterized by a single stock and a socio-economical interplay between two different types of traders has been developed and studied both from the qualitative and quantitative points of view. The model is strictly related to the microscopic Lux–Marchesi model for two categories of agents [215–217] and to kinetic models of opinion formation recently introduced in Chapter 6, Section 6.2. The model in [222] consists of a system of two kinetic equations for both the categories of agents, fundamentalists and chartists, and it is particularly rich, also taking into account psychological and behavioural components of the agents, like the way they interact with each other and perceive the risk. This is done by means of a suitable *value function* in agreement with the Prospect Theory by Kahneman and Tversky [188, 189]. The hypothesis, derived from this theory, implies an instantaneous reaction of the agents at a given time t, based on the ratio between the price and its derivative at time t.

7.4.1 The Lux–Marchesi model

The aim of the pioneering Lux–Marchesi model [216, 217] was to show that the scaling laws that are observed in financial markets can arise from mutual interaction between agents. In particular, Lux and Marchesi stressed the clear difference between the statistical properties of the model input, i.e. the normal noise that makes the stock fundamental price evolve in time, and the output, that is the price dynamics produced by the operations between agents.

In [216, 217], the agents are divided into two categories: the fundamentalists, who believe in the existence of a reference price S_F for the traded stocks (they sell stocks if the price $S(t) > S_F$ and they buy if $S(t) < S_F$), and the chartists, who are noise traders whose behaviour is dictated by herding and historical prices. While the total number of agents N remains fixed in time, the numbers of fundamentalists and chartists, n_F and n_C, are allowed to vary. This is done by assuming that at each time step each agent can change his category with a given transition probability. In this way, an internal dynamic between the two classes of agents is established.

The effect of the two classes of agents on the price is very different. While fundamentalists have a stabilizing effect on the market, as their operations drive the price towards the reference price, chartists have a destabilizing effect and can create bubbles and crashes. Furthermore, chartist agents are divided in two subcategories: optimists, who believe that the price will rise and hence always buy stocks, and pessimists, who believe that the price will decrease and so, on the contrary, always sell stocks. At each time step each agent can change his subcategory with a given transition probability.

In [222], the main features of the Lux–Marchesi model have been merged into the framework of kinetic theory, with the goal of introducing a kinetic description both for the behaviour of the microscopic agents and for the price, and then to exploit the tools provided by kinetic theory to get more insight into the way the microscopic dynamics of each trading agent can influence the evolution of the price. The main novelty in [222] was to introduce a mechanism of opinion formation for chartists, suitable for describing, for a given price $S(t)$ and price derivative $\dot{S}(t) = dS(t)/dt$, the microscopic dynamics of the investment propensity of chartists. The binary mechanism at the basis of this opinion formation was inspired by the opinion formation model presented in Chapter 6, Section 6.2. A remarkable feature of the proposed relations for the changes of opinion of chartists was the presence of the normalized value function $\Phi(\dot{S}(t)/S(t))$ in $[-1, 1]$ in the sense of Kahneman and Tversky that models the reaction of individuals towards potential gains and losses in the market [188, 189]. This permits the introduction of behavioural aspects in the market dynamics and consideration of the influence of psychology on the behaviour of financial practitioners.

7.4.2 A kinetic model for multiple agent interactions

The model of [222] describes a simple financial market characterized by a single stock or type of goods, and an interplay between two different trader populations, chartists and fundamentalists, which determine the price dynamics of such stock (type of goods). The goal was to introduce a kinetic description both for the behaviour of the microscopic agents and for the price, and then to exploit the tools provided by kinetic theory to get more insight about the way the microscopic dynamics of each trading agent can influence the evolution of the price, and be responsible for the appearance of *stylized* facts like *fat tails* and *lognormal* behaviour.

Similarly to the Lux–Marchesi model [216,217], the starting point is a population of two different kinds of trader, chartists and fundamentalists. Chartists are characterized by their number density ρ_C and the investment propensity (or opinion index) y of a single agent, whereas fundamentalists appear only through their number density ρ_F.

The value $\rho = \rho_F + \rho_C$ is invariant in time so that the total number of agents remains constant. We will assume for simplicity that $\rho = 1$.

Let us define $f(y,t)$, $y \in [-1,1]$, the distribution function of chartists with investment propensity y at time t. Positive values of y represent buyers, negative values characterize sellers and close to $y = 0$ we have undecided agents. Clearly,

$$\rho_C(t) = \int_{-1}^{1} f(y,t)\,dy.$$

Moreover, we define the mean investment propensity

$$Y(t) = \frac{1}{\rho_C(t)} \int_{-1}^{1} f(y,t) y\,dy. \tag{7.83}$$

For a given price $S(t)$ and price derivative $\dot{S}(t) = dS(t)/dt$, the microscopic dynamics of the investment propensity of chartists is characterized by a binary interaction, $(y, y_*) \to (y', y_*')$, defined by

$$\begin{aligned} y' &= (1 - \alpha_1 H(y) - \alpha_2) y + \alpha_1 H(y) y_* + \alpha_2 \Phi\left(\frac{\dot{S}(t)}{S(t)}\right) + D(y)\eta, \\ y_*' &= (1 - \alpha_1 H(y_*) - \alpha_2) y_* + \alpha_1 H(y_*) y + \alpha_2 \Phi\left(\frac{\dot{S}(t)}{S(t)}\right) + D(y_*)\eta_*. \end{aligned} \tag{7.84}$$

These interaction rules are reminiscent of the binary interactions between agents which measure the opinion change we introduced in Chapter 6, Section 6.1. Here, $\alpha_1 \in [0,1]$ and $\alpha_2 \in [0,1]$, with $\alpha_1 + \alpha_2 \leq 1$, measure the importance the individuals place on others' opinions and the actual price trend in forming expectations about future price changes. The random variables η and η_* are distributed accordingly, with zero mean and variance σ^2 and measure individual deviations from the average behaviour. The function $H(y) \in [0,1]$ is taken as concave and symmetric on the interval I, and characterizes the herding behaviour, whereas $D(y) \geq 0$ defines the diffusive behaviour, and will be also taken concave and symmetric on I. Simple examples of the herding and diffusion functions are given by

$$H(y) = a + b(1 - |y|), \qquad D(y) = (1 - y^2)^{\gamma},$$

with $0 \leq a + b \leq 1$, $a \geq 0, b > 0$, $\gamma > 0$. Here, a measures the herding influence on persuaded buyers and sellers and $a + b$ the influence on undecided agents. The parameter γ weights the concavity of D and thus how the amount of randomness decreases when moving from an undecided to a determined trader.

Other choices of concave functions in $[-1,1]$ are of course possible. Both functions should take into account that extremal positions $|y| \approx 1$ suffer less herding and fluctuations compared to indecisive positions $|y| \approx 0$. If symmetry is broken then the influence of herding and fluctuations is different according to whether you are a buyer or a seller. Note that in order to preserve the bounds for y it is essential that $D(y)$ vanishes in $y = \pm 1$.

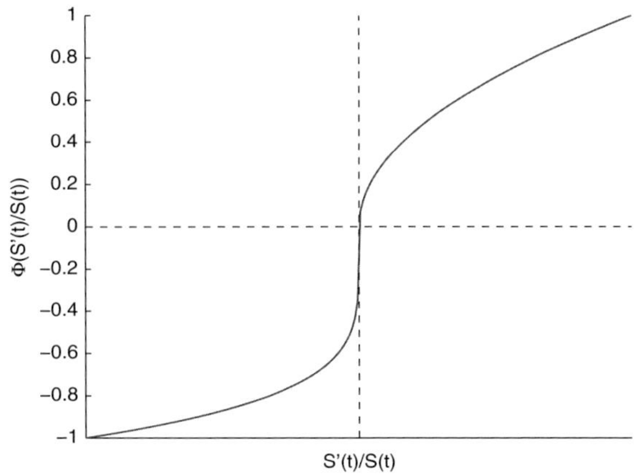

Fig. 7.6 An example of value function $\Phi(\dot{S}(t)/S(t))$.

For $b = 0$, $H(y)$ is constant, no herding effect is present and the mean investment propensity is preserved when the market influence is neglected ($\alpha_2 = 0$), as in the classical opinion models described in Chapter 6, Section 6.1. A remarkable feature of the above relations is the presence of the normalized value function $\Phi(\dot{S}(t)/S(t))$ in $[-1, 1]$ in the sense of Kahneman and Tversky [188, 189], which models the reaction of individuals towards potential gains and losses in the market [188]. This permits the introduction of behavioural aspects in the market dynamics and consideration of the influence of psychology on the behaviour of financial practitioners.

The value function is defined on deviations from a reference point, which is usually assumed equal to zero (but can also be considered positive or negative), and is normally concave for gains (implying risk aversion), commonly convex for losses (risk seeking) and is generally steeper for losses than for gains (loss aversion), see Figure 7.6.

It is worth observing that, for a given $D(y) \geq 0$, a suitable choice of the support of the random variable η avoids the dependence of the collisional kernel $B(y, y_*)$ on the variables y, y_*.
In fact, we have

$$
\begin{aligned}
y' &= (1 - \alpha_1 H(y) - \alpha_2)y + \alpha_1 H(y) y_* + \alpha_2 \Phi\left(\frac{\dot{S(t)}}{S(t)}\right) + D(y)\eta \\
&\leq (1 - \alpha_1 H(y) - \alpha_2)y + \alpha_1 H(y) + \alpha_2 + D(y)\eta.
\end{aligned}
$$

Then, to obtain $y' \leq 1$ for any $y \in [-1, 1]$, we have to choose η such that

$$
D(y)\eta \leq (1 - \alpha_1 - \alpha_2)(1 - y),
$$

which gives

$$
\eta \leq m_C(1 - \alpha_1 - \alpha_2), \quad m_C = \min_{y \in [-1,1]} \left\{ \frac{(1-y)}{D(y)}, D(y) \neq 0 \right\}.
$$

Analogously, we can ensure $y' \geq -1$ and thus it is enough to take

$$\eta \in [-m_C(1-\alpha_1-\alpha_2), m_C(1-\alpha_1-\alpha_2)]. \tag{7.85}$$

For example, if $D(y) = 1 - y^2$ we have $m_C = 1/2$. For this reason, in the rest of the section we will consider only random variables which satisfy this assumption.

Let us ignore for the moment the price evolution. The above binary interaction gives the following kinetic equation for the time evolution of chartists:

$$\frac{\partial f}{\partial t} = Q(f,f), \tag{7.86}$$

where, for any test function φ, the interaction operator Q can be conveniently written in weak form as

$$\int_{-1}^{1} \varphi(y) Q(f,f)\, dy = \left\langle \int_{[-1,1]^2} (\varphi(y') - \varphi(y)) f(y) f(y_*) dy_*\, dy \right\rangle. \tag{7.87}$$

Note that the mass density of chartists $\rho_C(t)$ is an invariant for the interaction.

Strategy exchange and price evolution

In addition to the change of investment propensity due to a balance between herding behaviour and the price-following nature of chartists, the model includes the possibility that an agent changes its strategy from chartist to fundamentalist, and vice versa.

Agents meet individuals from the other group, compare excess profits from both strategies and, with a probability depending on the pay-off differential, switch to the more successful strategy. When a chartist and a fundamentalist meet they characterize the success of a given strategy through the profits earned by comparing

$$X_C(y,t) = \psi(y)\left(\frac{\dot S(t)/\omega + D}{S(t)} - r\right), \quad X_F(t) = k\frac{|S_F - S(t)|}{S(t)}. \tag{7.88}$$

Here, $\psi(y) \in [-1,1]$ has the same sign as y and takes into account the change of sign in the profits according to the actual behaviour of the agent in the market, which relies on his investment propensity y. The simplest choice is $\psi(y) = \mathrm{sgn}(y)$.

The value D is the nominal dividend and r the average real return of the market, such that $r = D/S_F$, i.e. evaluated at its fundamental value S_F in a state of stable price $\dot S = 0$ the asset yields the same returns as other investments, or equivalently $X_C = X_F = 0$. The discount factor $k < 1$ is justified by the observation that X_F is an expected gain realized only after reversal of the fundamental value. Finally, $\omega > 0$ measures the frequency of the exchange rates.

A chartist characterized by an investment propensity y and a fundamentalist meet each other and, after comparing their strategies, they exchange strategies with a rate given by a suitable monotone function $B_{FC}(\cdot) \geq 0$. More precisely, a chartist switches to fundamentalist with a rate $B_{FC}(X_F - X_C)$, and a fundamentalist switches to chartist at a rate $B_{FC}(X_C - X_F)$.

For chartists we define the following linear strategy exchange operator:

$$Q_{FC}(f) = \omega\rho_F(t)f(y)(B_{FC}(X_C - X_F) - B_{FC}(X_F - X_C)).$$

Taking into account such strategy exchanges, we have the chartists–fundamentalists model

$$\begin{cases} \dfrac{\partial f}{\partial t} = Q(f,f) + \omega\rho_F(t)f(y)(B_{FC}(X_C - X_F) - B_{FC}(X_F - X_C)) \\ \dfrac{\partial \rho_F}{\partial t} = \omega\rho_F(t)\displaystyle\int_{-1}^{1} f(y)(B_{FC}(X_F - X_C) - B_{FC}(X_C - X_F))\,dy. \end{cases} \tag{7.89}$$

It can be immediately verified that the total number density $\rho_C + \rho_F$ is conserved in time.

Finally, we introduce the probability density $V(s,t)$ of a given price s at time t. The effective market price $S(t)$ is defined as the mean value

$$S(t) = \int_0^{\infty} V(s,t)s\,ds. \tag{7.90}$$

Following Lux and Marchesi [216, 217], the microscopic dynamics of the price is given by

$$s' = s + \beta(\rho_C t_C Y(t)s + \rho_F\gamma(S_F - s)) + \eta^* s, \tag{7.91}$$

where β represents the price speed evaluation and η^* is a random variable with zero mean and variance ς^2. In the above relation, chartists either buy or sell the same number t_C of units and γ is the reaction strength of fundamentalists to deviations from the fundamental value.

Thus, the chartists–fundamentalists system of equations (7.89) is complemented by the equation for the price distribution,

$$\frac{\partial V}{\partial t} = L(V), \tag{7.92}$$

where the operator L is linear, and in weak form it reads

$$\int_0^{\infty} L(V)(s)\varphi(s)\,ds = \left\langle \int_0^{\infty} b(s,\eta^*)V(s)(\varphi(s') - \varphi(s))\,ds \right\rangle, \tag{7.93}$$

with the transition rate $b(s,\eta^*) = \Psi(\eta^*)\chi(s' \geq 0)$.
As before, a suitable choice of the domain for the support of variable η^* ensures $s' \geq 0$. Assuming

$$\eta^* \in [-1 + \beta(\rho_C t_C + \rho_F\gamma), 1 - \beta(\rho_C t_C + \rho_F\gamma)], \quad \beta(\rho_C t_C + \rho_F\gamma) < 1,$$

permits us to express the transition rate in the simpler form

$$b(s,\eta^*) = \Psi(\eta^*).$$

Note that the expected value for the stock price satisfies the same differential equation as in [216, 217]:

$$\frac{dS(t)}{dt} = \beta\rho_C t_C Y(t)S(t) + \beta\rho_F\gamma(S_F - S(t)). \tag{7.94}$$

Booms, crashes and macroscopic stationary states

In order to study the macroscopic steady states and relate them to the value function Φ let us start by observing that the equilibrium states for the price satisfy

$$\rho_C t_C Y S + \rho_F \gamma (S_F - S) = 0,$$

and thus fall in one of the following categories:

(i) $\rho_F \neq 0, \quad S = \frac{\rho_F \gamma S_F}{\rho_F \gamma - \rho_C t_C Y}, \qquad \rho_F \gamma S_F - \rho_C t_C Y \geq 0;$
(ii) $\rho_F = 0, \quad Y = 0, \quad S$ arbitrary;
(iii) $\rho_F = 0, \quad S = 0, \quad Y$ arbitrary.

At equilibrium we require ρ_F, ρ_C and Y to be constants. In order for the number densities to be constants we require $Q_{FC} = 0$. For $\rho_F \neq 0$ and $\rho_C \neq 0$, thanks to the monotonicity of B_{FC}, we have $X_C = X_F$ or, equivalently, $S = S_F$. Note that Q_{FC} also vanishes when $\rho_F = 0$ or $\rho_C = 0$. These considerations reduce the set of possible equilibrium configurations to

(i) $\rho_F \neq 0, \quad S = S_F, \quad Y = 0;$
(ii) $\rho_F = 0, \quad Y = 0, \quad S$ arbitrary;
(iii) $\rho_F = 0, \quad S = 0, \quad Y$ arbitrary.

Finally, we consider the requirements for Y to be constant. In the case $Q_{FC} = 0$ the first moment equation reads

$$\begin{aligned} \frac{d}{dt} Y(t) = &- \alpha_1 \int_{-1}^{1} H(y) y f(y) dy - \alpha_2 \rho_C Y(t) \\ &+ \alpha_1 Y(t) \int_{-1}^{1} H(y) f(y) dy + \alpha_2 \rho_C \Phi \left(\frac{\dot{S}(t)}{S(t)} \right), \end{aligned}$$

which gives the steady-state condition

$$-\alpha_1 \int_{-1}^{1} H(y) y f(y) dy - \alpha_2 \rho_C Y + \alpha_1 Y \int_{-1}^{1} H(y) f(y) dy + \alpha_2 \rho_C \Phi \left(\frac{\dot{S}(t)}{S(t)} \right) = 0.$$

This gives a constraint for the value function Φ, precisely

$$\alpha_2 \rho_C \Phi \left(\frac{\dot{S}(t)}{S(t)} \right) = \alpha_1 \int_{-1}^{1} H(y) y f(y) dy + \alpha_2 \rho_C Y - \alpha_1 Y \int_{-1}^{1} H(y) f(y) dy,$$

which, in the simple case of H constant, reduces to

$$\alpha_2 \rho_C \left(\Phi \left(\frac{\dot{S}(t)}{S(t)} \right) - Y \right) = 0.$$

Now, using the fact that

$$\frac{\dot{S}(t)}{S(t)} = \beta \rho_C t_C Y(t) + \beta \rho_F \gamma \frac{(S_F - S(t))}{S(t)},$$

we can state:

Proposition 7.1 *The system of equations (7.89) in the case of H constant admits the following possible equilibrium configurations:*

(i) $\rho_F \neq 0, \quad S = S_F, \quad Y = 0, \quad \Phi(0) = 0$;

(ii) $\rho_F = 0, \quad Y = 0, \quad \Phi(0) = 0,$ *S arbitrary;*

(iii) $\rho_F = 0, \quad Y = Y_*,$ *with* $Y_* = \Phi(\beta t_C Y_*), \quad S = 0$.

Note that if the reference point for the value function $\Phi(0) \neq 0$, configurations (i) and (ii) are not possible for a constant H. This is in good agreement with the fact that an emotional perception of the market from the chartists acts as a source of instability for the market itself. In contrast, configuration (iii), corresponding to a market crash, can also be achieved for $\Phi(0) \neq 0$. The existence of a unique fixed point Y_* has to be guaranteed by the choice of Φ, β and t_C. Of course, if the reference point is set to zero, $\Phi(0) = 0$, we have $Y_* = 0$. It is easy to verify that these possible equilibrium configurations include the ones in the original Lux–Marchesi model [216].

In addition to the above equilibrium configurations, the model admits several other possible asymptotic behaviours in the form of booms and cycles. Some of the fundamental features of the model are summarized below.

Chartists alone ($\rho_F = 0, \rho_C = 1$) influence the price through their mean propensity to invest $Y(t)$, and at the same time the price trend influences their mean propensity through the value function $\Phi(\dot{S}(t)/S(t))$, since $\dot{S}(t)/S(t) = \beta Y(t) t_C$. Thus, except for the particular shape of the value function, if the mean propensity is initially (sufficiently) positive then it will continue to grow together with the price, and the opposite occurs if it is initially (sufficiently) negative.

The market goes towards a boom (exponential growth of the price) or a crash (exponential decay of the price) with

$$S(0)e^{-\beta t_C} \leq S(t) \leq S(0)e^{\beta t_C},$$

and agents tend to concentrate at $y = 1$ and $y = -1$ respectively, depending on the choices of H and Φ. This is in good agreement with the price-following nature of chartists. Fundamentalists alone ($\rho_F = 1, \rho_C = 0$) influence the price through their expectation of the fundamental price. So their effect is to drive the price towards the fundamental price. For a constant fundamental price S_F, the equilibrium state reached is characterized by $S = S_F$ and the trend is exponential.

The presence of fundamentalists acts in contrast to the chartists' pressure towards market booms or crashes. If their number is large enough, they are capable of driving the price towards the fundamental value. Otherwise, the chartists dynamics may dominate. In addition to booms and crashes, we now have the possibility of price cycles/oscillations around the fundamental value.

7.4.3 Fokker–Planck limit and kinetic asymptotic behaviour

Now we consider what happens at the kinetic scale. Due to the extreme difficulty of getting detailed information on the asymptotic behaviour of the kinetic coupled system, we will recover, for both distribution functions f and V, simplified Fokker–Planck models which preserve the main features of the original kinetic model. To keep notations simple, since we are mostly interested in the study of the equilibrium states we ignore the presence of the terms describing the change of strategy. However, they can easily be included in the scaling described below [222].

For this purpose we introduce a time scaling parameter ε and define

$$\tau = \varepsilon t, \quad \tilde{f}(y,\tau) = f(y,t), \quad \tilde{V}(s,\tau) = V(s,t).$$

To preserve the chartists dynamic in the limit, we must require that

$$\lim_{\alpha_1,\varepsilon\to 0} \frac{\alpha_1}{\varepsilon} = \tilde{\alpha_1}, \qquad \lim_{\alpha_2,\varepsilon\to 0} \frac{\alpha_2}{\varepsilon} = \tilde{\alpha_2}, \qquad \lim_{\sigma,\varepsilon\to 0} \frac{\sigma^2}{\varepsilon} = \lambda,$$

where λ is a positive constant.

Similarly for the price dynamic, we assume

$$\lim_{\beta,\varepsilon\to 0} \frac{\beta}{\varepsilon} = \tilde{\beta}, \qquad \lim_{\zeta,\varepsilon\to 0} \frac{\zeta^2}{\varepsilon} = \nu.$$

Performing similar computations as in Section 7.3.3 and omitting the details, we recover the following Fokker–Planck system [222]:

$$\begin{cases} \dfrac{\partial \tilde{f}}{\partial \tau} + \dfrac{\partial}{\partial y}\left[\left(\rho_C \tilde{\alpha_1} H(y)(\tilde{Y} - y) + \rho_C \tilde{\alpha_2}\left(\Phi\left(\dfrac{\dot{\tilde{S}}}{\tilde{S}}\right) - y\right)\right)\tilde{f}\right] \\ \qquad = \dfrac{\lambda \rho_C}{2}\dfrac{\partial^2}{\partial y^2}[(D_C{}^2(y))\tilde{f}], \\ \dfrac{\partial}{\partial \tau}\tilde{V} + \dfrac{\partial}{\partial s}\left[\tilde{\beta}\left(\rho_C \tilde{Y} t_C s + \rho_F \gamma(\tilde{S}_F - s)\right)\tilde{V}\right] = \dfrac{\nu}{2}\dfrac{\partial^2}{\partial s^2}\left(s^2 \tilde{V}\right). \end{cases} \tag{7.95}$$

To simplify the notations, in what follows we will omit the tildes in the variables f, V, Y and S.

If we now take $D_C(y) = 1 - y^2$ and $H(y) = 1$, we can explicitly compute the equilibrium state for chartists with a constant mean investment propensity $Y = Y_*$ as

$$f^\infty(y) = C_0(1+y)^{-2+Y_* \frac{(\tilde{\alpha_1}+\tilde{\alpha_2})}{2\lambda}}(1-y)^{-2-Y_* \frac{(\tilde{\alpha_1}+\tilde{\alpha_2})}{2\lambda}} \\ \exp\left(-\frac{(1 - Y_* y)(\tilde{\alpha_1} + \tilde{\alpha_2})}{\lambda(1-y^2)}\right), \tag{7.96}$$

where $C_0 = C_0(Y_*, \lambda/(\tilde{\alpha_1} + \tilde{\alpha_2}))$ is such that the mass of f^∞ is equal to ρ_C. Other choices of the diffusion function originate different steady states (see Chapter 6, Section 6.3).

Observe that in the case $Y_* \neq 0$ the distribution is not symmetric, and in the chartist population a predominant behaviour arises. Otherwise, when the reference point of the value function is set to zero we have a symmetric distribution with two peaks and mean value zero, and the macroscopic state of indecision is given, microscopically, by a polarization of the chartist population between two opposite kinds of behaviour (see Figure 7.7).

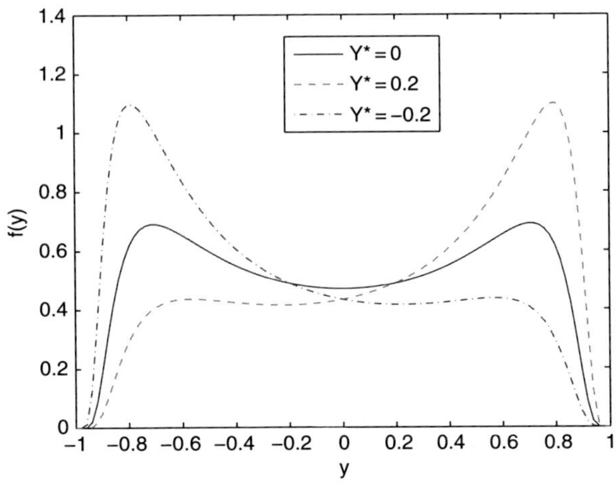

Fig. 7.7 Equilibrium distribution function of the chartist investment propensity for different values of $Y_* = 0, 0.2, -0.2$. Exact solutions with $\rho_C = 1$, $\beta = 0.1$, $t_C = 1$, $\lambda/(\tilde{\alpha_1} + \tilde{\alpha_2}) = 1$ and $f(y,0) = f^\infty(y)$.

In order to study the asymptotic behaviour for the price we must distinguish between the case $\rho_F \neq 0$ and $\rho_F = 0$.

Let us consider first the situation in which $\rho_F = 0$ (or, equivalently, $\rho_C = 1$). For this purpose, we introduce the scaling

$$V(s,\tau) = \frac{1}{s} v(\chi,\tau), \quad \chi = \log(s).$$

It is straightforward to show that $v(\chi,\tau)$ satisfies the following linear convection diffusion equation:

$$\frac{\partial}{\partial \tau} v(\chi,\tau) = \left[\frac{\nu}{2} - \tilde{\beta} Y t_C\right] \frac{\partial}{\partial \chi} v(\chi,\tau) + \frac{\nu}{2}\frac{\partial^2}{\partial \chi^2} v(\chi,\tau),$$

which, similarly to Section 7.3.3, admits the self-similar solution [104, 222]

$$v(\chi,\tau) = \frac{1}{(2\log(E(\tau)/S(\tau)^2)\pi)^{\frac{1}{2}}} \exp\left(-\frac{(\chi + \log(\sqrt{E(\tau)}/S(\tau)) - \log(S(\tau)))^2}{2\log(E(\tau)/S(\tau)^2)}\right),$$

with

$$E(\tau) = \int_0^\infty V(s,\tau) s^2 \, ds.$$

Then, reverting to the original variables gives the lognormal behaviour

$$V(s,\tau) = \frac{1}{s(2\log(E(\tau)/S(\tau)^2)\pi)^{\frac{1}{2}}} \exp\left(-\frac{(\log(s\sqrt{E(\tau)}/S(\tau)^2)^2}{2\log(E(\tau)/S(\tau)^2)}\right). \tag{7.97}$$

Note that $E(\tau)$ satisfies the differential equation

$$\frac{dE}{d\tau} = (2\tilde{\beta}Yt_C + \nu)E(\tau).$$

Thus, for a steady state characterized by (ii) in Proposition 7.1 we have $S(\tau) = S_0$, $Y = 0$ and $E(\tau) = e^{\nu\tau}E_0$.

Besides the above equilibrium state, equation (7.97) also characterizes the self-similar behaviour of the price distribution in the case of booms and crashes, when the price $S(\tau)$ grows arbitrarily or decays to zero. In particular, in the limit $S(\tau) \to 0$, point (iii) in Proposition 7.1, the distribution function $V(s,\tau)$ concentrates near zero.

Finally, we consider the microscopic behaviour of the model where both $\rho_C \neq 0$ and $\rho_F \neq 0$.

Recall now the Fokker–Planck equation for the price in (7.95) and consider the stationary case (i) in Proposition 7.1. The Fokker–Planck equation in such cases reads

$$\frac{\partial}{\partial\tau}V + \frac{\partial}{\partial s}\left[\left(\rho_F\gamma(S_F - s)\right)V\right] = \frac{\nu}{2}\frac{\partial^2}{\partial s^2}\left(s^2 V\right).$$

In this case, the steady state can be computed as in Chapter 5, Section 5.4.1:

$$V^\infty(s) = C_1(\Upsilon)\frac{1}{s^{1+\Upsilon}}e^{-\frac{(\Upsilon-1)S_F}{s}}, \tag{7.98}$$

where $\Upsilon = 1 + 2\rho_F\gamma/\nu$ and $C_1(\Upsilon) = ((\Upsilon - 1)S_F)^\Upsilon/\Gamma(\Upsilon)$, with $\Gamma(\cdot)$ being the usual Gamma function. Therefore, the stationary state is described by a Gamma-like distribution with Pareto power law tails.

Remark 7.2 *The presence of fundamentalists is then essential in order to obtain fat tails in the price distribution. Their presence forces the price to approach the mean value S_F in a way similar to the redistribution of wealth in the models proposed in [55, 105]. This feature seems to be essential for the development of power-law behaviours. The stationary state for price (7.98) has, in fact, the same structure as the stationary states for wealth (5.73).*

7.4.4 Numerical examples

In this section we consider different numerical simulations of the kinetic system. A Monte Carlo method of Nanbu type as described in Chapter 4 has been used for the simulations. In order to compute the kinetic behaviour of the price, we computed averages using a set of $N_s = 50000$ samples which can be thought of as possible realizations of the random variable s denoting the price. Since at the initial time the stock price S_0 is supposed to be known, all samples are initialized with the same value. We represent the initial chartists distribution with a set of $N_c = \rho_C(0)N$ sample agents with $N = 50000$. These do not represent real agents but simply a statistical realization of the random variable y. Such choices of N_s and N allows us to obtain results with a moderate effect of fluctuations without averaging.

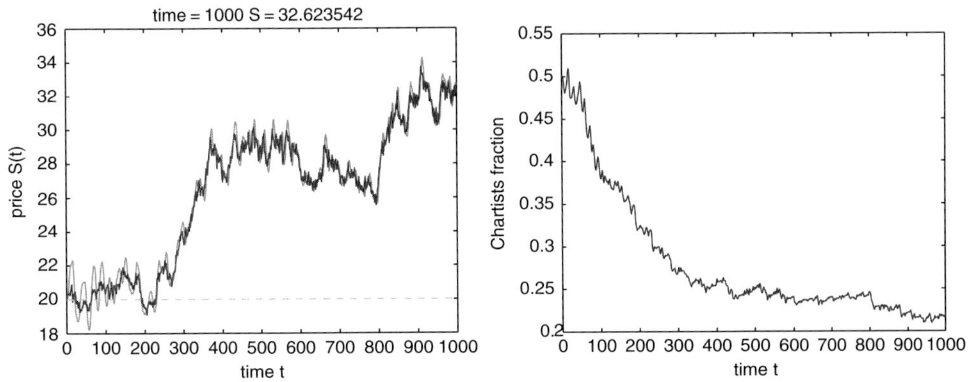

Fig. 7.8 A strong fundamentalist predominance forces the price to follow the fundamental price. The chartist population is characterized by the parameters $\alpha_1 = 0.25$ and $\alpha_2 = 0.7$. The figure on the right represents the variation of the chartists' fraction.

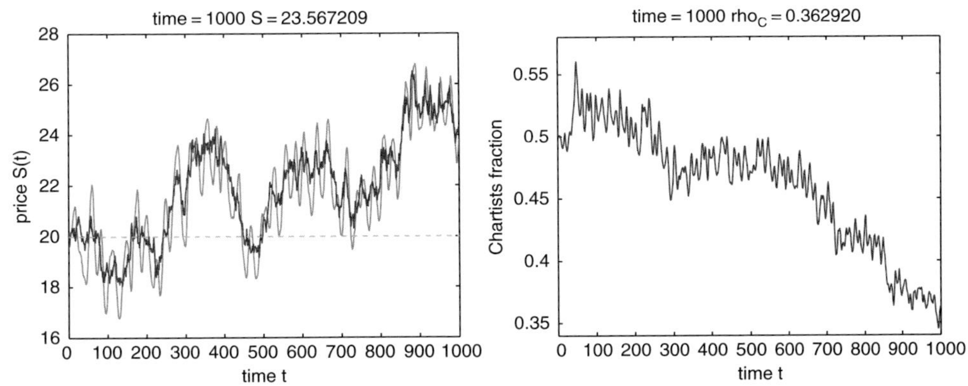

Fig. 7.9 The price follows the fundamental price but strong oscillations appear. The chartist population is characterized by the parameters $\alpha_1 = 0.6$ and $\alpha_2 = 0.4$. The figure on the right represents the variation of the chartists' fraction.

In all our computations we take the value function

$$\Phi(x) = \begin{cases} \min\{(x - R_0)^r, 1\}, & x > R_0, \\ -\min\{(R_0 - x)^l, 1\}, & x \leq R_0, \end{cases}$$

where R_0 is the reference point and $0 < l \leq r < 1$. In most numerical simulations we choose a non-symmetric value function with $R_0 = 0$, $r = 1/4$ and $l = 1/6$. We note that other choices of the reference point R and of the exponents r and l have a profound influence on the price dynamics without, however, changing the possible scenario that we illustrate in the subsequent examples.

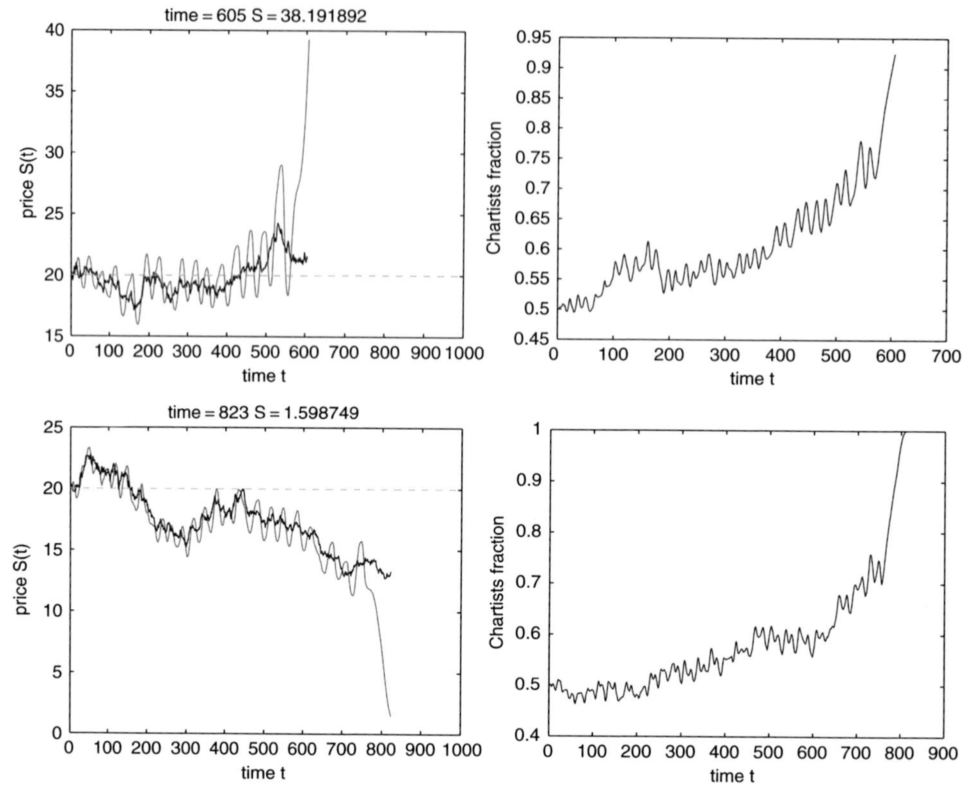

Fig. 7.10 A chartists predominance may lead to extreme behaviours like the case of a market boom (top) or a crash (bottom). The chartist population is characterized by the parameters $\alpha_1 = 0.65$ and $\alpha_2 = 0.3$. Top figure: $r = 1/6$, $l = 1/4$. Bottom figure: $r = 1/4$, $l = 1/6$. The figure on the right represents the variation of the chartists' fraction.

We consider the possible behaviours of the model in the realistic situation of a time-varying fundamental price. In our simulations we consider a temporal series of fundamental prices $S_F(t)$ with a stationary lognormal distribution based on a Gaussian with 0 mean and variance 0.01. Initially we take $S_F(0) = 20$. The remaining parameters are $\beta = 6$, $t_C = 0.02$, $\gamma = 0.1$, $\sigma = 0.8$, $\omega = 0.2$, $D = 0.004$, $k = 0.75$ and $\psi(y) = \text{sgn}(y)$. Here we consider a non-constant herding function of the form $H(y) = (1-|y|)$. We run different simulations for $T = 1000$ iterations with different values of α_1 and α_2, which measure respectively the herding and the market influence on the chartists. As expected from our theoretical analysis we can identify three main different behaviours:

- fundamentalists predominance (efficient market);
- no predominance (irrational market);
- chartists predominance (market boom or crash).

The predominance of fundamentalists originates damped oscillation of the price towards the fundamental value (see Figure 7.8), a balanced behaviour of the two populations is characterized by strong oscillation of the price around the fundamental value (see Figure 7.9), whereas the predominance of chartists leads the market towards a crash or a boom (see Figure 7.10) in agreement with the agents' tendency to overestimate or underestimate potential gains and losses.

7.5 Inhomogeneous models for the evolution of wealth

The goal of the models introduced in Chapter 5 and of the more realistic examples presented in this chapter is to follow the evolution of wealth distribution in a multiagent society, and overall to be able to describe its large-time behaviour, which in most cases is represented by an equilibrium distribution with a certain number of properties to be interpreted in terms of economic relevance.

As a matter of fact, in statistical physics, knowledge of the large-time behaviour of the density of energy and its approach to a universal profile are subsequently used to construct equations for hydrodynamics, which describe the space–time evolution of macroscopic observables.

A similar procedure could, in principle, be applied to the evolution of the density of wealth, in the case where this density depends on other important parameters (typically the space variable in a physical system).

The analogy with the evolution of a gas density depending on both the space and velocity variable has recently been developed in [135], in the case in which the behaviour of each agent in the system is identified by two main variables, the first given by his wealth, the second by his propensity to invest (in the hope of making a profit).

Similarly to what happens in a physical system, where each particle is identified by its position and velocity, and the position depends on the velocity through the classical equations of motion, in [135] it was assumed that the propensity still depends on the wealth through a suitable *equation of motion*. This relationship between propensity and wealth is largely formal, and could be modified in many ways. Nevertheless, it is quite interesting to observe how the behaviour of the hydrodynamics equations depends on the assumption of this *equation of motion*.

One important aspect of this technique is to justify it by solid arguments. In statistical mechanics the validity of the fluid dynamics is related to the gas density, or, equivalently, to the mean time between subsequent collisions [84]. In the present situation this validity is linked to the mean number of trades which occur in a fixed interval of time. This aspect of the matter has recently been discussed in a paper of Wang, Ding and Zhang [324], who introduced the concept of the statistical description of the velocity of money circulation. This concept is based on the *holding time of money* which is defined as the time interval between two transactions. Although this concept is kept in mind when economists think of the velocity, even though the term referring to this kind of time interval has been mentioned in several cases, it is somewhat new to them since there has been no explicit specification of it in economics. Recently, several efforts have been devoted to measuring the waiting time distributions in financial markets, see e.g. [220, 277]. In the process of money circulation, not only the amount of money each agent holds can be considered as random variable, but also the holding time between two transactions varies randomly. The theoretical investigation and the numerical simulations in [324] led to the conclusion that the velocity of money is proportional to the share for exchange, and, most importantly, inversely proportional to the number of agents, and independent of the average amount of money.

Using this result in a kinetic model of Boltzmann type shows that the velocity of relaxation to the steady distribution of wealth is inversely proportional to the velocity of money circulation, which justifies a hydrodynamical description when the same velocity is sufficiently high.

7.5.1 Inhomogeneous kinetic models

Taking into account the results of Chapter 5, collision-like kinetic models seem to share some common features. Almost all of these models identify a very important variable for the shape of the wealth distribution, which is usually called the saving propensity to trade or the saving rate, respectively. This parameter can either enter into the collision rule as a constant factor [91], or it can be chosen as a random quantity [96]. Other studies include the saving propensity as an independent variable [88], without questioning the relationship between wealth and saving.

These approaches to studying both wealth and saving distributions show that it could be reasonable to introduce other types of propensities into the game, not directly connected to the microscopic binary trade, which could be important in the evolution of wealth in a market of agents. Among others, one can assume that the evolution of the density of wealth is heavily dependent on the propensity to invest, and at the same time that this propensity is closely related to the amount of money each agent has to deal with.

In agreement with the classical theory of rarefied gases, where the particle density depends on the space variable x, the velocity variable w and time t, we study here the evolution of the distribution function as a function depending on the propensity $x \in [0, 1]$, wealth $w \in \mathbb{R}_+$ and time $t \in \mathbb{R}_+$, say $f = f(x, w, t)$. By analogy with the classical kinetic theory of rarefied gases, it is useful to emphasize the role of the different parameters by identifying the velocity with the wealth, and the position with the saving propensity. By doing this, one assumes at once that the variation of the distribution $f(x, w, t)$ with respect to the wealth parameter w will depend on *collisions* between agents, while the change of distributions in terms of the propensity x depends on the *transport* term, which contains the equation of motion, namely the law of variation of x with respect to time,

$$\frac{dx}{dt} = \mathcal{T}(x, w). \tag{7.99}$$

The time evolution of the distribution will obey a non-homogeneous Boltzmann-like equation, given by

$$\frac{\partial}{\partial t} f(x, w, t) + \mathcal{T}(x, w)\frac{\partial}{\partial x} f(x, w, t) = \frac{1}{\tau}\mathcal{Q}(f)(x, w, t). \tag{7.100}$$

In (7.100), $\mathcal{T}$ is the law of variation of the propensity to invest given in (7.99), while $\mathcal{Q}$ represents as usual the collision operator which describes the change of density due to binary trades. Finally, τ is a suitable relaxation time, depending on the velocity of money circulation [324]. Note that in physical applications where no forces are present, the transport term is simply $\mathcal{T}(x, w) = w$.

In any trading, savings come naturally [281]. In a real society or economy, saving propensity is a very inhomogeneous parameter, and interest in saving varies from person to person, according to their wealths. To move a step closer to the real situation, one has to introduce a saving factor widely distributed within the population [88, 96], and responsible for different outcomes in binary trades. The evolution of money in such trading can be written as a generalization of the usual trade given by (5.6), that is [88]:

$$\begin{aligned} v^* &= \gamma v + \varepsilon(\gamma, \mu) \left[(1-\gamma)v + (1-\mu)w \right], \\ w^* &= \mu w + (1 - \varepsilon(\gamma, \mu)) \left[(1-\gamma)v + (1-\mu)w \right]. \end{aligned} \tag{7.101}$$

Here, (γ, w) and (μ, v) denote the saving propensities and wealths of agents before collisions. In a single collision it is assumed that the agents maintain fixed saving propensities, so that the post-collision parameters are (γ, w^*) and (μ, v^*). Moreover, $\varepsilon(\gamma, \mu)$ denotes a random fraction, coming from the stochastic nature of the trading. Also, one can assume, as in [105], the trade

$$\begin{aligned} v^* &= (1-\gamma)v + \gamma w + \eta v, \\ w^* &= \gamma v + (1-\gamma)w + \tilde{\eta} w. \end{aligned} \tag{7.102}$$

In (7.102), the trade depends on a single saving rate $\gamma \in (0, 1)$, while the risks of the market are described by η and $\tilde{\eta}$, equally distributed random variables with zero mean and variance σ^2.

In all cases, the operator which describes the binary interaction is the usual bilinear operator defined by (5.8), rewritten below for convenience:

$$\begin{aligned} &\int_{\mathbb{R}_+} \mathcal{Q}(f, f)(w)\varphi(w)\, dw = \\ &\frac{1}{2} \left\langle \int_0^1 dy \int_{\mathbb{R}_+} \int_{\mathbb{R}_+} dv\, dw\, (\varphi(v^*) + \varphi(w^*) - \varphi(v) - \varphi(w)) f(y, v) f(x, w) \right\rangle . \end{aligned} \tag{7.103}$$

In the continuous trading limit $(\gamma \to 0, \sigma^2/\gamma \to \lambda)$, it has been shown in [105] (cf. Chapter 5, Section 5.4) that the collision operator (7.103) is well described by the Fokker–Planck collision operator

$$\mathcal{P}(w) = \frac{\lambda}{2} \frac{\partial^2}{\partial w^2}(w^2 f(w)) + \frac{\partial}{\partial w}(w - m(f)) f(w), \tag{7.104}$$

where $m(f)$ is the mean wealth of $f(w)$. The key parameter λ is obtained as the limit of the quotient of the variance and the saving rate. The homogeneous kinetic equation

$$\frac{\partial}{\partial t} f(w, t) = \mathcal{P}(w, t) \tag{7.105}$$

is such that both the mass and the mean wealth $m(f)$ are conserved in time. Moreover, for any initial density $f(w, t = 0) = f_0(w)$ with mass ρ and mean m, equation (7.105) has a unique stationary state, the so-called Maxwellian state $M_{\rho,m}(w)$ given by

$$M_{\rho,m}(w) = \rho \frac{((\Upsilon - 1)m)^{\Upsilon}}{\Gamma(\Upsilon)} \frac{1}{w^{1+\Upsilon}} \exp\left(-\frac{(\Upsilon - 1)m}{w}\right), \tag{7.106}$$

where

$$\Upsilon = 1 + \frac{2}{\lambda} > 1.$$

Therefore, the Maxwellian distribution exhibits a Pareto power law tail for large w's. In particular, higher moments of the equilibrium Maxwellian are given in terms of mass ρ and mean m. The second moment can easily be evaluated by considering that in equilibrium, i.e. as $t \to \infty$, one has

$$\begin{aligned}
0 =& \frac{\lambda}{2} \int_{\mathbb{R}_+} w^2 \frac{\partial^2}{\partial w^2} (w^2 M_{\rho,m}(w))\, dw + \int_{\mathbb{R}_+} w^2 \frac{\partial}{\partial w} [M_{\rho,m}(w)(w - m)]\, dw \\
=& \lambda \int_{\mathbb{R}_+} w^2 M_{\rho,m}(w)\, dw - 2 \int_{\mathbb{R}_+} w(w - m) M_{\rho,m}(w)\, dw \\
=& (\lambda - 2) \int_{\mathbb{R}_+} w^2 M_{\rho,m}(w)\, dw + 2m \int_{\mathbb{R}_+} w M_{\rho,m}(w)\, dw \\
=& (\lambda - 2) \int_{\mathbb{R}_+} w^2 M_{\rho,m}(w)\, dw + 2\rho m^2.
\end{aligned}$$

Thus, if $\lambda < 2$, the second moment of the Maxwellian is bounded, and

$$\int_{\mathbb{R}_+} w^2 M_{\rho,m}(w)\, dw = \frac{2}{2 - \lambda} \rho m^2. \tag{7.107}$$

In the rest of this chapter, we will assume that in a closed economy the Maxwellian distribution $M_{\rho,m}$, equilibrium solution of the Fokker–Planck equation (7.105), plays the same role as played by the Maxwell distribution in the kinetic theory of rarefied gases. However, contrary to what happens in classical kinetic theory, where the equilibrium Maxwellian has all moments bounded, in this case the number of moments bounded in the equilibrium depends on the parameter λ in front of the second-order term in (7.104).

7.5.2 The Euler equations

Section 7.5.1 described the main properties of the collision operator (7.104), such as the existence of a unique Maxwellian equilibrium with tails, and the consequent possibility of obtaining higher-order moments from the first two (mass and mean wealth). As in the classical kinetic theory of rarefied gases, these properties are the basis of the construction of a reasonable hydrodynamics for the evolution of the propensity. The underlying kinetic model is obtained by substituting the Fokker–Planck operator into the Boltzmann equation (7.100),

$$\frac{\partial}{\partial t}f(x,w,t)+\mathcal{T}(x,w)\frac{\partial}{\partial x}f(x,w,t)=\frac{1}{\tau}\mathcal{P}(f)(x,w,t). \tag{7.108}$$

The τ parameter (the analogue of the Knudsen number in the kinetic theory of gases [82]) represents a suitable relaxation time, and has to be assumed small in fluid-dynamical regimes. A clear understanding of the derivation of macroscopic equations in kinetic theory can be obtained through the use of the splitting method, very popular in the numerical approach to the Boltzmann equation [155, 249]. If at each time step we consider sequentially the transport and relaxation operators in the Boltzmann equation (7.108), during this short time interval we recover the evolution of the density from the joint action of the relaxation

$$\frac{\partial f}{\partial t}=\frac{1}{\tau}\,\mathcal{P}(x,w,t) \tag{7.109}$$

and transport

$$\frac{\partial f}{\partial t}+\mathcal{T}(x,w)\frac{\partial}{\partial x}f(x,w,t)=0. \tag{7.110}$$

As in classical kinetic theory, where the energy is conserved in collisions, the conservation of the mean wealth in the relaxation step is enough to guarantee that (7.109) pushes the solution towards the Maxwellian equilibrium with the same mass and mean of the initial datum. Then, if τ is sufficiently small, one can easily argue that the solution to (7.109) is *sufficiently close* to the Maxwellian, and this Maxwellian can be used in the transport step (7.110) to close the equations. In detail, since the Fokker–Planck operator (7.104) is both mass and momentum preserving, integrating equation (7.108) with respect to the wealth velocity w, using as test functions $\varphi(w)=1,w$ respectively, we obtain

$$\int_{\mathbb{R}_+}\left(\frac{\partial f}{\partial t}+\mathcal{T}(x,w)\frac{\partial}{\partial x}f(x,w,t)\right)\,dw=0, \tag{7.111}$$

and

$$\int_{\mathbb{R}_+}w\left(\frac{\partial f}{\partial t}+\mathcal{T}(x,w)\frac{\partial}{\partial x}f(x,w,t)\right)\,dw=0. \tag{7.112}$$

Let us fix the law $\mathcal{T}$ to be linearly dependent on w,

$$\mathcal{T}(x,w)=(w-\chi\bar{w})\mathcal{W}(x), \tag{7.113}$$

where χ is a positive constant and $\bar{w}$ represents a suitable fixed value of the wealth. Then, one obtains from (7.111), (7.112) the equations

$$\frac{\partial\rho}{\partial t}+\mathcal{W}(x)\frac{\partial}{\partial x}\Big[\rho\big(m-\chi\bar{w}\big)\Big]=0, \tag{7.114}$$

$$\frac{\partial(\rho m)}{\partial t}+\mathcal{W}(x)\frac{\partial}{\partial x}\left[\int_{\mathbb{R}_+}w^2f(x,w,t)\,dw-\chi\bar{w}\rho m\right]=0. \tag{7.115}$$

In (7.114) and (7.115), $\rho(x,t), m(x,t)$ are the macroscopic variables, the local density of agents with propensity x at time t, given by

$$\rho(x,t) = \int_{\mathbb{R}_+} f(x,w,t)\,dw, \tag{7.116}$$

and the local mean

$$m(x,t) = \frac{1}{\rho(x,t)} \int_{\mathbb{R}_+} w f(x,w,t)\,dw. \tag{7.117}$$

Equation (7.115) depends on the second moment of the density. Using the equilibrium Maxwellian (7.107), however, we can express this second moment in terms of the first two. By this relationship we finally obtain the following system of equations:

$$\frac{\partial \rho}{\partial t} + \mathcal{W}(x)\frac{\partial}{\partial x}\Big[\rho(m - \chi\bar{w})\Big] = 0, \tag{7.118}$$

$$\frac{\partial(\rho m)}{\partial t} + \mathcal{W}(x)\frac{\partial}{\partial x}\Big[\rho m\Big(\frac{2}{2-\lambda}m - \chi\bar{w}\Big)\Big] = 0, \tag{7.119}$$

which have to be solved on $(0,1)\times(0,T)$ with appropriate boundary and initial conditions. Using (7.118) we can rewrite the second equation as

$$\frac{\partial m}{\partial t} + \mathcal{W}(x)(m - \chi\bar{w})\frac{\partial m}{\partial x} + \frac{\lambda}{2-\lambda}\frac{1}{\rho}\frac{\partial}{\partial x}\big[\rho m^2\big] = 0. \tag{7.120}$$

The hydrodynamic equations (7.118), (7.119) can be written in symmetric hyperbolic form [114]. Multiplying (7.118) by m and subtracting it from (7.119) we obtain

$$\begin{aligned}\rho m_t + \mathcal{W}(x)\Big[&-(\rho_x m + \rho m_x)m \\ &+ \chi\bar{w} m\rho_x + (\rho_x m + 2\rho m_x)\frac{2}{2-\lambda}m - (\rho_x m + \rho m_x)\chi\bar{w}\Big] = 0.\end{aligned} \tag{7.121}$$

Define $u = (\rho, m)$. Then a direct computation shows that the system (7.118), (7.121) can be written in compact form as

$$A^0(t,x,u)\frac{\partial u}{\partial t} + A^1(t,x,u)\frac{\partial u}{\partial x} = 0. \tag{7.122}$$

In (7.122), A^0 and A^1 denote the symmetric matrices

$$A^0(t,x,u) = \begin{pmatrix} \frac{1}{\rho}\frac{\lambda}{2-\lambda}m^2 & 0 \\ 0 & \rho \end{pmatrix}, \tag{7.123}$$

$$A^1(t,x,u) = \begin{pmatrix} \rho(m-\chi\bar{w})\frac{\lambda}{2-\lambda}m^2 & \frac{\lambda}{2-\lambda}m^2 \\ \frac{\lambda}{2-\lambda}m^2 & (1+\frac{4}{2-\lambda})\rho m - \rho\chi\bar{w} \end{pmatrix}. \tag{7.124}$$

Note that, for all $\rho, m \in G$ belonging to a suitable set G, $A^0(t,x,u)$ is uniformly positive definite provided $\lambda < 2$. Due to their structure, suitable numerical methods are available [206].

7.5.3 Universality of the closure

One of the main advantages linked to the use of the Fokker–Planck collision operator (7.104) is that it is straightforward to recover its steady state, namely the Maxwellian (7.106). However, if we use a different collision operator like the Boltzmann operator (7.103), while it is possible to prove that in the case of conservative trades there exists a unique steady state (cf. Chapter 5), the explicit form of this Maxwellian is unknown. This problem is not present in the classical elastic kinetic theory of rarefied gases, where the Maxwellian is uniquely defined independently of the choice of the binary collision operator [84]. This fact causes a serious problem in the justification of the validity of the closure, which in principle has to be independent of the choice of the underlying microscopic model of collisions, except, eventually, for constant parameters. Using the results of [229], however, we can easily conclude that the closure law

$$\int_{\mathbb{R}_+} w^2 M_{\rho,m}(w)\,dw = C\rho m^2. \tag{7.125}$$

where $C = 2/(2-\lambda)$ in (7.107), has universal validity, and the type of binary trade used in the collision operator (7.103) is reflected only through the precise value of the constant C. Let us suppose once more that the (conservative) binary interactions are described by the rules

$$v^* = p_1 v + q_1 w, \qquad w^* = p_2 v + q_2 w, \tag{7.126}$$

where

$$\langle p_1 + p_2\rangle = 1, \quad \langle q_1 + q_2\rangle = 1. \tag{7.127}$$

We note that both trades (7.101) and (7.102) satisfy assumption (7.127). In this case, application of formula (7.103) with $\varphi(w) = w^n$ allows us to compute recursively the evolution of the principal moments

$$M_n(t) = \int_{\mathbb{R}_+} w^n f(w,t)\,dw$$

with $n \geq 2$ (see [229] for details). One obtains

$$\frac{d}{dt}M_n(t) = \frac{1}{2}\left\langle (p_1^n + p_2^n - 1) + (q_1^n + q_2^n - 1)\right\rangle M_n(t) +$$

$$\frac{1}{2}\sum_{k=1}^{n-1}\binom{n}{k}\left\langle p_1^k q_1^{n-k} + p_2^k q_2^{n-k}\right\rangle M_k(t)M_{n-k}(t). \tag{7.128}$$

Considering that the first moment is conserved, $M_1(t) = m$, equation (7.128) also furnishes a recursive computation of the principal moments of the stationary solution,

$$M_n = \rho\frac{\sum_{k=1}^{n-1}\binom{n}{k}\left\langle p_1^k q_1^{n-k} + p_2^k q_2^{n-k}\right\rangle M_k M_{n-k}}{2 - \left\langle p_1^n + p_2^n + q_1^n + q_2^n\right\rangle}. \tag{7.129}$$

In particular,

$$M_2 = \int_{\mathbb{R}_+} w^2 M_{\rho,m}(w)\, dw = \rho m^2 \frac{2\,\langle p_1 q_1 + p_2 q_2 \rangle}{2 - \langle p_1^2 + p_2^2 + q_1^2 + q_2^2 \rangle}, \tag{7.130}$$

which coincides with the law (7.125), in which

$$C = \frac{2\,\langle p_1 q_1 + p_2 q_2 \rangle}{2 - \langle p_1^2 + p_2^2 + q_1^2 + q_2^2 \rangle}. \tag{7.131}$$

If we consider the trade (7.102), where

$$\begin{aligned} p_1 &= 1 - \gamma + \eta, \quad q_1 = \gamma, \\ p_2 &= \gamma, \quad q_2 = 1 - \gamma + \tilde{\eta}, \end{aligned} \tag{7.132}$$

we obtain for C the value

$$C = \frac{2\gamma(1-\gamma)}{2\gamma(1-\gamma) - \sigma^2}. \tag{7.133}$$

Note that this value corresponds to the choice

$$\lambda = \frac{\sigma^2}{\gamma(1-\gamma)}$$

in (7.107). This result shows the meaning of the constant λ appearing in the Fokker–Planck equation (7.104) in terms of the underlying binary trade (7.102). We remark that also in this case the stationary state of the Boltzmann equation possesses tails. Consider now the trade (7.101), where for simplicity $\varepsilon(\gamma, \mu) = 1/2$,

$$\begin{aligned} p_1 &= (1+\gamma)/2, \quad q_1 = (1-\mu)/2, \\ p_2 &= (1-\gamma)/2, \quad q_2 = (1+\mu)/2. \end{aligned} \tag{7.134}$$

In this case

$$C = \frac{2(1-\gamma\mu)}{2 - (\gamma^2 + \mu^2)}. \tag{7.135}$$

This corresponds to the choice

$$\lambda = \frac{(\gamma-\mu)^2}{1-\gamma\mu}$$

in (7.107). Here, however, the constant λ is always strictly less than 2. This is related to the fact that, for pointwise collisions like the one defined by (7.134), the stationary solution has all moments bounded [229]. In conclusion, there is a universal validity of the closure of hydrodynamic equations, at least in the well-defined case of conservative economies.

7.5.4 Which law for the propensity to trade?

To proceed and to obtain (at least numerical) results on the time evolution of the macroscopic quantities, it is necessary to set the two variables x and w in relation.

In classical hydrodynamics, where the variables are position and velocity, this relation is obvious, since velocity is the time derivative of position. In the absence of forces, it corresponds to choose $\mathcal{T}(x, w) = w$. To find an analogue for our economic setting, namely a law for the propensity to trade, we resort to some arguments within the concepts of opinion formation, introduced in Chapter 6.

There, a class of kinetic models of opinion formation, based on two-body interactions involving both compromise and diffusion properties in exchanges between individuals, was introduced. In the quasi-invariant opinion limit, these models are described by partial differential equations of Fokker–Planck type (7.104).

The equilibrium state of the Fokker–Planck equation can be computed explicitly and, in the absence of internal points in which diffusion is missing, is in most cases well represented by a Beta distribution,

$$B(x; a, b) = \frac{x^a(1-x)^b}{\int_0^1 u^a(1-u)^b\, du} = \frac{\Gamma(a+b+2)}{\Gamma(a+1)\Gamma(b+1)}\, x^a(1-x)^b, \tag{7.136}$$

where $a, b > -1$, and Γ is the gamma function. In what follows, we assume that the stationary profile for the distribution of our propensity to trade follows a law of type (7.136). Taking into account that this stationary profile is stable, a highly reasonable hypothesis is to assume that the rate of variation of the propensity is proportional to the density of people having that propensity. In this case, in order to maintain the lower and upper bounds of $x(t)$ we assume the law

$$\mathcal{T}(x, w) = x^a(1-x)^b H(w), \tag{7.137}$$

where $a, b > 0$, and the coefficient $H(w)$ takes into account the dependence of the law of variation on the (relative) wealth. We note that, thanks to the positivity of a, b, people with propensity to trade close to zero or one are more stable in their propensity, while people with intermediate propensity have more inclination to change their ideas.

Lastly, the form of $H(w)$ can be deduced from an argument similar to that introduced in Section 7.2.1.

Consider the case of a rational investor seeking to maximize his utility from wealth after each trade. He can choose a combination of saving his current wealth and the return from a trade; thus, since η has zero mean, his expected post-trade wealth is

$$\langle w^* \rangle = w + x(v - w) = (1-x)w + xv.$$

The investor's choice of his propensity to invest can be interpreted as the choice of combinations of two options, w and v. The investor tries to maximize his expected utility from post-trade wealth $\langle u(w^*) \rangle$, where u is a utility function characterizing the investor's satisfaction from wealth, e.g. the Cobb–Douglas function (7.2),

$$U(w, v) = w^\alpha v^\beta, \; \alpha + \beta = 1.$$

The projection of this function onto the w–v-plane gives level curves of constant utility, the so-called *indifference curves*. The investor is indifferent to the different combinations of v and w on such a curve, or in other words, at each point on an indifference curve he has no preference for one combination over another. A rational investor will choose a value for x that maximizes his utility from post-trade wealth. Therefore it is clear that x should be a function of the investor's wealth w or his relative wealth $w - \chi\bar{w}$, if utility from wealth is measured with respect to the mean wealth $\chi\bar{w}$ as a reference value (χ denoting a suitable constant). In other words, the fundamental law of physics about position and velocity and their relation is replaced here by an economic law that relates a rational investor's propensity to invest and his wealth based on the principle of utility maximization.

In general, the optimal choice of x depends on the underlying utility function, and this can lead to quite complex and non-linear relationships between propensity to invest x and wealth w. However, recall that we are considering a regime where the time between trades τ is very small. If we assume that $\mathcal{T}$ is smooth enough, we can approximate it by a linear relation and ignore terms of higher order. Furthermore, empirical results [281] from economic literature suggest that typically individuals have decreasing absolute risk aversion.

A simple way to take these facts into account is to relate the time variation of the propensity to the relative (with respect to the mean) wealth. A way to cope with these demands is to introduce the following law:

$$\mathcal{T}(x, w) = \pm\vartheta x^a (1 - x)^b \big(w - \chi\bar{m}(t)\big), \tag{7.138}$$

where $\bar{m}(t) = \int_0^1 m(x,t)\,dx$ denotes the mean wealth at time t, and ϑ is a positive constant. Note that the choice $a = b = 1$ implies an exponential decay of $x(t)$ towards one of the two extremal points. Since $x(t) < 1$, its variation with respect to time can be controlled by a suitable choice of these parameters. Let us note further that the choice of a positive (negative) sign in (7.138) implies that individuals with a higher (lower) wealth will be more (less) willing to trade than individuals with lower (higher) wealth. Clearly, this is only one choice among many possibilities. However, it seems a promising, quite natural approach, which is at the same time flexible enough, and sufficiently easy to be tractable from a numerical point of view. Using this choice in (7.118) and (7.119), and absorbing ϑ into time, we arrive at the following system:

$$\frac{\partial \rho}{\partial t} + x^a(1-x)^b \frac{\partial}{\partial x}\Big[\rho\big(m - \chi\bar{m}\big)\Big] = 0, \tag{7.139}$$

$$\frac{\partial \rho m}{\partial t} + x^a(1-x)^b \frac{\partial}{\partial x}\Big[\rho m\Big(\frac{2}{2-\lambda}m - \chi\bar{m}\Big)\Big] = 0. \tag{7.140}$$

As before, using (7.139) we can rewrite the second equation as

$$\frac{\partial m}{\partial t} + x^a(1-x)^b(m - \chi\bar{m})\frac{\partial m}{\partial x} + \frac{\lambda}{2-\lambda}\frac{1}{\rho}\frac{\partial}{\partial x}\big[\rho m^2\big] = 0. \tag{7.141}$$

7.5.5 Numerical tests

To solve (7.139), (7.140) numerically, we use a standard finite element method. We choose quadratic Lagrangian elements on a uniform grid with 480 nodes. We use the initial conditions

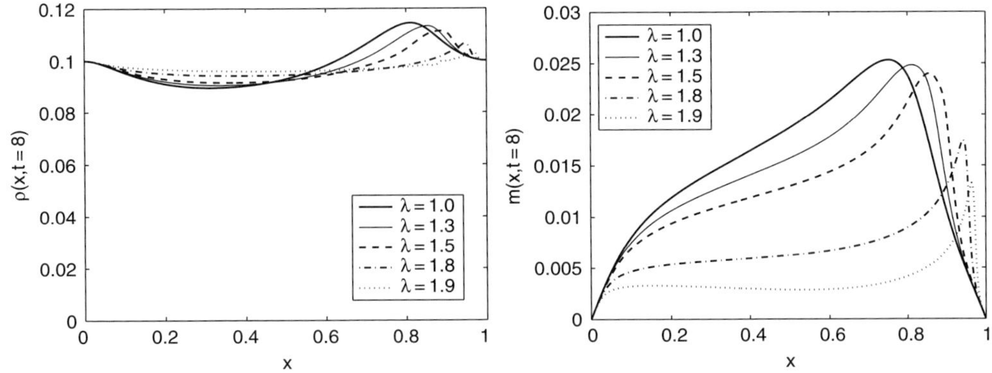

Fig. 7.11 Influence of different values for λ ($\lambda = 1, 1.3, 1.5, 1.8, 1.9$).

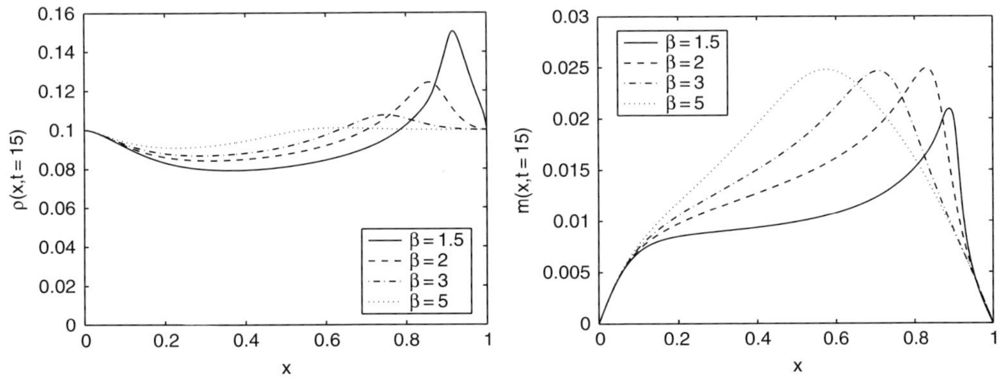

Fig. 7.12 Influence of different values for b ($b = 1.5, 2, 3, 5$).

$$\rho_0(x) = 0.1, \quad m_0(x) = x(1-x), \quad x \in (0,1).$$

At the boundaries we use homogeneous Neumann conditions for ρ and homogeneous Dirichlet conditions for m. If not mentioned otherwise, we choose $a = b = 2$, $\chi = 1$, $\lambda = 1$ and the final time is $T = 15$.

Figure 7.11 displays the numerical solution for different values of λ. Recall that a higher λ corresponds to tails with a lower Pareto index, and this corresponds to a society with a strong economy. We can observe the same effect here at the macroscopic level. As λ increases, the density of agents with a high propensity to trade increases while the wealth density is increasing for larger propensities and decreasing for smaller propensities, i.e. a large fraction of the total wealth is owned by a small group of agents.

Figure 7.12 shows the influence of different variants of law (7.138). For a value of $b = 1.5$ agents with wealth above the mean wealth increase their propensity to trade which leads to a peak formation in the density ρ close to $x = 1$. For higher values of b the propensity to invest grows slower when above the average wealth, therefore the peak is less pronounced.

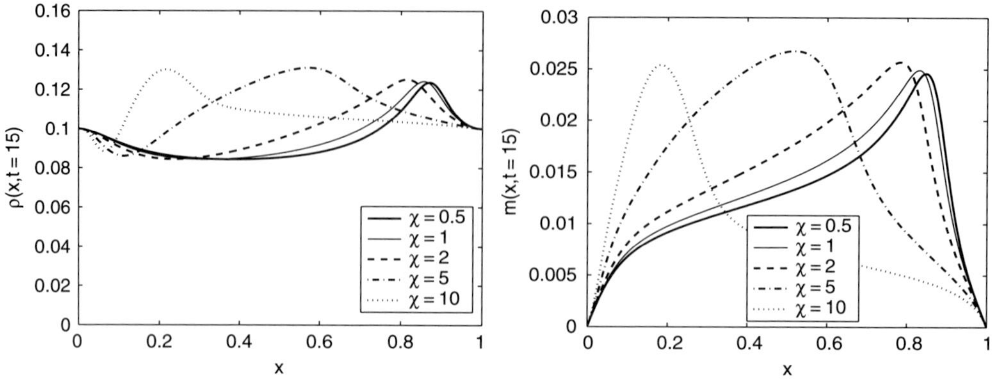

Fig. 7.13 Influence of different values for χ ($\chi = 0.5, 1, 2, 5, 10$).

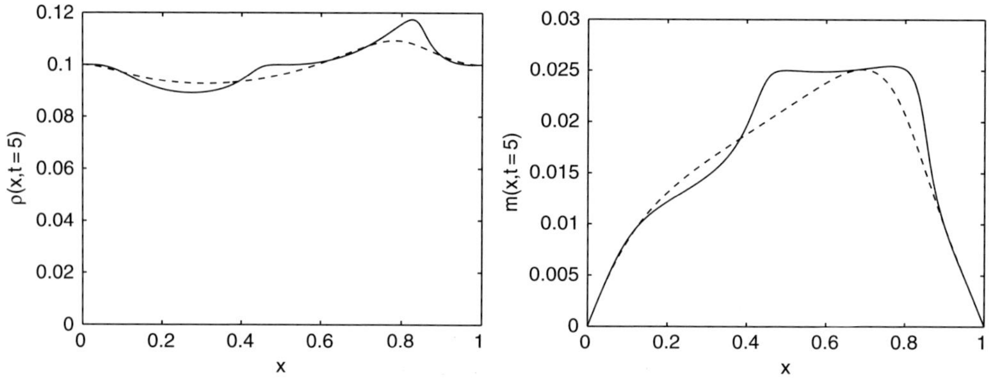

Fig. 7.14 Influence of different laws for propensity: solid lines correspond to law (7.142), dashed lines to (7.138).

The influence of the parameter χ can be observed in Figure 7.13. For high values of χ only very wealthy agents increase their propensity to trade, all others decrease it. This results in a peak formation at lower levels of x with agents saving most of their wealth.

The previous examples illustrate the influence of the law (7.99) in the evolution of the macroscopic quantities. Clearly, the proposed law (7.138) constitutes only a prototype towards a better understanding of the whole matter. Changing the law (7.138) allows us to clarify the role of the various parameters involved. By maintaining linearity with respect to the wealth parameter w, we consider in what follows the law

$$\tilde{\mathcal{T}}(x, w) = \nu(x)(w - \bar{m}(t)), \tag{7.142}$$

with $\nu(x) = cx^a(1/2-x)^\gamma(1-x)^b$. Law (7.142) assumes as hypothesis the (reasonable?) fact that in correspondence to some point (in this case $x = 1/2$) the propensity tends to stabilize. For numerical simulation we choose $a = b = 4$, $\gamma = 2$, $c = 750$, $T = 5$ and $\lambda = 1$, and the same initial and boundary conditions as above. Figure 7.14 shows the

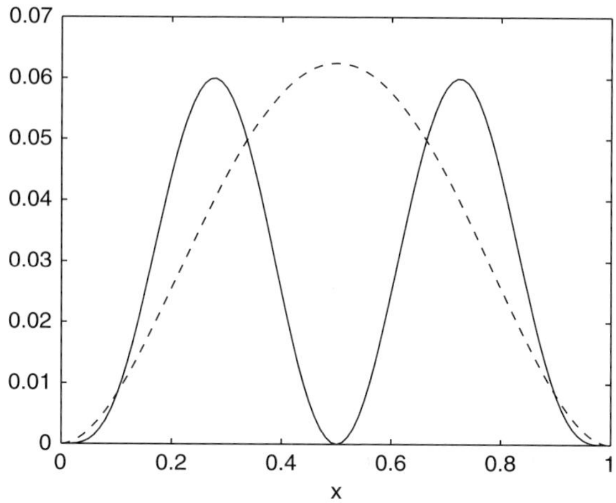

Fig. 7.15 Functions $\mathcal{W}(x)$ and $\nu(x)$ used in the numerical illustration.

plot of the densities ρ and m at time $T = 5$ resulting from the computations with the different laws. Finally, Figure 7.15 shows a comparison of $\mathcal{W}(x)$, $\nu(x)$ corresponding to the same values of a and b.

8
Modelling in life sciences

> Every resultant is either a sum or a difference of the co-operant forces; their sum, when their directions are the same – their difference, when their directions are contrary. Further, every resultant is clearly traceable in its components, because these are homogeneous and commensurable. It is otherwise with emergents, when, instead of adding measurable motion to measurable motion, or things of one kind to other individuals of their kind, there is a co-operation of things of unlike kinds. The emergent is unlike its components insofar as these are incommensurable, and it cannot be reduced to their sum or their difference.
>
> G. H. Lewes, *Problems of Life and Mind*, 1875.

Chapters 5, 6 and 7 contain a detailed description of various applications of the kinetic modelling to economic and social sciences. Although very important for the reflections that these areas have in our daily lives, this modelling covers only a small part of the potential applications of statistical mechanics and kinetic modelling of multiagent interacting systems. Having to make a selection of arguments, in this chapter we introduce, from the point of view of multiagent modelling, two main examples from life sciences, in which the methods presented in this book have been fruitfully employed to give both a valid alternative to the physical microscopic description, and, as a non-secondary by-product, to a new approach to numerical simulation.

The first example deals with a classic topic in genetics, which has involved both mathematicians and biologists in research in the last fifty years [329]. The second example deals with a kinetic version of flocking by Cucker and Smale [111], which describes the collective behaviour of a set of organisms, animals or devices. The pertinent literature in both subjects is very wide, and these examples are intended as significative prototypes for an introduction to the subjects.

8.1 The Luria–Delbrück distribution

The object of our study is known under the name of the Luria–Delbrück model [214]. This mathematical topic has a long history of nearly seventy years, beginning in 1943, when Salvador Luria, an Italian biologist, and Max Delbrück, a German physicist and biologist, published their article in *Genetics*, 'Mutations of bacteria from virus sensitivity to virus resistance'. This article proposed a model to describe the trend, within a population of bacteria, of the number of individuals who have a mutation that provides resistance to viruses (also called bacteriophages or phages).

Over the decades, numerous variations and mathematical formulations for this model have been proposed, primarily with the intention of obtaining a distribution for the number of cells which present a particular mutation. From the point of view of medical research the importance of these studies mainly lies in the fact that genetic mutations are the basis of the development of cancer, and to know what the probability is at a certain time of having a number of cells, within a population, which presents a particular mutation, can also be of great utility for the testing of drugs and experimentation with new treatments [28].

Fig. 8.1 Salvador Edward Luria[1]

The estimation of mutation rates is a challenging problem which has attracted the interest of generations of both geneticists and mathematicians. This problem, which is essentially a matter of evaluation of the growth of a population, had its origin in a series of classic experiments pioneered by Luria and Delbrück [214], which led to a seminal example of devising a mathematical model to estimate mutation rates. The model proposed by Luria and Delbrück assumed deterministic growth of mutant cells, which seemed too stringent an assumption to allow for efficient extraction of reliable information about mutation rates from experimental data. This shortcoming of the Luria and Delbrück model was, some years later, remedied by a slightly different mathematical formulation proposed by Lea and Coulson [204], who adopted the Yule stochastic birth process to mimic the growth of mutant cells. In the ensuing decades the Lea–Coulson formulation occupied so prominent a place in the study of mutation rates that the Lea–Coulson formulation is now commonly referred to as the Luria–Delbrück model. Noteworthy contributions to the study of the Lea–Coulson formulation are [8, 12, 21, 110, 184, 197, 218, 224, 245, 296] and numerous others cited in the essential review article by Zheng [329], which appeared at the end of the last century. In his paper, Zheng examines the various mathematical formulations, focusing on important practical issues closely linked with the distribution of the number of mutants. It appears evident that almost all of these formulations are prominently based on probabilistic arguments.

Very recently, the problem has been attacked from the multiagent system point of view in [191], by introducing a kinetic description of the density of mutation rates, in which the Luria–Delbrück distribution appears as a suitable limit of the underlying linear Boltzmann equation describing the growth of mutated cells. Following Luria and Delbrück's original model, mutation is, in the kinetic picture, a consequence of a specific *collision rule*, under which both the normal cells and the mutant cells grow deterministically but mutations occur randomly. The analysis in [191] easily extends to

[1]**Salvador Edward Luria**: born on 13 August 1912 in Torino (Italy), died on 6 February 1991 in Lexington, Massachusetts, (USA). He won the Nobel Prize in Physiology or Medicine in 1969, with Max Delbrück and Alfred Hershey, for their discoveries concerning the replication mechanism and the genetic structure of viruses.

cover the Lea and Coulson model [204], which is based on a different formulation under which the normal cells grow deterministically but the mutants grow stochastically [192].

As extensively discussed in Chapter 1, the core of the kinetic methodology is to identify the *microscopic interaction rule* which drives the underlying process of mutation, in order to write a kinetic equation of the Boltzmann type for the distribution density of the mutated cells. As is common in this approach, the construction of the kinetic equation emphasizes the differences between the microscopic interaction rules of the original Luria–Delbrück distribution and the Lea–Coulson modification. Even in this case, the kinetic description allows us to use in a natural and powerful way the Fourier description, which clarifies the nature of the assumption made by Lea and Coulson [204], namely the choice of a Yule process for the random growth of the mutated cells [192]. It is remarkable, in fact, that the kinetic description based on linear interactions is enormously flexible, and allows us to establish a direct connection between the *microscopic collision* and the consequent *macroscopic behaviour*. Once again, as happens in wealth distribution, where the characteristics of the binary trade reflect into the steady profile of wealth, or in opinion formation, where the details of the binary exchange of opinion produce the final distribution of opinion in the system, one can learn from microscopic dynamics the forthcoming macroscopic behaviour of mutant cells, and associate with the various models the corresponding mechanism of interaction.

8.1.1 The kinetic description of mutation rates

The most-used formulations of a mutation are generally expressed by (A) grey→ black, (B) grey→ grey + black. Here, *grey* stands for a normal cell and *black* a mutant cell. Note that under formulation (A), a mutation entails the loss of the normal cell. In formulation (B), the normal cell continues to live after a mutation. Consequently, the occurrence of a mutation does not decrease the rate at which mutations occur in subsequent time intervals. Assumption (B) was implicitly made by Lea and Coulson [204] and Armitage [12]. Since our goal is to present a kinetic description of the Lea–Coulson formulation, in what follows we will assume formulation (B).

The first mathematical description of the mutation process, which was subsequently modified by Lea and Coulson, goes back to Luria and Delbrück [214]. Because all subsequent formulations are simple variations of this formulation, we detail below its underlying assumptions.

The process starts at time $t = 0$ with one normal cell and no mutants. Normal cells are assumed to grow deterministically at a constant rate β_1. Therefore, the number of normal cells at time $t > 0$ is $N(t) = e^{\beta_1 t}$.

Mutants grow deterministically at a constant rate β_2. If a mutant is generated by a normal cell at time $s > 0$, then the clone spawned by this mutant will be of size $e^{\beta_2(t-s)}$ for any $t > s$. Mutations occur randomly at a rate proportional to $N(t)$. If μ denotes the per-cell per-unit-time mutation rate, then the standard assumption is that mutations occur in accordance with a Poisson process having the intensity function

$$\nu(t) = \mu N(t) = \mu e^{\beta_1 t}. \tag{8.1}$$

Consequently, the expected number of mutations occurring in the time interval $[0, t)$ is

$$\int_0^t \nu(s)\, ds = \frac{\mu}{\beta_1}\left(e^{\beta_1 t} - 1\right).$$

Let $X(t)$ denote the number of mutants existing at time t, and $f(n, t)$ the probability of having n mutants at time t. The previous assumptions imply that $X(t)$ can be expressed as

$$X(t) = \sum_{i=1}^{M(t)} \exp\{\beta_2(t - \tau_i)\}, \quad \text{if} \quad M(t) \geq 1, \tag{8.2}$$

while

$$X(t) = 0 \quad \text{if} \quad M(t) = 0\,.$$

Here, τ_i are the random times at which mutations occur, and $M(t)$ stands for the mutation process which is a Poisson process with intensity function $\nu(t)$ given in (8.1). The probabilistic description (8.2) allows for recovering the various quantities like mean, variance and cumulants (cf. the detailed historical description in [329]).

An interesting alternative to the probabilistic construction described above, is to use the typical approach of statistical mechanics, as adopted many times in this book. Within this approach, the study of the time evolution of the probability distribution $f(n, t)$ of mutants, together with a reasonable explanation of the growth process induced by this distribution, is achieved by means of kinetic collision-like models [191]. Assuming for simplicity a continuous parameter v for the number of mutants, the time evolution of the distribution $f(v, t)$ is therefore based on the assumption of collision-like trade events.

The interaction rule described by the previous probabilistic description indeed corresponds to assuming that, at time $t > 0$, the microscopic growth of mutants is driven by

$$v^* = v + \beta_2 v + \eta(t), \tag{8.3}$$

where $\eta \geq 0$ (absence of backward mutations) is a discrete random variable distributed according to a Poisson density $p(\nu(t))$ of intensity given by (8.1). Hence, for any given v, the number of mutants after the interaction can only assume the values

$$v_i^* = v + \beta_2 v + i, \quad i \in \mathbb{N}.$$

The interaction (8.3) then leads to the kinetic equation (in weak form) [191]

$$\frac{d}{dt}\int_{\mathbb{R}_+} f(v,t)\varphi(v)\, dv = \sum_{i=1}^{\infty} p_i(\nu(t)) \int_{\mathbb{R}_+} \left(\varphi(v_i^*) - \varphi(v)\right) f(v,t)\, dv, \tag{8.4}$$

where $\varphi(v)$ is a smooth function. The condition of having no mutants at time $t = 0$ translates into the initial condition

$$f(v, t = 0) = f_0(v) = \delta_0(v), \tag{8.5}$$

where $\delta_0(v)$ indicates as usual the Dirac delta function concentrated at $v = 0$.

Interaction (8.3) belongs to a class of linear interactions which can be fruitfully described in a unified way, which is sufficiently flexible to describe both the original Luria–Delbrück and Lea–Coulson formulations, and easier to understand from the modelling point of view.

In any of the aforementioned formulations of the growth of the mutant cells, the microscopic process is due to two simultaneous events. The first one describes the intrinsic growth of the mutants, which can be represented by a general interaction of type

$$v' = v + \mathcal{L}(v), \tag{8.6}$$

where $v \to \mathcal{L}(v) \geq 0$ is a map acting on v. Note that the choice $\mathcal{L}(v) = \beta_2 v$ gives the classical Luria–Delbrück relation (8.3). The map can be either deterministic or random. In the latter case, corresponding to the Lea–Coulson formulation, denoting by $\langle\cdot\rangle$ the mathematical expectation, in order to maintain the same mean growth of the deterministic interaction we will assume that

$$\langle \mathcal{L}(v)\rangle = \beta_2 v. \tag{8.7}$$

The second event describes the growth of the mutant cells due to the presence of the normal ones which can mutate according to assumption (B). In our picture this corresponds to assuming that the mutants are immersed in a background (of normal cells), which, due to their random mutation rule, undergo interactions which increase the number of mutants according to the time-dependent probability distribution $M(w, t)$. In the Luria–Delbrück setting the probability distribution of the background depends on time, and it is a Poisson process with intensity $\nu(t) = \mu N(t)$. Consequently,

$$\int_{\mathbb{R}_+} M(v,t)\, dv = 1\,; \quad \int_{\mathbb{R}_+} vM(v,t)\, dv = \mu e^{\beta_1 t}. \tag{8.8}$$

Coupling the interaction (8.3) with the interaction due to the background, one obtains the general microscopic interaction

$$v^* = v + \mathcal{L}(v) + w. \tag{8.9}$$

Following the arguments of Chapter 1, the effect of interactions (8.9) on the time variation of the mutant density can be quantitatively described by a linear Boltzmann-type equation in which the variation of the density $f = f(v, t)$ is due to a balance between gain and loss terms that, for the given number v, take into account all the interactions of type (8.9) which end up with the number v (gain term) as well as all the interactions which, starting from the number v, lose this value after interaction (loss term). The balance equation for the density of the number of mutant cells, written in weak form, reads

$$\frac{d}{dt}\int_{\mathbb{R}_+} \varphi(v) f(v,t)\, dv = \left\langle \int_{\mathbb{R}_+^2} \left(\varphi(v^*) - \varphi(v)\right) f(v,t) M(w,t)\, dv\, dw \right\rangle. \tag{8.10}$$

In (8.10), the presence of $\langle\cdot\rangle$ takes into account the possibility that $\mathcal{L}(v)$ could be a random quantity. Note that (8.8) implies that $f(v, t)$ remains a probability density if it is so initially:

$$\int_{\mathbb{R}_+} f(v,t)\,dv = \int_{\mathbb{R}_+} f_0(v)\,dv = 1. \tag{8.11}$$

Moreover, on the basis of (8.9), choosing $\varphi(v) = v$ shows that the total mean number of mutants satisfies the differential equation

$$\frac{dm(t)}{dt} = \int_{\mathbb{R}_+} \langle \mathcal{L}(v) \rangle f(v,t)\,dv + \mu e^{\beta_1 t}. \tag{8.12}$$

We note that the choice $\langle \mathcal{L}(v) \rangle = \beta_2 v$ implies

$$\frac{dm(t)}{dt} = \beta_2 m(t) + \mu e^{\beta_1 t}, \tag{8.13}$$

which, coupled with the initial condition $m(0) = 0$ induced by (8.5), leads to the explicit computation of the mean value of the Luria–Delbrück distribution ($\beta_1 \neq \beta_2$), for any time $t > 0$,

$$m(t) = \frac{\mu}{\beta_1 - \beta_2} \left(e^{\beta_1 t} - e^{\beta_2 t} \right).$$

This expression is independent of the particular form of the map $\mathcal{L}$, provided the map is such that condition (8.7) is satisfied. In other words, at the level of the growth of the mean value, the original formulation by Luria–Delbrück, and its modification considered by Lea–Coulson, produce the same time behaviour.

8.1.2 Preliminary results

As is usually done for this type of kinetic model, most of the analytical properties of the solution are obtained by means of Fourier analysis. The weak formulation (8.10) is, in fact, equivalent to the Fourier-transformed equation

$$\frac{\partial \widehat{f}(\xi,t)}{\partial t} = \hat{M}(\xi,t) \int_{\mathbb{R}_+} \langle e^{-i\xi \mathcal{L}(v)} \rangle e^{-i\xi v} f(v,t)\,dv - \hat{f}(\xi,t). \tag{8.14}$$

In (8.14) we used the fact that $M(v,t)$ is a probability density function, so that $\hat{M}(0,t) = 1$. The initial condition (8.5) turns into

$$\widehat{f_0}(\xi) = 1.$$

Note that, since $M(w,t)$ is a Poisson process of intensity $\nu(t) = \mu e^{\beta_1 t}$,

$$\hat{M}(\xi,t) = \exp\left\{ \mu e^{\beta_1 t} \left(e^{-i\xi} - 1 \right) \right\}.$$

The explicit evaluation of the gain term on the right-hand side of equation (8.14) requires the knowledge of $\mathcal{L}(v)$. In the case in which $\mathcal{L}(v) = \beta_2 v$, equation (8.14) becomes

$$\frac{\partial \widehat{f}(\xi,t)}{\partial t} = \hat{M}(\xi,t) \hat{f}((1+\beta_2)\xi) - \hat{f}(\xi,t). \tag{8.15}$$

Likewise, by assuming that $\mathcal{L}(v)$ is a Poisson variable of mean value $\beta_2 v$, thus satisfying (8.7),

$$\langle e^{-i\xi\mathcal{L}(v)}\rangle = \exp\left\{\beta_2 v\left(e^{-i\xi}-1\right)\right\} = \exp\left\{-iv\beta_2 i\left(e^{-i\xi}-1\right)\right\}, \tag{8.16}$$

and equation (8.14) becomes

$$\frac{\partial \widehat{f}(\xi,t)}{\partial t} = \hat{M}(\xi,t)\hat{f}\left(\xi+\beta_2 i\left(e^{-i\xi}-1\right),t\right)-\hat{f}(\xi,t). \tag{8.17}$$

Equations (8.15) and (8.17) differ only in the gain term on the right-hand side, where the Fourier transform of the probability density is evaluated at different points.

The existence of a solution to equation (8.15) and (8.17) can be seen easily using the same techniques as Chapter 2. In an analogous way, in order to show uniqueness of the solution, we use the Fourier-based metric (2.17). The goal is to prove that the operator on the right-hand side of (8.14) is Lipschitz continuous in this metric. Since in this case we deal with a right-hand side which is explicitly time dependent, we present the results in detail.

In order to show Lipschitz continuity, if $\hat{f}_1(\xi,t)$ and $\hat{f}_2(\xi,t)$ are two solutions to equation (8.15) corresponding to the initial data $\hat{f}_{0,1}(\xi)$ and $\hat{f}_{0,2}(\xi)$, it holds that

$$\begin{aligned}\frac{\partial}{\partial t}\frac{\hat{f}_1(\xi,t)-\hat{f}_2(\xi,t)}{|\xi|^s}&+\frac{\hat{f}_1(\xi,t)-\hat{f}_2(\xi,t)}{|\xi|^s}\\&=\frac{\hat{f}_1((1+\beta_2)\xi,t)-\hat{f}_2((1+\beta_2)\xi,t)}{|\xi|^s}\hat{M}(\xi,t).\end{aligned}$$

On the other hand, since $|\hat{M}(\xi,t)|\le 1$,

$$\begin{aligned}&\sup_{\xi\in\mathbb{R}}\left|\frac{\hat{f}_1((1+\beta_2)\xi,t)-\hat{f}_2((1+\beta_2)\xi,t)}{|\xi|^s}\hat{M}(\xi,t)\right|\\&\quad\le\sup_{\xi}\left|\frac{\hat{f}_1((1+\beta_2)\xi,t)-\hat{f}_2((1+\beta_2)\xi,t)}{|\xi|^s}\right|=(1+\beta_2)^s d_s(f_1(t),f_2(t)).\end{aligned} \tag{8.18}$$

Thanks to (8.18),

$$\left|\frac{\partial}{\partial t}\frac{\hat{f}_1(\xi,t)-\hat{f}_2(\xi,t)}{|\xi|^s}+\frac{\hat{f}_1(\xi,t)-\hat{f}_2(\xi,t)}{|\xi|^s}\right|\le(1+\beta_2)^s d_s(f_1(t),f_2(t)),$$

which, by Gronwall's inequality, implies

$$d_s(f_1(t),f_2(t))\le d_s(f_{0,1},f_{0,2})\exp\left\{\left((1+\beta_2)^s-1\right)\right)t\right\}, \tag{8.19}$$

which implies uniqueness.

The same method applies to equation (8.17). In this case, by choosing $s=1$, formula (8.18) reads

$$\begin{aligned}&\sup_{\xi}\left|\frac{\hat{f}_1\left(\xi+\beta_2 i\left(e^{-i\xi}-1\right),t\right)-\hat{f}_2\left(\xi+\beta_2 i\left(e^{-i\xi}-1\right),t\right)}{|\xi|}\right||\hat{M}(\xi,t)|\\&\quad\le\sup_{\eta}\left|\frac{\hat{f}_1(\eta,t)-\hat{f}_2(\eta,t)}{|\eta|}\right|\frac{|(\xi+\beta_2 i\left(e^{-i\xi}-1\right)|}{|i\xi|}\le(1+\beta_2)d_1(f_1(t),f_2(t)).\end{aligned} \tag{8.20}$$

The previous analysis indicates that one can obtain both existence and uniqueness of a solution to equation (8.14) for a wide choice of the map $\mathcal{L}(v)$.

The Fourier-transformed kinetic equation allows for a direct evaluation of the evolution equation for the cumulant-generating function, defined as

$$k(\xi, t) = \log \hat{f}(\xi, t). \tag{8.21}$$

As an example, a simple computation on equation (8.15) shows that $k(\xi, t)$ solves

$$\frac{\partial k(\xi, t)}{\partial t} = \exp\left\{k((1+\beta_2)\xi, t) - k(\xi, t)\right\} \hat{M}(\xi, t) - 1. \tag{8.22}$$

Generally, the cumulants can be extracted from the cumulant-generating function via differentiation (at zero) of $k(\xi, t)$. That is, the cumulants appear as the coefficients in the Maclaurin series of $k(\xi, t)$. In the case under consideration, the modulus of the first coefficient coincides with the mean value, and the second one with the variance. The evaluation of the subsequent derivatives of $k(\xi, t)$ appears extremely cumbersome, and the best one can do explicitly is to evaluate the first two coefficients of the Maclaurin series. On the other hand, since the explicit expression of the solution to the Boltzmann equation (8.14) is not known, the description of the solution though its cumulants is one of the main operative possibilities.

8.2 The quasi-invariant limit of the growth of mutant cells

Due to its particular structure, in general it is quite difficult to obtain analytic results, or to describe precisely properties of the solution to equation (8.14). As explained at the end of the previous section, it is cumbersome to deduce from equation (8.22) a recursive relation which allows for detailed computation of the cumulants. This difficulty is well known to people working on this topic, where explicit results are rarely present, independent of the different formulations. From this point of view, while alternative to standard probabilistic approaches, the kinetic picture does not produce evident simplifications.

By analogy with similar situations discussed in Chapter 1, Section 1.4, one possibility is to make use of particular asymptotics of the equation which result in simplified models of Fokker–Planck type, for which it is relatively easier to compute in many cases the relevant properties of the solution, if not its analytic expression.

In order to give, in the present case, a physical basis to these asymptotics, let us discuss in some detail the evolution equation for the mean, given by (8.13). This evolution equation has a universal value, since it does not depend on the particular choice of the map $\mathcal{L}$, but on its mean value. The scaling

$$\beta_1 \to \varepsilon\beta_1, \quad \beta_2 \to \varepsilon\beta_2, \quad \mu \to \varepsilon\mu, \quad t \to \varepsilon t = \tau, \tag{8.23}$$

is such that the mean value of the density $f(v, t)$ satisfies

$$\frac{dm_f(t)}{dt} = \varepsilon\beta_2 m_f(t) + \varepsilon\mu e^{\varepsilon\beta_1 t} = \varepsilon\left(\beta_2 m_f(t) + \mu e^{\beta_1 \tau}\right).$$

If we set $g_\varepsilon(v,\tau) = f(v,t)$, then $m_{g_\varepsilon}(\tau) = m_f(t)$, and the mean value of the density $g_\varepsilon(v,\tau)$ solves

$$\frac{dm_{g_\varepsilon}(\tau)}{d\tau} = \beta_2 m_{g_\varepsilon}(\tau) + \mu e^{\beta_1\tau}. \tag{8.24}$$

Note that equation (8.24) does not depend explicitly on the scaling parameter ε. In other words, we can reduce the growth of the mutants, waiting enough time to get the same law for the mean value of the density of mutants.

Using (8.23), we can write the kinetic equation 8.10 for $g_\varepsilon(v,\tau)$. This equation reads

$$\frac{d}{d\tau}\int_{\mathbb{R}_+} \varphi(v) g_\varepsilon(v,\tau)\,dv = \frac{1}{\varepsilon}\left\langle \int_{\mathbb{R}_+^2} (\varphi(v_\varepsilon^*) - \varphi(v)) g_\varepsilon(v,\tau) M_\varepsilon(w,\tau)\,dv\,dw \right\rangle. \tag{8.25}$$

In (8.25),

$$v_\varepsilon^* = v + \mathcal{L}_\varepsilon(v) + w,$$

and $\mathcal{L}_\varepsilon(v)$ denotes the map with mean growth $\varepsilon\beta_2$. Moreover, $M_\varepsilon(v,\tau)$ is the Poisson process with intensity $\varepsilon\mu$ and growth parameter $\varepsilon\beta_1$.

By means of equation (8.25) we can consequently investigate the situation in which most of the interactions produce a very small growth of mutants ($\varepsilon \to 0$), while at the same time the evolution of the density is such that (8.24) remains unchanged. We will call this limit the quasi-invariant growth limit. A similar procedure has been applied to various kinetic models in economics in Chapter 5 and opinion formation in Chapter 6.

Under the scaling (8.23), equation (8.14) for $g_\varepsilon(v,\tau)$ reads

$$\frac{\partial \hat{g}_\varepsilon(\xi,\tau)}{\partial\tau} = \frac{1}{\varepsilon}\left[\hat{M}_\varepsilon(\xi,\tau)\int_{\mathbb{R}_+} \langle e^{-i\xi\mathcal{L}_\varepsilon(v)}\rangle e^{-i\xi v} g_\varepsilon(v,\tau)\,dv - \hat{g}_\varepsilon(\xi,\tau)\right]. \tag{8.26}$$

Let us observe that equation (8.26) can be written equivalently as

$$\frac{\partial \hat{g}_\varepsilon(\xi,\tau)}{\partial\tau} = \frac{1}{\varepsilon}\int_{\mathbb{R}_+} \left(\langle e^{-i\xi\mathcal{L}_\varepsilon(v)}\rangle \hat{M}_\varepsilon(\xi,\tau) - 1\right) e^{-i\xi v} g_\varepsilon(v,\tau)\,dv, \tag{8.27}$$

where

$$\hat{M}_\varepsilon(\xi,\tau) = \exp\left\{\varepsilon\mu e^{\beta_1\tau}\left(e^{-i\xi} - 1\right)\right\}$$

is the Fourier transform of the Poisson process $M_\varepsilon(v,\tau)$. By assuming that $\mathcal{L}(v)$ is a Poisson variable of mean value $\beta_2 v$, so that (8.16) holds, (8.27) becomes

$$\frac{\partial \hat{g}_\varepsilon(\xi,\tau)}{\partial\tau} = \frac{1}{\varepsilon}\int_{\mathbb{R}_+} \left(\exp\left\{\varepsilon\left(\beta_2 v + \mu e^{\beta_1\tau}\right)\left(e^{-i\xi} - 1\right)\right\} - 1\right) e^{-i\xi v} g_\varepsilon(v,\tau)\,dv,$$

which, expanding the exponential function in the integral in Taylor series gives

$$\begin{aligned}\frac{\partial \hat{g}_\varepsilon(\xi,\tau)}{\partial\tau} &= \left(e^{-i\xi} - 1\right)\int_{\mathbb{R}_+} \left(\beta_2 v + \mu e^{\beta_1\tau}\right) e^{-i\xi v} g_\varepsilon(v,\tau)\,dv + \varepsilon R(\xi,\tau)\\ &= \left(e^{-i\xi} - 1\right)\left(i\beta_2\frac{\partial \hat{g}_\varepsilon(\xi,t)}{\partial\xi} + \mu e^{\beta_1\tau}\hat{g}_\varepsilon(\xi,\tau)\right) + \varepsilon R_\varepsilon(\xi,\tau).\end{aligned} \tag{8.28}$$

The remainder term R is such that

$$|R_\varepsilon(\xi,\tau)| \le \frac{1}{2}|e^{-i\xi}-1|^2 \int_{\mathbb{R}_+} \left|\beta_2 v + \mu e^{\beta_1 \tau}\right|^2 g_\varepsilon(v,\tau)\,dv. \tag{8.29}$$

Standard computations then show that, for any fixed time $\tau > 0$, the remainder term remains uniformly bounded with respect to ε. Therefore, letting $\varepsilon \to 0$ we obtain that the limit function $g(v,\tau)$ satisfies the equation

$$\frac{\partial \hat{g}(\xi,\tau)}{\partial \tau} = \left(e^{-i\xi}-1\right)\left(i\beta_2 \frac{\partial \hat{g}(\xi,\tau)}{\partial \xi} + \mu e^{\beta_1\tau}\hat{g}(\xi,\tau)\right). \tag{8.30}$$

We remark that equation (8.30) is the evolution equation for the Fourier transform of the Lea–Coulson distribution function [329]. Consequently, the Lea–Coulson formulation of the Luria–Delbrück distribution is obtained as the quasi-invariant growth limit of the kinetic model (8.17), in which the self-growth of the mutated cells follows a Poisson process of intensity proportional to the number v of mutated cells. Under this assumption, differently from what happens in the original Luria–Delbrück distribution, the distribution function $g(v,\tau)$ takes values only on the positive natural numbers.

Equation (8.30) can easily be transformed back to the space of probability distributions to get, for $n \ge 0$, the evolution equation for the probabilities $g(n,\tau)$ of having n mutant cells at time τ. These probabilities obey the recursive equation

$$\frac{\partial g(n,\tau)}{\partial \tau} = \beta_2\left((n-1)g(n-1,\tau) - ng(n,\tau)\right) + \mu e^{\beta_1\tau}\left(g(n-1,\tau)-g(n,\tau)\right), \tag{8.31}$$

where $g(-1,\tau) = 0$. Note that, if we assume as in Section 8.1.1 that at time $\tau = 0$ there are no mutants, the initial conditions are given by

$$g(0,\tau=0) = 1, \quad g(n,\tau=0) = 0 \quad \text{if} \quad n \ge 1.$$

Consider now that, for a given $\varepsilon > 0$, the Taylor expansion we used to obtain formula (8.28), can be used to express equation (8.30) like a kinetic equation of type (8.27) plus a remainder term which now depends on the solution $g(v,\tau)$ and its moments,

$$\begin{aligned} \frac{\partial \hat{g}(\xi,\tau)}{\partial \tau} &= \left(e^{-i\xi}-1\right)\left(i\beta_2 \frac{\partial \hat{g}(\xi,\tau)}{\partial \xi} + \mu e^{\beta_1\tau}\hat{g}(\xi,\tau)\right) \\ &= \frac{1}{\varepsilon}\left[\hat{M}_\varepsilon(\xi,\tau)\int_{\mathbb{R}_+} \langle e^{-i\xi\mathcal{L}_\varepsilon(v)}\rangle e^{-i\xi v} g(v,\tau)\,dv - \hat{g}(\xi,\tau)\right] - \varepsilon R(\xi,\tau), \end{aligned} \tag{8.32}$$

where now the remainder term takes the form

$$R(\xi,\tau) = \frac{1}{2}\left(e^{-i\xi}-1\right)^2 \int_{\mathbb{R}_+} \theta(\varepsilon,v)\left(\beta_2 v + \mu e^{\beta_1\tau}\right)^2 e^{-i\xi v} g(v,\tau)\,dv, \tag{8.33}$$

with $0 \le \theta(\varepsilon,v) \le 1$. Since the mean and the variance of the Lea–Coulson distribution remain bounded in time [329], formula (8.33) shows that

$$\frac{R(\xi,\tau)}{|\xi|} \le \int_{\mathbb{R}_+} \left|\beta_2 v + \mu e^{\beta_1 \tau}\right|^2 g(v,\tau)\,dv = K(\tau) < +\infty.$$

If $\hat{g}$ and $\hat{g}_\varepsilon$ denote the solutions to (8.26) and (8.30), respectively, let us define

$$h(\xi,\tau) = \frac{\hat{g}(\xi,\tau) - \hat{g}_\varepsilon(\xi,\tau)}{|\xi|}.$$

Using equation (8.32), and proceeding as in the computations of formula (8.18), we find that $h(\xi,\tau)$ satisfies

$$\partial_\theta h(\xi,\tau) + \frac{1}{\varepsilon}\, h(\xi,\tau) \le \frac{1+\varepsilon\beta_2}{\varepsilon}\, \|h\|_\infty(\theta) + \varepsilon K(\tau). \tag{8.34}$$

This is equivalent to

$$\partial_\tau \left(h(\xi,\tau)e^{\tau/\varepsilon}\right) \le \frac{1+\varepsilon\beta_2}{\varepsilon}\, \|h(\cdot,\tau)e^{\tau/\varepsilon}\|_\infty + \varepsilon K(\tau)e^{\tau/\varepsilon}.$$

Integrating from 0 to τ, we get

$$h(\xi,\tau)e^{\tau/\varepsilon} \le h(\xi,0) + \int_0^\tau \left(\frac{1+\varepsilon\beta_2}{\varepsilon}\, \|h(\cdot,t)e^{t/\varepsilon}\|_\infty + \varepsilon K(t)e^{t/\varepsilon}\right) dt.$$

Hence, if $H_\varepsilon(\tau) = \|h(\cdot,\tau)e^{\tau/\varepsilon}\|_\infty$,

$$H_\varepsilon(\tau) \le H_\varepsilon(0) + \int_0^\tau \frac{1+\varepsilon\beta_2}{\varepsilon}\, H_\varepsilon(t)\,dt + \int_0^\theta \varepsilon K(t)e^{t/\varepsilon}\,dt.$$

Now, by the generalized Gronwall inequality,

$$u(\tau) \le \phi(\tau) + \int_0^\tau \lambda(t)u(t)\,dt$$

implies

$$u(\tau) \le \phi(0)\exp\left\{\int_0^\tau \lambda(t)\,dt\right\} + \int_0^\tau \exp\left\{\int_t^\tau \lambda(t)\,dt\right\} \frac{d\phi}{dt}\,dt.$$

Applying this inequality with $\lambda(\tau) = (1+\varepsilon\beta_2)/\varepsilon$ and $\phi(\tau) = H_\varepsilon(0) + \int_0^t \varepsilon K(t)e^{t/\varepsilon}\,dt$, we obtain

$$H_\varepsilon(\tau) \le \exp\left\{\frac{1+\varepsilon\beta_2}{\varepsilon}\right\}\left(H_\varepsilon(0) + \varepsilon\int_0^\tau K(t)e^{\beta_2 t}dt\right),$$

namely

$$\|h(\cdot,\tau)\|_\infty \le e^{\beta_2\tau}\left(\|h(\cdot,0)\|_\infty + \varepsilon\int_0^\tau K(t)e^{\beta_2 t}dt\right).$$

Therefore, by choosing the same initial data for equations (8.26) and (8.30), so that $\|h(\cdot,0)\|_\infty = 0$, we obtain that, for $\tau > 0$,

$$d_1(g,g_\varepsilon)(\tau) = \sup_\xi \frac{|\hat{g}(\xi,\tau) - \hat{g}_\varepsilon(\xi,\tau)|}{|\xi|} \le \varepsilon e^{\beta_2\tau}\int_0^\theta K(t)e^{\beta_2 t}dt.$$

Letting $\varepsilon \to 0$ shows that the solution to the kinetic model (8.26) converges to the solution of the evolution equation for the Lea–Coulson formulation (8.30). The previous computations can be put in the form of a rigorous Theorem.

Theorem 8.1 *Let the probability density f_0 be defined as in (8.5). Then, choosing as parameters $\varepsilon\beta_1$, $\varepsilon\beta_2$ and $\varepsilon\mu$, as $\varepsilon \to 0$, the weak solution to the kinetic equation for the scaled density $g_\varepsilon(v,\tau) = f(v,t)$, with $\tau = \varepsilon t$, converges to a probability density $g(w,\tau)$. This density is a weak solution of the Fokker–Planck-like equation (8.30), whose solution is the Fourier transform of the Luria–Delbrück distribution in the Lea–Coulson formulation.*

The same strategy applies to the classical Luria–Delbrück distribution, which appears as the weak limit of the kinetic equation (8.15). In this case, using the same mathematical tools, one can prove that the solution to (8.15) converges to the solution to the Fokker–Planck-like equation

$$\frac{\partial \hat{g}(\xi,\tau)}{\partial \tau} = \beta_2 \xi \frac{\partial \hat{g}(\xi,\tau)}{\partial \xi} + \mu e^{\beta_1 \tau} \left(e^{-i\xi} - 1 \right) \hat{g}(\xi,\tau). \tag{8.35}$$

Once again, the limit procedure leading to equation (8.35) illustrates the powerful role of the Fourier-based metric (2.17) introduced in Chapter 2.

8.2.1 The Bartlett formulation

As extensively discussed in Section 8.1.1, in the Lea–Coulson formulation of the Luria–Delbrück distribution, the growth of mutants is assumed to be a random process, while the growth of normal cells is assumed to be completely deterministic. In the Bartlett formulation the growth of both normal and mutant cells is assumed to be fully stochastic [23, 329]. In this formulation the Luria–Delbrück model is represented by a two-dimensional birth process $(X_1(t); X_2(t)), t \geq 0$, where $X_1(t)$ and $X_2(t)$ represent the population size at time t of the normal cells and that of the mutant cells, respectively. The formulation is called fully stochastic because it models the growth of both the normal cells and the mutant cells by stochastic growth processes.

At a kinetic level, the Bartlett formulation belongs to a class of linear kinetic equations for the distribution density $f = f(v,w,t)$ in which v and w represent the number of mutant and normal cells, respectively, which are subject to the interaction rules

$$v^* = v + \mathcal{L}_2(v) + w\,; \quad w^* = w + \mathcal{L}_1(w). \tag{8.36}$$

As for the simpler interaction (8.9), the two maps $\mathcal{L}_1$ and $\mathcal{L}_2$ can be either deterministic or random. According to our standard assumptions,

$$\langle \mathcal{L}_1(w) \rangle = \beta_1 w\,; \quad \langle \mathcal{L}_2(v) \rangle = \beta_2 v.$$

Resorting to the scaling of Section 8.1.1, the effect of interactions (8.36) on the time variation of the density $f = f(v,w,t)$ can be quantitatively described by a linear Boltzmann-type equation, that in weak form reads

$$\frac{d}{dt} \int_{\mathbb{R}_+^2} \varphi(v,w) f(v,w,t)\, dv\, dw = \langle \int_{\mathbb{R}_+^2} \left(\varphi(v^*,w^*) - \varphi(v,w) \right) f(v,w,t)\, dv\, dw \rangle. \tag{8.37}$$

In (8.37), the presence of $\langle \cdot \rangle$ takes into account the possibility that $\mathcal{L}_i(v), i = 1,2$, could be random quantities. Note that, by choosing $\varphi(v,w) = 1$, one easily recovers that $f(v,t)$ remains a probability density if it is so initially,

$$\int_{\mathbb{R}^2_+} f(v,w,t)\,dv\,dw = \int_{\mathbb{R}^2_+} f_0(v,w)\,dv\,dw = 1. \tag{8.38}$$

Moreover, on the basis of (8.36), choosing $\varphi(v,w) = v$ (respectively $\varphi(v,w) = w$) shows that the total mean numbers $m_1(t) = \int_{\mathbb{R}^2_+} wf(v,w,t)$ and $m_2(t) = \int_{\mathbb{R}^2_+} vf(v,w,t)$ of normal cells and mutants satisfy the differential system

$$\frac{dm_1(t)}{dt} = \int_{\mathbb{R}^2_+} \langle \mathcal{L}_1(w)\rangle f(v,w,t)\,dv = \beta_1 m_1(t), \tag{8.39}$$

$$\frac{dm_2(t)}{dt} = \int_{\mathbb{R}^2_+} (\langle \mathcal{L}_2(w)\rangle + w) f(v,w,t)\,dv = \beta_2 m_2(t) + m_1(t). \tag{8.40}$$

Note that, by solving (8.39) with initial condition $m_1(0) = \mu$ and substituting the result into (8.40), one obtains that the mean value $m_2(t)$ of mutants solves (8.13). Therefore, the evolution of the mean of mutants in the Bartlett formulation coincides with the evolution of the mean in the original Luria–Delbrück model.

Let us now suppose that the two maps $\mathcal{L}_i(v)$, $i = 1, 2$, are independent random quantities. Then, by choosing $\varphi(v,w) = e^{-i(\xi v + \eta w)}$, it is easily seen that the Fourier transform of the solution to (8.37),

$$\hat{f}(\xi,\eta,t) = \int_{\mathbb{R}^2_+} e^{-i(\xi v+\eta w)} f(v,w,t)\,dv\,dw,$$

satisfies the equation

$$\frac{\partial \widehat{f}(\xi,\eta,t)}{\partial t} = \int_{\mathbb{R}^2_+} \langle e^{-i\xi(\mathcal{L}_2(v)+w)}\rangle \langle e^{-i\eta \mathcal{L}_1(w)}\rangle e^{-i(\xi v+\eta w)} f(v,w,t)\,dv - \hat{f}(\xi,\eta,t). \tag{8.41}$$

If now $\mathcal{L}_i(v)$, $i = 1, 2$ are Poisson distributed with means $\beta_i v, i = 1, 2$, so that

$$\langle e^{-i\xi \mathcal{L}_2(v)}\rangle = e^{\beta_2 v(e^{-i\xi}-1)}, \quad \langle e^{-i\eta \mathcal{L}_1(w)}\rangle = e^{\beta_1 w(e^{-i\eta}-1)},$$

equation (8.41) reduces to

$$\frac{\partial \hat{f}(\xi,\eta,t)}{\partial t} = \hat{f}\left(\xi + i\beta_2(e^{-i\xi}-1), \xi+\eta+i\beta_1(e^{-i\eta}-1), t\right) - \hat{f}(\xi,\eta,t), \tag{8.42}$$

which represents the kinetic equation relative to the Bartlett formulation. Note that the initial conditions on mass and mean value move to

$$\hat{f}(0,0,t=0) = 1, \qquad \left.\frac{\partial \hat{f}(\xi,\eta,t=0)}{\partial \xi}\right|_{\xi=\eta=0} = 0, \qquad \left.\frac{\partial \hat{f}(\xi,\eta,t=0)}{\partial \eta}\right|_{\xi=\eta=0} = \mu. \tag{8.43}$$

Let us apply to the parameters the same scaling as in (8.23). Then proceeding as in Section 8.2 we find that the function $\hat{g}(\xi,\eta,\tau) = \hat{f}(\xi,\eta,t)$, where $\tau = \varepsilon t$, solves the equation

$$\frac{\partial \hat{g}(\xi,\eta,\tau)}{\partial \tau} = i\beta_2(e^{-i\xi}-1)\frac{\partial \hat{g}}{\partial \xi} + \left(\xi + i\beta_1(e^{-i\xi}-1)\right)\frac{\partial \hat{g}}{\partial \eta}, \tag{8.44}$$

with initial conditions given as in (8.43). Equation (8.44) differs from equation (95) in Zheng [329] in the coefficient of $\partial\hat{g}/\partial\eta$. Reverting to the physical space, we obtain, for

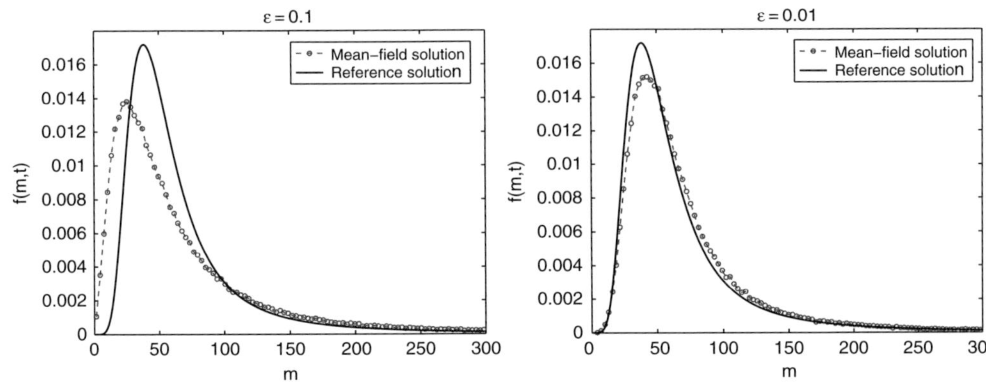

Fig. 8.2 Luria–Delbrück setting. Distribution of mutant cells at $t = 6.7$ with $\beta = 2.5$ for $\epsilon = 0.1$ (top) and $\epsilon = 0.01$ (bottom) in the kinetic model (8.10), with $\mathcal{L}(v) = \beta v$. The reference solution is computed using Lemma 2 in [329].

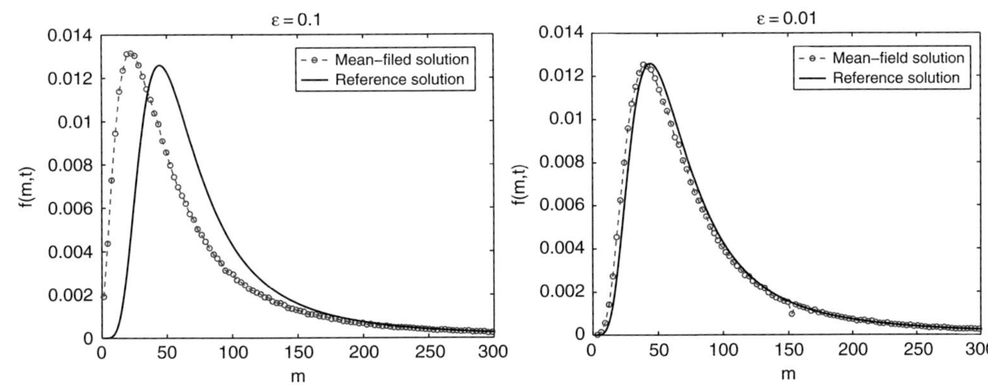

Fig. 8.3 Lea–Coulson setting. Distribution of mutant cells at $t = 6.7$ with $\beta = 2.8$ for $\epsilon = 0.1$ (top) and $\epsilon = 0.01$ (bottom) in the kinetic model (8.10), with $\mathcal{L}(v)$ a Poisson process of mean βv. The reference solution is computed using Lemma 2 in [329] and numerical quadrature.

$n, m \geq 0$, the evolution equation for the probabilities $g(n, m, \tau)$ of having n mutant cells and m normal cells at time τ. These probabilities obey the recursive equation

$$\frac{\partial g(n,m,\tau)}{\partial \tau} = \beta_2 \left((n-1)g(n-1,m,\tau) - ng(n,m,\tau)\right) +$$

$$\beta_1 \left((m-1)g(n,m-1,\tau) - mg(n,m,\tau)\right)\left(g(n-1,\tau) - g(n,\tau)\right), \tag{8.45}$$

where $g(-1,\tau) = 0$. If we assume, as in Section 8.1.1, that at time $\tau = 0$ there are no mutants, the initial conditions are given by

$$g(0,\tau = 0) = 1, \quad g(n,\tau = 0) = 0 \quad \text{if} \quad n \geq 1.$$

8.2.2 Numerical examples

We compare the continuous distribution of mutants obtained using the different kinetic models in the generalized Fokker–Planck limit and some standard methods for computing the approximated discrete distribution in the Luria–Delbrück and Lea–Coulson settings (see Lemma 2, page 11, in [329]).

As usual, for the numerical solution of the kinetic models we adopt a standard Monte Carlo simulation method. Here, the main difference is the necessity of generating Poisson samples at each time step. This can easily be achieved by standard algorithms; see, for example, [195, 227].

The test cases considered were proposed in [329]. We start from initial conditions where no mutants are present: $f_0(m) = \delta(0)$ and $N_0 = 1$. The parameters are $\mu = 10^{-7}$, $\alpha - \mu = 3$ and the final computation time is $t = 6.7$. To reduce fluctuations, the total number of simulation samples is 5×10^5. First we consider the Luria–Delbrück case for $\beta = 2.5$, and then the Lea–Coulson case for $\beta = 2.8$.

In Figure 8.2 we report the solution obtained simulating the kinetic model (8.25) for different values of the scaling parameter ϵ, and for the map $\mathcal{L}_\varepsilon(v) = \varepsilon\beta_2 v$. The results show the convergence of the mutant distribution prescribed by the model towards the classical Luria–Delbrück solution computed as in [329].

Similarly, we report in Figure 8.3 the solution of the kinetic model (8.25) for different values of the scaling parameter ϵ and for the random map $\mathcal{L}_\varepsilon(v)$ such that $\langle \mathcal{L}_\varepsilon(v) \rangle = \varepsilon\beta_2 v$. As expected, convergence of the mutant distribution towards the Lea–Coulson solution computed as in [329] is observed.

8.3 Self-organized systems and swarming models

In biology and physics, the main goal of simulations of self-organized systems is to be able to interpret and predict different swarming or multiagent aggregation behaviours, on the basis of microscopic interactions between agents. There has been a large amount of literature on flocking, herding and schooling. Much of it is descriptive, most of the remaining proposes models, which are then studied via computer simulations (see for example [149, 313]).

Fig. 8.4 Steven Smale[2]

Among others, the recent work of Cucker and Smale [111], connected with the emergent behaviours of flocks in a population of birds, engendered a noticeable resonance in the mathematical community.

The model proposed by Cucker and Smale is a particle model, in which the velocity of a bird is influenced by all the other birds' velocities according to a decreasing function of their mutual space distance. If this influence is strong enough, it is shown that all birds tend exponentially fast to move

[2]**Steven Smale**: born on 15 July 1930 in Flint, Michigan, (USA). He was awarded the Fields Medal in 1966 for his work on topology in higher dimensions. In 2007, Smale was awarded the Wolf Prize in mathematics for his ground breaking contributions that have played a fundamental role in shaping differential topology, dynamical systems, mathematical economics and other subjects in mathematics.

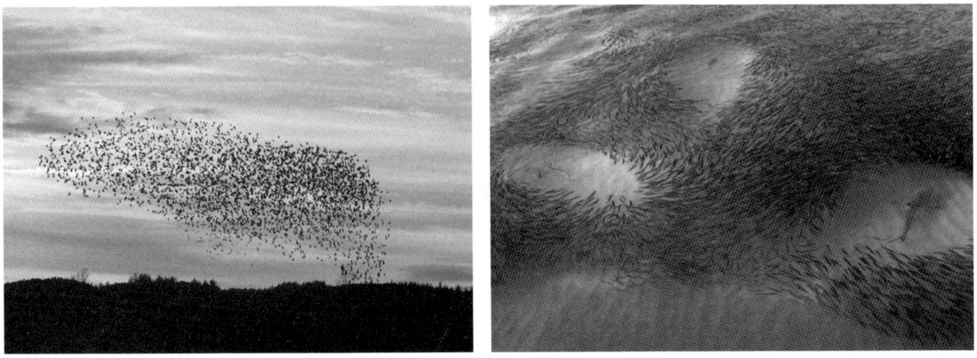

Fig. 8.5 A flock of birds (left) and a school of fishes attacked by predators (right).

with their global mean velocity, independently of the initial configuration, while their mutual distances tend to remain fixed. This situation is called *unconditional flocking*.

Naturally occurring synchronized motion has inspired several directions of research within the control community. A well-known application is related to formation flying missions and missions involving the coordinated control of several autonomous vehicles [198, 262]. There are several current projects dealing with formation flying and coordinated control of satellites, like the DARWIN project of the European Space Agency (ESA) with the goal of launching a space-based telescope aiding in the search for possible life-supporting planets, or the PRISMA project led by the Swedish Space Corporation (SSC) which will be the first real formation flying space mission [115]. One of the fundamental problems is to ensure that, with a minimal amount of fuel expenditure, the spacecraft fleet keep remaining in flight (flock), without losing mutual radio contact, and eventually scattering [181].

Our first aim in this section is to introduce a microscopic description of the binary interactions between birds which is inspired by the model of Cucker and Smale [111]. Making use of a collisional mechanism between individuals similar to the change in velocity of the bird population introduced in [111], we derive a dissipative spatially dependent Boltzmann-type equation which describes the behaviour of the flock in terms of a density $f = f(x, v, t)$, which depends both on position x and velocity v. This kinetic equation differs from most of the analogous ones we have introduced in the previous chapters, because the interactions here are dependent on both position and velocity. The underlying equation is reminiscent of the modification of the Boltzmann equation due to Povzner [265]. As usual, and similarly to the asymptotic procedure we described in Chapter 1, Section 1.4, one obtains in the *grazing limit procedure* a simpler equation which retains all the properties of the underlying Boltzmann equation, and in addition can be studied in detail. This equation has been derived as the mean field limit of the Cucker and Smale model by Ha and Tadmor [170]. We refer to [73,109,314] for some recent reviews on microscopic and kinetic modelling of swarming dynamics.

8.3.1 The Cucker–Smale model

In [111] Cucker and Smale studied the phenomenon of flocking in a population of birds whose members are moving in the physical space $\mathbb{R}^3$. The goal was to prove that under certain communication rates between the birds, the state of the flock converges to one in which all birds fly with the same velocity. The main hypothesis justifying the behaviour of the population is that every bird adjusts its velocity to a weighted average of the relative velocities with respect to the other birds.

For a system of N individuals this model is described by the following dynamical system:

$$\begin{cases} \dfrac{dx_i}{dt} = v_i, \\ \dfrac{dv_i}{dt} = \dfrac{\gamma}{N} \displaystyle\sum_{j=1}^{N} a_{ij}(v_j - v_i) , \end{cases} \tag{8.46}$$

for $i = 1, \ldots, N$, where the weights a_{ij}, named *communication rates*, quantify the way the birds influence each other, independently of their total number N, and γ measures the interaction strength. In [111], it is assumed that the communication rate is a function of the distance between birds, namely

$$a_{ij} = \frac{1}{(1 + \|x_i - x_j\|^2)^{\beta}}, \tag{8.47}$$

for some $\beta \geq 0$. For $x, v \in \mathbb{R}^N$, denote

$$\Gamma(x) = \frac{1}{2} \sum_{i \neq j} \|x_i - x_j\|^2, \qquad \Lambda(v) = \frac{1}{2} \sum_{i \neq j} \|v_i - v_j\|^2. \tag{8.48}$$

Then, under suitable restrictions on β and γ, and certain initial configurations (see [111] for details), it is proven that there exists a constant B_0 such that $\Gamma(x(t)) \leq B_0$ for all $t > 0$, while $\Lambda(v(t))$ converges towards zero as $t \to \infty$, and the vectors $x_i - x_j$ tend to a limit vector $\widehat{x}_{ij}$, for all $i, j \leq N$.

In particular, and rather remarkably, when $\beta < 1/2$, no restrictions on γ and initial configurations are needed [112, Theorem 1] and [111]. In this case, called unconditional flocking, the behaviour of the population of birds is perfectly specified. All birds tend to fly exponentially fast with the same velocity, while their relative distances tend to remain constant. This result has recently been improved in several works [169, 170, 283], where other weights or communication rates are studied and sharper rates of convergence are obtained in certain cases.

Let us now consider the time-discretized counterpart of (8.46) at time $t_n = n\Delta t$ with $n \in \mathbb{N}$ and $\Delta t > 0$; for the i-th individual, we have

$$v_i(t_n + \Delta t) - v_i(t_n) = \frac{\gamma \Delta t}{N} \sum_{i=1}^{N} a_{ij} \left(v_j(t_n) - v_i(t_n)\right). \tag{8.49}$$

Equation (8.49) can be fruitfully rephrased in a different way, which will be helpful in the following. Suppose that we are looking at two birds only, numbered i and j. Assume that their velocities are modified in time according to the rule

$$v_i(t_n + \Delta t) = (1 - \gamma \Delta t\, a_{ij}) v_i(t_n) + \gamma \Delta t\, a_{ij} v_j(t_n),$$
$$v_j(t_n + \Delta t) = \gamma \Delta t\, a_{ij} v_i(t_n) + (1 - \gamma \Delta t\, a_{ij}) v_j(t_n). \tag{8.50}$$

Then momentum is preserved after the interaction,

$$v_i(t_n + \Delta t) + v_j(t_n + \Delta t) = v_i(t_n) + v_j(t_n),$$

while the energy increases or decreases according to the value of γ,

$$v_i^2(t_n + \Delta t) + v_j^2(t_n + \Delta t) = v_i^2(t_n) + v_j^2(t_n) - 2\gamma \Delta t a_{ij}\, (1 - \gamma \Delta t\, a_{ij})\, (v_i(t_n) - v_j(t_n))^2.$$

For $\gamma \Delta t < 1$, the energy is dissipated. Note that in this case the relative velocity is decreasing, since

$$|v_i(t_n + \Delta t) - v_j(t_n + \Delta t)| = |1 - 2\gamma \Delta t\, a_{ij}||v_i(t_n) - v_j(t_n)| < |v_i(t_n) - v_j(t_n)|, \tag{8.51}$$

and the velocities of the two birds tend towards the mean velocity $(v_i + v_j)/2$. In the case $\gamma \Delta t < 1/2$, the interaction (8.50) is similar to a binary interaction between molecules of a dissipative gas (see [77] and the references therein).

In the general case of a population of N birds, the binary law (8.50) is taken into account together with the assumption that the i-th bird modifies its velocity by giving the same weight to all the other velocities. As a consequence of this,

$$v_i(t_n + \Delta t) = \frac{1}{N} \sum_{j=1}^{N} \{(1 - \gamma \Delta t\, a_{ij}) v_i(t_n) + \gamma \Delta t\, a_{ij} v_j(t_n)\}, \tag{8.52}$$

which is a different way to obtain formula (8.49), as a sum of binary interactions.

8.3.2 A Boltzmann description of Cucker–Smale flocking

Equations (8.50) describe a linear binary interaction between particles, in which the post-collision velocities depend both on the position and velocity of the colliding particles. To treat them in the usual way, we introduce a density distribution of particles, depending on spatial position, velocity and time evolution, which interact according to (8.50).

Let $f(x, v, t)$ denote the density of birds in the position $x \in \mathbb{R}^n$ with velocity $v \in \mathbb{R}^n$ at time $t \geq 0$, $n \geq 1$. The kinetic model for the evolution of $f = f(x, v, t)$ can easily be derived by considering that the change in time of $f(x, v, t)$ depends both on transport (birds fly freely if they do not interact with others) and interactions with other birds. Discarding other effects, this change in density depends on a balance between the gain and loss of birds with velocity v due to binary interactions. Let us assume that two birds with positions and velocities (x, v) and (y, w) modify their velocities after the interaction according to (8.50),

$$v^* = (1 - \gamma H(x - y)) v + \gamma H(x - y) w,$$
$$w^* = \gamma H(x - y) v + (1 - \gamma H(x - y)) w, \tag{8.53}$$

where now the communication rate function H takes the form

$$H(x) = \frac{1}{(1+|x|^2)^\beta}, \quad x \in \mathbb{R}^n, \tag{8.54}$$

and $\gamma < 1/2$. Note that, as usual in collisional kinetic theory, the change in velocities due to binary interactions does not depend on time. It is clear that the binary relations (8.53) have much in common with the binary consensus model (6.1) introduced in Chapter 6.

A first important consequence of the interaction mechanism given by (8.53) is that the support of the allowed velocities cannot increase. In fact, since $0 \le a \le 1$ and $0 < \gamma < \frac{1}{2}$,

$$|v^*| = |(1-\gamma H(x-y))v + \gamma H(x-y)w| \le (1-\gamma H)|v| + \gamma H|w| \le \max\{|v|, |w|\}.$$

Inserting the mechanism of collisions of type (8.53) into the binary collision operator leads to the integro-differential equation of Boltzmann type

$$\left(\frac{\partial f}{\partial t} + v \cdot \nabla_x f\right)(x,v,t) = Q(f,f)(x,v,t), \tag{8.55}$$

that in weak form, namely for all smooth functions $\varphi(x,v)$ and $t \ge 0$, reads

$$\begin{aligned} \frac{\partial}{\partial t}\int_{\mathbb{R}^{2d}} \varphi(x,v) f(x,v,t)\, dv\, dx + \int_{\mathbb{R}^{2n}} (v \cdot \nabla_x \varphi(x,v)) f(x,v,t)\, dv\, dx \\ = \sigma \int_{\mathbb{R}^{4n}} (\varphi(x,v^*) - \varphi(x,v)) f(x,v,t) f(y,w,t) dv\, dx\, dw\, dy. \end{aligned} \tag{8.56}$$

In the classical kinetic theory of rarefied gases, the Russian mathematician Povzner [265] proposed a modified Boltzmann collision operator considering a smearing process for the pair collisions. This modified Povzner collision operator looks as follows:

$$Q_P(f,f)(x,v) = \int_{\mathbb{R}^d}\int_{\mathbb{R}^d} B\left(\frac{1}{J} f(x,v_*) f(y,w_*) - f(x,v) f(y,w)\right) dw\, dy. \tag{8.57}$$

In (8.57), $B = B(x-y, v-w)$ is the collision kernel and J the Jacobian of the velocity transformations

$$\begin{aligned} v^* &= (I - A(|x-y|))v + A(|x-y|)w, \\ w^* &= A(|x-y|)v + (I - A(|x-y|))w, \end{aligned} \tag{8.58}$$

where A is a $d \times d$ matrix and I the identity matrix. These last relations imply the conservation of momentum. We note that, differently from interactions (8.53), in Povzner's equation the matrix A is such that energy is also preserved in a collision and the Jacobian of the transformation is equal to 1. It is clear that the right-hand side of equation (8.56) can be viewed like the weak form of a Povzner-type equation with dissipative interactions, where the matrix $A(|x-y|) = \gamma H(x-y)I$, $B \equiv \sigma$ and $J = (1 - 2\gamma H(x-y))^d$. It is remarkable that, while the Povzner equation was first introduced for purely mathematical reasons and usually ignored by physicists, related kinetic equations can be fruitfully introduced to model multiagent systems in biology and ecology. The form (8.56) is easier to handle, and it is the starting point for exploring the evolution of macroscopic quantities (moments).

Remark 8.1 *Binary interactions of type (8.53), despite their linearity with respect to the velocity variable, are deeply different from all the other interactions introduced so far, in which the post-collision quantities depend on the pre-collision quantities only, except for the additional difficulty that this dependence can be assumed to be random. Here, both position and velocity enter into the game to determine the new velocities. This peculiarity is typical of problems which involve processes of decision (typically orientation) in which relative positions play an essential rule.*

8.3.3 Mean-field kinetic equation for flocking

While the time evolution of the population density is described in detail by the Boltzmann equation (8.55), a precise description of the phenomenon of swarming is mainly related to the large-time behaviour of the solution. On the other hand, this large-time behaviour has to depend mainly on the type of collisions (8.53), and not on the size of the parameter γ which determines the *strength* of the interaction itself. In this situation, an accurate description can also be furnished by resorting to simplified models, which turn out to be valid exactly for large times. We made use of a similar asymptotic procedure in Chapter 6, Section 6.2.3, where the computations were fully justified, and in this chapter, Section 8.2. For this reason, we do not insist here on all the details, and we outline only the main physical assumptions which justify the procedure.

Let us assume that the parameter γ, which measures the intensity of the velocity change in the binary interactions, is small ($\gamma \ll 1$). Then, in order that these interactions produce a variation of the collision integral, one needs to introduce a further scaling. In the aforementioned cases discussed in Chapters 6 and 8, the asymptotic analysis was performed by a suitable scaling of time. Here, we obtain the same effect by scaling the collision frequency. The most interesting case comes from the choice $\sigma\gamma = \lambda$, where λ is a fixed positive constant. In this case, by expanding $\varphi(x, v^*)$ in Taylor's series of $v^* - v$ up to the second order, the weak form of the collision integral takes the form

$$\begin{aligned}\sigma \int_{\mathbb{R}^{4d}} &(\varphi(x,v^*) - \varphi(x,v)) f(x,v,t) f(y,w,t)\, dx\, dv\, dy\, dw \\ &= \gamma\sigma \int_{\mathbb{R}^{4d}} (\nabla_v \varphi(x,v) \cdot (w-v))\, H(x-y) f(x,v,t)\, f(y,w,t)\, dx\, dv\, dy\, dw \\ &\quad + \underbrace{\frac{\gamma^2}{2}\sigma \int_{\mathbb{R}^{4d}} \left[\sum_{i,j=1}^{d} \frac{\partial^2 \varphi(x,\tilde{v})}{\partial v_i^2} (w_j - v_j)^2 \right] H(x-y)^2 f(x,v) f(y,w)\, dx\, dv\, dy\, dw}_{=I},\end{aligned} \tag{8.59}$$

with $\tilde{v} = \theta v + (1-\theta) v^*$, $0 \le \theta \le 1$. If the collisions are nearly grazing, $\gamma \ll 1$, while $\gamma\sigma = \lambda$, we can cut the expansion (8.59) after the first-order term. In fact, since the second moment of the solution to the Boltzmann equation is non-increasing,

$$\int_{\mathbb{R}^{2d}} |v|^2 f(x,v,t)\, dx\, dv \le \int_{\mathbb{R}^{2d}} |v|^2 f_0(x,v)\, dx\, dv,$$

and $a(x) \le 1$,

$$|I| \le 2\|\varphi(x,v)\|_{C_0^2} \int_{\mathbb{R}^{2d}} |v|^2 f_0(x,v)\, dx\, dv. \tag{8.60}$$

Hence, we have a uniform-in-time upper bound for the remainder term of order $\gamma^2\sigma = \lambda\gamma \ll 1$. It follows that, in the regime of small γ and high collision frequency, so that $\sigma\gamma = \lambda$, the Boltzmann collision operator $Q(f,f)$ is approximated by the dissipative operator $I_\gamma(f,f)$, in strong divergence form,

$$I_\gamma(f,f) = \lambda\nabla_v \cdot \{f(x,v,t)\left[(H(x)\nabla_v W(v)) * f\right](x,v,t)\}, \tag{8.61}$$

where $W(v) = \frac{1}{2}|v|^2$ and $*$ is the (x,v)-convolution. Notice that, as remarked by McNamara and Young in their pioneering paper [233], the operator $I_\gamma(f,f)$ maintains the same dissipation properties as the full Boltzmann collision operator.

The target equation corresponds to a nonlinear-type friction equation for the density $f = f(x,v,t)$, which reads

$$\frac{\partial f}{\partial t} + v \cdot \nabla_x f = \lambda\nabla_v \cdot \left[\xi(f)(x,v,t) f(x,v,t)\right], \tag{8.62}$$

where

$$\begin{aligned}\xi(f)(x,v,t) &= \left[(H(x)\nabla_v W(v)) * f\right](x,v,t) \\ &= \int_{\mathbb{R}^{2d}} \frac{v-w}{(1+|x-y|^2)^\beta} f(y,w,t)\, dy\, dw.\end{aligned} \tag{8.63}$$

Remark 8.2 *Equation (8.62) has been derived as the mean-field limit of the discrete and finite-dimensional model (8.49) of Cucker and Smale by Tadmor and Ha [170]. The comparison between the two derivations shows that mathematically the kinetic description presented here is more difficult to justify rigorously than the corresponding one in [170]. On the other hand, the kinetic description in terms of binary interactions is more natural, since it describes the total variation of velocity in terms of subsequent variations of velocity consequent to binary interactions, in which the two agents adapt their velocity by looking at each other's position and velocity. This fact has been used recently in [3] for the derivation of efficient algorithms for the numerical solution of (8.55) and (8.62).*

8.3.4 Alternative formulations and model improvements

In this section we present some alternative formulations of the Boltzmann equation describing the binary interaction dynamics for alignment, and some improvements to the modelling such as the introduction of a perception cone and a roosting force.

The collision operator (8.57) has much in common with a classical Boltzmann operator for Maxwell molecules, in the sense that the collision frequency is independent of the velocity and position of individuals. An alternative Povzner–Boltzmann kinetic approximation is obtained with the interaction operator

$$Q_P(f,f) = \int_{\mathbb{R}^d}\int_{\mathbb{R}^d} H(x-y)\left(\frac{1}{J} f(x,v_*) f(y,w_*) - f(x,v) f(y,w)\right) dw\, dy, \tag{8.64}$$

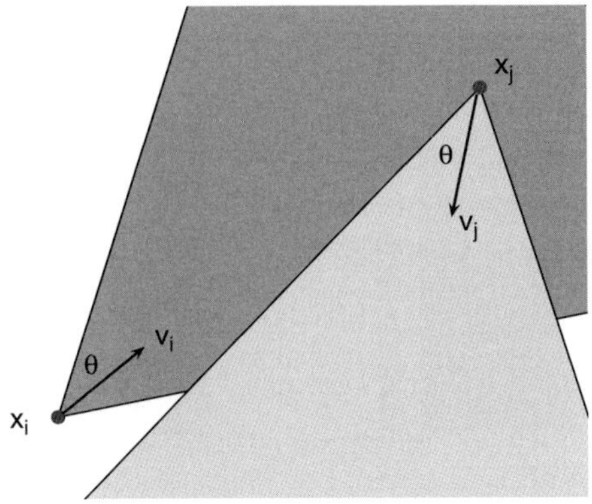

Fig. 8.6 One of the possible configurations in the interaction with the visual cone.

where now

$$\begin{aligned} v^* &= (1-\gamma)v + \gamma w \\ w^* &= \gamma v + (1-\gamma)w, \end{aligned} \tag{8.65}$$

and therefore $J = (1-2\gamma)^d$. From the modelling viewpoint, the function H is interpreted here as the frequency of interactions instead of the strength of the same interactions. Note that the approach is similar to the one introduced in Section 6.2.1 for opinion models with bounded confidence.

Clearly the two formulations (8.57) and (8.64) are not equivalent in general. It is easy to verify that formally we obtain the same mean-field limit (8.62). Note, however, that now the second-order term in the expansion (8.59) is slightly different and reads

$$I = \int_{\mathbb{R}^{4d}} \left[\sum_{i,j=1}^{d} \frac{\partial^2 \phi(x,\tilde{v})}{\partial v_i^2} (w_j - v_j)^2 \right] H(x-y) f(x,v,t) f(y,w,t) dv dx dw dy.$$

Since $H(x-y) < 1$, $H(x-y) > H(x-y)^2$, and in practice we may expect a slower convergence to the mean-field dynamic for small values of γ.

For interacting animals like birds, fishes and insects the visual perception of the single individual plays a fundamental role [108, 146]. In [76], the authors introduce into the dynamic a further rule: the *visual cone*. A *visual cone* identifies the area in which interaction is possible and a blind area where there can be no interaction. Mathematically speaking, the visual cone depends on an angle, θ, that gives us the visual width. Together with position and velocity, the visual area can be described as follows:

$$\Sigma(x,v,\theta) = \left\{ y \in \mathbb{R}^d : \frac{(x-y)\cdot v}{|(x-y)|\,|v|} \geq \cos(\theta/2) \right\}. \tag{8.66}$$

Note that the introduction of the visual cone breaks the typical symmetry of the interaction (see Figure 8.6).

The drawback of this choice is that a single individual that has no one in his visual cone, never changes his direction. For real situations this assumption is clearly

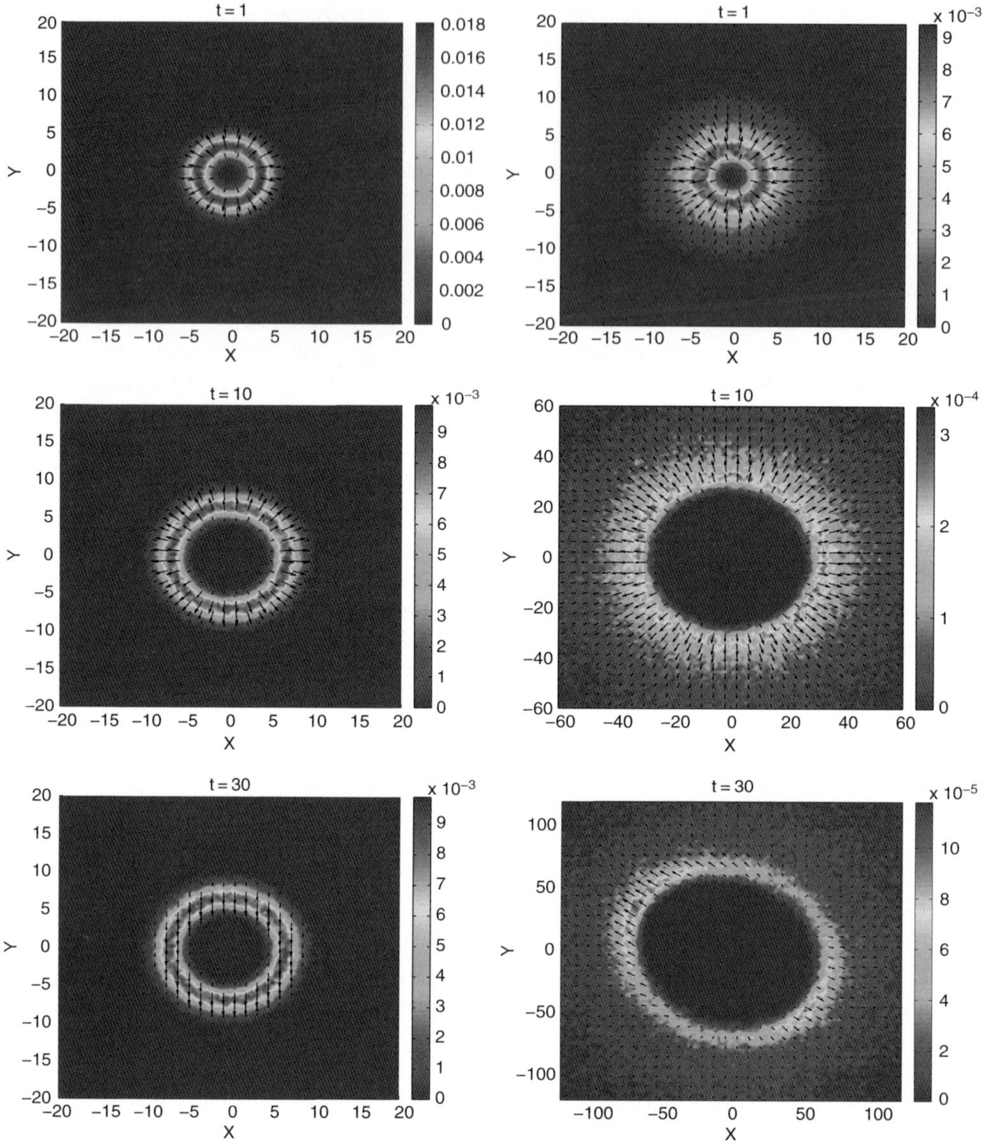

Fig. 8.7 Two-dimensional flocking with the Cucker–Smale alignment dynamic. On the left without visual cone, on the right with visual cone. The perception limitation splits the flock into two groups.

too strong, since many other stimuli are received from the surroundings. We cannot ignore other perceptions like hearing, smell and visual memory. For example, fish use their visual perception mostly at large/medium distances whereas at medium/short distances they rely on their lateral line.

These observations lead naturally to improving the idea of a visual cone by introducing a *perception cone* as follows: we assign two different weights measuring the strength/probability of the interaction. A weight p_1 in the case of strong perception and p_2 for weak perception, with $0 \le p_2 \le p_1 \le 1$. Note that taking $p_1 = 1$ and $p_2 = 0$ we have the standard visual cone. For example, in the simulation section we consider a function H_α, where $\alpha = (\theta, p_1, p_2)$, with the following form:

$$H_\alpha(x_i, x_j, v_i) = \{p_1\chi(x_j \in \Sigma_i) + p_2\chi(x_j \in \mathbb{R}^d \setminus \Sigma_i)\}\, H(|x_i - x_j|), \tag{8.67}$$

where $\chi(\cdot)$ is the indicator function and $\Sigma_i = \Sigma(x_i, v_i, \theta)$ is defined according to (8.66).

Related efforts to improve the dynamics also consider different ingredients such as *topological interactions*, where individuals interact only with the closest individuals and with a limited number of them, see [19, 109]. Another improvement concerns the introduction of a term describing a *roosting force* [76]. In fact, flocking phenomena tend to stay localized in a particular area. This can be modelled by means of a force which acts orthogonally to the single velocity, giving each particle a tendency to move towards the origin. A similar effect has been subsequently introduced in [2] by means of a *boundary force*.

To illustrate the influence of the perception cone, we consider a flock of N individuals with position $(x, y) \in \mathbb{R}^2$, initially distributed as

$$f_0(x, y, v_x, v_y) = \frac{1}{4\pi^2\sigma^2\nu^2} e^{-(x^2+y^2)/(2\sigma^2)} \left(e^{-(v+v^0)^2/(2\nu^2)} + e^{-(v-v^0)^2/(2\nu^2)}\right),$$

where $v^0 = (10, 10)$ and σ and ν are respectively the variance of the space and velocity variables. In the following simulations we use $N = 100000$ particles, and we compare the evolution of the same initial data, with and without the visual cone. As in the one-dimensional case, every bird interacts according to the law (8.67). In Figure 8.7 we plot the evolution of this initial data, comparing for each time step the two cases. The parameters for the visual cone case are $\theta = 4/3\pi$, $p_1 = 1$ and $p_2 = 0.04$. To reconstruct the probability density function in space we use a 100×100 grid for each figure. Moreover, we add to the plots the velocity flux field to get an idea of the flock direction.

8.3.5 Models including repulsion and attraction

Almost all the classical models that describe swarming phenomena are based on the simple definition of three zones around each individual, the so called *three-zones model* [11, 273].

Let us briefly summarize the three-zones assumptions. We define three areas: a velocity alignment zone, a short-range repulsion zone and an attraction zone. Each interaction between two individuals in the system is evaluated on the respective position in the model (see Figure 8.8):

- *Repulsion zone:* when too close to others, the individual tends to move away from that area.
- *Alignment zone:* where the individual tries to identify the possible direction of the group and to align with it.

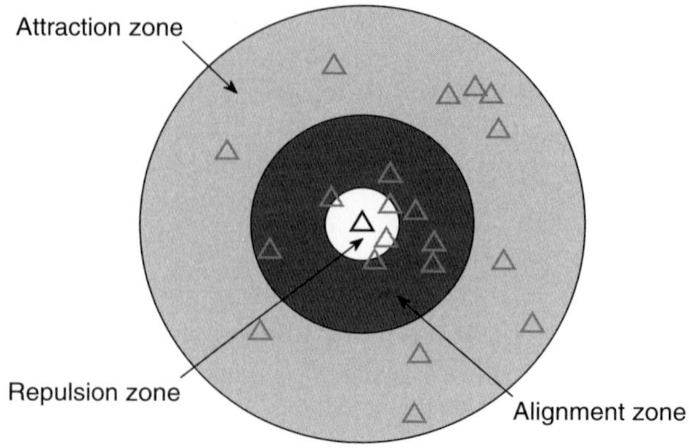

Fig. 8.8 Sketch of the three-zone model.

- *Attraction zone:* when an individual is too far from the group, he wants to get closer.

Typically, different interaction models are taken in the different zones or attention is focused on a specific zone, like the alignment/consensus dynamic in the Cucker and Smale model. The corresponding dynamical system has the following general form:

$$\begin{cases} \dfrac{dx_i}{dt} = v_i \\ \dfrac{dv_i}{dt} = S(v_i) + \dfrac{1}{N}\displaystyle\sum_{j=1}^{N} [A(x_i, x_j)(v_j - v_i) + R(x_i, x_j)] \end{cases} \tag{8.68}$$

for $i = 1, \ldots, N$, where $S(v_i)$ characterizes a self-propelling term, $A(x_i, x_j)(v_j - v_i)$ the alignment process between individuals with positions and velocities given by (x_i, v_i) and (x_j, v_j), and the term $R(x_i, x_j)$ the attraction-repulsion dynamic. For example, for the Cucker–Smale model described in Section 8.3.1 we have $A(x_i, x_j) = H(x_i - x_j)$.

The microscopic model introduced by D'Orsogna, Bertozzi et al. [128] considers self-propelling, attraction and repulsion dynamics and reads

$$\begin{cases} \dfrac{dx_i}{dt} = v_i \\ \dfrac{dv_i}{dt} = \gamma(\alpha - \beta|v_i|^2)v_i - \dfrac{\gamma}{N}\displaystyle\sum_{j \neq i} \nabla_{x_i} U(|x_j - x_i|) \end{cases} \tag{8.69}$$

for $i = 1, \ldots, N$, where α, β are non-negative parameters, $U : \mathbb{R}^d \longrightarrow \mathbb{R}$ is a given potential modelling the short-range repulsion and long-range attraction, and γ measures the strength of the interaction. The term associated with α models the *self-propulsion* of individuals, whereas the term corresponding to β represents a *friction* following Rayleigh's law. The combination and the balance of these two terms result in the tendency of the system (if we ignore the effect due to the repulsion and attraction term) to reach the asymptotic speed $|v| = \sqrt{\alpha/\beta}$, although not influencing the orientation of the velocities. A typical choice for U is the Morse potential which is radial and given by

$$U(x) = k(|x|), \quad k(r) = -C_A e^{-r/\ell_A} + C_R e^{-r/\ell_R}, \tag{8.70}$$

where C_A, C_R and ℓ_A, ℓ_R are the strengths and the typical lengths of attraction and repulsion, respectively. The most relevant situations for biological applications are determined for $C = C_R/C_A > 1$ and $\ell = \ell_R/\ell_A < 1$, which correspond to long-range attraction and short-range repulsion.

Let us assume that two individuals with positions and velocities (x, v) and (y, w) modify their velocities after the interaction according to

$$v^* = \mathcal{C}(x, v; y, w); \quad w^* = \mathcal{C}(y, w; x, v), \tag{8.71}$$

where $\mathcal{C}$ is a suitable *interaction rule*. This leads to the following integro-differential equation of Boltzmann type,

$$\left(\frac{\partial f}{\partial t} + v \cdot \nabla_x f\right)(x, v, t) = Q(f, f)(x, v, t), \tag{8.72}$$

where

$$\begin{aligned}\int_{\mathbb{R}^{2n}} & Q(f, f)\varphi(x, v)\, dx\, dv \\ &= \sigma \int_{\mathbb{R}^{4n}} (\varphi(x, v^*) - \varphi(x, v)) f(x, v, t) f(y, w, t) dv\, dx\, dw\, dy.\end{aligned} \tag{8.73}$$

In (8.73), v^* is given by (8.71). As before, the role of the *interaction operator* Q is to provide a stochastic description of the interactions happening between each pair of individuals. The binary law given by (8.71) describes the details of the microscopic interaction, and can be very general.

Following the same approach as described in Section 8.3.1 if we consider the self-propelling, attraction, repulsion model (8.69) we have the binary interaction rule

$$\mathcal{C}(x, v; y, w) = v + \gamma\left[(\alpha - \beta|v|^2)v - \nabla U(|x - y|)\right]. \tag{8.74}$$

the second order, the weak form of the interaction integral takes the form

$$\begin{aligned}\int_{\mathbb{R}^{4d}} & (\varphi(x, v^*) - \varphi(x, v)) f(x, v, t) f(y, w, t)\, dx\, dv\, dy\, dw \\ = \gamma & \int_{\mathbb{R}^{4d}} \left(\nabla_v \varphi(x, v) \cdot \left[(\alpha - \beta|v|^2)v - \nabla U(|x - y|)\right]\right) f(x, v, t)\, f(y, w, t)\, dx\, dv\, dy\, dw \\ + \frac{\gamma^2}{2} & \int_{\mathbb{R}^{4d}} \left[\sum_{i,j=1}^{d} \frac{\partial^2 \varphi(x, \tilde{v})}{\partial v_i^2} \left[(\alpha - \beta|v|^2)v - \nabla U(|x - y|)\right]_j^2\right] f(x, v) f(y, w)\, dx\, dv\, dy\, dw.\end{aligned} \tag{8.75}$$

For $\gamma \ll 1$, i.e. for the strength of the interaction being very small, such that $\sigma\gamma = \lambda$ is constant and $\sigma\gamma^2 \ll 1$, a regime that is expected at large time, we may approximate the interaction operator with the sole first-order term, i.e.,

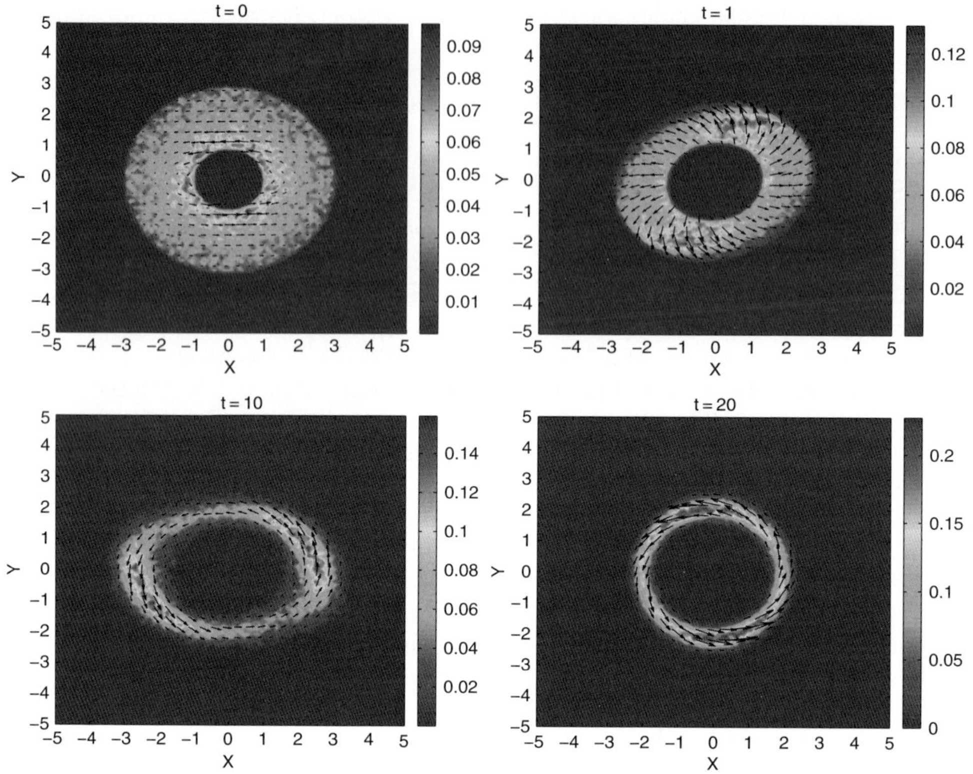

Fig. 8.9 Flock at several time steps. Parameters in the attraction–repulsion potential are such that $C = C_R/C_A = 30$, $l = l_R/l_A = 0.3$, $\alpha = 0.7$, $\beta = 0.05$. Equilibrium is reached after $t = 10$.

$$\frac{\partial}{\partial t}\int_{\mathbb{R}^{2d}} \varphi(x,v) f(x,v,t)\, dv\, dx + \int_{\mathbb{R}^{2d}} (v\cdot\nabla_x\varphi(x,v)) f(x,v,t)\, dv\, dx$$
$$\approx \lambda \int_{\mathbb{R}^{4d}} (\nabla_v\varphi(x,v)\cdot\left[(\alpha-\beta|v|^2)v - \nabla U(|x-y|)\right])\, f(x,v,t)\, f(y,w,t)\, dx\, dv\, dy\, dw,$$

or, in the strong form, we derive the corresponding nonlinear friction-type equations

$$\frac{\partial f}{\partial t} + v\cdot\nabla_x f = \lambda\left[(\nabla_x U * \rho)\cdot\nabla_v f - \nabla_v\cdot((\alpha-\beta|v|^2)vf)\right], \tag{8.76}$$

where

$$\rho(x,t) = \int_{\mathbb{R}^d} f(x,v,t)\, dv, \tag{8.77}$$

and $*$ is the x-convolution.

Finally, we show a numerical simulation performed with the binary interaction algorithm derived in [3] of the D'Orsogna–Bertozzi model et al. model reproducing the typical flock dynamics at kinetic level. We consider the bidimensional case taking into account $N = 100000$ particles. According to the interaction described in (8.69), we consider the long-range attraction and short-range repulsion, starting from initial data where the particles are uniformly distributed on a bidimensional torus, with a circular motion. The evolution shows how the attraction and repulsion forces stretch the flock reaching a condition of equilibrium in a stable circular motion.

8.4 Systems interacting with few individuals

The aim of this section is to present different levels of description for the dynamic of a large group of agents influenced by a small number of external agents [4]. A similar setting in opinion formation has been studied in Section 6.5. In a biological context, this corresponds to the behaviour of a flock or a school of fish attacked by one or more predators, or the movement of a herd of sheep guided by a sheepdog.

From the modelling viewpoint this involves considering a microscopic dynamic described by classical flocking models interacting with a set of few individuals characterized in way similar to what was done in [29, 102]. Moreover, motivated by the analysis in [19], we endow the classical dynamic of interaction both with a *metric* as well as a *topological interaction rule.*

We start from the microscopic dynamic, given by an ODE system, and we derive two other different levels of description: the mesoscopic (or kinetic) level through a *mean-field limit* and the macroscopic level through a suitable *hydrodynamic approximation.* Here, differently from the first-order macroscopic models proposed in [102], we obtain second-order models for the corresponding continuum dynamic.

8.4.1 Microscopic model

We are interested in the study of a dynamical system composed of N individuals and N^p external agents with the following general structure:

$$\begin{cases} \dfrac{dx_i}{dt} = v_i \\ \dfrac{dv_i}{dt} = \dfrac{1}{N^*} \displaystyle\sum_{j \in \Sigma_{N^*}(x_i)} F(x_i, v_i, x_j, v_j) + \dfrac{1}{N_p} \sum_{k=1}^{N_p} F^p(x_i, v_i, p_k) \\ \dfrac{dp_h}{dt} = \varphi_h(t, \mathbf{p}, \mathcal{A}\rho^N(p_h, t)) \end{cases} \tag{8.78}$$

where $i = 1, \ldots, N$, $h = 1, \ldots, N_p$, $(\mathbf{x}, \mathbf{v})_i = (x_i, v_i)$ lives in $\mathbb{R}^{2d}$, $d \geq 1$, $i = 1, \ldots, N$ and $(\mathbf{p})_h = p_h \in \mathbb{R}^{nd}$, with $n = 1, 2$; $h = 1, \ldots, N_p$, and $N_p \ll N$. The function F describes the interactions inside the swarm, and F^p depicts the interaction with each external agent p_h.

Moreover, we endow the model with a *topological* rule of interaction. Each agent will interact only with a fix number of agents of their species,

$$\Sigma_{N^*}(x_i) = \{\text{the } N^* \text{ closest neighbours respect to } i\}.$$

If $N^* = N$, each agent interacts with all the others and the *topological interaction* coincides with standard global interaction. Functions $\varphi_h : [0, +\infty) \times (\mathbb{R}^{nd})^{N_p} \times \mathbb{R} \longrightarrow \mathbb{R}$

describe the evolution in time of each external individual and they depend on the discrete density ρ^N defined as the empirical measure

$$\rho^N(x,t) = \frac{1}{N}\sum_{i=1}^{N} \delta(x - x_i(t)).$$

According to [102], we define $\mathcal{A}$ as a convolution operator $\mathcal{A}\rho^N(x,t) = (\rho^N * \eta)(x,t)$, where η is a smooth kernel with compact support.

8.4.2 Kinetic model

To obtain a mesoscopic description of the system (8.78), we proceed through a mean-field limit following [73]. Similarly, we can recover the same mean-field kinetic equation by deriving first its corresponding Boltzmann model and then considering the asymptotic quasi-invariant limit. Note that, since the limit is done only for the first set of equations which describes the evolution of the swarm, as a result we obtain a hybrid model composed of one kinetic equation and the system of ODEs governing the external agents.

The basic idea of the mean-field limit is to derive, through a weak argument, a single evolutionary equation for f^N, the empirical measures defined as

$$f^N(x,v,t) = \frac{1}{N}\sum_{i=1}^{N} \delta(x - x_i(t))\delta(v - v_i(t)).$$

Next, one computes the limit $f^N \longrightarrow f$ for $N \longrightarrow \infty$ and performs a rigorous derivation of the limiting kinetic equation. A well-posedness theory for this asymptotic derivation has been developed in [67] for a general set of swarming models.

Moreover, we assume that for all N the ratio N^*/N is fixed and equal to some constant μ. This supposition allows us to define the *topological density* as $\rho^* = \mu\bar{\rho}$ where $\bar{\rho} = \int_{R^d}\int_{R^d} f(y,v,t)\,dy\,dv$ is constant in time. Then, in the mean-field limit, an analogue of the topological set of interaction $\Sigma_{N^*}(x)$ is described by a characteristic function on the ball $\mathcal{B}(x,R^*)$, with centre x and radius $R^*(x,t)$ such that we have

$$R^*(x,t) = \min\left\{R \textit{ s.t. } R > 0, \int_{\mathbb{R}^d}\int_{\mathcal{B}(x,R)} f(y,v,t)\,dy\,dv = \rho^*\right\}.$$

The resulting model reads

$$\begin{cases} \dfrac{\partial f}{\partial t} + v\cdot\nabla_x f = -\nabla_v\cdot(\mathcal{E}^*(x,v)f) - \nabla_v\cdot(\mathcal{E}^p(x,v)f), \\ \dfrac{dp_h}{dt} = \varphi_h(t,\mathbf{p},\mathcal{A}\rho(p_h,t)), \end{cases} \tag{8.79}$$

where $h = 1,\ldots,N_p$, $\rho(x,t) = \int_{R^d} f(x,v,t)dv$ and

$$\mathcal{E}^*(x,v) = \frac{1}{\mu}\int_{\mathbb{R}^d}\int_{\mathcal{B}(x,R^*)} F(x,v,y,w)\,dy\,dw, \quad \mathcal{E}^p(x,v) = \frac{1}{N_p}\sum_{k=1}^{N_p} F^p(x,v,p_k).$$

Remark 8.3 *If we consider the case of the combined* Cucker–Smale *and* D'Orsogna–Bertozzi et al. *models and* $p_h = (x_h^p, v_h^p)$, *we can explicitly compute* $\mathcal{E}^*$ *and* $\mathcal{E}^p$ *as*

$$\mathcal{E}^*(x,v) = \frac{1}{\mu}\int_{\mathcal{B}(x,R^*)}\left(\int_{\mathbb{R}^d} H(x-y)(w-v)f(y,w)dw - \nabla_x U(|x-y|)\rho(y)\right)dy \\ +(\alpha-\beta|v|^2)v, \tag{8.80}$$

and

$$\mathcal{E}^p(x,v) = \frac{1}{N_p}\sum_{k=1}^{N_p} H^p(x-x_k^p)(v_k^p - v) - \frac{1}{N_p}\sum_{k=1}^{N_p}\nabla_x U^p(|x-x_k^p|). \tag{8.81}$$

8.4.3 Hydrodynamic approximation

Lastly, we will detail a possible macroscopic description of the system. From a numerical point of view, this corresponds to reducing the dimensionality of the problem in such a way that simulations become affordable. Any macroscopic description of a kinetic equation depends upon the local equilibria. In the kinetic theory of rarefied gases, this is a well-studied task, which connects the Boltzmann equation with the Euler and Navier–Stokes systems of fluid dynamics [80]. In a situation like the present one, the determination of the local equilibrium state of the system is, in general, a very difficult task. One usually resorts to approximate equilibrium states which are physically reasonable and simplify the mathematical computations. Here we follow the approach introduced in [100] and subsequently used in [73] (cf. also [118–120]). Of course, however, a different equilibrium state would provide a different set of macroscopic equations.

First let us define the momentum and the temperature of the system as

$$\rho u(x,t) = \int v f dv, \quad T(x,t) = \int |v-u|^2 f dv.$$

In order to obtain a system of equations which describes the evolution of the mass density ρ and the momentum ρu we integrate the kinetic equation in (8.79) against dv and $v dv$.

According to [100], we impose the momentum closure assuming that the fluctuations are negligible, i.e. that the temperature $T(x,t) = 0$, and the velocity distribution is *monokinetic*: $f(x,v,t) = \rho(x,t)\delta(v - u(x,t))$.

The previous assumptions lead to the hydrodynamic system

$$\begin{cases} \dfrac{\partial \rho}{\partial t} + div_x(\rho u) = 0 \\ \dfrac{\partial \rho u}{\partial t} + \nabla_x \cdot (\rho u \otimes u) = \mathcal{F}^*(x,u)\rho(x,t) + \mathcal{F}^p(x,u)\rho(x,t) \\ \dfrac{dp_h}{dt} = \varphi_h(t, \mathbf{p}, \mathcal{A}\rho(p_h,t)) \end{cases}, \tag{8.82}$$

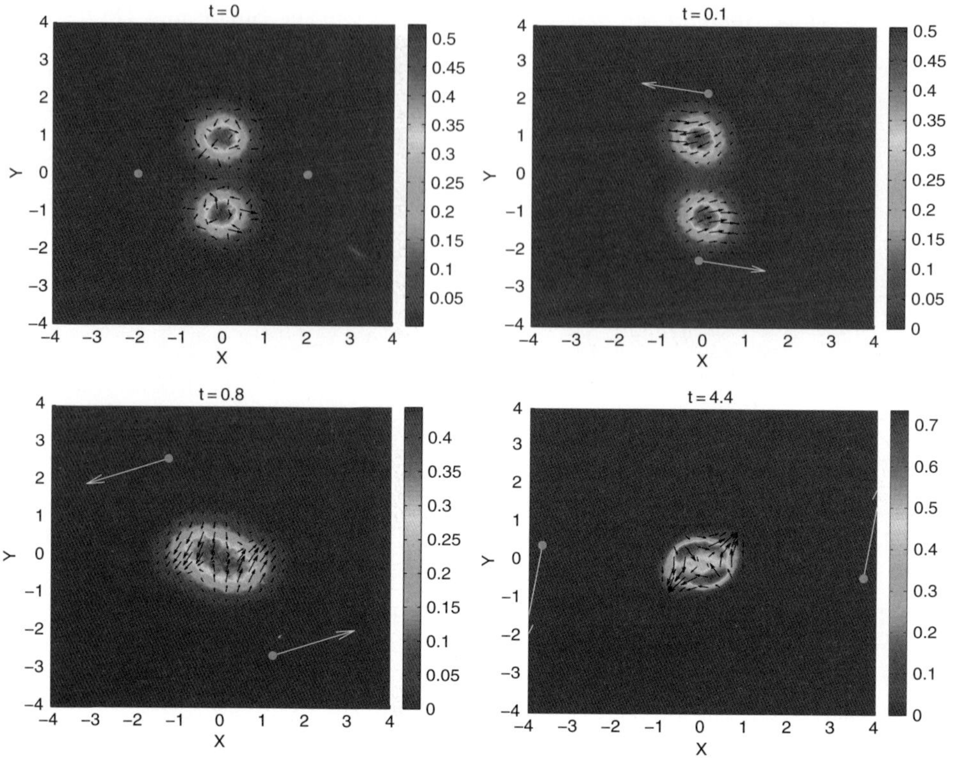

Fig. 8.10 Shepherd dogs. Simulation at different times with parameters $a = 2.5$, $b = 0.1$ and $\gamma = 0.45$.

where $h = 1, \ldots, N_p$ and, in the particular case of the Cucker–Smale and D'Orsogna–Bertozzi et al. model with $p_h = (x^p_h, v^p_h)$, we have

$$\mathcal{F}^*(x,u) = \frac{1}{\mu}\int_{\mathcal{B}(x,R^*)} (H(x-y)(u(y,t)-u(x,t)) - \nabla_x U(|x-y|))\,\rho(y,t)\,dy \\ +(\alpha - \beta|u|^2)u, \tag{8.83}$$

$$\mathcal{F}^p(x,u) = \frac{1}{N_p}\sum_{k=1}^{N_p} H^p(x-x^p_k)(v^p_k - u(x,t)) - \frac{1}{N_p}\sum_{k=1}^{N_p} \nabla_x U^p(|x-x^p_k|). \tag{8.84}$$

8.4.4 Numerical examples

We show here two numerical examples taken from [102] for two different biological dynamics. In order to solve the kinetic model the simulations have been performed using the asymptotic binary interaction algorithm proposed in [3]. We recall that the algorithm is based on a stochastic routine which uses N_s sample particles and that the overall cost is $O(N_s)$.

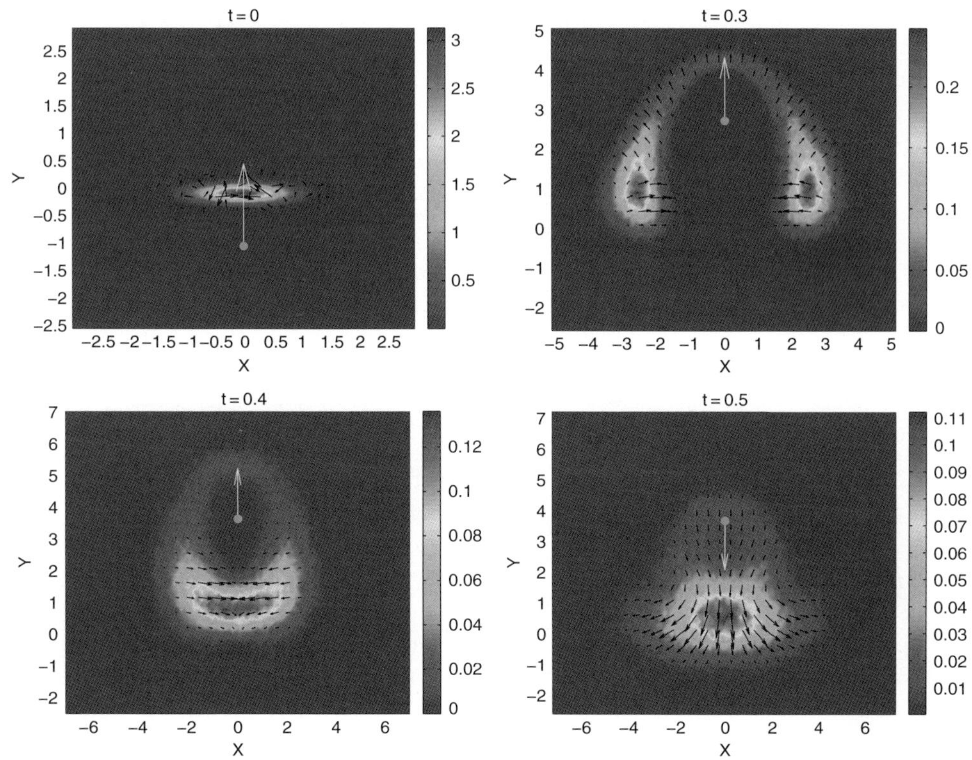

Fig. 8.11 Swarm attacked by a predator. Simulation at different times with parameters $a = 4$, $b = 2$ and $\gamma = 0.45$.

The dynamics considered for the swarm in both cases are characterized by

$$F(x_i, v_i, x_j, v_j) = H(x_i - x_j)(v_j - v_i) - \nabla_{x_i} U(|x_i - x_j|),$$

and only a *repulsion* dynamic with respect to p_h in F^p is considered, given by $U^p(r) = r^c/c$ with $c > 0$, and where $r = |x_i - p_h|$.

The simulation represents the evolution of a swarm controlled by two leaders $p_1, p_2 \in \mathbb{R}^{2n}$, $n = 1$, interacting with the swarm accordingly to the following functions:

$$\varphi_h(t, \mathbf{p}, s_h) = V_p \frac{s_h^{\perp}}{\sqrt{1 + |s_h|^2}}; \quad V_p = 300, \quad r_p = 5, \quad h = 1, 2,$$

$$\eta(x) = \frac{3}{\pi r_p^6} (\max\{0, r_p^2 - |x|^2\})^2, \quad s_h = \mathcal{A}\rho(x_h^p, t) = (\rho *_x \nabla\eta)(x_h^p, t).$$

As we can see in Figure 8.10, the action of the leaders is able to force the flocking of the particles.

Finally, we consider the evolution of a swarm which undergoes the action of a predator. The predator is modelled by the evolution of $p = (x^p, v^p) \in \mathbb{R}^{2n}$, $n = 2$, and its evolution is led by the following potential:

$$\varphi(t, p, s) = (v^p, V_p s); \quad V_p = 1500, \quad r_p = 5,$$

$$\eta(x) = \frac{3}{\pi r_p^6}(\max\{0, r_p^2 - |x|^2\})^2, \quad s = (\mathcal{A}\rho)(x^p, t) = (\rho * \nabla\eta)(x^p, t).$$

We report the results in Figure 8.11. It is evident how the predator attack splits the flock into two groups which subsequently merge together again.

8.5 Final remarks

With the brief discussion on the modelling of swarming we conclude our introduction to the mathematical modelling of interacting multiagent systems. In the last section we presented some important topics that in this monograph were only mentioned. Among them, the macroscopic description of these phenomena plays a basic role. It was our choice to remain mostly with the kinetic description.

In recent years, mathematical modelling and simulations of kinetic equations applied to socio-economic and life sciences have gained increasing importance, the number of relevant papers has become enormously high and it is very difficult to summarize in a few pages even the most significant contributions. This situation was the basis of our choice to present in detail only some selected arguments, together with a solid basis for the numerical simulations particularly adapted to the kinetic modelling.

The reader who wants to deepen his or her knowledge on the theoretical and applied aspects of the mathematical modelling introduced in this book is referred to some recent monographs and collections [27, 85, 89, 98, 121, 225, 242, 252, 299, 300, 317].

We hope we were able to convey in these few pages our passion for these topics and that this book will serve to bring other researchers to the fascinating world of kinetic theory. We close the book by quoting Clifford Truesdell [311]. In writing this book we didn't have such an ambitious goal, but we are pleased to see that after more than 30 years Truesdell's hope has turned into reality.

> The cognoscenti of Maxwell's second kinetic theory are few, too few even to make up a club. *A fortiori* they are not yet professionalised, not yet incarcerated within and shielded by a carapace of union rules. Thus they lie in peril. I think this treatise is the first on the subject who deserve the qualification 'mathematical'. In these times, dark as they are for the mathematics of natural science, I fear it may be the last. Earnestly I hope, on the contrary, that it will open a new day in the kinetic theory, promote a new floraison, to the point that this fascinating branch of rational mechanics will at last surrender its mysteries.
>
> Clifford Truesdell, Prologue to *Fundamentals of Maxwell's Kinetic Theory of a Simple Monatomic Gas*, 1980

Appendix A
Basic arguments on Fourier transforms

A.1 Definitions

We will briefly introduce here the basic properties of Fourier transforms, with the aim of facilitating the understanding of most of the mathematical arguments introduced in the previous chapters, which are based on these transforms. Since most of the functions we dealt with were probability density functions, we will always refer to non-negative integrable functions, unless otherwise stated. The Fourier transform of a function $f(v)$ has been defined in Chapter 1, Section 1.2, through (1.17) and (1.18). We will repeat this definition here for convenience. The function

$$\hat{f}(\xi) = \int_{\mathbb{R}} e^{-iv\xi} f(v)\, dv \tag{A.1}$$

is called the Fourier transform of $f(v)$. Note that, if $f(v)$ is non-negative, taking $\xi = 0$ in (A.1) shows that

$$\hat{f}(\xi) = \int_{\mathbb{R}} f(v)\, dv = \|f\|_{L^1}. \tag{A.2}$$

The probability density functions introduced in Appendix B provide easy-to-handle examples.

Example A.1 Let us consider the uniform probability density function given by

$$r(v) = \begin{cases} 1 \text{ if } -1/2 \le v \le 1/2, \\ 0 \text{ otherwise.} \end{cases} \tag{A.3}$$

$r(v)$ is also called a normalized boxcar function. The Fourier transform of this density is

$$\hat{r}(\xi) = \int_{\mathbb{R}} e^{-iv\xi} r(v)\, dv = \int_{-1/2}^{1/2} e^{-iv\xi}\, dv = \frac{e^{i\xi/2} - e^{-i\xi/2}}{i\xi} = \frac{2}{\xi} \sin\frac{\xi}{2},$$

when $\xi \neq 0$. When $\xi = 0$,

$$\hat{r}(0) = \int_{-1/2}^{1/2} 1 \cdot dv = 1 = \lim_{\xi \to 0} \frac{2}{\xi} \sin\frac{\xi}{2}.$$

■

Example A.2 For a given $\lambda > 0$, let us consider the exponential probability density function given by

$$f(v) = \begin{cases} \lambda e^{-\lambda v} & \text{if } v \geq 0, \\ 0 & \text{if } v < 0. \end{cases}$$

The Fourier transform of this density is

$$\hat{f}(\xi) = \int_{\mathbb{R}} e^{-iv\xi} f(v)\, dv = \int_0^{+\infty} e^{-iv\xi} \lambda e^{-\lambda v}\, dv = \lambda \int_0^{+\infty} e^{-v(i\xi+\lambda)}\, dv = \frac{\lambda}{\lambda + i\xi}.$$

■

A.2 Properties of the Fourier transform

A.2.1 Linearity

For any given constants c and d, let us build the new function $h(v) = cf(v) + dg(v)$ as a linear combination of two functions $f(v)$ and $g(v)$. Then the Fourier transform of h is

$$\hat{h}(\xi) = \int_{\mathbb{R}} e^{-iv\xi} h(v)\, dv = \int_{\mathbb{R}} e^{-iv\xi} [cf(v) + dg(v)]\, dv =$$

$$c \int_{\mathbb{R}} e^{-iv\xi} f(v)\, dv + d \int_{\mathbb{R}} e^{-iv\xi} g(v)\, dv = c\hat{f}(\xi) + d\hat{g}(\xi). \qquad \text{(A.4)}$$

A.2.2 Time shifting

Let us build a new function $h(v) = f(v - v_0)$ by shifting a function $f(v)$ by a positive quantity v_0. An easy way to check the effect of this shift is to note that if the original function $f(v)$ has a discontinuity at the point $v = c$, then this discontinuity moves to the point $v - v_0 = c$ for the new function $h(v)$, that is at the point $v = v_0 + c$, which is at a distance c to the right of the original jump. The Fourier transform of $h(v)$ is

$$\hat{h}(\xi) = \int_{\mathbb{R}} e^{-iv\xi} h(v)\, dv = \int_{\mathbb{R}} e^{-iv\xi} f(v - v_0)\, dv =$$

$$\int_{\mathbb{R}} e^{-i(v_*+c)\xi} f(v_*)\, dv_* = e^{-ic\xi} \int_{\mathbb{R}} e^{-iv_*\xi} f(v_*)\, dv_* = e^{-ic\xi} \hat{f}(\xi). \qquad \text{(A.5)}$$

A.2.3 Scaling

For any given positive constant c, we consider the function $h(v) = f(v/c)$ obtained by scaling the variable by a factor c. Then the Fourier transform of h is

$$\hat{h}(\xi) = \int_{\mathbb{R}} e^{-iv\xi} h(v)\, dv = \int_{\mathbb{R}} e^{-iv\xi} f(v/c)\, dv = c \int_{\mathbb{R}} e^{-iv_* c\xi} f(v_*)\, dv_* = c\hat{f}(c\xi). \qquad \text{(A.6)}$$

Example A.3 To acquire practice with the previous properties, let us evaluate the Fourier transform of the function

$$R(v) = \begin{cases} b \text{ if } c - a/2 \le v \le c + a/2, \\ 0 \text{ otherwise.} \end{cases}$$

Note that, if $r(v)$ is the uniform density defined in (A.3),

$$R(v) = b\, r\left(\frac{v-c}{a}\right).$$

By combining the properties of linearity, shifting and scaling described above, we can easily determine the Fourier transform of $R(v)$. Let us first consider the case in which $b = 1$ and $c = 0$. We define $R_1(v) = r(v/a)$. Then, according to the scaling property (A.6),

$$\hat{R}_1(\xi) = b\hat{r}(b\xi) = b\,\frac{2}{b\xi}\sin\frac{b\xi}{2} = \frac{2}{\xi}\sin\frac{b\xi}{2}.$$

Next, we consider $c \neq 0$, and define $R_2(v) = r\left(\frac{v-c}{a}\right) = R_1(v-c)$. By the translation property (A.5) we obtain

$$\hat{R}_2(\xi) = e^{-i\xi c}\hat{R}_1(\xi) = e^{-i\xi c}\frac{2}{\xi}\sin\frac{b\xi}{2}.$$

Last, by the linearity property (A.4), since $R(v) = b\,R_2(v)$,

$$\hat{R}(\xi) = b\,\hat{R}_2(\xi) = be^{-i\xi c}\frac{2}{\xi}\sin\frac{b\xi}{2}.$$

■

A.2.4 Differentiation

Let us build a new function $h(v) = f'(v)$ by differentiating the old function $f(v)$. Then the Fourier transform of $h(v)$ is

$$\hat{h}(\xi) = \int_{\mathbb{R}} e^{-iv\xi}\, h(v)\, dv = \int_{\mathbb{R}} e^{-iv\xi}\, f'(v)\, dv.$$

Integrating by parts, and assuming that $f(\pm\infty) = 0$, gives

$$\hat{h}(\xi) = -\int_{\mathbb{R}} (-i\xi)e^{-iv\xi}\, f(v)\, dv = i\xi\hat{f}(\xi). \tag{A.7}$$

Example A.4 Property (A.7) can be fruitfully used in all cases in which $h(v)$ is the derivative of a product of the old function $f(v)$ and entire powers of the variable v.

For example, if $h(v) = (vf(v))'$, and $vf(v) = 0$ vanishes at infinity, integration by parts gives

$$\hat{h}(\xi) = \int_{\mathbb{R}} (-i\xi)e^{-iv\xi}\,(vf(v))'\,dv = \int_{\mathbb{R}} (i\xi)e^{-iv\xi}\,vf(v)\,dv =$$

$$\xi \int_{\mathbb{R}} (-iv)e^{-iv\xi}\,f(v)\,dv = -\xi \int_{\mathbb{R}} \frac{\partial e^{-iv\xi}}{\partial \xi}\,f(v)\,dv = -\xi \frac{\partial}{\partial \xi} \int_{\mathbb{R}} e^{-iv\xi}\,f(v)\,dv = -\xi \frac{\partial \hat{f}}{\partial \xi}.$$

■

Example A.5 Example A.4 indicates that the differentiation property is quite useful when one uses the Fourier transform to solve differential, or, more generally, partial differential equations. To see how this property works, let us consider the Fokker–Planck equation (1.103) introduced in Chapter 1, which has a stationary solution given by the Gaussian density (1.104). This stationary solution solves the differential equation

$$\frac{\partial^2 h}{\partial v^2} + \frac{\partial}{\partial v}(vh) = 0. \tag{A.8}$$

The solution to (A.8) is easily found by writing it in the equivalent form

$$\frac{\partial}{\partial v}\left(\frac{\partial h}{\partial v} + vh\right) = 0,$$

and then looking for functions which satisfy the differential equation

$$\frac{\partial h}{\partial v} + vh = 0.$$

Now, either $h(v) = 0$, or dividing by h we obtain

$$\frac{1}{h}\frac{\partial h}{\partial v} + v = \frac{\partial}{\partial v}\left(\log h + \frac{v^2}{2}\right) = 0,$$

which implies

$$h_\infty(v) = ce^{-v^2/2}.$$

By fixing the mass equal to one, the (unique) solution to (A.8) with this mass is the Gaussian function

$$M(v) = \frac{1}{\sqrt{2\pi}}e^{-v^2/2}. \tag{A.9}$$

Owing to the differentiation property, and to Example A.4, we obtain that equation (A.8) in Fourier transform reads

$$-\xi^2\hat{h}(\xi) - \xi\frac{\partial \hat{h}}{\partial \xi} = 0. \tag{A.10}$$

If $\xi\hat{h}(\xi) \neq 0$, we can divide both sides of (A.10) to write it in the equivalent form

$$\frac{1}{\hat{h}}\frac{\partial \hat{h}}{\partial \xi} = -\xi,$$

which can be integrated to give

$$\hat{h}_\infty(\xi) = ce^{-\xi^2/2}.$$

By fixing the mass equal to one, one concludes that the (unique) solution to (A.10) is the function

$$\hat{M}(\xi) = e^{-\xi^2/2}. \tag{A.11}$$

This example shows that the Fourier transform of a Gaussian density is an exponential function of the same type. ■

A.2.5 Convolutions

An important property of Fourier transforms, which is at the basis of the analysis of kinetic models with binary collisions discussed in most of the chapters of this book, deals with the convolution property. Indeed, the Fourier transform translates between convolution and multiplication of functions. If $f(v)$ and $g(v)$ are integrable functions with Fourier transforms $\hat{f}(\xi)$ and respectively $\hat{g}(\xi)$, and we build the new integrable function $h(v)$ given by the convolution product of f and g,

$$h(v) = f * g(v) = \int_{\mathbb{R}} f(v-w)g(w)\,dw,$$

then the Fourier transform of the convolution is given by the product of the Fourier transforms. In fact, by Fubini's theorem

$$\hat{h}(\xi) = \int_{\mathbb{R}} e^{-iv\xi}\, f * g(v)\,dv = \int_{\mathbb{R}}\int_{\mathbb{R}} e^{-iv\xi}\, f(v-w)g(w)\,dv\,dw =$$

$$\int_{\mathbb{R}}\int_{\mathbb{R}} e^{-i(v-w)\xi}\, f(v-w)e^{-iw\xi}g(w)\,dv\,dw =$$

$$\int_{\mathbb{R}}\left(\int_{\mathbb{R}} e^{-i(v-w)\xi}\, f(v-w)\,dv\right) e^{-iw\xi}g(w)\,dw = \hat{f}(\xi)\hat{g}(\xi).$$

Appendix B
Important probability distributions

In this short appendix we list some important distributions and some basic properties that are encountered in this book. First, we record an elementary lemma.

Lemma B.1 *Suppose that X is a continuous random variable, having probability density function $f(x)$.*

i) If a is a real number, then the probability density function of $X + a$ is $f(x - a)$.
ii) If b is a positive number, then the probability density function of bX is $b^{-1}f(x/b)$.

B.1 Uniform distribution

Let a and b be real numbers with $a < b$. The *uniform distribution* on the interval $[a, b]$ is the continuous distribution whose probability density function f is given by

$$f(x) := \begin{cases} (b-a)^{-1} & \text{if } a \leq x \leq b, \\ 0 & \text{otherwise.} \end{cases}$$

To say that the random variable X has this distribution, we usually write $X \sim U(a, b)$. The concept of uniform distribution easily extends to higher dimensions. Given any subset $\Omega \in \mathbb{R}^n$, where $n \geq 1$, the random variable X is a uniform random variable on Ω if the probability density function of X is given by

$$f(x) := \begin{cases} \mathfrak{m}(\Omega)^{-1} & \text{if } x \in \Omega, \\ 0 & \text{otherwise.} \end{cases}$$

Uniform random variables are considered in Chapter 1, Section 1.3.2, and in Chapter 5, Section 5.5.

B.2 Beta distribution

Let a and b be positive numbers. The *Beta distribution with parameters a and b* is the continuous distribution whose probability density function f is given by

$$f(x) := \begin{cases} \frac{\Gamma(a+b)}{\Gamma(a)\Gamma(b)} x^{a-1}(1-vx)^{b-1} & \text{if } 0 \leq x \leq 1, \\ 0 & \text{otherwise.} \end{cases} \tag{B.1}$$

In (B.1), Γ is the *Gamma function*

$$\Gamma(a) = \int_0^{+\infty} t^{a-1} e^{-t}\, dt.$$

The Gamma function satisfies the recursion $\Gamma(a) = (a-1)\Gamma(a-1)$. In particular, if a is a positive integer, then $Gamma(a) = (a-1)!$. To say that the random variable X has this distribution, we usually write $X \sim Beta(a, b)$. If $a = b = 1$, $Beta(1, 1)$ coincides with the uniform distribution $U(0, 1)$.

B.3 Normal distribution

Let μ be a real number and σ a positive number. The *normal distribution with mean μ and variance σ^2* is the continuous distribution whose probability density function is given by

$$f(x) = \frac{1}{\sqrt{2\pi}\sigma} \exp\left(-(x-\mu)^2/(2\sigma^2)\right), \qquad -\infty < x < +\infty. \tag{B.2}$$

It is common to write $X \sim N(\mu, \sigma^2)$ to say that the random variable X has this distribution. If the random variable $Z \sim N(0, 1)$, i.e. if Z has the *standard normal distribution*, then, by Lemma B.1, $\sigma Z + \mu \sim N(\mu, \sigma^2)$.

In the kinetic theory of rarefied gases, and in this book, the normal probability density function is usually referred as a Maxwellian density. The Fourier transform of the normal probability density function (B.2) reads

$$\hat{f}(\xi) = \exp\left(-i\mu\xi - \frac{1}{2}\sigma^2\xi^2\right). \tag{B.3}$$

The normal (Maxwellian) density is the natural equilibrium distribution of the Boltzmann equation in which the elastic binary interactions are energy preserving [82].

B.4 Exponential distribution

Let λ be a positive number. The *exponential distribution with parameter* λ is the continuous distribution whose probability density function is given by

$$f(x) := \begin{cases} \lambda e^{-\lambda x} & \text{if } x \geq 0, \\ 0 & \text{if } x < 0. \end{cases} \tag{B.4}$$

We write $X \sim Exp(\lambda)$ to say that the random variable X has this distribution. The exponential distribution with parameter λ has mean value $1/\lambda$ and variance $1/\lambda^2$. If $X \sim Exp(1)$, and $b > 0$, then $bX \sim Exp(1/b)$. The Fourier transform of the exponential probability density function (B.4) is

$$\hat{f}(\xi) = \left(1 + i\frac{\xi}{\lambda}\right)^{-1}. \tag{B.5}$$

The exponential density function often appears as equilibrium profile for kinetic models of wealth distribution treated in Chapter 5.

A related probability density function is given by

$$f(x) := \frac{1}{2}\lambda\, e^{-\lambda|x|} \qquad -\infty < x < +\infty. \tag{B.6}$$

The Fourier transform of the density (B.6) can easily be evaluated by means of the Fourier transform of the exponential distribution, to give

$$\hat{f}(\xi) = \left(1 + \left(\frac{\xi}{\lambda}\right)^2\right)^{-1}. \tag{B.7}$$

The probability density function (B.6) appears in Chapter 1, Section 1.3.3, in connection with the Rosenau approximation of hydrodynamics.

B.5 Gamma distribution

Let a and b be positive numbers. The *Gamma distribution with parameters* a *and* b is the continuous distribution whose probability density function f is given by

$$f(x) := \begin{cases} b^a x^{a-1} e^{-bx}/\Gamma(a) & \text{if } x \geq 0, \\ 0 & \text{if } x < 0. \end{cases} \tag{B.8}$$

We write $X \sim \mathit{Gamma}\,(a, b)$ to say that the random variable X has this distribution. The Gamma distribution has mean a/b and variance a/b^2. If $X \sim \mathit{Gamma}\,(a, b)$, and $c > 0$, then $cX \sim \mathit{Gamma}\,(a, b/c)$. Whenever $a = 1$, the Gamma distribution $\mathit{Gamma}\,(1, b)$ coincides with the exponential distribution $\mathit{Exp}\,(b)$.

The Fourier transform of the Gamma density function (B.8) is

$$\hat{f}(\xi) = \left(1 + i\frac{\xi}{b}\right)^{-a}. \tag{B.9}$$

Gamma densities have the typical profile of the steady states of the kinetic models for wealth distribution discussed in Chapter 5.

In the same chapter, the *Inverse Gamma distribution of shape parameter* a *and scale parameter* $a - 1$ was considered [295]:

$$f(x) := \begin{cases} (a-1)^a e^{-\frac{(a-1)}{x}}/(\Gamma(a)x^{a+1}) & \text{if } x \geq 0, \\ 0 & \text{if } x < 0. \end{cases} \tag{B.10}$$

This density is peaked around the mean value 1 and has heavy tails, since it decays at infinity like $x^{-(a+1)}$.

B.6 Bernoulli distribution

A Bernoulli random variable models the toss of a (possibly unfair) coin, or any random experiment that has exactly two possible outcomes. Let p be a number in the interval $[0, 1]$. The *Bernoulli distribution with parameter* p is the discrete distribution

$$f(x) = p\delta_0(x - \alpha) + (1 - p)\delta_0(x - \beta),$$

where α and β are the values of the two possible outcomes. We write $X \sim \mathit{Bernoulli}\,(p)$ to say that the random variable X has this distribution.

The indicator random variable of any event B has a Bernoulli distribution with $p = P(B)$.

B.7 Poisson distribution

Let λ be a positive number. The *Poisson distribution with parameter* λ is the discrete probability function

$$f(x) = \sum_{k=1}^{+\infty} \frac{\lambda^k e^{-\lambda}}{k!} \delta_0(x - k). \tag{B.11}$$

We write $X \sim \mathit{Poisson}\,(\lambda)$ to say that the random variable X has this distribution. For the Poisson distribution, the mean and the variance are both equal to λ.

The Fourier transform of the Poisson distribution has the form

$$\hat{f}(\xi) = \exp\left(\lambda(e^{-i\xi} - 1)\right). \tag{B.12}$$

The Poisson distribution appears in Chapter 2, Section 2.5, and in Chapter 8, and it is related to growth processes.

List of figures

1.1 Daniel Bernoulli[1] 3

1.2 Ludwig Eduard Boltzmann[2] 4

1.3 James Clerk Maxwell[3] 6

1.4 Sir Geoffrey Ingram Taylor[4] 12

1.5 Subrahmanyan Chandrasekhar[5] 17

1.6 Cedric Villani[6] 30

1.7 Mark Kac[7] 35

1.8 Carlo Cercignani[8] 43

2.1 Josiah Willard Gibbs[9] 48

2.2 Graph of the function (2.27) at subsequent times. 59

3.1 John von Neumann[10] 80

3.2 Convergence of the Monte Carlo approximation of π. 81

3.3 Plot of the points (X_n, X_{n+1}). Uncorrelated pseudo-random sequence (left), and the presence of correlations (right). 84

3.4 Transformation of N uniformly distributed random numbers U_n using the relation $X_n = -\log(U_n)$. 87

3.5 Acceptance–rejection method: η has been sampled from $w(x)$. It will be accepted with probability equal to the ratio $f(\eta)/w(\eta)$. 89

3.6 Standard normal via rejection using the proposal density $g(x) = \frac{1}{2}e^{-|x|}$. 91

3.7 Convergence of the Box–Muller algorithm using the LCG generator (left) against Matlab 7 `randn` function (right). 95

3.8 Sampling from the surface of a sphere using $2N$ uniformly random sequences $(U_1)_i$, $(U_2)_i$ and the transformations $\Phi_i = 2\pi(U_1)_i$, $\Theta_i = \arccos(1 - 2(U_2)_i)$. 96

3.9 Importance sampling to estimate the integral of $g(x) = 4\sqrt{1 - x^2}$ using $f(x) = 2(2 - x)/3$. 103

3.10 Left: the functions $\Psi_1(x)$ and $W_1(x)$. Right: reconstruction of a standard normal density (−) from $N = 500$ samples using piecewise constant (o) and B-spline functions (∗) with $\Delta x = 0.4$. 109

3.11 Solution of the heat equation with $D = 1$ at $t = 0.4$ by estimating the expectation of the heat kernel with $M = 100$, $N = 120$ (left) and by the random walk method with $M = 100$, $N = 12 \times 10^3$ and simple piecewise constant reconstruction (right). The initial data is the sum of two square waves $u_0(x) = 0.5$, $x \in [-2.5, -0.5]$, $u_0(x) = 1$, $x \in [0.5, 2.5]$ and $u_0(x) = 0$ elsewhere. 112

3.12 Solution of convection–diffusion at $t = 0.5$ (left) and $t = 2$ (right) with $a = 1$ and $D = 0.05$ using $N = 12000$ samples and $M = 100$ grid points in the reconstruction. The initial data is the sum of two square waves $u_0(x) = 0.5$, $x \in [-4, -2]$, $u_0(x) = 1$, $x \in [-1, 1]$ and $u_0(x) = 0$ elsewhere. 116

3.13 Random gradient representation of the standard normal with zero mean and unit variance. Left: simple histogram representation using 50 cells of 100 samples from its derivative. Right: plot of the function $u_N(x)$ computed using (3.80). 118

4.1 Stanislaw Marcin Ulam[11] 123

4.2 Monte Carlo solution of the Goldstein–Taylor model for $\rho(x, t)$ at time $t = 0.38$ with $\sigma = 1$ and $c = 1$. The figure refers to $N = 15000$ samples and 100 grid points in the reconstruction. The initial data is $\rho_0(x) = 1$ for $x < 0$ and $\rho_0(x) = 0$ for $x > 0$, with $f_0^+(x) = f_0^-(x)$. The continuous line represents the reference solution obtained by finite differences. 125

4.3 Standard Monte Carlo. Solution at $t = 10$ for $F(\rho) = \rho^2/2$. Left: $N = 1000$ and $\Delta t = 0.1$. Right: $N = 10000$, $\Delta t = 0.01$. The limit case is $\varepsilon = 0$ with $c = 1$. The solution is reconstructed on $M = 100$ grid points. The initial data is a normal with zero mean and unit variance. 126

4.4 Random gradient method. Solution at $t = 10$ for $F(\rho) = \rho^2/2$. Left: $N = 100$ and $\Delta t = 0.1$. Right: $N = 500$, $\Delta t = 0.05$. The limit case is $\varepsilon = 0$ with $c = 1$. The initial datum is a normal density with zero mean and unit variance. 128

4.5 Solution of the BGK model at $t = 0.05$ for the Lax shock tube problem ($\rho = 0.445$, $U_x = 0.698$, $U_y = U_z = 0$, $p = 3.528$ for $x < 0.5$ and $\rho = 0.5$, $U_x = U_y = U_z = 0$, $p = 0.571$ for $x > 0.5$) and Maxwellian initial data using 200 space cells and 500 particles per cell with $\Delta t = 0.1$. Left: density for $\varepsilon = 0.01$. Right: density for $\varepsilon = 0.0005$. The dashed line is the corresponding solution of the Euler equation (4.16). 132

4.6 McKean graphs for $k = 0$, $k = 1$, $k = 2$ and $k = 3$. 137

4.7 Generation of samples in the algorithm. 138

4.8 The asymptotic preserving Monte Carlo diagram. Here, f is the solution of the kinetic equation, $f_N^{\Delta t}$ the corresponding Monte Carlo approximation characterized by Δt and by the use of N samples, M is the Maxwellian equilibrium and M_N its approximation with N samples. 140

4.9 Solution at $t = 0.005$ for $\alpha = 0$ (left) and $\alpha = 1$ (right). The dotted line represents the initial datum. The results were obtained with $N = 50000$ particles, $\delta = 0.1$ at time $t = 0.005$. The values of h used in the computations were $h = 2 \times 10^{-6}$ for $\alpha = 0$ and $h = 5 \times 10^{-3}$ for $\alpha = 1$. 148

5.1 Vilfredo Pareto[12] 160

5.2 The cumulative probability distribution of wealth[13] in the United Kingdom. Low capitals fit an exponential curve, and high capitals fit a Pareto power law. Reprinted from [130], © 2001, with permission from Elsevier. 163

5.3 The CPT model: regimes for the formation of Pareto tails. 168

5.4 Winner takes all model: Evolution of the fraction of agents with zero wealth (left) and blow up (right). Here, N is the total number of agents used in the Monte Carlo simulation. 171

5.5 CPT kinetic market model. Asymptotic behaviour of the Fokker–Planck model and the Boltzmann model for $\alpha = 2.0$ and $\beta^2 = 0.1$ (top), $\beta^2 = 0.01$ (bottom). The figures on the right are in log-log scale. The Boltzmann solution has been computed with a Monte Carlo simulation using $N = 10000$ agents [105]. 186

5.6 Histogram of steady-state distribution for $\gamma_1 = \gamma_2 = 0.875$ (left) and for $\gamma_1 = \gamma_2 = 0.99$ (right). 204

5.7 Cumulative wealth distribution for $\gamma_1 = \gamma_2 = 0.875$ (left) and for $\gamma_1 = \gamma_2 = 0.99$ (right). 204

5.8 Histogram of steady-state distribution for $\gamma_1 = 0.875$ and $\gamma_2 = 0.99$ (left) in comparison with the histogram for the union of two disjunct populations with the same parameters (right). 205

5.9 Cumulative wealth distribution for $\gamma_1 = 0.875$ and $\gamma_2 = 0.99$ (left) in comparison with the cumulative wealth distribution for the union of two disjunct populations with the same parameters (right). 205

6.1 Left: Various realizations of the diffusion function. Right: Effect of diffusion parameter λ. 250

6.2 $P(w) = \chi(|w - \bar{w}| \leq \rho)$ with $\rho = 0.5$ (left) and $\rho = 0.8$ (right), $\lambda = 0.003$ and $p = q = 0.5$. 250

6.3 Steady profile g corresponding to the choices of Maxwellian localization $w_- = -0.8$ and $w_+ = 0.8$. $p = q = 0.5$ (left) and $p = 0.6, q = 0.4$ (right). 251

6.4 Steady profile g for $p = q = 0.5$, corresponding to different choices of Maxwellian localization $w_- = -0.3$ and $w_+ = 0.8$ (left), $w_- = -0.8$ and $w_+ = 0.6$ (right). 251

6.5 Paul Felix Lazarsfeld[14] 252

6.6 Illustration of stationary solutions (6.96) for different values of α and parameters given by (6.97). The solid line corresponds to $\alpha = 1$, the dashed one to $\alpha = 0.75$ and the dashed-dotted one to $\alpha = 0.5025$. 256

6.7 Difference between approximate solution g_2 and exact solution $g_{2,\infty}$ in L^1-Norm. 257

7.1 Daniel Kahneman[15] 261

7.2 Francis Ysidro Edgeworth[16] 263

7.3 Oskar Morgenstern[17] 274

7.4 Distribution function at $t = 50, 200, 500$. The continuous line is the lognormal Fokker–Planck solution. The right plot is in log-log scale. 281

7.5 Test 3. The corresponding Lorentz curves. The Gini coefficients are $G = 0.1$, $G = 0.2$ and $G = 0.3$ respectively. 282

7.6 An example of value function $\Phi(\dot{S}(t)/S(t))$. 285

7.7 Equilibrium distribution function of the chartist investment propensity for different values of $Y_* = 0, 0.2, -0.2$. Exact solutions with $\rho_C = 1$, $\beta = 0.1$, $t_C = 1$, $\lambda/(\tilde{\alpha_1} + \tilde{\alpha_2}) = 1$ and $f(y, 0) = f^\infty(y)$. 291

7.8 A strong fundamentalist predominance forces the price to follow the fundamental price. The chartist population is characterized by the parameters $\alpha_1 = 0.25$ and $\alpha_2 = 0.7$. The figure on the right represents the variation of the chartists' fraction. 293

7.9 The price follows the fundamental price but strong oscillations appear. The chartist population is characterized by the parameters $\alpha_1 = 0.6$ and $\alpha_2 = 0.4$. The figure on the right represents the variation of the chartists' fraction. 293

7.10 A chartists predominance may lead to extreme behaviours like the case of a market boom (top) or a crash (bottom). The chartist population is characterized by the parameters $\alpha_1 = 0.65$ and $\alpha_2 = 0.3$. Top figure: $r = 1/6$, $l = 1/4$. Bottom figure: $r = 1/4$, $l = 1/6$. The figure on the right represents the variation of the chartists' fraction. 294

7.11 Influence of different values for λ ($\lambda = 1, 1.3, 1.5, 1.8, 1.9$). 305

7.12 Influence of different values for b ($b = 1.5, 2, 3, 5$). 305

7.13 Influence of different values for χ ($\chi = 0.5, 1, 2, 5, 10$). 306

7.14 Influence of different laws for propensity: solid lines correspond to law (7.142), dashed lines to (7.138). 306

7.15 Functions $\mathcal{W}(x)$ and $\nu(x)$ used in the numerical illustration. 307

8.1 Salvador Edward Luria[18] 309

8.2 Luria–Delbrück setting. Distribution of mutant cells at $t = 6.7$ with $\beta = 2.5$ for $\epsilon = 0.1$ (top) and $\epsilon = 0.01$ (bottom) in the kinetic model (8.10), with $\mathcal{L}(v) = \beta v$. The reference solution is computed using Lemma 2 in [329]. 321

8.3 Lea–Coulson setting. Distribution of mutant cells at $t = 6.7$ with $\beta = 2.8$ for $\epsilon = 0.1$ (top) and $\epsilon = 0.01$ (bottom) in the kinetic model (8.10), with $\mathcal{L}(v)$ a Poisson process of mean βv. The reference solution is computed using Lemma 2 in [329] and numerical quadrature. 321

8.4 Steven Smale[19] 322

8.5 A flock of birds (left) and a school of fishes attacked by predators (right). 323

8.6 One of the possible configurations in the interaction with the visual cone. 329

8.7 Two-dimensional flocking with the Cucker–Smale alignment dynamic. On the left without visual cone, on the right with visual cone. The perception limitation splits the flock into two groups. 330

8.8 Sketch of the three-zone model. 332

8.9 Flock at several time steps. Parameters in the attraction–repulsion potential are such that $C = C_R/C_A = 30$, $l = l_R/l_A = 0.3$, $\alpha = 0.7$, $\beta = 0.05$. Equilibrium is reached after $t = 10$. 334

8.10 Shepherd dogs. Simulation at different times with parameters $a = 2.5$, $b = 0.1$ and $\gamma = 0.45$. 338

8.11 Swarm attacked by a predator. Simulation at different times with parameters $a = 4$, $b = 2$ and $\gamma = 0.45$. 339

References

[1] Adams R.A. and Fournier J.J.F. (2003). *Sobolev Spaces.* Amsterdam: Academic Press.

[2] Agueh M., Illner R., and Richardson A. (2011). Analysis and simulations of a refined flocking and swarming model of Cucker–Smale type. *Kinet. Relat. Models*, **4**:1–16.

[3] Albi G. and Pareschi L. (2013). Binary interaction algorithms for the simulation of flocking and swarming dynamics. *SIAM, Mult. Model. Simul.*, **11**: 1–29.

[4] Albi G. and Pareschi L. (2013). Modeling self-organized systems interacting with few individuals: From microscopic to macroscopic dynamics. *App. Math. Letters*, **24**: 397–401.

[5] Aletti G., Naldi G. and Toscani G. (2007). First-order continuous models of opinion formation. *SIAM J. Appl. Math.*, **67**/3: 837–853.

[6] Alsmeyer G., Iksanov A. and Rösler U. (2009). On distributional properties of perpetuities. *J. Theor. Probab.*, **22**: 666–682.

[7] Amoroso L. (1925). Ricerche intorno alla curva dei redditi. *Ann. Mat. Pura Appl.* Ser. 4, **21**: 123–159.

[8] Angerer W.P. (2001). An explicit representation of the Luria–Delbrück distribution. *J. Math. Biol.*, **42**: 145–174.

[9] Angle J. (1986). The surplus theory of social stratification and the size distribution of personal wealth. *Social Forces*, **65**/2: 293–326.

[10] Angle J. (2006). The inequality process as a wealth maximizing process. *Physica A*, **367**: 388–414.

[11] Aoki I. (1982). A simulation study on the schooling mechanism in fish. *Bulletin of The Japanese Society of Scientific Fisheries*, **48**: 1081–1088.

[12] Armitage P. (1952). The statistical theory of bacterial populations subject to mutation. *J. Royal Statist. Soc. B*, **14**: 1–40.

[13] Arnold A., Carrillo J.A., Desvillettes L., Dolbeault J., Jüngel A., Lederman C., Markowich P.A., Toscani G. and Villani C. (2004). Entropies and equilibria of many-particle systems: An essay on recent research. *Monatsh. Math.*, **142**: 35–43.

[14] Babovsky H. (1986). On a simulation scheme for the Boltzmann equation. *Math. Methods Appl. Sci.*, **8**: 223–233.

[15] Babovsky H. and Illner R. (1989). A convergence proof for Nanbu's simulation method for the full Boltzmann equation. *SIAM J. Num. Anal.*, **26**: 45–65.

[16] Bhatnagar P.L., Gross E.P. and Krook M. (1954). A model for collision processes in gases. I. Small amplitude processes in charged and neutral one-component systems. *Phys. Rev.*, **94**: 511–525.

[17] Bal G., Papanicolaou G. and Ryzhik L. (2002). Radiative transport limit for the random Schroedinger equation. *Nonlinearity*, **15**: 513–529.

[18] Baldassarri A., Marini Bettolo Marconi U. and Puglisi A. (2002). Kinetic models of inelastic gases. *Mat. Mod. Meth. Appl. Sci.*, **12**: 965–983.

[19] Ballerini M., Cabibbo N., Candelier R., Cavagna A., Cisbani E., Giardina I., Lecomte V., Orlandi A., Parisi G., Procaccini A., Viale M. and Zdravkovic V. (2008). Interaction ruling animal collective behavior depends on topological rather than metric distance: Evidence from a field study. *PNAS*, **105**: 1232–1237.

[20] Barenblatt G.I. (1952). On some unsteady motions of a liquid and a gas in a porous medium. *Prinkl. Math. Meh.*, **16**: 67–78.

[21] Barenblatt G.I. (1996). *Scaling, Self-Similarity, and Intermediate Asymptotics.* Cambridge: Cambridge University Press.

[22] Baringhaus L. and Grübel R. (1997). On a class of characterization problems for random convex combination. *Ann. Inst. Statist. Math.*, **49**/3: 555–567.

[23] Bartlett M.S. (1966). *An Introduction to Stochastic Processes with Special Reference to Methods and Applications* (2nd edn). London: Cambridge University Press.

[24] Bassetti F. and Toscani G. (2010). Explicit equilibria in a kinetic model of gambling. *Phys. Rev. E*, **81**: 066115.

[25] Bassetti F., Ladelli L. and Matthes D. (2011). Central limit theorem for a class of one-dimensional kinetic equations. *Prob. Theor. Rel. Fields*, **150**: 77–109.

[26] Bauer W.F. (1958). The Monte Carlo method. *J. of the Society for Industrial and Applied Mathematics*, **6**/4: 438–451.

[27] Bellomo N. (2008). *Modeling Complex Living Systems: Kinetic Theory and Stochastic Game Approach.* Boston: Birkhauser.

[28] Bellomo N. and Delitala M. (2008). From the mathematical kinetic, and stochastic game theory to modelling mutations, onset, progression and immune competition of cancer cells. *Physics of Life Reviews*, **5**: 183–206.

[29] Bellomo N. and Soler J. (2012). On the mathematical theory of the dynamics of swarms viewed as complex systems. *Math. Mod. and Meth. in Appl. Sci.*, **22**: 1140006.

[30] Ben-Avraham D., Ben-Naim E., Lindenberg K. and Rosas A. (2003). Self-similarity in random collision processes. *Phys. Rev. E*, **68**: R050103.

[31] Ben-Naim E. (2005). Opinion dynamics, rise and fall of political parties. *Europhys. Lett.*, **69**: 671.

[32] Ben-Naim E. and Krapivski P. (2000). Multiscaling in inelastic collisions. *Phys. Rev. E*, **61**: R5–R8.

[33] Ben-Naim E.and Krapivski P. (2002). Nontrivial velocity distributions in inelastic gases. *J. Phys. A*, **35**: L147–L152.

[34] Ben-Naim E., Krapivski P. and Redner S. (2003). Bifurcations and patterns in compromise processes. *Physica D*, **183**: 190–204.

[35] Ben-Naim E., Krapivski P., Vazquez R. and Redner S. (2003). Unity and discord in opinion dynamics. *Physica A*, **330**: 99–106.

[36] Benedetto D., Caglioti E. and Pulvirenti M. (1997). A kinetic equation for granular media. *Mat. Mod. Numer. Anal.*, **31**: 615–641.

[37] Benedetto D., Caglioti E., Carrillo J.A. and Pulvirenti M. (1998). A non-Maxwellian distribution for one-dimensional granular media. *J. Stat. Phys.*, **91**: 979–990.

[38] Bird G.A. (1970). Direct simulation of the Boltzmann equation. *Phys. Fluids*, **13**: 2676–2681.

[39] Bird G.A. (1994). *Molecular Gas Dynamics and the Direct Simulation of Gas Flows*. Oxford: Clarendon Press.

[40] Bisi M., Carrillo J.A. and Toscani G. (2005). Contractive metrics for a Boltzmann equation for granular gases: Diffusive equilibria. *J. Stat. Phys.*, **118**: 301–331.

[41] Bisi M., Carrillo J.A. and Toscani G. (2006). Decay rates in probability metrics towards homogeneous cooling states for the inelastic Maxwell model. *J. Stat. Phys.*, **124**: 625–653.

[42] Bisi M., Spiga G. and Toscani G. (2009). Kinetic models of conservative economies with wealth redistribution. *Commun. Math. Sci.*, **7**/4: 901–916.

[43] Bobylev A.V. (1975). The method of the Fourier transform in the theory of the Boltzmann equation for Maxwell molecules. *Dokl. Akad. Nauk SSSR*, **225**: 1041–1044.

[44] Bobylev A.V. (1988). The theory of the nonlinear spatially uniform Boltzmann equation for Maxwellian molecules. *Sov. Sci. Rev. c*, **7**: 111–233.

[45] Bobylev A.V. and Cercignani C. (2003). Self-similar asymptotics for the Boltzmann equation with inelastic and elastic interactions. *J. Stat. Phys.*, **110**: 333–375.

[46] Bobylev A.V. and Gamba I. (2006). Boltzmann equations for mixtures of Maxwell gases: Exact solutions and power like tails. *J. Stat. Phys.*, **124**: 497–516.

[47] Bobylev A.V., Carrillo J.A. and Gamba I. (2001). On some properties of kinetic and hydrodynamic equations for inelastic interactions. *J. Stat. Phys.*, **98**: 743–773; Erratum: *J. Stat. Phys.*, **103**: 1137–1138.

[48] Bobylev A.V., Cercignani C. and Toscani G. (2003). Proof of an asymptotic property of self-similar solutions of the Boltzmann equation for granular materials. *J. Stat. Phys.*, **111**: 403–417.

[49] Bobylev A.V., Cercignani C. and Gamba I. (2008). Generalized kinetic Maxwell models of granular gases, in G. Capriz, P. Giovine and P.M. Mariano (eds), *Mathematical Models of Granular Matter*, Lecture Notes in Mathematics Vol. 1937. New York: Springer-Verlag.

[50] Bobylev A.V., Cercignani C. and Gamba I. (2009). On the self-similar asymptotics for generalized non-linear kinetic Maxwell models. *Comm. Math. Phys.*, **291**: 599–644.

[51] Bobylev A.V. and Nanbu K. (2000). Theory of collision algorithms for gases and plasmas based on the Boltzmann equation and the Landau–Fokker–Planck equation. *Phys. Rev. E*, **61**: 4576–4586.

[52] Bolley F. and Carrillo J.A. (2007). Tanaka theorem for inelastic Maxwell models. *Comm. Math. Phys.*, **276**: 287–314.

[53] Bossy M.and Talay D. (1997). A stochastic particle method for the McKean–Vlasov and the Burgers equation. *Math. Comp.*, **66**/217: 157–192.

[54] Boltzmann L. (1995). Weitere Studien über das Wärmegleichgewicht unter Gasmolekülen. *Sitzungsberichte der Akademie der Wissenschaften*, **66**: 275–370, in *Lectures on Gas Theory*. Berkeley: University of California Press (1964). Translated by S.G. Brush. Reprint of the 1896–1898 Edition. Reprinted by Dover Publ.

[55] Bouchaud J.F. and Mézard M. (2000). Wealth condensation in a simple model of economy. *Physica A*, **282**: 536–545.

[56] Boudin L. and Salvarani F. (2009a). The quasi-invariant limit for a kinetic model of sociological collective behavior. *Kinetic Rel. Mod.*, **2**: 433–449.

[57] Boudin L. and Salvarani F. (2009b). A kinetic approach to the study of opinion formation. *ESAIM: Math. Mod. Num. Anal.*, **43**: 507–522.

[58] Boudin L., Mercier A. and Salvarani F. (2012). Conciliatory and contradictory dynamics in opinion formation. *Physica A*, **391**: 5672–5684.

[59] Bracewell R.N. (2000). *The Fourier Transform and Its Applications* (3rd edn). Boston: McGraw-Hill.

[60] Broadwell J.E. (1964). Shock structure in a simple discrete velocity gas. *Phys. Fluids*, **7**: 1243–1247.

[61] Brugna C., Demichelis S., and Toscani G. (2013). Kinetic models for the trading of goods. *J. Stat. Phys.*, **151**: 549–566

[62] Burda Z., Johnston D., Jurkiewicz J., Kamiński M., Nowak M.A., Papp G. and Zahed I. (2002). Wealth condensation in Pareto macroeconomies. *Phys. Rev. E*, **65**: 026102.

[63] Caflisch R.E. (1988). Monte Carlo and quasi-Monte Carlo methods. *Acta Numerica*, **7**: 1–49.

[64] Caflisch R.E., Dimarco G. and Pareschi L. (2010). Direct simulation Monte Carlo schemes for Coulomb interactions in plasmas. *Comm. App. Ind. Math.*, **1**: 72–91.

[65] Caflisch R.E. and Pareschi L. (1999). An implicit Monte Carlo method for rarefied gas dynamics I: The space homogeneous case. *J. Comput. Phys.*, **154**: 90–116.

[66] Caglioti E. and Villani C. (2002). Homogeneous cooling states are not always good approximations to granular flows. *Arch. Rat. Mech. Anal.*, **163**: 329–343.

[67] Canizo J.A., Carrillo J.A. and Rosado J. (2011). A well-posedness theory in measures for some kinetic models of collective motion. *Math. Models Methods Appl. Sci.*, **21**: 515–539.

[68] Carleman T. (1957). *Problèmes mathématiques de la théorie cinetique des gaz.* Uppsala: Almqvist and Wiksells.

[69] Carlen E.A., Carvalho M.C. and Gabetta E. (2000). Central limit theorem for Maxwellian molecules and truncation of the Wild expansion. *Commun. Pure Appl. Math.*, **53**: 370–397.

[70] Carlen E.A., Carvalho M.C., Le Roux J., Loss M. and Villani C. (2010). Entropy and chaos in the Kac model. *Kinetic Rel. Mod.*, **3**: 85–122.

[71] Carlen E.A., Gabetta E. and Toscani G. (1999). Propagation of smoothness and the rate of exponential convergence to equilibrium for a spatially homogeneous Maxwellian gas. *Commun. Math. Phys.*, **305**: 521–546.

[72] Carlen E.A. and Salvarani F. (2002). On the optimal choice of coefficients in a truncated Wild sum and approximate solutions for the Kac equation. *J. Stat. Phys.*, **109**: 261–277.

[73] Carrillo J.A., Fornasier M., Toscani G. and Vecil F. (2010). Particle, kinetic and hydrodynamic models of swarming, in G. Naldi, L. Pareschi and G. Toscani (eds), *Mathematical Modeling of Collective Behavior in Socio-economic and Life Sciences*, Modelling and Simulation in Science and technology, 297–336. Boston, MA: Birkhäuser.

[74] Carrillo J.A., Cordier S. and Toscani G. (2009). Over-populated tails for conservative-in-the-mean inelastic Maxwell models. *Discrete and Continuous Dynamical Systems A*, **24**/1: 59–81.

[75] Carrillo J.A., Jüngel A., Markowich P.A., Toscani G. and Unterreiter A. (2001). Entropy dissipation methods for degenerate parabolic systems and generalized Sobolev inequalities. *Monatsh. Math.*, **133**: 1–82.

[76] Carrillo J.A., Klar A., Martin S. and Tiwari S. (2010). Self-propelled interacting particle systems with roosting force. *Math. Models Methods Appl. Sci*, **20**:1533–1552.

[77] Carrillo J.A. and Toscani G. (2007). Contractive probability metrics and asymptotic behavior of dissipative kinetic equations. *Riv. Mat. Univ. Parma*, **6**/7: 75–198.

[78] Case K.M. and Zweifel P.F. (1967). *Linear Transport Theory*. Reading: Addison-Wesley.

[79] Castellano C., Fortunato S. and Loreto V. (2009). Statistical physics of social dynamics. *Rev. Mod. Phys.*, **81**: 591–646.

[80] Cercignani C. (1975). *Theory and Application of the Boltzmann Equation*. Edinburgh: Scottish Academic Press.

[81] Cercignani C. (1982). H-theorem and trend to equilibrium in the kinetic theory of gases. *Arch. Mech.*, **34**: 231–241.

[82] Cercignani C. (1988). *The Boltzmann Equation and its Applications* Springer Series in Applied Mathematical Sciences, Vol. 67. New York: Springer-Verlag.

[83] Cercignani C. (1990). *Mathematical Methods in Kinetic Theory*. (2nd edn). New York: Plenum Press.

[84] Cercignani C., Illner R. and Pulvirenti M. (1994). *The Mathematical Theory of Dilute Gases* Springer Series in Applied Mathematical Sciences, Vol. 106. New York: Springer-Verlag.

[85] Cercignani C. (2000). *Rarefied Gas Dynamics: from Basic Concepts to Actual Calculations* Cambridge Texts in Applied Mathematics. Cambridge: Cambridge University Press.

[86] Chaikin P.M. and Lubensky T.C. (2007). *Principles of Condensed Matter Physics* (4th print edn). Cambridge: Cambridge University Press.

[87] Chakrabarti A.S. and Chakrabarti B.K. (2009). Statistical theories of income and wealth distribution. *Economics, Open-Access, Open-Assessment E-Journal*, http://www.economics-ejournal.org/economics/journalarticles/2010-4.

[88] Chakrabarti B.K. and Chatterjee A. (2004). Ideal gas-like distributions in economics: Effects of saving propensity, in H. Takayasu (ed.), *Applications of Econophysics*. Tokyo: Springer-Verlag.

[89] Chakrabarti B.K., Chakraborti A. and Chatterjee A. (2006). *Econophysics and Sociophysics: Trends and Perspectives*. Berlin: Wiley VCH.

[90] Chakraborti A. (2002). Distribution of money in model markets of economy. *Int. J. Mod. Phys. C*, **13**/10: 1315–1321.

[91] Chakraborti A. and Chakrabarti B.K. (2000). Statistical mechanics of money: How saving propensity affects its distributions. *Eur. Phys. J. B.*, **17**: 167–170.

[92] Chakraborti A. and Patriarca M. (2008). Gamma-distribution and wealth inequality. *Pramana J. Phys.*, **71**/2: 233–243.

[93] Chandrasekhar S. (1950). *Radiative Transfer*. Oxford: Clarendon Press; Reprinted (1960), New York: Dover Publications.

[94] Chatterjee A. and Chakrabarti B.K. (2007a). Ideal-gas-like market models with savings: Quenched and annealed cases. *Physica A*, **382**: 36–41.

[95] Chatterjee A. and Chakrabarti B.K. (2007b). Kinetic exchange models for income and wealth distributions. *Eur. Phys. J. B*, **60**: 135–149.

[96] Chatterjee A., Chakrabarti B.K. and Manna S.S. (2004). Pareto law in a kinetic model of market with random saving propensity. *Physica A*, **335**: 155–163.

[97] Chatterjee A., Chakrabarti B.K. and Stinchcombe R.B. (2005). Master equation for a kinetic model of trading market and its analytic solution. *Phys. Rev. E*, **72**: 026126.

[98] Chatterjee A., Sudhakar Y. and Chakrabarti B.K. (2005). *Econophysics of Wealth Distributions*, New Economic Windows Series. Milan: Springer-Verlag.

[99] Chorin A.J. (1973). Numerical study of slightly viscous flows. *J. Fluid Mech.*, **57**: 785–796.

[100] Chuang Y.-L., D'Orsogna M.R., Marthaler D., Bertozzi A.L. and Chayes L. (2007). State transitions and the continuum limit for a 2D interacting, self-propelled particle system. *Physica D*, **232**: 33–47.

[101] Coelho R., Richmond P., Barry J. and Hutzler S. (2008). Double power law in income and wealth distributions. *Physica A*, **387**: 3847–3851.

[102] Colombo R. and Lécureux-Mercier M. (2012). An analytical framework to describe the interactions between individuals and a continuum. *J. Nonlinear Sci.*, **22**: 39–61.

[103] Comincioli V., Della Croce L. and Toscani G. (2009). A Boltzmann-like equation for choice formation. *Kinetic Rel. Mod.*, **2**: 135–149.

[104] Cordier S., Pareschi L. and Piatecki C. (2009). Mesoscopic modeling of financial markets. *J. Stat. Phys.*, **134**: 161–184.

[105] Cordier S., Pareschi L. and Toscani G. (2005). On a kinetic model for a simple market economy. *J. Stat. Phys.*, **120**: 253–277.

[106] Coron F. and Perthame B. (1991). Numerical passage from kinetic to fluid equations. *SIAM J. Num. Anal.*, **28**: 26–42.

[107] Couzin I.D., Krause J., Franks N.R. and Levin S. (2005). Effective leadership and decision making in animal groups on the move. *Nature*, **433**: 513–516.

[108] Couzin I.D., Krause J., James R., Ruxton G.D. and Franks N.R. (2002). Collective memory and spatial sorting in animal groups. *J. Theor. Biol.*, **218**: 1–11.

[109] Cristiani E., Piccoli B. and Tosin A. (2010). Modeling self-organization in pedestrians and animal groups from macroscopic and microscopic viewpoints, in G.

Naldi, L. Pareschi and G. Toscani (eds), *Mathematical Modeling of Collective Behavior in Socio-economic and Life Sciences*, Modelling and Simulation in Science and Technology, 337–364. Boston, MA: Birkhäuser.

[110] Crump K.S. and Hoel D.G. (1974). Mathematical models for estimating mutation rates in cell populations. *Biometrika*, **61**: 237–252.

[111] Cucker F. and Smale S. (2007). Emergent behavior in flocks. *IEEE Trans. Automat. Control*, **52**: 852–862.

[112] Cucker F. and Smale S. (2007). On the mathematics of emergence. *Japan J. Math.*, **2**: 197-227.

[113] D'Addario R. (1932). Intorno alla curva dei redditi di Amoroso. *Riv. Italiana Statist. Econ. Finanza*, **4**/1.

[114] Dafermos C. (1999). *Hyperbolic Conservation Laws in Continuum Physics*. New York: Springer-Verlag.

[115] D'Amico S., Ardaens J.S., De Florio S. and Montenbruck O. (2008) Autonomous formation flying: Tandem-x, Prisma and beyond, in *Proceedings of the 5th International Workshop on Satellite Constellation & Formation Flying*. Evpatoria, Ukraine, July 2–4.

[116] Defuant G., Neau D., Amblard F. and Weisbuch G. (2000). Mixing beliefs among interacting agents. *Advances in Complex Systems*, **3**: 87–98.

[117] Defuant G., Amblard F., Weisbuch G. and Faure T. (2002). How can extremism prevail? A study based on the relative agreement interaction model. *J. Art. Soc. Soc. Sim.*, **5**/4.

[118] Degond P. and Motsch S. (2007). Macroscopic limit of self-driven particles with orientation interaction. *C. R. Acad. Sci. Paris*, Ser. I, **345**: 555–560.

[119] Degond P. and Motsch S. (2008). Large scale dynamics of the Persistent Turning Walker model of fish behavior. *J. Stat. Phys.*, **131**:989–1021.

[120] Degond P. and Motsch S. (2011). A macroscopic model for a system of swarming agents using curvature control, *J. Stat. Phys.*, **143**:685–714.

[121] Degond P., Pareschi L. and Russo G. (2004). *Modelling and Computational Methods for Kinetic Equations*, Modeling and Simulation in Science, Engineering and Technology Series. Boston: Birkhäuser.

[122] Dell'Antonio G. (1983). Large-time small coupling behavior of a quantum particle in a random potential. *Ann. I.H.P. Nonlinear Anal.*, **39**: 339–384.

[123] Deshpande S.M. (1986). *A second order accurate kinetic theory based method for inviscid compressible flows.* Technical Paper 2613. Hampton: NASA

[124] Desvillettes L. (1995). About the regularizing properties of the non-cut-off Kac equation. *Comm. Math. Phys.*, **168**: 417–440.

[125] Desvillettes L., Furioli G. and Terraneo E. (2009). Propagation of Gevrey regularity for solutions of Boltzmann equation for Maxwellian molecules. *Trans. Amer. Math. Soc.*, **361**: 1731–1747.

[126] Devroye L. (1986). *Non-uniform Random Variate Generations*, New York: Springer-Verlag.

[127] Dimarco G. and Pareschi L. (2011). Exponential methods for kinetic equations. *SIAM J. Num. Anal*, **49**: 2057–2077.

[128] D'Orsogna M.R., Chuang Y.L., Bertozzi A.L. and Chayes L.S. (2006). Self-propelled particles with soft-core interactions: Patterns, stability and collapse. *Physical Review Letters*, **96**: 104–302.

[129] Drăgulescu A. and Yakovenko V.M. (2000). Statistical mechanics of money. *Eur. Phys. Jour. B*, **17**: 723–729.

[130] Drăgulescu A. and Yakovenko V.M. (2001). Exponential and power-law probability distributions of wealth and income in the United Kingdom and the United States. *Physica A*, **299**: 213–221.

[131] Düring B., Markowich M., Pietschmann J.F. and Wolfram M.T. (2009). Boltzmann and Fokker–Planck equations modelling opinion formation in the presence of strong leaders. *Proc. R. Soc. A*, **465**: 3687–3708.

[132] Düring B., Matthes D. and Toscani G. (2008a). Exponential and algebraic relaxation in kinetic models for wealth distribution, in N. Manganaro et al. (eds), *Proceedings of the 14th Conference on Waves and Stability in Continuous Media*. Hackensack, NJ: World Sci. Publ.

[133] Düring B., Matthes D. and Toscani G. (2008b). Kinetic equations modeling wealth redistribution: A comparison of approaches. *Phys. Rev. E*, **78**: 056103.

[134] Düring B., Matthes D. and Toscani G. (2009). A Boltzmann type approach to the formation of wealth distribution curves. *Riv. Mat. Univ. Parma*, **8**: 199–261.

[135] Düring B. and Toscani G. (2007). Hydrodynamics from kinetic models of conservative economies. *Physica A*, **384**: 493–506.

[136] Düring B. and Toscani G. (2008). International and domestic trading and wealth distribution. *Commun. Math. Sci.*, **6**/4: 1043–1058.

[137] Durrett R. and Liggett T. (1983). Fixed points of the smoothing transformation. *Z. Wahrsch. Verw. Gebiete*, **64**: 275–301.

[138] Edgeworth F.Y. (1881). *Mathematical Psychics: An Essay on the Application of Mathematics to the Moral Sciences*. London: Kegan Paul.

[139] Ermakov S.M., Nekrutkin V.V. and Sipin A.S. (1989) *Random Processes for Classical Equations of Mathematical Physics*. Dordrecht: Kluwer Academic Publishers.

[140] Ernst M.H. (1981). Nonlinear model-Boltzmann equations and exact solutions. *Phys. Rep. (Review Section of Phys. Lett.)*, **78**/1: 1–171.

[141] Ernst M.H. and Brito R. (2002a). High energy tails for inelastic Maxwell models. *Europhys. Lett.*, **58**: 182–187.

[142] Ernst M.H. and Brito R. (2002b). Scaling solutions of inelastic Boltzmann equation with over-populated high energy tails. *J. Stat. Phys.*, **109**: 407–432.

[143] Evans L.C. (1998). *Partial Differential Equations*, Grad. Stud. Math., Vol. 19. Providence, RI: Amer. Math. Soc.

[144] Fannjiang A., Jin S. and Papanicolaou G. (2003). High frequency behavior of the focusing nonlinear Schroedinger equation with random inhomogeneities. *SIAM J. Appl. Math.*, **63**: 1328–1358.

[145] Feller W. (1957). *An Introduction to Probability Theory and Its Applications*. (2nd edn), New York: John Wiley & Sons.

[146] Fernández-Juricic E., Erichsen J.T. and Kacelnik A. (2004). Visual perception and social foraging in birds. *Trends in Ecology & Evolution*, **19**: 25–31.

[147] Ferrero J.C. (2005). The monomodal, polymodal, equilibrium and nonequilibrium distribution of money, in Chatterjee A., Yarlagadda S. and Chakrabarti B.K. (eds), *Econophysics of Wealth Distributions.* Milan: Springer-Verlag.

[148] Fishman G.S. (1985). *Monte Carlo. Concepts, Algorithms and Applications.* New York: Springer-Verlag.

[149] Flierl G., Grünbaum D., Levin S. and Olson D. (1999). From individuals to aggregations: the interplay between behavior and physics. *J. Theor. Biol.*, **196**:397–454.

[150] Friedman D.D. (1990). *Price Theory: An Intermediate Text.* Cincinnati: South-Western Publishing Co..

[151] Fristedt B. and Gray L. (1997). *A Modern Approach to Probability Theory.* Boston: Birkhäuser.

[152] Furioli G., Pulvirenti A., Terraneo E. and Toscani G. (2009). Strong convergence towards self-similarity for one-dimensional dissipative Maxwell models. *J. Funct. Anal.*, **257**: 2291–2324.

[153] Furioli G., Pulvirenti A., Terraneo E. and Toscani G. (2010). Convergence to self-similarity for the Boltzmann equation for strongly inelastic Maxwell molecules. *Annales de l'Institut Henri Poincaré (C) Analyse Non Linéaire*, **27**: 719–737.

[154] Gabetta E., Markowich P.A. and Unterreiter A. (1999). A note on the entropy production of the radiative transfer equation. *Appl. Math. Lett.*, **12**: 111–116.

[155] Gabetta E., Pareschi L. and Toscani G. (1997). Relaxation schemes for nonlinear kinetic equations. *SIAM J. Numer. Anal.*, **34**: 2168–2194.

[156] Gabetta E., Toscani G. and Wennberg B. (1995). Metrics for probability distributions and the trend to equilibrium for solutions of the Boltzmann equation. *J. Stat. Phys.* **81**: 901–934.

[157] Galam S., Gefen Y. and Shapir Y. (1982). Sociophysics: A new approach of sociological collective behavior. *J. Math. Sociology*, **9**: 1–13.

[158] Garibaldi U., Scalas E. and Viarengo P. (2007). Statistical equilibrium in simple exchange games II. The redistribution game. *Eur. Phys. Jour. B*, **60**: 241–246.

[159] Goldie C.M. and Maller R.A. (2000). Stability of perpetuities. *Ann. Probab.*, **28**: 1195–1218.

[160] Goldstein S. (1951). On diffusion by discontinuous movementsand on the telegraph equation. *Q. J. Mechanics Appl. Math.*, **4**: 129–156.

[161] Gómez-Serrano J., Graham C. and Le Boudec J-Y. (2012). The bounded confidence model of opinion dynamics. *Math. Models Methods Appl. Sci.*, **22**: 1–46.

[162] Goudon T., Junca S. and Toscani G. (2002). Fourier-based distances and Berry–Esseen like inequalities for smooth densities. *Monatsh. Math.*, **135**: 115–136.

[163] Graham C. and Meleard S. (1997). Stochastic particle approximations for generalized Boltzmann models and convergence estimates. *Annals of Prob.*, **25**: 115–132.

[164] Gradshteyn I.S. and Ryzhik I.M. (2000). *Table of Integrals, Series, and Products* (6th edn). San Diego: Academic Press.

[165] Gray R.M. (2009). *Probability, Random Processes and Ergodic Properties.* New York: Springer-Verlag.

[166] Guala S. (2008). Taxes in a simple wealth distribution model by inelastically scattering particles. arXiv: 0807.4484v1.

[167] Gupta A.K. (2006a). Money exchange model and a general outlook. *Physica A*, **359**: 634–640.

[168] Gupta A.K. (2006b). Models of wealth distributions: A perspective, in Chakrabarti B.K., Chakraborti A. and Chatterjee A. (eds) *Econophysics and Sociophysics: Trends and Perspectives*. Weinheim: Wiley VHC.

[169] Ha S.Y. and Liu J.G. (2009). A simple proof of the Cucker–Smale flocking dynamics and mean-field limit. *Commun. Math. Sci.* **7**: 297–325.

[170] Ha S.Y. and Tadmor E. (2008). From particle to kinetic and hydrodynamic descriptions of flocking. *Kinetic Rel. Mod.*, **1**: 415–435.

[171] Halton J.H. (1970). A retrospective and perspective survey of the Monte Carlo method. *SIAM Review*, **12**: 1–63.

[172] Hammersley J.M. and Handscomb D.C. (1964). *Monte Carlo Methods*. London: Methuen & Co., and New York: John Wiley & Sons.

[173] Haselgrove C. (1961). A method for numerical integration. *Math. Comp.*, **15**: 323–327.

[174] Hayes B. (2002). Follow the money. *American Scientist*, **90**: 400–405.

[175] Hegselman R. and Krause U. (2002). Opinion dynamics and bounded confidence models, analysis and simulation. *J. Art. Soc. Soc. Sim.*, **5**: 2.

[176] Hill I. and Taylor R. (2001). Recent trends in dividends payments and share buy-backs. *Economic Trends*, **567**: 42–44.

[177] Hoffmann-Jœrgensen J. (1994). *Probability with a View Towards Statistics* Vol. 1. New York: Chapman & Hall.

[178] Hopf E. (1950). The partial differential equation $u_t + uu_x = u_{xx}$. *Comm. on Pure and Appl. Math.*, **3**: 201–230.

[179] Hughes B. (1995). *Random Walks and Random Environments*. Vol. 1. Oxford: Clarendon Press.

[180] Ispolatov S., Krapivsky P.L. and Redner S. (1998). Wealth distributions in asset exchange models. *Eur. Phys. Jour. B*, **2**: 267–276.

[181] Jadbabaie A., Lin J. and Morse A.S. (2003). Coordination of groups of mobile autonomous agents using nearest neighbor rules. *IEEE Trans. on Autom. Control*, **48**: 988–1001.

[182] Jaynes E.T. (1965). Gibbs vs Boltzmann entropies. *American J. of Phys.*, **33**: 391–398.

[183] Jin S. and Xin Z. (1995). The relaxation schemes for systems of conservation laws in arbitrary space dimensions. *Comm. Pure Appl. Math.*, **48**: 235–276.

[184] Jones M.E., Thomas S.M. and Rogers A. (1994). Luria–Delbrück fluctuation experiments: Design and analysis. *Genetics*, **136**: 1209–1216.

[185] Kac M. (1956). Foundations of kinetic theory, in *Proceedings of the Third Berkeley Symposium on Mathematical Statistics and Probability, 1954–1955* 3: 171–197. Berkeley and Los Angeles: University of California Press.

[186] Kac M. (1959). *Probability and Related Topics in the Physical Sciences*. London, New York: Interscience.

[187] Kac M. (1974). A stochastic model related to the telegrapher's equation. *Rocky Mountain Journal of Mathematics*, **4**: 497–509.

[188] Kahneman D. and Tversky A. (1979). Prospect theory: An analysis of decision under risk. *Econometrica*, **47**/2: 263–292.

[189] Kahneman D. and Tversky A. (2000). *Choices, Values, and Frames.* Cambridge: Cambridge University Press.

[190] Kalos M.H. and Withlock P.A. (1986). *Monte Carlo Method. Volume I: Basics.* New York: Wiley & Sons.

[191] Kashdan E. and Pareschi L. (2012). Mean field dynamics and the continuous Luria–Delbrück distribution. *Mathematical Biosciences*, **240**/2: 223–230.

[192] Kendall D.G. (1949). Stochastic processes and population growth. *Journal of the Royal Statistical Society, B*, **11**: 230–282.

[193] Kielek Z. (1988). An application of the convolution iterates to evolution equations in Banach space. *Univ. Iagel. Acta Math.*, **27**: 247–257.

[194] Klar A. and Wegener R. (1996). A kinetic model for vehicular traffic derived from a stochastic microscopic model. *Transp. Theory Stat. Phys.*, **25**: 785–798.

[195] Knuth D.E. (1969). *Seminumerical Algorithms. The Art of Computer Programming*, Vol. 2. New York: Addison Wesley.

[196] Knuth D.E. (1997). *The Art of Computer Programming*, Vol. 1. Reading, MA: Addison-Wesley.

[197] Koch A.L. (1982). Mutation and growth rates from Luria–Delbrück fluctuation tests. *Mutat. Res.*, **95**: 129–143.

[198] Krogstad T.R. (2009) *Attitude Synchronization in Spacecraft Formations: Theoretical and Experimental Results.* PhD Thesis, NTNU, Trondheim.

[199] Kullback S. and Leibler R.A. (1951). On information and sufficiency. *Annals of Mathematical Statistics*, **22**: 79–86.

[200] Laha R.G. and Rohatgi V.K. (1979). *Probability Theory.* New York: John Wiley & Sons.

[201] Landau L. (1936). Die kinetische Gleichung für den Fall Coulombscher Wechselwirkung. *Phys. Z. Sowjet.*, **10**: 154. The transport equation in the case of Coulomb interactions, in D. Ter Haar (ed.) (1981). *Collected Papers of L.D. Landau*, 163–170. Oxford: Pergamon Press.

[202] Lasry J. M. and Lions P. L. (2007). Mean field games. *Japanese J. of Mathematics*, **2**: 229–260.

[203] Lazarsfeld P.F., Berelson B.R. and Gaudet H. (1944). *The people's choice: How the voter makes up his mind in a presidential campaign.* New York: Duell, Sloan & Pierce.

[204] Lea D.E. and Coulson C.A. (1949). The distribution of the numbers of mutants in bacterial populations. *J. Genetics*, **49**: 264–285.

[205] Leveque R.J. (1992). *Numerical Methods for Conservation Laws.* Basel: Birkhäuser.

[206] Leveque R.J. (2002). *Finite Volume Methods for Hyperbolic Problems*, Cambridge Texts in Applied Mathematics. Cambridge: Cambridge University Press.

[207] Levy M., Levy H. and Solomon S. (1994). A microscopic model of the stock market: Cycles, booms and crashes. *Economics Letters*, **45**: 103–111.

[208] Levy M., Levy H. and Solomon S. (2000). *Microscopic Simulation of Financial Markets: From Investor Behaviour to Market Phenomena.* San Diego: Academic Press.

[209] Li Hailiang and Toscani G. (2004). Long-time asymptotics of kinetic models of granular flows. *Arch. Ration. Mech. Anal.*, **172**: 407–428.

[210] Liggett T.M. (1999). *Stochastic Interacting Systems, Contact, Voter and Exclusion Processes.* Berlin: Springer-Verlag.

[211] Lighthill J.M. (1958). *Fourier Analysis and Generalized Functions.* Cambridge: Cambridge University Press.

[212] Liu Q. (2000). On generalized multiplicative cascades. *Stochastic Process. Appl.*, **86**: 263–286.

[213] Liu J.S. (2001). *Monte Carlo Strategies in Scientific Computing*, Springer Series in Statistics. New York: Springer-Verlag.

[214] Luria S.E. and Delbrück M. (1943). Mutations of bacteria from virus sensitivity to virus resistance. *Genetics*, **28**: 491–511.

[215] Lux T. (1998). The socio-economic dynamics of speculative markets, interacting agents, chaosand the fat tails of return distributions. *Journal of Economic Behavior & Organization*, **33**: 143–165.

[216] Lux T. and Marchesi M. (1999). Scaling and criticality in a stochastic multiagent model of a financial market. *Nature*, **397**/11: 498–500.

[217] Lux T. and Marchesi M. (2000). Volatility clustering in financial markets, a microscopic simulation of interacting agents. *International Journal of Theoretical and Applied Finance*, **3**: 675–702.

[218] Ma W.T., Sandri G. vH. and Sarkar S. (1992). Analysis of the Luria–Delbrück distribution using discrete convolution powers. *J. Appl. Prob.*, **29**: 255–267.

[219] Madras N. (2002). *Lectures on Monte Carlo Methods.* Fields Institute Monographs, Vol. 16. Providence, RI: American Mathematical Society.

[220] Mainardi F., Raberto M., Gorenflo R. and Scalas E. (2000). Fractional calculus and continuous-time finance II: The waiting-time distribution. *Physica A*, **287**: 468–481.

[221] Malcai O., Biham O., Solomon S. and Richmond P. (2002). Theoretical analysis and simulations of the generalized Lotka–Volterra model. *Phys. Rev. E*, **66**: 031102.

[222] Maldarella D. and Pareschi L. (2012). Kinetic models for socio-economic dynamics of speculative markets. *Physica A*, **391**: 715–730.

[223] Mandelbrot B. (1960). The Pareto–Lévy law and the distribution of income. *International Economic Review*, **1**: 79–106.

[224] Mandelbrot B. (1974). A population birth-and-mutation process, I: Explicit distributions for the number of mutants in an old culture of bacteria. *J. Appl. Prob.*, **11**: 437–444.

[225] Mantegna R.N. and Stanley H.E. (1999). *An Introduction to Econophysics: Correlations and Complexity in Finance.* Cambridge: Cambridge University Press.

[226] Marsaglia G. (1985). A current view of random number generators, keynote address, in *Proceedings, Computer Science and Statistics: 16th Symposium on the Interface.* Amsterdam: North-Holland.

[227] Marsaglia G. (1986). The incomplete Γ function as a continuous Poisson distribution. *Computers & Mathematics with Applications*, **12**: 1187–1190.

[228] Maslach C., Stapp J. and Santee, R.T. (1985) Individuation: Conceptual analysis and assessment. *J. Personality Social Psychology*, **49**: 729–738.

[229] Matthes D. and Toscani G. (2008a). On steady distributions of kinetic models of conservative economies. *J. Stat. Phys.*, **130**: 1087–1117.

[230] Matthes D. and Toscani G. (2008b). Analysis of a model for wealth redistribution. *Kinetic Rel. Mod.*, **1**: 1–22.

[231] McKean H.P. Jr. (1966). Speed of approach to equilibrium for Kac's caricature of a Maxwellian gas. *Arch. Rat. Mech. Anal.*, **21**: 343–367.

[232] McKean H.P. Jr. (1967). An exponential formula for solving Boltmann's equation for a Maxwellian gas. *J. Combinatorial Theory*, **2**: 358–382.

[233] McNamara S. and Young W.R. (1993). Kinetics of a one-dimensional granular medium in the quasi-elastic limit. *Phys. Fluids A*, **5**: 34–45.

[234] Meerschaert M.M. and Scheffler H.P. (2001). *Limit Distributions for Sums of Independent Random Vectors: Heavy Tails in Theory and Practice.* New York: Wiley Interscience.

[235] Metropolis N. (1987). The beginning of the Monte Carlo method, *Los Alamos Science*, **15**: 125–131.

[236] Metropolis N. and Ulam S. (1949). The Monte Carlo method. *J. of the American Statistical Association*, **44**/247: 335–341.

[237] Mohanty P.K. (2006). Generic features of the wealth distribution in an ideal-gas-like market. *Phys. Rev. E*, **74**: 011117.

[238] Morgenstern D. (1954). General existence and uniqueness proof for spatially homogeneous solutions of the Maxwell–Boltzmann equation in the case of Maxwellian molecules. *Proc. Natl. Acad. Sci. USA*, **40**: 719–721.

[239] Morgenstern O. and von Neumann J. (1947). *Theory of Games and Economic Behavior.* 2d ed. Princeton: Princeton University Press.

[240] Morgenstern O. (1950). *On the Accuracy of Economic Observations.* Princeton: Princeton University Press.

[241] Myers T.G. (1998). Thin films with high surface tension. *SIAM Reviews*, **40**: 441–462.

[242] Naldi G., Pareschi L. and Toscani G. (2010). *Mathematical Modeling of Collective Behavior in Socio-Economic and Life Sciences.* Boston: Birkhauser.

[243] Nanbu K. (1980). Direct simulation scheme derived from the Boltzmann equation. *J. Phys. Soc. Japan*, **49**: 2042–2049.

[244] Ochrombel R. (2001). Simulation of Sznajd sociophysics model with convincing single opinions. *Int. J. Mod. Phys. C*, **12**: 1091–1092.

[245] Pakes A.G. (1993). Remarks on the Luria–Delbrück distribution. *J. Appl. Prob.*, **30**: 991–994.

[246] Papoulis A. (1991). *Probability, Random Variables, and Stochastic Processes* (3rd edn). New York: McGraw-Hill.

[247] Pareschi L. (2005). Hybrid multiscale methods for kinetic and hyperbolic problems. *ESAIM: Proceedings*, **15**: 87–120.

[248] Pareschi L. and Russo G. (2000). Asymptotic preserving Monte Carlo methods for the Boltzmann equation. *Transp. Theo. Stat. Phys.*, **29**: 415–430.

[249] Pareschi L. and Russo G. (2001). Time relaxed Monte Carlo methods for the Boltzmann equation. *SIAM J. Sci. Comput.*, **23**: 1253–1273.

[250] Pareschi L. and Russo G. (2001). An introduction to Monte Carlo methods for the Boltzmann equation. *ESAIM: Proceedings*, **10**: 35–75.

[251] Pareschi L., Russo G. and Toscani G. (2004). A kinetic approximation to Hele–Shaw flow. *C. R., Math., Acad. Sci. Paris, Ser. I 338*, **2**: 177–182.

[252] Pareschi L., Russo G. and Toscani G. (2006). *Modelling and Numerics of Kinetic Dissipative Systems*. New York: Nova Science Publ.

[253] Pareschi L. and Seaid M. (2004). A new Monte Carlo approach for conservation laws and relaxation systems. *Lecture Notes in Computer Science*, **3037**: 281–288.

[254] Pareschi L. and Toscani G. (2006). Self-similarity and power-like tails in non-conservative kinetic models. *J. Stat. Phys.*, **124**: 747–779.

[255] Pareschi L. and Wennberg B. (2001). A recursive Monte Carlo method for the Boltzmann equation in the Maxwellian case. *Monte Carlo Methods Appl.*, **7**: 349–358.

[256] Pareschi L., Trazzi S. and Wennberg B. (2008). Adaptive and recursive time relaxed Monte Carlo methods for rarefied gas dynamics. *SIAM J. Sci. Comput.*, **31**: 1379–1398.

[257] Pareto V. (1897). *Cours d'Économie Politique*. Lausanne and Paris.

[258] Patriarca M., Chakraborti A. and Germano G. (2006). Influence of saving propensity on the power-law tail of the wealth distribution. *Physica A*, **369**: 723–736.

[259] Patriarca M., Chakraborti A. and Kaski K. (2004). A statistical model with a standard Gamma distribution. *Phys. Rev. E*, **70**: 016104.

[260] Patriarca M., Heinsalu E. and Chakraborti A. (2010). Basic kinetic wealth-exchange models: common features and open problems. *Eur. Phys. J. B*, **73**: 145-153.

[261] Patriarca M., Chakraborti A., Heinsalu E. and Germano G. (2007). Relaxation in statistical many-agent economy models. *Europ. Phys. J. B*, **57**: 219–224.

[262] Perea L., Gómez G. and Elosegui P. (2009). Extension of the Cucker–Smale control law to space flight formations. *AIAA Journal of Guidance, Controland Dynamics*, **32**: 527–537.

[263] Pianegonda S., Iglesias J.R., Abramson G. and Vega J.L. (2003). Wealth redistribution with finite resources. *Physica A*, **322**: 667–675.

[264] Platkowski T. and Illner R. (1988). Discrete velocity models of the boltzmann equation: A survey on the mathematical aspects of the theory. *SIAM Review*, **30**: 213–255.

[265] Povzner A.Y. (1962). The Boltzmann equation in kinetic theory of gases. *Amer. Math. Soc. Transl.*, Ser. 2 **47**: 193–216.

[266] Press W.H., Teukolsky S.A., Vetterling W.T. and Flannery B.P (2007). *Numerical Recipes: The Art of Scientific Computing* (3rd edn). New York: Cambridge University Press. Chapter 7. Random Numbers.

[267] Pullin D.I. (1979). Generation of normal variates with given sample mean and variance. *J. Sta. Comput. Simul.*, **9**: 303–309.

[268] Pulvirenti A. and Toscani G. (2004). Asymptotic properties of the inelastic Kac model. *J. Stat. Phys.*, **114**: 1453–1480.

[269] Rachev S.T. and Rüschendorf L. (1998). *Mass Transportation Problems.* Probability and its Applications. New York: Springer-Verlag.

[270] Repetowicz P., Hutzler S. and Richmond P. (2005). Dynamics of money and income distributions. *Physica A*, **356**: 641–654.

[271] Rash W. (1997). *Politics on the Nets: Wiring the Political Process.* New York: Freeman.

[272] Rey T. and Toscani G. (2013). Large-time behavior of the solutions to Rosenau type approximations to the heat equation. *SIAM J. Appl. Math.*, **73**: 1416–1438.

[273] Reynolds C.W. (1987). Flocks, herds and schools: A distributed behavioral model. *ACM SIGGRAPH Computer Graphics*, **21**: 25–34.

[274] Risken H. (1996). *The Fokker–Planck Equation. Methods of Solution and Applications.* Berlin: Springer-Verlag.

[275] Rjasanow S. and Wagner W. (2005). *Stochastic Numerics for the Boltzmann Equation.* Springer Series in Computational Mathematics. Berlin: Springer-Verlag.

[276] Rodino L. (1993). *Linear Partial Differential Operators in Gevrey Spaces.* River Edge, NJ: World Scientific Publishing Co. Inc..

[277] Romanovsky M. and Oks E. (2001). Time intervals distribution of stock transactions and time correlation of stock indices in the model space. *Physica A*, **299**: 168–174.

[278] Rosenau P. (1992). Tempered diffusion: A transport process with propagating fronts and inertial delay. *Physical Review A*, **46**: 12–15.

[279] Rubinstein B.Y. (1981). *Simulation and the Monte Carlo Method.* New York: Wiley & Sons.

[280] Samorodnitsky G. and Taqqu M.S. (1994) *Stable Non-Gaussian Random Processes: Stochastic Models with Infinite Variance.* New York: Chapman and Hall.

[281] Samuelson P.A. (1980). *Economics.* Auckland: McGraw-Hill Int.

[282] Scalas E., Garibaldi U. and Donadio S. (2006). Statistical equilibrium in the simple exchange games I. Methods of solution and application to the Bennati–Dragulescu–Yakovenko (BDY) game. *Eur. Phys. J. B*, **53**: 267–272.

[283] Shen J. (2008). Cucker–Smale flocking under hierarchical leadership. *SIAM J. Appl. Math.*, **68**: 694–719.

[284] Sherman A. and Mascagni M. (1994). A gradient random walk method for two-dimensional reaction–diffusion equations. *SIAM J. Sci. Comput.*, **15**: 1280–1293.

[285] Schochet S. and Tadmor E. (1992). The regularized Chapman–Enskog expansion for scalar conservation laws. *Arch. Ration. Mech. Anal.*, **119**: 95–107.

[286] Silver J., Slud E. and Takamoto K. (2002). Statistical equilibrium wealth distributions in an exchange economy with stochastic preferences. *J. Econ. Theory*, **106**: 417–435.

[287] Slanina F. (2004). Inelastically scattering particles and wealth distribution in an open economy. *Phys. Rev. E*, **69**: 046102.

[288] Slanina F. and Lavička H. (2003). Analytical results for the Sznajd model of opinion formation. *Eur. Phys. J. B*, **35**: 279–288.

[289] Solomon S. (1998). Stochastic Lotka–Volterra systems of competing auto-catalytic agents lead generically to truncated Pareto power wealth distribution, truncated Levy distribution of market returns, clustered volatility, booms and crashes, in A-P.N. Refenes, A.N. Burgess and J.E. Moody (eds) *Computational Finance 97*. Dordrecht: Kluwer Academic Publishers.

[290] Spohn H. (1977). Derivation of a transport equation for electrons moving through random impurities. *J. Stat. Phys.*, **17**: 385–412.

[291] Stanley H.E., Afanasyev V., Amarala L.A.N., Buldyreva S.V., Goldberger A.L., Havlina S., Leschhorna H., Maassa P., Mantegna R.N., Penga C.-K., Prince P.A., Salinger M.A., Stanley M.H.R. and Viswanathana G.M. (1996). Anomalous fluctuations in the dynamics of complex systems: From DNA and physiology to econophysics. *Physica A*, **224**/12: 302–321.

[292] Stauffer D. (2002). Percolation and Galam theory of minority opinion spreading. *Int. J. Mod. Phys. C*, **13**: 975–978.

[293] Stauffer D. and De Oliveira P.M.C. (2002). Persistence of opinion in the Sznajd consensus model, computer simulation. *Eur. Phys. J. B*, **30**: 587–592.

[294] Stauffer D. and Meyer-Ortmanns H. (2004). Simulation of consensus model of Deffuant et al. on a Barabasi–Albert network. *Int. J. Mod. Phys. B*, **15**: 241–246.

[295] Steutel F.W. and van Harn K. (2004). *Infinite Divisibility of Probability Distributions on the Real Line.* New York: Marcel Dekker, Inc.

[296] Stewart F.M., Gordon D.M. and Levin B.R. (1990). Fluctuation analysis: The probability distribution of the number of mutants under different conditions. *Genetics*, **124**: 175–185.

[297] Strang G. (1968). On the construction and comparison of difference schemes. *SIAM J. Num. Anal.*, **5**: 506–517.

[298] Sznajd-Weron K. and Sznajd J. (2000). Opinion evolution in closed community. *Int. J. Mod. Phys. C*, **11**: 1157–1166.

[299] Takayasu H. (2004). *Application of Econophysics.* Tokyo: Springer-Verlag.

[300] Takayasu H. (2005). *Practical Fruits of Econophysics.* Tokyo: Springer-Verlag.

[301] Tanaka H. (1973). An inequality for a functional of probability distributions and its applications to Kac's one-dimensional model of a Maxwellian gas. *Z. Wahrsch. Verw. Gebiete*, **27**: 47–52.

[302] Tanaka H. (1978). Probabilistic treatment of the Boltzmann equation of Maxwellian molecules. *Z. Wahrsch. Verw. Gebiete*, **46**/1: 67–105.

[303] Taylor G.I. (1921). Diffusion by continuous movements. *Proc. London Math. Soc.*, **20**: 196–211.

[304] Thomas G.E. and Stamnes K. (1999). *Radiative Transfer in the Atmosphere and Ocean.* Cambridge: Cambridge University Press.

[305] Topaz C.M. and Bertozzi A.L. (2004). Swarming patterns in a two-dimensional kinematic model for biological groups. *SIAM J. Appl. Math.*, **65**: 152–174.

[306] Topaz C.M., Bertozzi A.L. and Lewis M.A. (2006). A nonlocal continuum model for biological aggregation. *Bulletin of Mathematical Biology*, **68**: 1601–1623.

[307] Toscani G. (2000). One-dimensional kinetic models of granular flows. *RAIRO Modél Math. Anal. Numér.*, **34**: 1277–1291.

[308] Toscani G. (2006). Kinetic models of opinion formation. *Commun. Math. Sci.*, **4**: 481–496.

[309] Toscani G. (2009). Wealth redistribution in conservative linear kinetic models with taxation. *Europhysics Letters*, **88**: 10007.

[310] Toscani G. and Villani C. (1999). Probability metrics and uniqueness of the solution to the Boltzmann equation for a Maxwell gas. *J. Stat. Phys.*, **94**: 619–637.

[311] Truesdell C. and Muncaster R.G. (1980). *Fundamentals of Maxwell's Kinetic Theory of a Simple Monatomic Gas: Treated as a Branch of Rational Mechanics.* New York: Academic Press.

[312] Vazquez J.L. (2007). *The Porous Medium Equation: Mathematical Theory.* Oxford: Oxford University Press.

[313] Vicsek T., Czirók A., Ben-Jacob E., Cohen I. and Shochet O. (1995). Novel Type of Phase Transition in a System of Self-Driven Particles. *Phys. Rev. Lett.*, **75**: 1226–1229.

[314] Vicsek T. and Zafeiris A. (2012). Collective motion. *Phys. Rep.*, **517**:71–140.

[315] Villani C. (1998a). *Contribution à l'étude mathématique des équations de Boltzmann et de Landau en théorie cinétique des gaz et des plasmas.* PhD Thesis, Univ. Paris-Dauphine.

[316] Villani C. (1998b). On a new class of weak solutions to the spatially homogeneous Boltzmann and Landau equations. *Arch. Rational Mech. Anal.*, **143**: 273–307.

[317] Villani C. (2002). A Review of Mathematical Topics in Collisional Kinetic Theory, in S. Friedlander and D. Serre (eds), *Handbook of Mathematical Fluid Dynamics*, Vol. 1, New York: Elsevier.

[318] Villani C. (2003a). Cercignani's conjecture is sometimes true and always almost true. *Commun. Math. Phys.*, **234**: 455–490.

[319] Villani C. (2003b). *Topics in Optimal Transportation*, Graduate Studies in Mathematics 58. Providence: American Mathematical Society.

[320] Voit J. (2005). *The Statistical Mechanics of Financial Markets.* Berlin: Springer-Verlag.

[321] Yakovenko V.M. (2009a). Statistical mechanics approach to econophysics, in R.A. Meyers (ed.), *Encyclopedia of Complexity and System Science.* New York: Springer-Verlag.

[322] Yakovenko V.M. and Rosser J.B. (2009b). Colloquium: Statistical mechanics of money, wealth and income. *Rev. Mod. Phys.*, **81**: 1703–1725.

[323] Wagner W. (1992). A convergence proof for Bird's direct simulation Monte Carlo method for the Boltzmann equation. *J. Stat. Phys.*, **66**: 1101–1044.

[324] Wang Y., Ding N. and Zhang L. (2003). The circulation of money and holding time distribution. *Physica A*, **324**: 665–677.

[325] Wasserstein L.N. (1969). Markov processes over denumerable products of spaces describing large systems of automata. *Problems of Information Transmission*, **5**: 47–52.

[326] Weidlich W. (2000). *Sociodynamics: a Systematic Approach to Mathematical Modelling in the Social Sciences.* Amsterdam: Harwood Academic Publishers.

[327] Weisbuch G., Deffuant G., Amblard F. and Nadal J.P. (2002). Meet, discuss and segregate! *Complexity*, **7**: 55–63.

[328] Wild E. (1951). On Boltzmann's equation in the kinetic theory of gases. *Proc. Camb. Phyl. Soc.*, **47**: 602–609.

[329] Zheng Qi (1999). Progress of a half century in the study of the Luria–Delbrück distribution. *Math. Biosciences*, **162**: 1–32.

[330] Zolotarev V.M. (1983). Probability Metrics. *Theory Prob. Appl.*, **28**: 278–302.

[331] Zender C. (2001). *Radiative Transfer in the Earth System.* Univ. California at Irvine. `http://dust.ess.uci.edu/ess_rdn/rt.pdf`.

Index

acceptance–rejection methods, 88
  convex combination, 92
  normal distribution, 90
antithetic variates, 104
asymptotic preserving Monte Carlo, 139

Beta distribution, 190
BGK model
  Monte Carlo method, 131
BGK operator, 47
binary interaction models, 49
  boundary conditions, 38
  Dirac delta form, 28
  grazing limit, 30
  in $\mathbb{R}$, 26
  in $\mathbb{R}_+$, 36
  Kac model, 35
  market models, 164
  Monte Carlo methods, 133
  opinion formation, 226
  random process, 27
  steady state, 32, 39
  strong form, 131
  symmetric interactions, 28
  weak form, 27
binomial distribution, 82
Bird's method, 135
Boltzmann equation, 4
  binary collision, 5
  collision term, 5
  conservations, 8
  entropy, 48
  free transport, 5
  macroscopic quantities, 5
  Maxwellian equilibrium, 5
  Maxwellian molecules, 5, 43
  space homogeneous, 6
bounded confidence model, 226
Box–Muller method, 93
Broadwell model, 154
  Monte Carlo method, 155
Burgers' equation, 122

Carleman model, 46
  Monte Carlo method, 153
Cauchy distribution, 121
cell mutation models
  Bartlett model, 319
  cumulant generating function, 315
  kinetic equation, 311
  Lea–Coulson distribution, 317
  Lea–Coulson model, 309, 312
  Luria–Delbrück distribution, 319
  Luria–Delbrück model, 308, 310, 312
  Monte Carlo, 322
  mutation rates, 309
  Poisson distribution, 311
  quasi-invariant limit, 315
central limit theorem, 41, 82
  centred p–stable law, 42
  Gaussian law, 41
Chebyshev inequality, 72
collision kernel, 5
collision operator, 5
confidence interval, 82

Dirac delta function, 10, 60
dissipative gas, 7
dissipative models, 61
  grazing limit, 62
drift equation, 57
  entropy, 61
  random process, 61
Duhamel formula, 9

econophysics, 159
Edgeworth box, 260, 261
  Cobb–Douglas utility function, 262
  Fokker–Planck equation, 269
  kinetic equation, 265
Euler equations, 298

financial market models
  boom and crash, 295
  Chakraborti–Chakrabarti model, 297
  Corder–Pareschi–Toscani model, 297
  Euler equations, 298
  finite element method, 304
  Fokker–Planck equations, 297
  fundamental price, 282
  heterogeneous traders, 296
  hydrodynamic approximation, 301
  inhomogeneous models, 295
  investment propensity, 296, 302
  market risk, 272
  Monte Carlo, 292
  quasi-invariant limit, 297
  saving propensity, 297
  steady state, 298
  value function, 293

wealth distribution, 280, 298
financial markets models, 272
bond, 273
booms and crashes, 288
chartists, 283
demand curve, 274
dividends, 273, 276
Fokker–Planck equations, 279, 289
fundamentalists, 283
herding, 285
heterogeneous traders, 281
investment propensity, 284
Levy–Levy–Solomon model, 272
Lux–Marchesi model, 282
market clearance, 275
Monte Carlo, 281
opinion formation, 284
portfolio, 272
price, 276, 278
price behaviour, 287
quasi-invariant limit, 279
self-similar solution, 280
steady state, 280
stock, 273
stock return, 276
strategy exchange, 286
value function, 283, 285
von Neumann–Morgenstern utility, 273
Fourier metric, 55
Fourier transform, 10, 54
fractional diffusion, 114

gambling process, 190
exact solutions, 192
Gamma distribution, 163
generalized functions, 60
Gini coefficient, 194, 281
Goldstein–Taylor model, 11
random process, 13
steady state, 16
Goldstein-Taylor model
Monte Carlo method, 124
telegrapher's equation, 153
growth process, 69
grazing limit, 72
Poisson distribution, 73

heterogeneous traders, 297
Cordier–Pareschi–Toscani model, 199
Fokker–Planck equation, 200
steady state, 202
Hyperbolic relaxation systems, 125
Jin–Xin model, 125
Monte Carlo method, 127
random gradient method, 127

importance sampling, 102
individual preferences, 205, 297
Dirac delta, 209
Fokker–Planck equation, 209
Pareto tails, 206
steady state, 208
inverse transform methods, 86
exponential distribution, 87
geometric distribution, 87
Pareto distribution, 88

Kac equation, 50
Monte Carlo method, 141

Lévy stable distribution, 114
Laplace transform, 11
Levy–Levy–Solomon model, 260
linear kinetic models, 11
lognormal distribution, 280
Lux–Marchesi model, 261

matching moments, 104
McKean graphs, 136
Monte Carlo method, 79
doubling/removing technique, 111
convection–diffusion equations, 115
diffusion problems, 112
evolutionary PDEs, 110
integration, 96
kinetic equations, 123
multi-dimensional problems, 149
nonlinear diffusion, 144
reconstruction, 108
variance reduction, 99
multiagent system, 5

Nanbu's method, 134
nonlinear diffusion
porous media equation, 156
nonlinear drift equation, 68

observable quantities, 9
opinion formation models, 227
bounded confidence, 249
bounded confidence interval, 230
compromise propensity, 228
external factors, 244
Fokker–Planck equations, 237, 247, 254
individual deviations, 228
leadership, 252
linear models, 244
Maxwellian case, 231
quasi-invariant limit, 234
steady state, 241, 248, 255
opinion leadership, 252

Pareto distribution, 160
Pareto index, 162
particle density, 8
particle systems, 40
Poisson distribution, 122
Povzner–Boltzmann equation, 149, 325
Bird's method, 151
Nanbu's method, 150

Prospect theory, 261
pseudo-random numbers, 83
  LCG methods, 84
  non-uniform distribution, 85

radiative transfer, 17
  Monte Carlo method, 131
  random process, 18
  steady state, 21
random gradient method, 116
random variables, 83
random walk method, 113
recursive Monte Carlo method, 136
Rosenau approximation, 21, 51
  heat equation, 25
  Monte Carlo method, 143
  random process, 22
  steady state, 24

sample mean, 82
sample variance, 82
splitting methods, 119
stratified sampling, 99
sum of random variables, 95
swarming models
  Cucker–Smale model, 322, 324
  D'Orsogna–Bertozzi et al. model, 332
  Euler equations, 337
  external agents, 335
  flocking behaviour, 322
  hydrodynamic approximation, 337
  leadership, 335
  mean-field model, 327, 334, 336
  milling, 335
  Monte Carlo, 331, 338
  Morse potential, 333
  perception cone, 329
  Povzner–Boltzmann equation, 325, 333
  predator attack, 339
  Rayleigh's law, 332
  roosting force, 329
  shepherd dogs, 338
  three-zones model, 331
  topological interactions, 329

taxation, 210, 212

utility function, 260
  Cobb–Douglas utility, 262
  von Neumann–Morgenstern utility, 274

Wasserstein metric, 44
weak solution, 9
wealth distribution, 6, 159, 162
wealth distribution models, 163
  bimodal distribution, 203, 204
  Chakraborti–Chakrabarti model, 167
  Chatterjee–Chakrabarti–Manna model, 206
  conservative models, 165
  Cordier–Pareschi–Toscani model, 168, 184
  exact solutions, 189
  Fokker–Planck equations, 183
  heterogeneous traders, 199
  individual preferences, 205
  market risk, 168
  Monte Carlo, 185, 203
  non-conservative models, 181
  redistribution model, 213
  saving propensity, 167
  self-similar solutions, 186
  Slanina's model, 182
  steady state, 176
  taxation model, 212
  winner takes all model, 169
wealth redistribution, 210
  optimal taxation, 222
  steady state, 217
Wild convolution, 50
Wild sum, 44, 49
  coefficients, 53